5th International Conference on Advanced Materials Science and Technology (ICAMST 2017)

AA006151

IOP Conference Series: Materials Science and Engineering
Volume 367

Makassar, Indonesia
19 – 20 September 2017

ISBN: 978-1-5108-6482-5
ISSN: 1757-8981

Printed from e-media with permission by:

Curran Associates, Inc.
57 Morehouse Lane
Red Hook, NY 12571

CURRAN ASSOCIATES INC.
proceedings
.com

Some format issues inherent in the e-media version may also appear in this print version.

Copyright© (2017) by the Institute of Physics
All rights reserved. The material featured in this book is subject to
IOP copyright protection, unless otherwise indicated.

Printed by Curran Associates, Inc. (2018)

For permission requests, please contact the Institute of Physics
at the address below.

Institute of Physics
Dirac House, Temple Back
Bristol BS1 6BE UK

Phone: 44 1 17 929 7481
Fax: 44 1 17 920 0979

techtracking@iop.org

Additional copies of this publication are available from:

Curran Associates, Inc.
57 Morehouse Lane
Red Hook, NY 12571 USA
Phone: 845-758-0400
Fax: 845-758-2633
Email: curran@proceedings.com
Web: www.proceedings.com

Table of contents

Volume 367

The 5th International Conference on Advanced Materials Sciences and Technology (ICAMST 2017)

19–20 September 2017, Makassar, Indonesia

Accepted papers received: 16 May 2018
Published online: 12 June 2018

Preface

Papers

IOP Conf. Series: Materials Science and Engineering **367** (2018) 011001 doi:10.1088/1757-899X/367/1/011001

Preface

The Consortium of Materials Science and Technology (KoSaTeM) and Materials Research Society of Indonesia (MRS-Id) hold The 5th International Conference on Advanced Materials Science and Technology (ICAMST) on 19th-20th September 2017 at the Grand Clarion Hotel & Convention Centre, Makassar, Indonesia.

Leading Indonesian and ASEAN and International scholars and experts were invited to attend the conference, to exchange views and discuss issues relating to those most important and timely issues in the mentioned topics. The conference aimed to help participants to develop a more in-depth understanding of recent developments in the materials sciences and technology while also provided a forum for analysis and formulation of policy recommendations.

In this special occasion, the 5th ICAMST 2017 involved high level scientists as plenary/keynote speakers. The plenary speakers were: **Prof. Dr. Seiichi Miyazaki** (Nagoya University, Japan), Assoc. **Prof. Dr. Ariando** (National University of Singapore, Singapore), **Prof. Mohamad Deraman** (Universiti Kebangasaan Malaysia, Malaysia), and **Assoc. Prof. Dr. Risa Suryana** (Sebelas Maret University, Indonesia). There were about 80 papers presented and submitted at the conference.

Content from this work may be used under the terms of the Creative Commons Attribution 3.0 licence. Any further distribution of this work must maintain attribution to the author(s) and the title of the work, journal citation and DOI.
Published under licence by IOP Publishing Ltd

IOP Conf. Series: Materials Science and Engineering **367** (2018) 011002 doi:10.1088/1757-899X/367/1/011002

List of Committee ICAMST 2017

Chairman	:	Prof. Dr. Eko Hadi Sujiono, UNM, Indonesia
Co-Chairmen	:	Prof. Dr.Eng. Khairurrijal, ITB, Indonesia Assoc. Prof. Dr.Eng. Kuwat Triyana, UGM, Indonesia Prof. Dr. Heru Susanto, UNDIP, Indonesia Assoc. Prof. Dr.Eng. Risa Suryana, UNS, Indonesia Prof. Dr. Sutikno, UNNES, Indonesia Assoc. Prof. Dr. Markus Diantoro, UM, Indonesia
Scientific Committee		Assoc. Prof. Dr. Ariando, NUS, Singapore Prof. Dr. Mohamad Deraman, UKM, Malaysia Prof. Dr. Amun Amri, UNRI, Indonesia Prof. Dr. Darminto, ITS, Indonesia Prof. Dr. Hadi Nur, UTM, Malaysia Prof. Dr. Dahlang Tahir, UNHAS, Indonesia Assoc.Prof. Dr. Nurlela Rauf, UNHAS, Indonesia Assoc.Prof. Paulus Lobo Gareso, UNHAS, Indonesia Assoc.Prof. Subaer, Ph.D., UNM, Indonesia
Secretary	:	Helmy Pratiwi
Treasurer	:	Samnur
Information and Documentation	: :	Jasdar Agus Andi Muhammad Ridham
Paper Publication Division	:	Nasrul Ihsan Vicran Zharvan
Event Division	: :	Yasser Djawad Ichsan Invanni
Transportation and Accommodation Division	:	Imran Dany Wijaya
Refreshment Division	: :	Muhammad Yusriadi Dahlan, Andi Chaerunnisa, Rezki Aulia Imran

Content from this work may be used under the terms of the Creative Commons Attribution 3.0 licence. Any further distribution of this work must maintain attribution to the author(s) and the title of the work, journal citation and DOI.
Published under licence by IOP Publishing Ltd

IOP Conf. Series: Materials Science and Engineering **367** (2018) 011003 doi:10.1088/1757-899X/367/1/011003

List of Participant ICAMST 2017

No.	Name	Email	Institution
1	Madi Sumadi	madi@ugm.ac.id	Laboratorium Penelitian dan Pengujian Terpadu (LPPT) Universitas Gajah Mada (UGM)
2	Nasrul Ihsan	n.ihsan@unm.ac.id	Universitas Negeri Makassar (UNM)
3	Yanurita Dwihapsari	yanuritadh@gmail.com	Institut Teknologi Sepuluh Nopember Surabaya (ITS)
4	MARKUS DIANTORO	markus.diantoro.fmipa@um.ac.id	State University of Malang (Universitas Negeri Malang), Department of Physics, Faculty of Mathematics and Natural Sciences, Jl. Semarang 5 Malang 65145, Indonesia
5	Khairurrijal Khairurrijal	krijal@fi.itb.ac.id	Institut Teknologi Bandung (ITB)
6	Heru Susanto	heru.susanto@che.undip.ac.id	Department of Chemical Engineering, Faculty of Engineering, Diponegoro University
7	Musyarofah Musyarofah	musyarofah13@mhs.physics.its.ac.id	Institute of Technology Sepuluh Nopember (ITS)
8	Rianti Amalia	riantiamalia175@gmail.com	Universitas Indonesia (UI)
9	Prantasi Harmi Tjahjanti	prantasiharmi@umsida.ac.id	Universitas Muhammadiyah Sidoarjo
10	Frida Ermawati	frida.ermawati@unesa.ac.id	Physics Department, Faculty of Mathematics and Natural Sciences, Universitas Negeri Surabaya
11	Kamaruddin Kamaruddin	21205040@s.itb.ac.id	Institut Teknologi Bandung
12	Asep Suhendi	suhendi@telkomuniversity.ac.id	Telkom University
13	Febrian Dedi Sastrawan	dedi.hagi@gmail.com	IKIP PGRI Jember
14	Shibghatullah Muhammady	mshibghatullah@gmail.com	Department of Physics, Faculty of Mathametics and Natural Sciences, Institut Teknologi Bandung
15	slamet priyono	slam013@lipi.go.id	Research Center for Physics-Indonesia Institute of Sciences (LIPI) Komplek Puspiptek Gd 442 Serpong Tangerang Selatan Banten Indonesia
16	Zahrotul Jannah	zahrotuljannah1@mhs.unesa.ac.id	Physic Department
17	Yohanes Edi Gunanto	yohanes.gunanto@uph.edu	University of Pelita Harapan
18	Ismudiati Puri Handayani	dekpuri@gmail.com	Engineering Physics, Telkom University
19	Edy Supriyanto	edysupriyanto@unej.ac.id	University Of Jember
20	Nasrul Ihsan	nas.ihsan@gmail.com	State University of Makassar
21	Sutikno Madnasri	smadnasri@yahoo.com	Semarang State University
22	Sarah Zulfa Khairunnisa	sarahzulfakhairunnisa@gmail.com	Institut Teknologi Bandung
23	Kusuma masudah	masudahkusuma@gmail.com	Universitas Pesantren Tinggi Darul Ulum
24	Dhyah Annur	dhyahannur@gmail.com	Indonesia Institute of Science-Development and Application Unit for

Content from this work may be used under the terms of the Creative Commons Attribution 3.0 licence. Any further distribution of this work must maintain attribution to the author(s) and the title of the work, journal citation and DOI.
Published under licence by IOP Publishing Ltd

			Biocompatible Implant Materials in Orthopedics
25	Zahrotul Jannah	zjannah14@gmail.com	Physich
26	Aprilia Erryani	apri011@lipi.go.id	Pusat Penelitian Metalurgi dan Material, Lembaga Ilmu Pengetahuan Indonesia
27	Nasikhudin Nasikhudin	nasikhudin.fmipa@um.ac.id	Universitas Negeri Malang
28	Lukman Noerochim	lukman@mat-eng.its.ac.id	Institut Teknologi Sepuluh Nopember
29	H. Husain	husain.physics@unm.ac.id	Institut teknologi Sepuluh Nopember Surabaya.
30	Slamet Priyono	pslamet046@gmail.com	Research Center for Physics-Indonesia Institute of Sciences - Komplek PUSPIPTEK Gd 442 Serpong Tangerang Selatan Banten
31	Nur Aini Fauziyah	aini.zierra@gmail.com	Departement of Physics, Faculty of Mathematics and Sciences - Institute of Technology Sepuluh Nopember, Surabaya 60111 Indonesia
32	Lydia Anggraini	lydia.anggra@president.ac.id	President University
33	Euis Sustini	euis@fi.itb.ac.id	Institut Teknologi Bandung (ITB)
34	MOCH. WISNU ARIF SEKTIONO	moch.afandy@gmail.com	Institut Teknologi Bandung (ITB)
35	Ayu Kurnia Agustina	ayu.kurnia.a@mail.ugm.ac.id	Universitas Gadjah Mada
36	Linda Armitasari	linda.armitasari@mail.ugm.ac.id	UNIVERSITAS GADJAH MADA
37	Nurul Fatriani	nurulfatrianii@gmail.com	Universitas Negeri Malang
38	Rusliana Fatayati	rusliana.fatayati@mail.ugm.ac.id	Universitas Gadjah Mada
39	Ryan Nur Iman	ryanfisuin14sgd@gmail.com	Universitas Islam Negeri Sunan Gunung Djati Bandung
40	Retno Dwi Wulandari	retnodwiwulandari12@gmail.com	Institut Teknologi Bandung
41	Askar Triwiyanto	askartri@president.ac.id	School of Mechanical Engineering, President University
42	Grasia Meliolla	grasia.meliolla@gmail.com	Pusat Mikroelektronika ITB
43	Fauziah A.	fauziah.physics@gmail.com	Institut Teknologi Bandung
44	Musyarofah Musyarofah	musyarofah.wongfisika@gmail.com	Institute Technology of Sepuluh Nopember
45	Franciska Pramuji Lestari	franciskapramuji@gmail.com	Research Center for Metallurgy and Materials - Indonesian Institute of Sciences (LIPI)
46	Putri Nur Anggraini	putr014@lipi.go.id	Research Center for Electronics and Telecommunications - Indonesian Institute of Science (P2ET- LIPI)
47	Supardianningsih -	supardianningsih@mail.ugm.ac.id	Universitas Gadjah Mada
48	Ni'matul Awalin	alinvella@gmail.com	Institut Teknologi Sepuluh Nopember (ITS)
49	Natalita Nursam	natalita.mn@gmail.com	Research Center for Electronics and Telecommunication, Indonesian Institute of Science (P2ET LIPI)
50	Aditya Rianjanu	aditya.rianjanu@mail.ugm.ac.id	Departement of Physics Universitas Gadjah Mada

51	Fatimah Noor	fatimah@fi.itb.ac.id	Institut Teknologi Bandung
52	Yasser Djawad	yasserdj@gmail.com	Universitas Negeri Makassar
53	Kamsul Abraha	kamsul@ugm.ac.id	Universitas Gadjah Mada
54	Setia Budi	setiabudi@unj.ac.id	Department of Chemistry, Faculty of Mathematics and Sciences, Universitas Negeri Jakarta
55	umi nuraini	uminuraini36@gmailcom	Institut Teknologi Sepuluh Nopember (ITS)
56	Anawati Anawati	anawati@sci.ui.ac.id	Departemen Fisika, FMIPA UI
57	Seiichi Miyazaki	miyazaki@nuee.nagoya-u.ac.jp	Graduate School of Engineering
58	Umi Nuraini	uminuraini36@gmail.com	Institut Teknologi Sepuluh Nopember (ITS)
59	Nur Aisyah Humairah	humairah.amjad27@gmail.com	Universitas Sulawesi Barat
60	Sulardjaka Sulardjaka	sulardjaka@undip.ac.id	Diponegoro University
61	Hamzah Fansuri	h.fansuri@chem.its.ac.id	Institut Teknologi Sepuluh Nopember
62	Putri Rahmawati	putri.rahma0396@gmail.com	Inorganic Material Research Group, Chemistry Department, Faculty of Mathematics and Natural Sciences, Sebelas Maret University
63	Heri Kiswanto	h.kiswanto@mail.ugm.ac.id	Universitas Gadjah Mada
64	Rendy Muhamad Iqbal	rendiqbale@gmail.com	Department of Chemistry, Institut Teknologi Sepuluh Nopember, Surabaya
65	SUMAR HADI SURYO	sumarhs.undip@gmail.com	Lecturer in Diponegoro University
66	Ahzab Sandi	sandyunm@gmail.com	Institut Teknologi bandung
67	Eko Hadi Sujiono	e.h.sujiono@unm.ac.id	Universitas Negeri Makassar, Makassar 90224. Indonesia
68	Akmal Zulfi	akmal.zulfi@yahoo.com	Institut Teknologi Bandung (ITB)
69	Hartatiek Hartatiek	hartatiek.fmipa@um.ac.id	Jurusan Fisika FMIPA Universitas Negeri Malang
70	Ida Hamidah	idahamidah@upi.edu	Universitas Pendidikan Indonesia
71	Novita Wahyuni	novita.wahyuni95@gmail.com	Institut Teknologi Bandung
72	Dhewa Edikresnha	dhewa@staff.itb.c.id	Institut Teknologi Bandung
73	Robeth Viktoria Manurung	rvmanurung@gmail.com	Pusat Penelitian Elektronika dan Telekomunikasi - Lembaga Ilmu Pengetahuan Indonesia
74	Fardiman Ruli	fardi2020@gmail.com	Department of Physics, Faculty of Mathematics and Natural Science, Universitas Indonesia
75	Suvina V	suvina@jainuniversity.ac.in	Jain university
76	Sahrul Saehana	sahrulsaehana@gmail.com	Universitas Tadulako
77	Windri Handayani	windri.h@sci.ui.ac.id	Departemen Biologi, FMIPA, Universitas Indonesia
78	Alimin Alimin	alkimi_unhalu@yahoo.com	UNIVERSITAS HALU OLEO - KENDARI
79	Budi Mulyanti	bmulyanti@upi.edu	Department of Electrical Engineering Education, Faculty of Engineering and Technology Education, Universitas

IOP Conf. Series: Materials Science and Engineering **367** (2018) 011003 doi:10.1088/1757-899X/367/1/011003

			Pendidikan Indonesia, Bandung, Indonesia
80	La Agusu	la_agusu@yahoo.com	Department of Physics, Universitas Halu Oleo
81	santi rosanti	sanderosdr@gmail.com	Departemen Fisika Universitas Indonesia
82	Rani Deliana Panggabean	rani1945panggabean@gmail.com	Universitas Gadjah Mada (UGM)
83	Arie Listyarini	me.aurora.2646@gmail.com	Department of Physics, Universitas Indonesia
84	Bunda Amalia	bundy.amalia@gmail.com	Departemen Fisika FMIPA UI
85	Zulkarnain Jalil	zjalil@unsyiah.ac.id	University of Syiah Kuala
86	Silviana Simbolon Silviana	simbolonsilvi@gmail.com	Pusat Penelitian Fisika LIPI (Research Center for Physics, Indonesian Intitute of Sciences)
87	Fitri Suryani Arsyad	fsarsyad90@gmail.com	Unversitas Sriwijaya
88	Banu Nursanni	bnursanni@ymail.com	Postgraduate Program of Materials Science Study, Department of Physics, Faculty of Mathematics and Sciences, Universitas Indonesia
89	Muhammad Nurdin	mnurdin06@yahoo.com	Department of Chemistry, Universitas Halu Oleo, Southeast Sulawesi, Indonesia
90	Naila Mubarok	naila_mubarok@yahoo.com	indonesian
91	Risa Suryana	rsuryana@staff.uns.ac.id	Department of Physics, Faculty of Mathematics and Natural Sciences, Universitas Sebelas Maret, Indonesia
92	hamdan akbar notonegoro	hamzah29@yahoo.com	Universitas Sultan Ageng Tirtayasa
93	Yudyanto Yudyanto	yudyanto.fmipa@um.ac.id	Fisika FMIPA Universitas Negeri Malang
94	Karthikeyan Shunmugavel	karthisrivi@gmail.com	Assistant Professor
95	Wijayanti Dwi Astuti	wijayanti.dwi.a@ugm.ac.id	Universitas Gadjah Mada (UGM)
96	Nurnadiyah Syuhada	nurnadiyahsyuhada.ns@gmail.com	Advanced Functional Material Laboratory, Institut Teknologi Bandung
97	Rhesti Suhanto	rhestinsuhanto@gmail.com	Department of Engineering Physics, Faculty of Industrial Technology, Institut Teknologi Bandung, Jl. Ganesha 10, Bandung 40132, Indonesia
98	hendrik hendrik	hend027@lipi.go.id	Research Center for Metallurgy and Materials Indonesia Institute of Sciences
99	Ahmad Taufiq	ahmad.taufiq.fmipa@um.ac.id	Universitas Negeri Malang
100	Subandi Subandi	subandi.fmipa@um.ac.id	Universitas Negeri Malang
101	Muhammad Fauzan	alfauzan.rifqi@ugm.ac.id	Department of Electrical Engineering and Informatics, Vocational Collage, Universitas Gadjah Mada
102	Sunaryono Sunaryono	sunaryono.fmipa@um.ac.id	Universitas Negeri Malang

103	Kiagus Dahlan	kiagusdahlan@gmail.com	Division of Biophysics, Department of Physics, Faculty of Mathematics and Natural Sciences, Bogor Agricultural University
104	Arif Hidayat	arif.hidayat.fmipa@um.ac.id	Universitas Negeri Malang
105	nandang Mufti	nandang.mufti.fmipa@um.ac.id	Universitas Negeri Malang
106	Munasir Munasir	munasir_physics@unesa.ac	Universitas Negeri Surabaya
107	Munasir Munasir	munasir_physics@unesa.ac.id	Universitas Negeri Surabaya
108	Nike Khusnah	nike.fk@gmail.com	Universitas Brawijaya
109	Heru Susanto	heru.susanto@undip.ac.id	Department of Chemical engineering, Diponegoro University
110	Silviana Simbolon	m.yunus1603@gmail.com	Pusat Penelitian Fisika (Lembaga Ilmu Penelitian Indonesia)
111	Andi Indra Wulan Sari Ramadani	wulanramadhan.physics@yahoo.com	Polytechnic Institute of Nuclear Technology, National Nuclear Energy Agency (BATAN)
112	Irhamsyah Andi	irhamsyah@unm.ac.id	Universitas Negeri Makassar
113	Darmawati Darwis	darmadarwis@yahoo.com	CEntre of Organic Electronics Tadulako University
114	Samsul Hidayat	samsul.hidayat.fmipa@um.ac.id	Jurusan Fisika FMIPA UM
115	Setyawan P. Sakti	sakti@ub.ac.id	Brawijaya University
116	Askar Triwiyanto	ascarg@gmail.com	President University
117	Hasniah Aliah	sawitriasti@gmail.com	UIN SUNAN GUNUNG DJATI BANDUNG
118	Kuwat Triyana	triyana@ugm.ac.id	Universitas gadjah Mada
119	Budhy Kurniawan	bkuru@fisika.ui.ac.id	Universitas Indonesia (UI)
120	Budhy Kurniawan	bkuru@fisika.ui.ac.id	Universitas Indonesia (UI)
121	Abdulloh Fuad	abdulloh.fuad.fmipa@um.ac.id	Universitas Negeri Malang (UM)
122	Prof. Ariando	phyarian@nus.edu.sg	National University of Singapore
123	Prof. Mohammad Deraman	mderaman113@gmail.com	Universiti Kebangsaan Malaysia
124	Prof. Seiichi Miyazaki		Nagoya University Japan

IOP Conf. Series: Materials Science and Engineering **367** (2018) 011004 doi:10.1088/1757-899X/367/1/011004

Photographs

1) Registration

Content from this work may be used under the terms of the Creative Commons Attribution 3.0 licence. Any further distribution of this work must maintain attribution to the author(s) and the title of the work, journal citation and DOI.
Published under licence by IOP Publishing Ltd

IOP Conf. Series: Materials Science and Engineering **367** (2018) 011004 doi:10.1088/1757-899X/367/1/011004

2) Open Ceremony

IOP Conf. Series: Materials Science and Engineering **367** (2018) 011004 doi:10.1088/1757-899X/367/1/011004

3) Photo Session

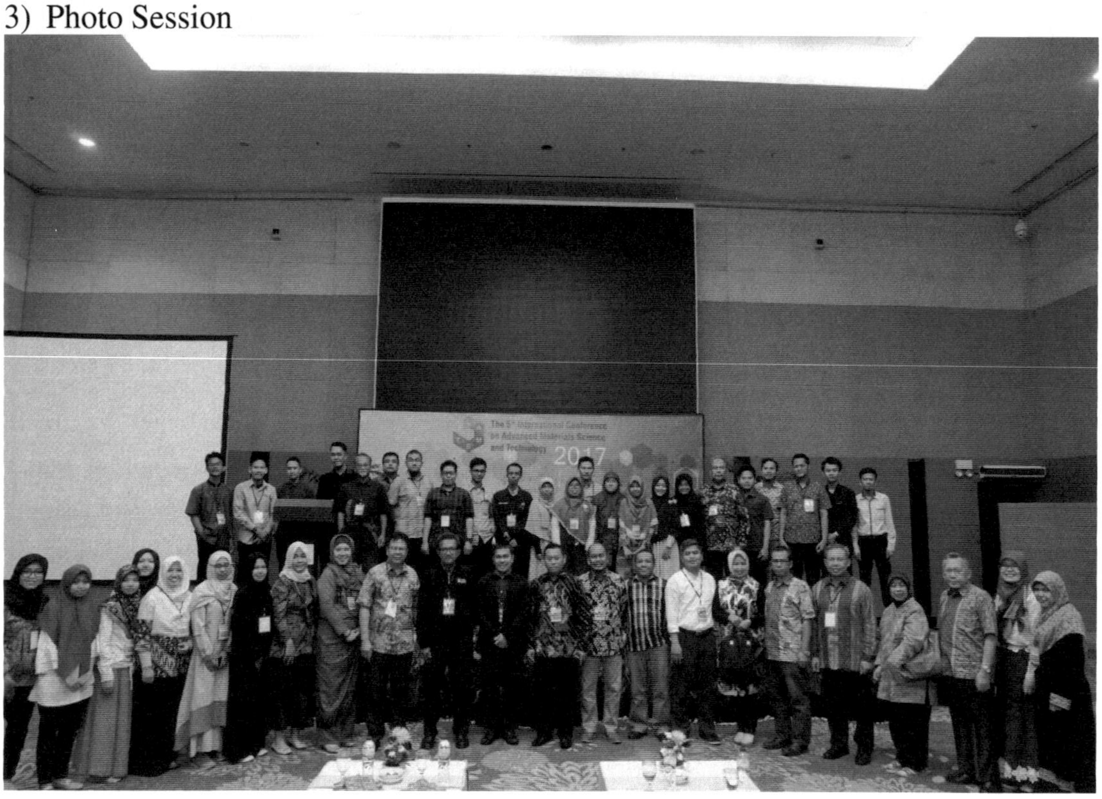

IOP Conf. Series: Materials Science and Engineering **367** (2018) 011004 doi:10.1088/1757-899X/367/1/011004

4) Plenary Session

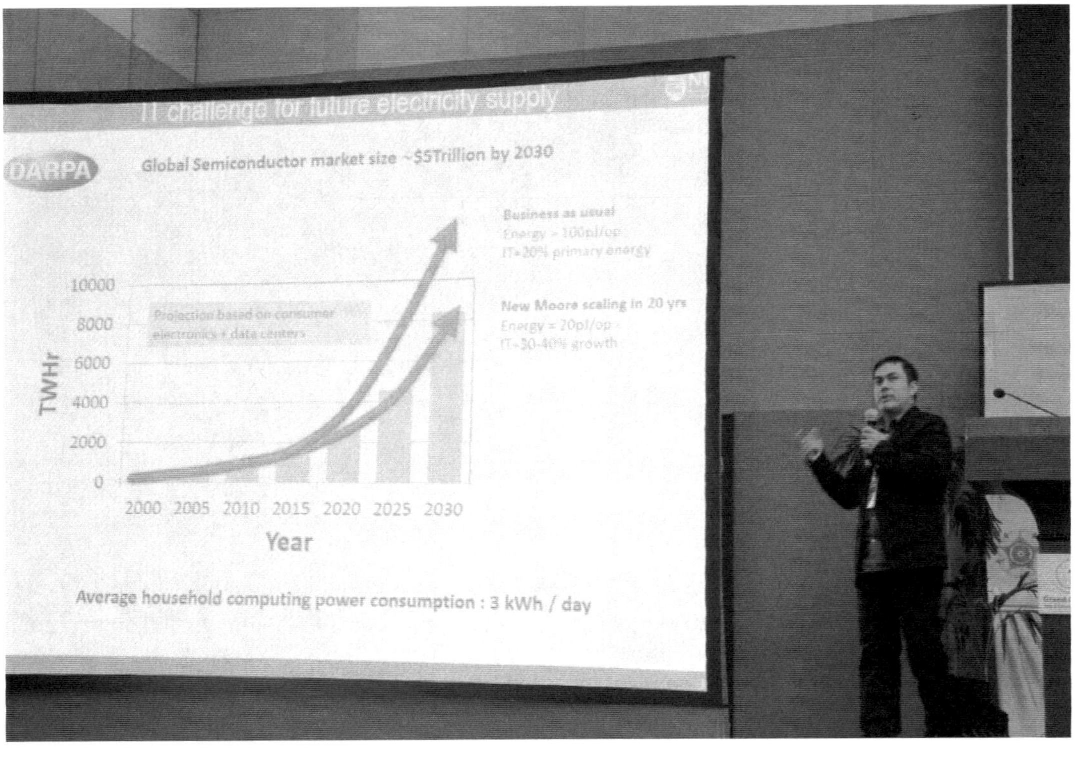

5) Presentation

IOP Conf. Series: Materials Science and Engineering **367** (2018) 011004 doi:10.1088/1757-899X/367/1/011004

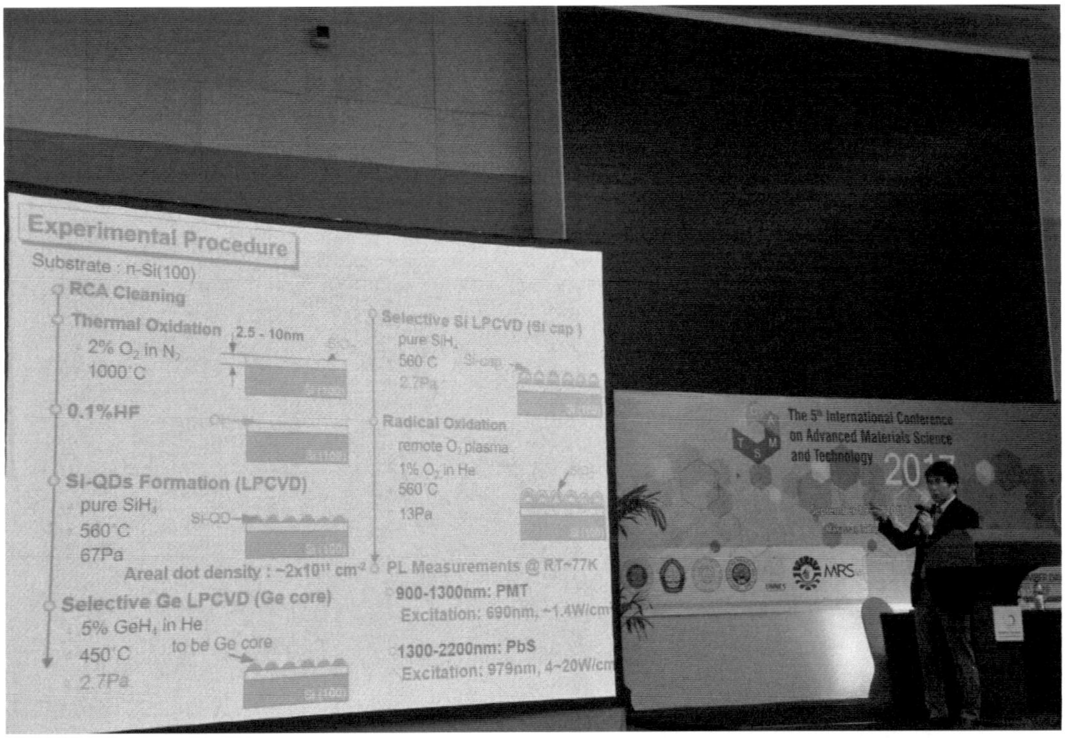

IOP Conf. Series: Materials Science and Engineering **367** (2018) 011004 doi:10.1088/1757-899X/367/1/011004

6) Participants

IOP Conf. Series: Materials Science and Engineering **367** (2018) 011004 doi:10.1088/1757-899X/367/1/011004

7) Parallel Session

IOP Conf. Series: Materials Science and Engineering **367** (2018) 011004 doi:10.1088/1757-899X/367/1/011004

IOP Conf. Series: Materials Science and Engineering **367** (2018) 011004 doi:10.1088/1757-899X/367/1/011004

8) Posters Session

IOP Conf. Series: Materials Science and Engineering **367** (2018) 011005 doi:10.1088/1757-899X/367/1/011005

Peer review statement

All papers published in this volume of *IOP Conference Series: Materials Science and Engineering* have been peer reviewed through processes administered by the proceedings Editors. Reviews were conducted by expert referees to the professional and scientific standards expected of a proceedings journal published by IOP Publishing.

Content from this work may be used under the terms of the Creative Commons Attribution 3.0 licence. Any further distribution of this work must maintain attribution to the author(s) and the title of the work, journal citation and DOI.
Published under licence by IOP Publishing Ltd

Effect of Zinc on Crystal Structure and Magnetic Properties of $Co_{1-x}Zn_xFe_2O_4$ Nanoparticles Synthesized by Coprecipitation Method

H. Kiswanto[1], A. Puspitasari[1], E. Suharyadi[1], T. Kato[2] and S. Iwata[3]

[1]Department of Physic, Universitas Gadjah Mada, Yogyakarta, Indonesia
[2]Departmen of Electronics, Nagoya University, Japan
[3]Institute for Material and Systems for Sustainability, Nagoya University, Japan

E-mail : esuharyadi@ugm.ac.id

Abstract. The effects of zinc (Zn) on the crystal structure and magnetic properties of $Co_{1-x}Zn_xFe_2O_4$ nanoparticles (NPs) were investigated. $Co_{1-x}Zn_xFe_2O_4$ NPs have been successfully synthesized by co-precipitation method with synthesis temperature of 70 °C and various of x from 0.2 to 0.8. The structural and magnetic properties were determined and characterized in detail by X-ray Diffractometer (XRD), Transmission Electron Microscope (TEM), Fourier Transform Infrared (FTIR) and vibrating sample magnetometer (VSM). X-ray Diffraction analysis showed that the samples had the single phase cubic spinel structure. The crystallite size of the samples was calculated by using Scherrer's formula within the range of 9.3 nm to 11.2 nm. The increase in lattice parameter from 8.179 Å to 8.212 Å by increasing zinc content as a result of the different ionic radii. The peak of FTIR spectra observed around 478 cm^{-1} and 586 cm^{-1} was assigned to vibration of octahedral and tetrahedral sites which confirmed the presence of the metal-oxide stretching band. The decrease in saturation magnetization (M_s) from 34.3 $emu{\cdot}g^{-1}$ to 20.3 $emu{\cdot}g^{-1}$ and the decrease in coercivity (H_c) from 674 Oe to 158 Oe are significantly affected by the increase of zinc content.

Keywords. Coercivity, coprecipitation method, magnetization, nanoparticles, Scherrer's formula.

1. Introduction

The investigation of nanoparticles ferrites recently become interesting because of its broad application in various fields. Nanoparticles ferrites have been used in many electronic devices because of their high electrical resistivity, mechanical hardness, and chemical stability. According to its structure, ferrites are classified into three types [1], i.e., spinel, garnets, and hexagonal ferrites. Study of spinel ferrite to be fascinating due to its interesting properties such as superparamagnetism, quantum magnetic tunneling and surface spin canting effect [2].

Among those spinel ferrites, we are interested in cobalt ferrite (CoFe2O4) and zinc ferrite ($ZnFe_2O_4$). Cobalt ferrite is known as a hard magnetic material with high coercivity, large magnetic anisotropy, high mechanical hardness and chemical stability [3]. On the other hand, zinc ferrite is a soft magnetic material with high resistivity, and high chemical stability [4]. Hard ferrite is suitable for inductance application, while soft ferrite is suitable for high-frequency applications. The combination

Content from this work may be used under the terms of the Creative Commons Attribution 3.0 licence. Any further distribution of this work must maintain attribution to the author(s) and the title of the work, journal citation and DOI.
Published under licence by IOP Publishing Ltd

of both properties will complement each other, and hence cobalt-zinc mixed ferrite becomes useful for inductance and high-frequency applications. They are used for power inductors, cores of transformers, electric motor, generator, etc. Additionally, Cobalt ferrite has inverse spinel, but zinc ferrite has normal spinel [5]. It could be inferred that studying cobalt-zinc ferrite become very interesting due to their different properties.

Deraz *et al.* [6] prepared $Co_{1-x}Zn_xFe_2O_4$ by combustion route found that increasing zinc concentration influenced the lattice constant, unit cell volume, ionic radii, the distance between magnetic ion and bond lengths on the tetrahedral and octahedral sites. Hassadee *et al.* [7] prepared $Co_{1-x}Zn_xFe_2O_4$ ($0 \leq x \leq 0.5$) by ceramic processing revealed that substitution of Co^{2+} with Zn^{2+} lead to increase the lattice parameter. Ali *et al.* [8] synthesized Co_{1-x} $Zn_xFe_2O_4$ ($x = 0.0 - 0.3$) by sol/gel method reported that the zinc substitution increases in saturation magnetization. They observed that the structural and magnetic properties could be modified by varying the concentration. Previous research has been done work with high-temperature synthesis, but lack of investigation in low temperature. The properties of $Co_xZn_{1-x}Fe_2O_4$ are influenced by the shape, size, and structure which depend on the synthesis process [9-11]. In this present work, the $Co_{1-x}Zn_xFe_2O_4$ NPs is synthesized by varying x from 0.2 to 0.8 using co-precipitation methods due to an easy in a homogeneous mixing the main components. The purpose of this study is to investigate the effect of Zn concentration on crystal structure and magnetic properties of $Co_xZn_{1-x}Fe_2O_4$.

2. Materials and methods

The precursor was used in the synthesis were $CoCl_2 \cdot 6H_2O$, $ZnSO_4 \cdot 6H_2O$, $FeCl_3 \cdot 6H_2O$ (Merck KGaA, Germany), NaOH and HCl. At first, we prepared two different solutions, i.e. solution A, and solution B. The solution A was a mixture of 0.2 M solution of $CoCl_2 \cdot 6H_2O$ mixed $ZnSO_4.6H_2O$ and prepared in 50 mL of distilled water. Separately, the solution B was prepared using 0.4 M of $FeCl_3 \cdot 6H_2O$ in 50 mL distilled water. Then solution A and B were mixed at room temperature, stirred, followed by added 3.37 mL of HCl (37 %). This mixture was drop-by-drop into a base solution (0.6 M solution of NaOH dissolved in 50 mL distilled water) and kept 70 °C under constant stirring. The solutions were maintained at 80 °C for 60 min. Then cooled to room temperature and precipitated. The fine particles were collected by using magnetic separation. These particles were washed several times to remove salt or impurities and dried at 90 °C for 4 h to obtain a powder.

The powder samples were determined and analyzed by Rigaku MiniFlex 600 Benchtop X-Ray Diffractometer using monochromatic Cu-Ka radiation ($\lambda = 1.5406$ Å). XRD patterns were recorded in the range of 20° to 80°. The average crystallite size was calculated by applying peak broadening of (311) using Scherrer's formula. The morphology and microstructure were observed by a Transmission Electron Microscope (JEOL TEM 1400) at 200 kV. Infra-Red Spectroscopy measurement in the range of 400 cm^{-1} to 4000 cm^{-1} was measured by Fourier Transform Infra-Red (Shimadzu FTIR Spectrometer). Magnetic properties measurement was carried out by using a VSM (Riken Denshi Co., Ltd) at room temperature.

3. Results and discussion

Figure 1 shows XRD patterns of the sample. The pattern shows that all the characteristic peaks are a spinel structure. All the peaks perfectly match with the crystalline phase of cubic spinel structure as compared with JCPDS data. All XRD peaks have been matched with JCPDS file no. 82-1049 ($ZnFe_2O_4$) and 22-1086 ($CoFe_2O_4$) [7]. The estimation of the crystallite size is based on the full width half maximum (FWHM) of the highest intensity of (311) plane. The crystallite size of the samples was estimated using Scherrer's formula. The lattice parameter was also calculated from the same plane of the spinel structure using Bragg's equation.

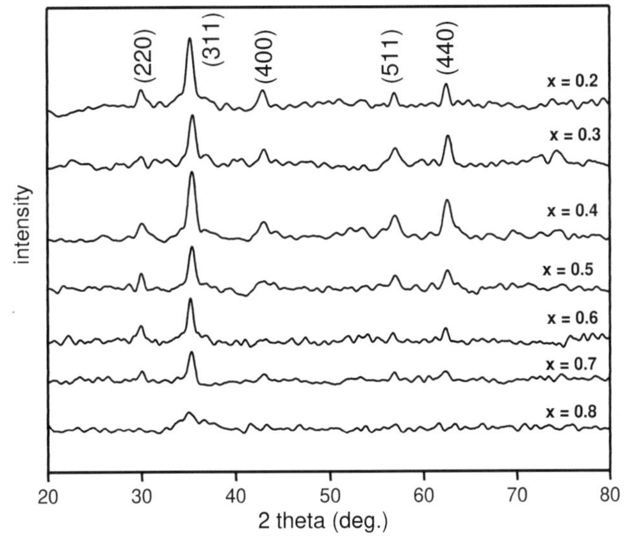

Figure 1. X-Ray Diffraction pattern for $Co_{1-x}Zn_xFe_2O_4$ with x varied from 0.2 to 0.8

Figure 2 shows the relation between zinc concentration on the crystallite size and lattice parameter. The crystallite size decreases with the increase in zinc concentration. It shows that the lattice parameter varies from 8.179 Å to 8.248 Å. It was observed that lattice parameter increases as zinc content increase. It could be explained that the ionic radii of zinc are higher than cobalt. The ionic radii of Zn^{2+} (0.88 Å) is larger than Co^{2+} (0.83 Å) [6]. Hence the lattice parameter increases as the increase in zinc content. Thus, the presence of zinc modifies in structural properties such as lattice parameter, ionic radii and the bond length on tetrahedral and octahedral sites of cubic spinel structure [8].

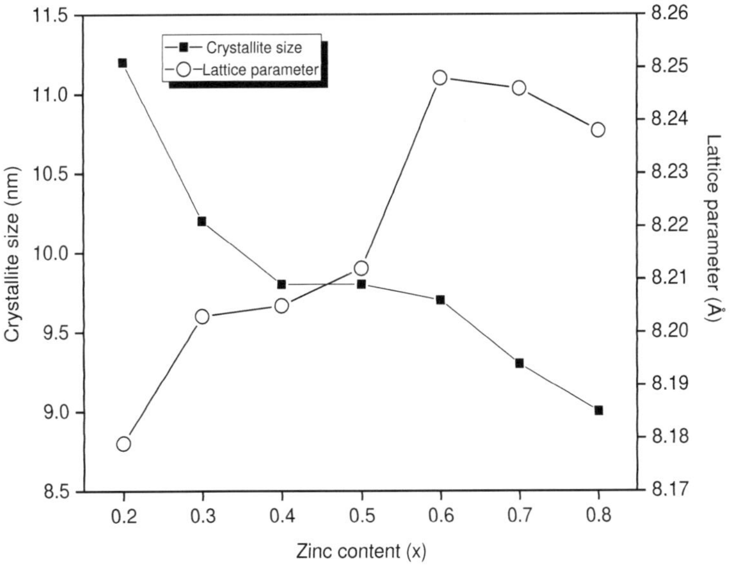

Figure 2. Crystallite size and lattice parameter of the $Co_xZn_{1-x}Fe_2O_4$

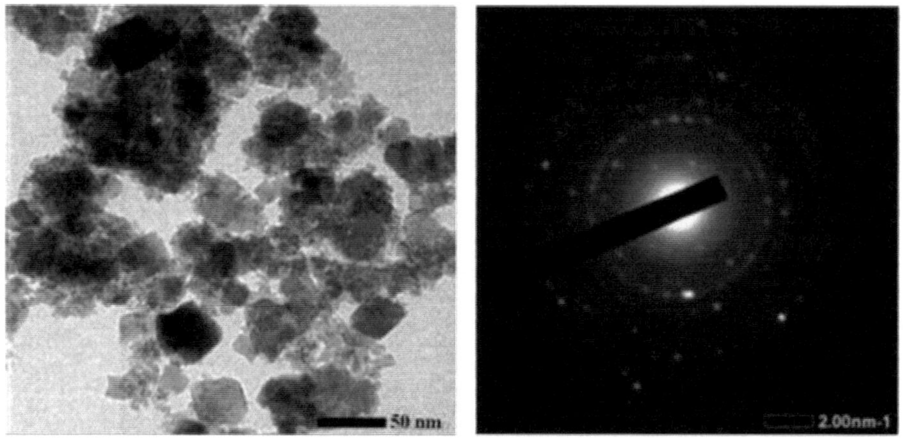

Figure 3. TEM Images of $Co_{0.5}Zn_{0.5}Fe_2O_4$ (a) Morphology (b) SAED pattern

The TEM images are shown in the Figure 3. The selected sample is x = 0.2 with 20 % zinc concentration because it has the highest crystallinity and low non-ferrite phase. The indicator that sample has good crystallinity as well as regular spherical shape and size boundaries of the particles. However, the powders sample can easily agglomerate due to the nanosize effect and intrinsic magnetism. To obtain a dispersed sample, the yielded samples were ground. The SAED pattern (Figure 3.b) shows that the sample is being polycrystalline. The crystallite size of nanoparticles has a range value from 9 nm to 11 nm.

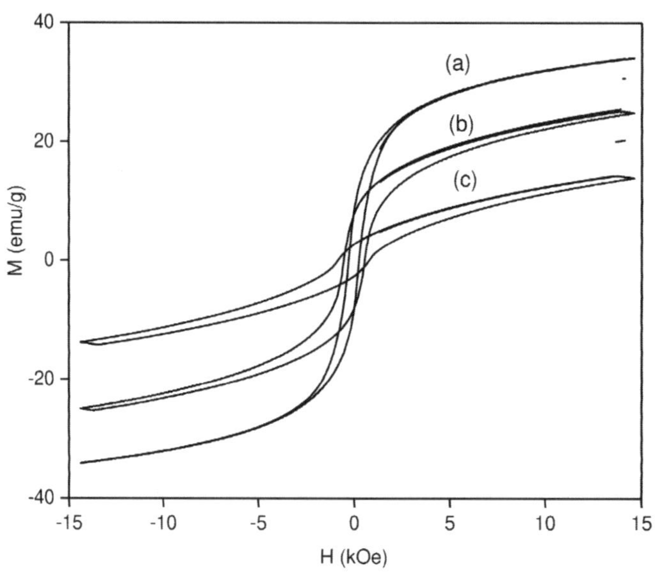

Figure 4. Hysteresis loop of $Co_{1-x}Zn_xFe_2O_4$ for (a) x = 0.2, (b) x = 0.5, and (c) x = 0.8

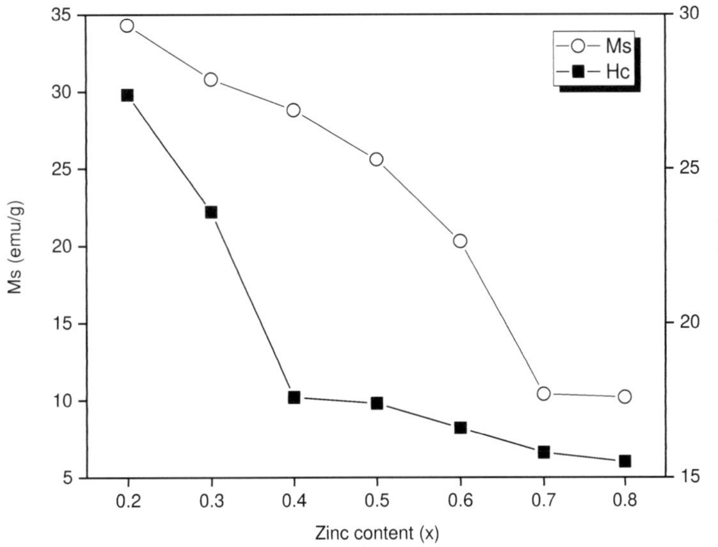

Figure 5. M_s and H_c of $Co_{1-x}Zn_xFe_2O_4$ NPs

Figure 4 shows the hysteresis curve of the sample. The M_s reaches the maximum value of 34.3 emu·g^{-1} for x = 0.2 and then decrease. The M_s and H_c decrease by increasing zinc content as shown in Figure 5. The decrease in M_s due to the replacement of cations because of the changed preferential occupancy in the case of nanoparticles, a change in the distribution of Co^{2+} and Zn^{2+} in the two sites. Since zinc ions substitute for cobalt ions, the magnetic coupling weakens because of that zinc does not have an impact on the magnetic moment. The reducing of magnetic properties could be understood as a result of the relatively high orbital contribution of Co^{2+} to the magnetic moment.

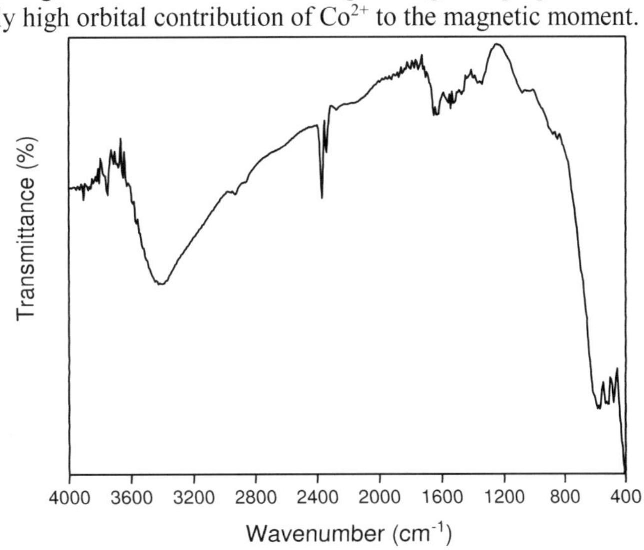

Figure 6. FTIR spectra of $Co_{0.5}Zn_{0.5}Fe_2O_4$

FTIR spectra of the sample are presented in Figure 6. There are some main broads of metal-oxygen bands are shown by the FTIR spectra. The wave number observed at 400 cm^{-1} to 500 cm^{-1} are identified as a stretching vibration of metal-oxygen from the tetrahedral site. The band in the range of 500 cm^{-1} to 600 cm^{-1} are recognized as octahedral metal stretching. We observed the band at 478 cm^{-1},

509 cm^{-1}, 586 cm^{-1} and 393 cm^{-1} are assigned as $M_{tetra} \leftrightarrow O$. The absorption bands observed at 3404 cm^{-1} prove the presence of adsorbed water on the surface of the ferrite nanoparticles.

4. Conclusions

The samples of $Co_{1-x}Zn_xFe_2O_4$ NPs have been successfully synthesized by co-precipitation methods. The XRD patterns show a cubic spinel structure for all sample. The increase in lattice parameter from 8.179 Å to 8.212 Å by increasing the zinc content as a result of Zn^{2+} substituted for Co^{2+} in the crystal structure. The crystallites size was found within the range of 9.3 nm to 11.2 nm. The decrease in magnetization saturation from 34.3 emu·g^{-1} to 20.3 emu·g^{-1} and the decrease in coercivity from 674 to 158 were affected by the increase of the zinc content. The peak observed around 478 cm^{-1} and 586 cm^{-1} of FTIR spectra were assigned to vibration of octahedral and tetrahedral sites which confirm the presence of a metal-oxide stretching band in ferrite.

Acknowledgments

This work was supported by Nanofabrication Consortium Project of Nagoya University, Ministry of Culture, Sports, Science, and Technology (MEXT), Japan and Grant of "Penelitian Berbasis Kompetensi", Ministry of Research Technology and Higher Education Republic of Indonesia (Kemenristek DIKTI).

References

[1] Panda R K 2016 Studies on Electric and Magnetic Properties of Cobalt Ferrite and its Modified Systems (Doctoral dissertation) *National Institute Technology of Rourkela*

[2] Bhowmik R N *et al* 2003 Magnetic enhancement of $Co_{0.2}Zn_{0.8}Fe_2O_4$ spinel oxide by mechanical milling *Physical Review B* **68** p 134433

[3] Kim Y I, Kim D, and Lee C S 2003 Synthesis and characterization of $CoFe_2O_4$ magnetic nanoparticles prepared by temperature-controlled coprecipitation method *Physica B: Condensed Matter* **337** pp 42–51

[4] Bhise B V *et al* 1991 X-ray infrared and magnetization studies on Mn substituted Ni-Zn ferrites *Journal of materials science letters* **10** pp 922–924

[5] Manouchehri S *et al* 2010 Preparation of Superparamagnetic of $Co_{0.5}Zn_{0.5}Fe_2O_4$ at Room Temperature by Co-precipitation Method and Investigation of Its Physical Properties *International Journal of Nanoscience and Nanotechnology* **6** pp 15–22

[6] Deraz NM and Alarifi A 2012 Structural, morphological and magnetic properties of nano-crystalline zinc substituted cobalt ferrite system *Journal of analytical and applied pyrolysis* **94** pp 41–47

[7] Hassadee A, Jutarosaga T, and Onreabroy W 2012 Effect of zinc substitution on structural and magnetic properties of cobalt ferrite *Procedia Engineering* **32** pp 597–602

[8] Ali M B *et al* 2016 Effect of zinc concentration on the structural and magnetic properties of mixed Co–Zn ferrites nanoparticles synthesized by sol/gel method *Journal of Magnetism and Magnetic Materials* **398** pp 20–25

[9] Deraz N M 2008 Production and characterization of pure and doped copper ferrite nanoparticles *Journal of Analytical and Applied Pyrolysis* **82** pp 212–222

[10] Alarifi A, Deraz N M, and Shaban S 2009 Structural, morphological and magnetic properties of NiFe2O4 nano-particles *Journal of Alloys and Compounds* **486** pp 501–506

[11] Deraz N M and Shaban S 2009 Optimization of catalytic, surface and magnetic properties of nanocrystalline manganese ferrite *Journal of Analytical and Applied Pyrolysis* **86** pp 173–179

Hydrothermal Synthesis of Reduced Graphene Oxide Using Urea as Reduction Agent: Excellent X-band Electromagnetic Absorption Properties

L. Agusu[1], L. O. Ahmad[2], Alimin[2], M. Nurdin[3], and Herdianto[3], S. Mitsudo[3], and H. Kikuchi[3]

[1]Department of Physics, Faculty of Mathematics and Sciences, Universitas Halu Oleo, Kampus Bumi Tridharma Anduonohu Kendari 93232, Indonesia
[2]Department of Chemistry, Faculty of Mathematics and Sciences, Universitas Halu Oleo, Kampus Bumi Tridharma Anduonohu Kendari 93232, Indonesia.
[3]Research Center for Development of Far-Infrared Region, University of Fukui, 3-9-1 Bunkyo, Fukui 910-8507, Japan

E-mail: laagusu@uho.ac.id

Abstract. We report a strong absorption of microwave energy at X-band (8 GHz to 12 GHz) by N-doped graphene. Attachment of nitrogen on the layered structure of GO improves the reflection loss of GO slab (2.0 mm, thickness) from −10 dB to −25.0 dB with a sharp bandwidth ~0.3 GHz. As for the broader bandwidth of about 1.4 GHz, reflection loss is −10.5 dB. This significant absorption may take place by improvement of magnetic property of NG through high magnetic coupling of localized spins induced by a defect on the surface of graphene. N atoms play as the electron trapper, easily influenced by self-magnetic moments and incoming electromagnetic fields to produce electric and/or magnetic losses. Here, urea acts as the reducing agent and N atoms donor for graphene oxide in hydrothermal process at a temperature of 190 °C.

Keywords. Electronic adsorption, hydrothermal synthesis, reduction agent, and urea.

1. Introduction

Graphene has been produced using mechanical [1] or thermal exfoliation, chemical vapor deposition [2], and epitaxial growth [3]. Mechanical or thermal exfoliation process was limited to involving a small number of the product. CVD and epitaxial growth need a substrate for deposition graphene, restrict the applicability, and yet, the substrate may influence the properties of graphene. Due to the applicability of graphene in a wide range of applications and promising new technologies, people were searching for a cheaper, more straightforward, more efficient and better yielding method of processing graphene, that can be scaled up massively compared to current techniques, and be financially suitable for industrial or commercial applications. One of the most effective ways of synthesized graphene on a large scale could be by the chemical reduction of graphene oxide. Graphene oxide itself has been popularly produced by hummer method that involved oxidation process of graphite followed by exfoliation process.

Content from this work may be used under the terms of the Creative Commons Attribution 3.0 licence. Any further distribution of this work must maintain attribution to the author(s) and the title of the work, journal citation and DOI.
Published under licence by IOP Publishing Ltd

As for chemical reduction of graphene oxide, in many cases, the active reduction agents were used [4–6]. Unfortunately, due to environmental safety aspects, strong acids cannot be utilized in the large-scale production facility. Using metals like Fe, Zn, and Al as reducing agents have also attracted significant interest in recent years [7]. However, to some extent, it was limited by the problem of a metal impurity.

Improvement of the magnetic property of graphene, in another side, was being an exciting issue in graphene research. Magnetism can take place in graphene via zig-zag structure, defect, or doped with the active substance. Doping graphene with sulfur or nitrogen [8, 9], for example, has proven enhancement of magnetic property of carbon or graphene. The existence of dopant on graphene surface can also change the electronic property of graphene.

In this report, doping of graphene oxide with nitrogen will be facilitated using urea as reduction agent through the hydrothermal process. It shows an excellent absorption of electromagnetic energy at X-band (8 GHz to 12 GHz).

2. Materials and methods

Graphene oxide (GO) was produced from fine graphite powder (particle size < 20 μm, Sigma Aldrich) following a modified of the Hummer's method [10]. Typically, a 2 g of graphite powder was added, under stirring, to 98 mL of concentrated H_2SO_4 in an ice bath. Under robust agitation, $KMnO_4$ (8.0 g) and 4 g $NaNO_3$ were added wisely until the complete suspension formed. Stirring process was continued for additional 24 h at ~35 °C, followed by addition of 15 mL H_2O_2 where color changed from dark brown to yellow. After that, the mixture was centrifuged, filtered, washed with 10 ml HCl 37 % and several times with H_2O to remove the remaining metal ions. The resulting solid, graphite oxide was dried in air at 105 °C for 12 h. The resultant of graphite oxide was then diluted to 200 mL of aquades, stirred overnight and subsequently sonicated for 120 min at 53 kHz. Finally, GO was filtered and dried for 24 h at 105 °C.

Nitrogen-doped graphene (NG) was produced *in situ* via the hydrothermal reduction of GO in the presence of urea with a fixed weight ratio of urea and GO (3:1). The 1.8 g of urea and 0.6 g of GO were mixed in 160 mL of de-ionized water under stirring process. The mixture, then, was transferred to a 200 mL teflon lined hydrothermal chamber and heated for 12 h at 190 °C. The resulting powders were characterized by multiple techniques as described below.

Crystalline structure and phase purity of the as-prepared samples (NG and GO) were examined using powder X-ray diffraction (XRD). XRD pattern was recorded with graphite-monochromated Cu-Kα (γ = 1.54056 Å) radiation on Rigaku RINT UltimaIII ranging from 10° to 70° (2θ) with a scanning speed of 2° per min. The surface morphology of the resulting products was investigated by a scanning electron microscopy (SEM) TESCAN. Transmission electron microscope (TEM) images were collected from TEM device of JEOL JEM 1400 USA. Fourier transforms infrared spectroscopy (FTIR) spectra were recorded on a Fourier transform infrared spectrometer (Bruker Vertex V70). Reflection loss of samples at X-band (8 GHz to 12 GHz) were measured using vector network analyzer (VNA) Advantest R3770 Network Analyzer

3. Results and discussion

3.1. Crystal structure

XRD patterns are shown in Figure 1a. GO diffraction spectrum exhibits a usual sharp peak of (002) crystal phase at 11.0° that corresponds to an interlayer distance of about ~0.80 nm. On the contrary, XRD peak of NG shows a board shape centering at ~24.78° that is a layer by layer distance of about 0.36 nm. Shifting of this peak from a strong peak of (100) GO crystal phase at 26.8°, reflecting a self-healing of graphitic crystal structure related to nitrogen embedding. The (002) reflections, corresponding to the coherent and parallel stacking of graphene like-sheets, become broader, signifying the increase of amorphous nature. However, the (100) reflection corresponds to the honeycomb structure which is formed by sp^2 hybridized carbons [11]. The heteroatom (N) doping in graphene facilitates defect site formation and destruction in the carbon lattice which, then, results in low crystallinity.

FTIR spectrum of GO and NG are shown in Figure 1b. In the range of GO, the peaks located at 3418 and 1111 cm^{-1} are attributed to the stretching vibration of hydroxyl groups [12]. The peaks at 1720 and 1219 cm^{-1} are assigned to the stretching vibrations of C=O and C–O–C in conjugated graphene, respectively. After hydrothermal reaction, the peak at 3418 cm^{-1} decreases accordingly. A disappearing peak at 1720 cm^{-1} can be assigned to a complete reduction of GO taking place where C=O groups are suppressed completely. More importantly, the peak at 1592 cm^{-1}, corresponding to the in-plane vibration of C=C, shifts to 1586 cm^{-1}. The disappearance is due to the superposition of the vibrations of C=C and C=N, which gives the evidence of the embedding of nitrogen-containing groups.

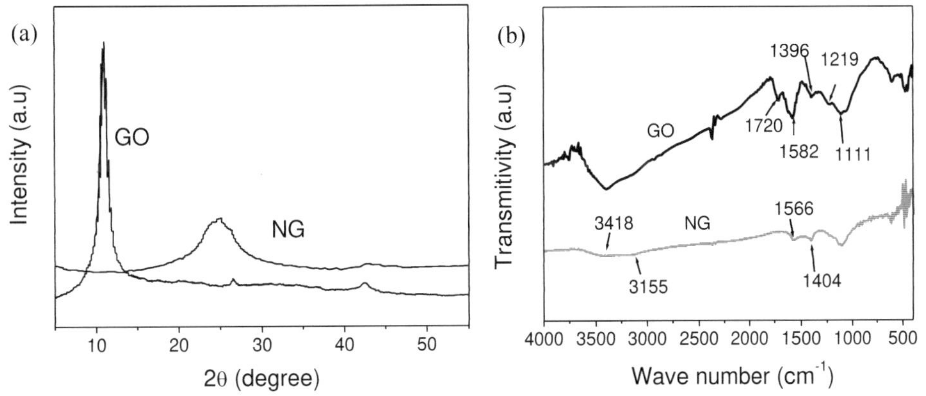

Figure 1. Spectroscopic measurement data of GO and NG, (a) XRD diffraction pattern and (b) FTIR spectra

3.2. Surface morphology

Morphology of GO and NG are presented in Figure 2. As for GO, the clump layered GO is shown in Figure 2a and Figure 2c. Morphology of NG surface in Figure 2b and Figure 2d is resulting from the considerable folding of an individual graphene sheet. The dark spots appeared on Figure 2c can be attributed to the evidence of destructive parts of carbon lattice.

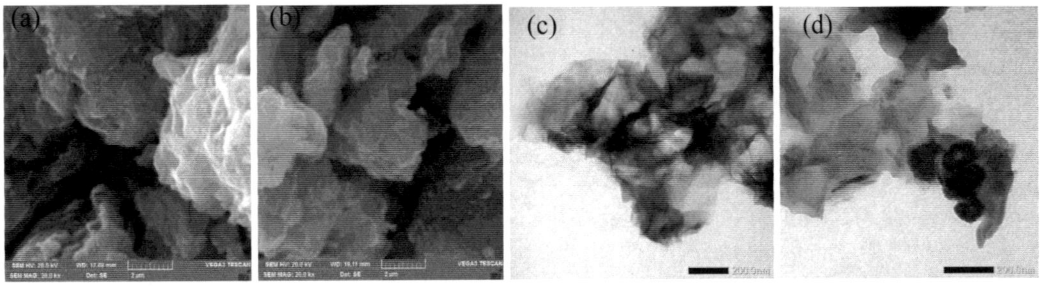

Figure 2. Images of SEM and low magnification TEM; SEM images of (a) GO and (b) NG; TEM images of (c) GO and (d) NG

3.3. Absorption capability

Absorption capabilities of GO and NG through measurement of reflection loss (R_L) of sample backed by copper plate are shown in Figure 3. Sample thickness of 1.0 and 2.0 mm are involved in analyses at X-band with frequency ranging from 8 GHz to 12 GHz. Attachment of nitrogen on the layered structure of GO improves absorption capability by a factor of 2.5 compared to absorption of GO only. In the case of only 2 mm NG thickness, R_L is -25.0 dB (bandwidth ~ 0.3 GHz) corresponding to power absorption of 91.8 %, or R_L is –10.6 dB (bandwidth ~1.5 GHz) related to 65.4% of absorption of incoming microwave power. These achievements are very significant compared to another result [13]. They used graphene having a thickness of 4 mm. However, R_L was only in the order of comparable to graphene having a thickness of ~ –4.3 dB. A similar absorption, in [13], was achieved by 2 mm of Fe-doped graphene. Significant absorption of microwave energy may occur from self-doping effect took place by lattice defects in graphene and also some carriers induced into graphene through replacement of oxygen sites by nitrogen atoms [14]. A defect in graphene can only produce a small quantity of carrier, and hence, it would not substantially change the dielectric effect and absorption capability as shown in [15]. This significant absorption may take place by improvement of magnetic property of NG.

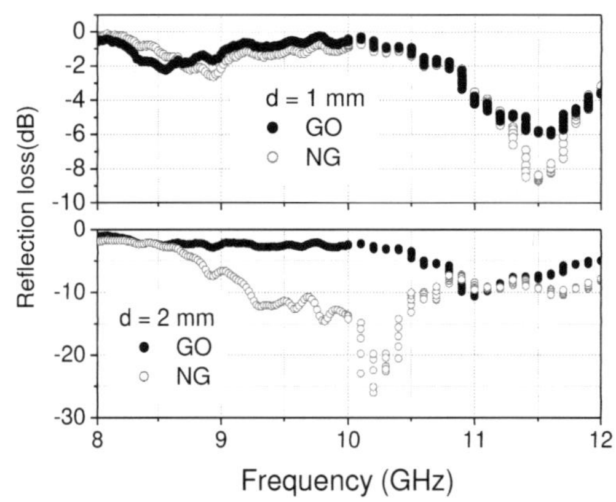

Figure 3. Reflection loss measurements data of GO and NG at different thickness, (a) 1.0 mm and (b) 2.0 mm.

3.4. Discussions

Magnetic properties of graphene can be investigated through intra-molecular and inter-molecular magnetic interaction. Intramolecular magnetic interaction arises from magnetic exchange interaction between unpaired electrons in a molecule. The typical interaction producing magnetism is between unpaired electron *d* or *f* orbital and graphene π electron. However, the unusual interaction creating the magnetic moment was shown by calculation which took place by unpaired 2*p* electrons of nitrogen atoms at a low concentration of nitrogen atoms [16]. This mechanism is one of the reasonable sources of magnetism responsible for strong absorption of microwave energy. Introducing nitrogen doping would possibly perform the spin symmetry breaking, resulting in energy degeneracy at some doping configurations. The spin symmetry breaking would cause spin-polarized effects, which induce a magnetic response in graphene.

Intermolecular magnetic interaction arises when ferromagnetic (FM) or antiferromagnetic (AFM) molecules approach each other. Defects in the structure of graphene can facilitate evidence of localized spin in the form of the armchair and/or zigzag structure. The localized spin on the armchair and zig-zag structure interact each other to form a ferromagnetic and antiferromagnetic state, respectively. Experimental studies have been performed to understand intermolecular interactions [17]. Unfortunately, the intermolecular ferromagnetic spin couplings are rather weak. Thermal vibration effect marked it. Recently, the magnetic property of pure graphene was demonstrated experimentally at room temperature. The room temperature magnetism was ascribed to the ferromagnetic order of spin magnetic moment arrangement at the graphene layer edges [18]. Evidence of self-magnetism in graphene film plays a vital role in that experiment. In the case of NG, number defects can be multiplied in compare to undoped graphene. Figure 4 shows indication of defect in high-resolution TEM image of NG with three probable formations of N atoms on the structure of graphene (pyrrolic, pyridinic, and graphitic attachments).

Other detail and clear explanations can be found in [13]. The magnetic coupling between the magnetic moments may appear via Ruderman–Kittel–Kasuya–Yosida (RKKY) interactions by delocalized π electrons, which is expected to decay as D^{-r}, where D is the distance between the magnetic moments and r is the decay exponent. In this case, the increasing of localized magnetic moments of graphene because of doping, the gap between the magnetic moments will decrease. As for the decay exponent r, enhancing of electron-electron interaction may reduce the value of r, and hence, increasing magnetic coupling. Instead of RKKY interactions, the bound magnetic polaron (BMP) may responsible for the existence of ferromagnetism in graphene. Ferromagnetic coupling is derived from shallow donor electrons that form BMPs, which overlap to create a spin-split impurity band. The N atoms act as BMP and trapping the free electrons. The electrons trapped in these BMPs tend to get easily polarized under the influence of the magnetic field and, resulting in ferromagnetism. It should be noted, here, that an appropriate amount of N atoms should be required for establishment of ferromagnetic graphene at room temperature.

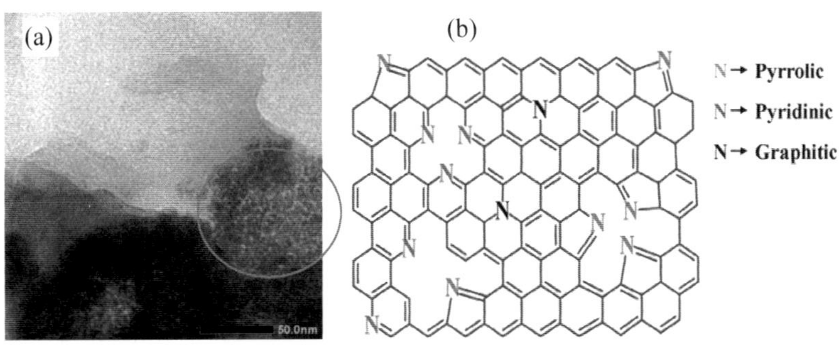

Figure 4. Evidence of defect on NG, (a) TEM image and (b) Three possible attachment of N atoms on graphene surface (pyrrolic, pyridinic, and graphitic)

4. Conclusions

N-doped graphene has been synthesized successfully by the hydrothermal process at 190 °C and 12 h using urea as reduction agent and N atom donors. Absorption capability of NG was investigated through measurement of reflection loss of metal backed graphene slab at X-band. In the case of only 2 mm of NG thickness, R_L is –25.0 dB (bandwidth ~ 0.3 GHz) corresponding to power absorption of 91.8 % or –10.6 dB (bandwidth ~1.5 GHz) comparable to absorption of 65.4 % of incoming microwave power. These achievements are very significant compared to –4.3 dB reflection loss of 4 mm graphene slab in [13]. The defect of graphene has been identified via a moderate magnification of TEM image. Defect and attachment of N atoms play an important role in the improvement of electric and/or magnetic losses which responsible for excellent absorption of X-band microwave energy.

Acknowledgments

This work was partially supported by the Fundamental Research Grant in 2017 from the Ministry of Research and Higher Education Republic of Indonesia, contract no. 065/ADD/SP2H/LT/DRPM/VIII/2017.

References

[1] Novoselov K S *et al* 2004 Electric Field Effect in Atomically Thin Carbon Films *Science* **306** 666–669
[2] Yang W *et al* 2013 Epitaxial growth of single-domain graphene on hexagonal boron nitride *Nature Materials* **12** 792–797

[3] Chen G, Wen W G, and Wu D J Preparation and characterization of graphite nanosheets from ultrasonic powdering technique 2004 *Carbon* **42** 753–759
[4] Stankovich S *et al* 2007 Synthesis of graphene-based nanosheets via chemical reduction of exfoliated graphite oxide *Carbon* **45** 1558–1565
[5] Wang G *et al* 2008 Facile Synthesis and Characterization of Graphene Nanosheets *J. Phys. Chem. C* **112** 8192–8195
[6] Muszynski R, Seger B, and Kamat P V 2008 Decorating Graphene Sheets with Gold Nanoparticles *J. Phys. Chem. C* **112** 5263–5266
[7] Fan Z J *et al* 2011 Facile synthesis of graphene nanosheets via Fe reduction of exfoliated graphite oxide *ACS Nano.* **5** 191–198
[8] Gao H *et al* 2012 Synthesis of S-doped graphene by liquid precursor *Nanotechnology* **23** p 275605
[9] Wang H, Maiyalagan T, and Wang X 2012 Review on Recent Progress in Nitrogen-Doped Graphene: Synthesis, Characterization, and Its Potential Applications *ACS Catal.* **2** 781−794
[10] Agusu L *et al* 2017 Transmission characteristic of graphene/TiO_2 paper measured at Ka-band *AIP Conference Proceedings* **1801** p 050006
[11] Geng D *et al* 2011 Nitrogen doping effects on the structure of graphene *Applied Surface Science* **257** 9193-9198
[12] Zhang Y *et al* 2015 Nitrogen-doped graphene as a cathode material for dye-sensitized solar cells: effects of hydrothermal reaction and annealing on electrocatalytic performance *RSC Adv.* **5** p 10430
[13] Zhao X *et al* 2013 Excellent microwave absorption property of Graphene-coated Fe nanocomposites *Scientific Reports* **3** p 3421
[14] Huard B *et al* 2008 Evidence of the role of contacts on the observed electron-hole asymmetry in graphene *Phys. Rev. B* **78** p 121402
[15] Wang Z, Qin S, and Wang C 2014 *Eur. Phys. J. B* **87** p 88
[16] Liu Y *et al* 2013 Realization of ferromagnetic graphene oxide with high magnetization by doping graphene oxide with nitrogen *Scientific Reports* **3** p 2566
[17] Rajca A 1994 Organic Diradicals and Polyradicals: From Spin Coupling to Magnetism? *Chem. Rev.* **94** 871-893
[18] Wang C and Diao D 2017 Self-magnetism induced large magnetoresistance at room temperature region in graphene nanocrystallited carbon film *Carbon* **112** 162-168

Contribution of Relaxation Effect to the Permittivity of $Mg_{1-x}Zn_xTiO_3$ Ceramics

F. U. Ermawati[1], Z. A. I. Supardi[1], S. Suasmoro[2], S. Pratapa[2] and T. Hübert[3]

[1]Physics Department, Faculty of Mathematics and Natural Sciences, Universitas Negeri Surabaya (UNESA), Surabaya 60231, Indonesia
[2]Physics Department, Faculty of Mathematics and Sciences, Institut Teknologi Sepuluh Nopember (ITS), Surabaya 60111, Indonesia
[3]Bundesanstalt für Materialforschung und –prüfung (BAM) Unter den Eichen 87, 12205 Berlin, Germany

E-mail : frida.ermawati@unesa.ac.id

Abstract. This work reported the investigation on the effect of relaxation to the permittivity (ε'_r) characteristic of $Mg_{1-x}Zn_xTiO_3$ ceramics for $x = 0 - 0.5$ (MZT0 – MZT0.5) measured from 1 Hz to 330 MHz. Within that frequency range, the relaxation effect that consists of the space charge (SC) and the dipolar (D) polarization mechanisms were identified. The contribution of the D relaxation in MZT0 – MZT0.2 systems extents overall from about 100 Hz to 330 MHz, while that in MZT0.3 – MZT0.5 systems is from 50 kHz to 330 MHz. The remaining frequencies, *i.e.* from 1 to 90 Hz for MZT0 – MZT0.2 and from 1 Hz to 50 kHz for MZT0.3 – MZT0.5, are attributed to the SC relaxation. The D polarization mechanism provides constant ε'_r values which vary from (15.4 – 17.0) ± 0.3 throughout the samples. Contribution of the SC polarization mechanism to the characteristic is supported by the simultaneous presence of different content and level of resistivity of the secondary phase of $(Mg_{1-\alpha}Zn_\alpha)_2TiO_4$ in MZT0 – MZT0.2 systems and of $(Zn_{1-\alpha}Mg_\alpha)_2TiO_4$ in MZT0.3 – MZT0.5, along with the presence of the main $Mg_{1-x}Zn_xTiO_3$ phase, as a result of the variation of zinc content in the systems.

Keywords. Dielectric ceramic, dipolar polarization, $Mg_{1-x}Zn_xTiO_3$, permittivity, and space charge polarization.

1. Introduction

Magnesium titanate ($MgTiO_3$) with ilmenite structure has been a favorite dielectric ceramic candidate for using as electronic components in various applications in communication systems operating at microwave frequencies such as resonators, filters, and antennas for global positioning systems [1, 2], as well as type I capacitors [3, 4]. These are due to characteristics of the dielectric met the important criteria for using in the microwave frequencies, *i.e.* a high relative permittivity (ε_r) to reduce the size of the components, especially when miniaturization of microwave circuits has been a concern in wireless communication industry, a high-quality factor ($Q \times f$) to increase the frequency selectivity, and a good temperature stability (τ_f) to ensure the performance stability at high temperatures [4].

Polarization processes in ceramic dielectrics due to the application of electric field occurs via relaxation and resonance effects. Dielectric relaxation refers to the relaxation response of a dielectric

Content from this work may be used under the terms of the Creative Commons Attribution 3.0 licence. Any further distribution of this work must maintain attribution to the author(s) and the title of the work, journal citation and DOI.
Published under licence by IOP Publishing Ltd

medium to an external, oscillating electric field. Such response, which is due to space charge (SC) and dipolar (D) polarizations, is often described in terms of permittivity as a function of frequency in the range below 10^{10} Hz; and for ideal systems this response can be described by the Debye equation [5]. On the other hand, dielectric sonance results from the distortion related to ionic and electronic polarizations that shows the behavior of the resonance type at optical frequencies since the resonances of electrons in atoms and of ions in crystals occur at optical frequencies (about 10^{15} Hz and 10^{13} Hz respectively) [5]. Information on how the relaxation and the resonance effects contribute to the nature of permittivity of dielectric material is therefore advantageous. As far the author's concern, however, the number of reports to discuss this matter is very limited.

This paper is devoted to reporting the study on the characteristic of real part of the relative permittivity (ε'_r) of $Mg_{1-x}Zn_xTiO_3$ dielectric ceramic systems measured from 1 Hz to 330 MHz, to identify the relaxation effect within that frequency range in the form of SC and D polarization mechanisms, and to investigate the contribution of those polarization mechanisms to the ε'_r characteristic. Analyses were related to the structural data and the electrical property of the identified phases in the systems.

2. Materials and methods

The six different compositions of $Mg_{1-x}Zn_xTiO_3$ dried powders for x = 0, 0.1, 0.2, 0.3, 0.4 and 0.5 (MZT0 – MZT0.5) were synthesized from Mg, Ti and Zn metal powders (Merck©) following the dissolved-metals-mixing procedure reported in [6, 7]. The MZT0 – MZT0.5 green bodies were prepared using a uniaxial die in the form of a disc and to press at 100 MPa using a Weber-hand hydraulic press. A Carbolite RHF-1400 furnace was occupied to sinter the bodies. Sintering was performed in the atmospheric environment and at the heating rate of 10 °C/min. The bulk density was measured following the Archimedes technique. Structural characterization of the sintered ceramics was performed using X'Pert Diffractometer (Philips) with CuK_α radiation. Dielectric property in the form of real part of the relative permittivity (ε'_r) were measured from 1 Hz to 32 MHz using SI 1260 Impedance/Gain-Phase Analyzer (Solartron Analytical), followed by from 1 kHz to 330 MHz using HP 4291A RF Impedance/Material Analyzer and HP 16453A Dielectric Test Fixture (Agilent). All the measurements were performed at room temperature (23 ± 2) °C.

3. Results and discussion

3.1. *The Origin of Space Charge and Dipolar Polarization Mechanisms*

Polarization processes in ceramic dielectric systems are related to the resonance and relaxation effects. The first effect is in the form of electronic and ionic polarizations that takes place at optical frequencies (about 10^{15} Hz and 10^{13} Hz, respectively). The second effect, *i.e.* the effect that was under investigation in this work, consists of the dipolar (D) and the space charge (SC) polarizations that occur at lower frequencies, below 10^{10} Hz [5]. The SC relaxation involves a limited transport of charge carriers, in this case, in the form of distribution of thermally excited electrons above 0 K in the Fermi-Dirac statistic. Such transport will be stopped at a potential barrier, possibly in the form of a grain boundary or an inter-phase boundary. The applied electric field polarizes the distribution of thermally excited electrons following the orientation of crystal lattices. Meanwhile, the D polarization is due to the presence of permanent electric dipoles in the crystal lattices, could be the oscillation of cations or cation-anion bonds in the system. In the case of the MZT0 – MZT0.5 dielectric ceramics under investigation, the intended permanent dipoles are therefore addressed to the oscillation of Ti^{4+} and Mg^{2+} cations in MZT0 system and of Ti^{4+} and $(Mg_{1-x}Zn_x)^{2+}$ cations in MZT0.1 – MZT0.5 systems, or the oscillation of TiO_6 and MgO_6 bonds in the crystal lattices of MZT0 and of TiO_6 and $(Mg_{1-x}Zn_x)O_6$ bonds in the crystal lattices of MZT0.1 – MZT0.5 as a result of the ABO_3 stoichiometry of the ilmenite structure of the systems, where A and B are cations while O is anion.

3.2. Dielectric Characteristic: Real Part of Relative Permittivity

Figure 1 provides characteristics of the real part of relative permittivity (ε_r') of MZT0 – MZT0.5 ceramic systems after the systems were sintered at 1400 °C for 8 h and that were measured from 1 Hz to 330 MHz using SI 1260 Impedance/Gain-Phase Analyzer (Solartron Analytical), HP 4291A RF Impedance/Material Analyzer and HP 16453A Dielectric Test Fixture (Agilent).

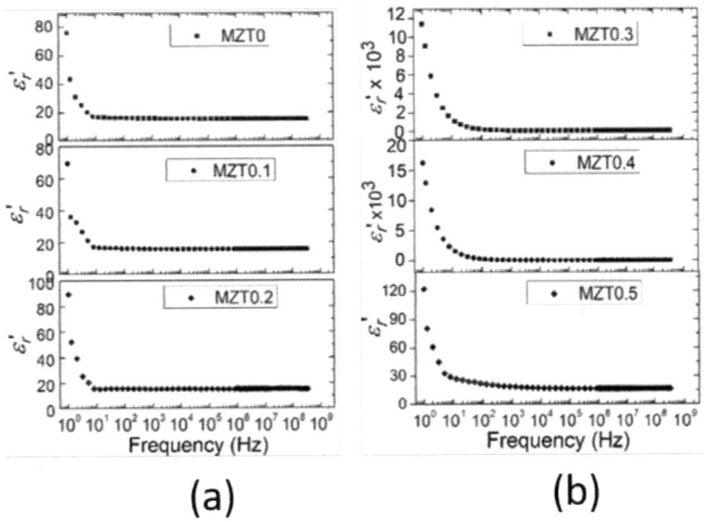

Figure 1. Variation of ε_r' with frequency from 1 Hz to 330 MHz measured from: (a) MZT0 – MZT0.2 and (b) MZT0.3 – MZT0.5 ceramic systems.

A dielectric relaxation response usually consists of relative permittivity (ε_r) recorded from two different zones. The first zone located in the low-frequency region is signified by an exponential decrease of ε_r-values and high loss factor (tan δ). These exponential decrease and the high loss factor characteristics were considered due to the presence of SC polarization as a result of the accumulation of mobile charges at structural surfaces and interfaces. The second zone is located at the frequencies which are higher than that of the first zone. Such zone is marked by a relatively constant ε_r due to the presence of permanent dipoles in the crystal structure. All the features described in the two different zones are recorded on the curves in Figure 1. As shown, the ε_r' curves reach the maximum values at 1 Hz and then decline exponentially before being constant thereafter. The decline exponential part is therefore justified as the SC polarization contribution, while the constant and the frequency independent part is contributed by the D polarization. Figure 1 also depicts that in MZT0 – MZT0.2 curves, the ε_r' values at 1 Hz are all similar, *i.e.* (76.0 ± 1.0) for MZT0, (69.4 ± 1.0) for MZT0.1 and (89.5 ± 1.2) for MZT0.2. Another similarity of the ε_r' values at 1 Hz are also shown in MZT0.3 – MZT0.5 curves, *i.e.* (11,372.0 ± 110.0) for MZT0.3, (16,261.0 ± 130.0) for MZT0.4 and (122.0 ± 1.5) for MZT0.5. The three later values, on average, are 1000 times higher than the three former values. This significant difference is believed mainly due to the population number difference of the charge carriers that are successfully passing through the potential barriers existed in the systems. In this case, the number of existing populations in MZT0.3 – MZT0.5 ceramics are about 1000 times higher than that in MZT0 – MZT0.2 ceramics. In a polycrystalline and mixed-phase ceramic, the SC polarization is predominantly arising from the difference between the conductivity of the phases present. Analysis on the conductivity of the phases present in MZT0 – MZT0.5 ceramics is provided in the following paragraphs and the data was supported by the phase identification result on the XRD spectra shown in Figure 2 and the Rietveld analysis in Figure 3.

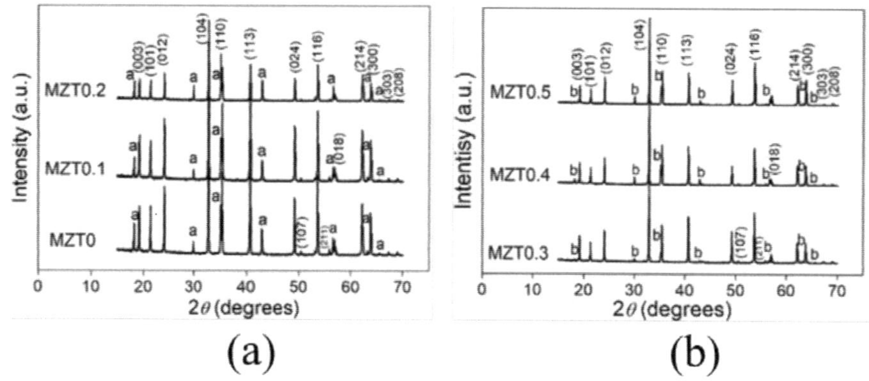

Figure 2. The XRD spectra (CuK_α radiation) of: (a) MZT0 – MZT0.2 and (b) MZT0.3 – MZT0.5 ceramics sintered at 1400 °C for 8 h. The indexed peaks = $Mg_{1-x}Zn_xTiO_3$, a = $(Mg_{1-\alpha}Zn_\alpha)_2TiO_4$ and b = $(Zn_{1-\alpha}Mg_\alpha)_2TiO_4$.

Figure 2 shows the XRD spectra of MZT0 – MZT0.5 ceramics after sintering at 1400 °C for 8 h. On each spectrum, two different phases are identified: 1) $MgTiO_3$ (PDF No. 06-0494) as the main phase, *i.e.* peaks with *hkl* and 2) the peaks labeled 'a' and 'b' as the secondary phase. The phase identification analysis on these secondary 'a' and 'b' phases reveals that the 2θ position and the variation of relative intensity of the 'a' peaks recorded in the MZT0 – MZT0.2 spectra corresponds to the 2θ position and the variation of relative intensity of Mg_2TiO_4 phase (PDF No.25-1157). Meanwhile the 2θ position and the variation of relative intensity of the 'b' phase detected in the MZT0.3 – MZT0.5 spectra matches the 2θ position and the variation of relative intensity of Zn_2TiO_4 (PDF No. 77-0014). Based on the Rietveld analysis data in Figure 3, the intended main phases are $Mg_{1-x}Zn_xTiO_3$ solid solution that can be indexed as $MgTiO_3$, while the secondary phases are $(Mg_{1-\alpha}Zn_\alpha)_2TiO_4$ and $(Zn_{1-\alpha}Mg_\alpha)_2TiO_4$ solid solutions that can be indexed as Mg_2TiO_4 and Zn_2TiO_4 respectively; in this case, the α-values are unknown. The porosity data in Figure 3B was 1 – *RD*, where *RD* is the fraction of relative density. The relative density is the ratio between the bulk density and the theoretical density. The theoretical density is the output of the Rietveld analysis.

Kumar [8] reported that Mg_2TiO_4 dielectric exhibits the characteristic of AC conductivity which varies from 5.0×10^{-6} $(\Omega\cdot cm)^{-1}$ at 75 kHz to 1.5×10^{-4} $(\Omega\cdot cm)^{-1}$ at 20 MHz. Meanwhile, Obradovic [9] measured the characteristic of resistivity of Zn_2TiO_4 dielectric of 1.02 Ωm at 4 MHz. This value corresponds to a conductivity of approximately 0.98 $(\Omega\cdot m)^{-1}$. These two-conductivity data confirmed that the $(Mg_{1-\alpha}Zn_\alpha)_2TiO_4$ phases detected in the MZT0 – MZT0.2 ceramics are very resistive as compared to the $(Zn_{1-\alpha}Mg_\alpha)_2TiO_4$ phases recorded in the MZT0.3 – MZT0.5 ceramics. Filipovic [10] measured the electrical resistivity of 0.38 MΩ·m at 5 MHz from the $MgTiO_3$ dielectric. This value is equivalent to a conductivity of about 2.63 $(\mu\Omega\cdot m)^{-1}$. Based on this literature, MZT0.3 – MZT0.5 ceramics were justified to be more conductive than MZT0 – MZT0.2 ceramics. Potential barriers in conductive dielectrics are much lower than the height of the same barriers in resistive dielectrics. As a result, the motion of charge carriers passing through the barriers in MZT0.3 – MZT0.5 ceramics should be easier and smoother than when the similar charge carriers going through the barriers in MZT0 – MZT0.2 ceramics. The population number of the charge carriers at the boundaries in MZT0.3 – MZT0.5 ceramics are therefore much greater than the similar number found at the boundaries in MZT0 – MZT0.2 ceramics. Bigger population numbers mean a larger polarization and, hence, a higher permittivity value. For that reason, at 1 Hz, the ε'_r values in MZT0.3 – MZT0.5 ceramics should be higher than the values in MZT0 – MZT0.2 ceramics; that is what is shown in Figure 1.

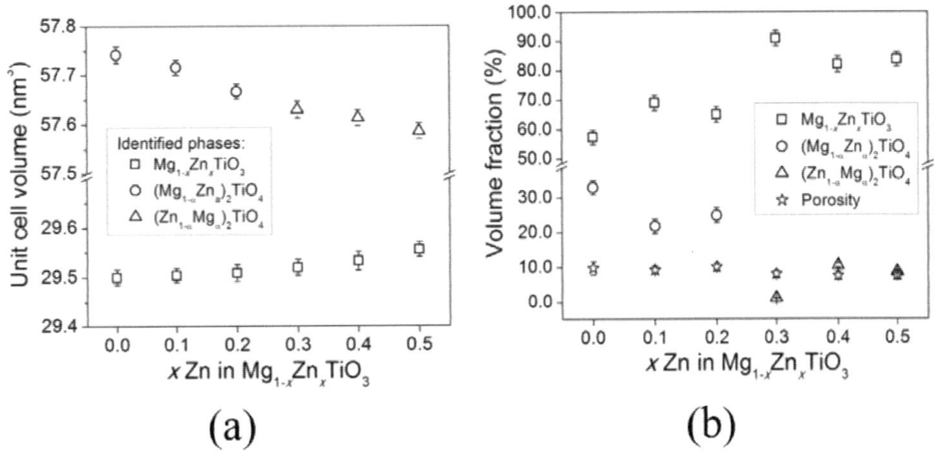

Figure 3. (a) Unit cell volume of the ($Mg_{1-x}Zn_xTiO_3$), ($(Mg_{1-\alpha}Zn_\alpha)_2TiO_4$ and ($(Zn_{1-\alpha}Mg_\alpha)_2TiO_4$) solid solutions. (b) The volume fraction of the three phases and of the porosity present in the systems.

As the frequency in Figure 1 increases, the SC polarization effect to the ε'_r values of the ceramics become less important and the role was replaced by the D polarization. Based on the above analysis, it is obvious that the values of ε'_r at 1 Hz in MZT0.3 – MZT0.5 ceramics will decay within a wide frequency range, *i.e.* by order of about 1 kHz of the similar frequency range observed in MZT0.1 – MZT0.2 ceramics. The D interaction response to the ε'_r values is frequency independent, as shown in Figure 1. The region between SC and D polarizations is called the transition area. Table 1 lists the location of the SC and the D polarization regions and their transition area, as well as the frequency independent ε'_r values recorded in the D polarization region recapitulated from Figure 1. As observed, in general, the ε'_r values in the D polarization region increase with the zinc content.

Table 1. Space charge and dipolar polarization regions and their transition area and the frequency independent ε'_r values in MZT0 – MT0.5 dielectric ceramic systems.

Sample	The SC polarization region (± 0.1 %)	The transition area (± 0.1 %)	The D polarization region (± 0.1 %)	The ε'_r values in the D polarization region
MZT0	(1.0 – 12.2) Hz	13.0 – 80.0 Hz	120.0 Hz – 330.0 MHz	15.4 ± 0.3
MZT0.1	(1.0 – 7.0) Hz	8.0 – 80.0 Hz	127.0 Hz – 330.0 MHz	16.2 ± 0.3
MZT0.2	(1.0 – 10.4) Hz	20.0 – 30.5 Hz	80.0 Hz – 330.0 MHz	15.1 ± 0.3
MZT0.3	1.0 Hz – 15.7 kHz	(32.0 – 80.0) kHz	120.0 kHz – 330.0 MHz	16.3 ± 0.3
MZT0.4	1.0 Hz – 18.3 kHz	(20.0 – 30.0) kHz	32.5 kHz – 330 MHz	16.2 ± 0.3
MZT0.5	1.0 Hz – 1.5 kHz	(2.0 – 4.0) kHz	5.0 kHz – 330 MHz	17.0 ± 0.3

4. Conclusions

Dielectric characteristic of the real part of the relative permittivity ε'_r in $Mg_{1-x}Zn_xTiO_3$ (x = 0 – 0.5) ceramic systems from 1 Hz to 330 MHz has been analyzed based on the phase composition data and the electrical properties of the identified phases. Contribution of the space charge and the dipolar polarization mechanisms to the characteristic were confirmed and the ε'_r values were justified. Variation of the zinc content in the systems has classified the systems into two different sub-systems, each with different conductivity level.

Acknowledgments
This work was partially funded by the research grant provided by the Universitas Negeri Surabaya under the contract number 891/UN38/HK/LT/2017.

References

[1] Hamieh, T., Kawtharani, F., Kassas, A., Quercioli, R., Houivet, D., Bernard, J., ... & Reda, M. 2013. Ultrafine Grinding of $MgTiO_3$ Based Ceramic Influencing the Material Properties. *J Phys Chem Biophys*, *3* (122), 2161-0398.

[2] Roopas Kiranet al 2012 Effect of Grain Size on the Microwave Dielectric Characteristics of High-Energy Ball-Milled Zinc Magnesium Titanate Ceramics *Journal of the American Ceramic Society* **95** pp 1973–1979

[3] Wang H *et al* 2012 Sintering behavior and microwave dielectric properties of $MgTiO_3$ ceramics doped with B_2O_3 by sol-gel method *Journal of Materials Science & Technology* **28** pp 751–755

[4] Xue X, Yu H and Xu G 2013 Phase composition and microwave dielectric properties of Mg-excess $MgTiO_3$ ceramics *Journal of Materials Science: Materials in Electronics* **24** pp 1287–129

[5] Moulson A J and Herbert J M 2003 *Electroceramics: materials, properties, applications* (John Wiley & Sons)

[6] Ermawati F U, Suasmoro S and Pratapa S 2015 A Simple Dissolved Metals Mixing Route to Prepare Nanostructured $Mg_{0.8}Zn_{0.2}TiO_3$ Solid Solution *In Advanced Materials Research* **1112** (Trans Tech Publications) pp 47–52

[7] Ermawati F U *et al* 2016 Preparation and structural study of $Mg_{1-x}Zn_xTiO_3$ ceramics and their dielectric properties from 1 Hz to 7.7 GHz *Journal of Materials Science: Materials in Electronics* **27** pp 6637–6645

[8] Kumar S *et al* 2009 Structural and electrical properties of Mg_2TiO_4 *Journal of the Ceramic Society of Japan* **117** pp 689–692

[9] Obradovic N, Mitrovic N and Pavlovic V 2009 Structural and electrical properties of sintered zinc-titanate ceramics *Ceramics International* **35** pp 35–37

[10] Filipović S *et al* 2009 Influence of mechanical activation on structural and electrical properties of sintered $MgTiO_3$ ceramics *Science of Sintering* **41** pp 117–123

Stabilization of PVA/Chitosan/TiO$_2$ Nanofiber Membrane with Heat Treatment and Glutaraldehyde Crosslink

Nasikhudin[1,3], M. Diantoro[3], A. Kusumaatmaja[1,2], and K. Triyana[1,2]

[1]Departement of Physics, Universitas Gadjah Mada, Sekip Utara BLS.21 Yogyakarta, 55281 Indonesia
[2]Nanomaterials Research Group Universitas Gadjah Mada, Jl. Kaliurang Km.4 Sekip Utara Yogyakarta, 55281 Indonesia
[3]Departement of Physics, Faculty of Mathematics and Natural Sciences, Universitas Negeri Malang, Indonesia

E-mail: triyana@ugm.ac.id

Abstract. PVA/chitosan/TiO$_2$ nanofiber membrane have been prepared by electrospinning method. PVA/Chitosan/TiO$_2$ nanofibers are poor stability in water. In this paper, the stabilization of PVA/chitosan/TiO$_2$ nanofiber membrane was improved by heat treatment and GA crosslink. The structure and morphology of PVA/chitosan/TiO$_2$ nanofibers were characterized by FTIR, SEM, and EDAX. The result showed that the stabilization methods affected the morphology of nanofiber membranes and their stability in water. Heat treatment reduces fiber diameter and membrane pore size, in contrast, GA crosslink increase fiber diameter and membrane pore size. Stability in water of nanofiber membranes was examined by immersing in the water for 24 h. The SEM result shows that after immersing in water for 24 h, PVA/Chitosan/TiO$_2$ nanofibers without stabilization rapidly shrink and dissolve after dipped in water and finally form transparent gel shape material with a grain of TiO$_2$ nanoparticle. PVA/Chitosan/TiO$_2$ nanofibers with heat treatment don't dissolve after immersing but present a certain degree of aggregation and swelling, which indicates that heat treatment can improve the stability of the fiber but still low. Meanwhile, PVA/Chitosan/TiO$_2$ nanofibers with GA crosslink do not dissolve and change morphology and fiber diameter after immersing 24 h. It indicates that PVA/Chitosan/TiO$_2$ nanofibers with GA crosslink have excellent stability in water.

Keywords. Chitosan, GA crosslink, heat treatment, nanofibers, PVA, Stabilization, and TiO$_2$.

1. Introduction

Polyvinyl alcohol (PVA) as a hydrophilic polymer with partial crystallization has a lot of advantages such as chemical resistance, excellent biocompatibility, high melting point, non-toxic, simple processing, and high water permeability [1]. On the other hand, chitosan (CS) is an organic polymer which has the properties of non-toxic, biocompatible, biodegradable, biofunctional, and hydrophilic [2]. Moreover, CS has been used as an antimicrobial and antiviral material in the fields of biotechnology, pharmaceutics, wastewater treatment, cosmetics, agriculture, food science, and textiles because of its advantageous in biological properties [3]. It has been reported that the combination between PVA and CS can be produced nanofibers by electrospinning [4].

Content from this work may be used under the terms of the Creative Commons Attribution 3.0 licence. Any further distribution of this work must maintain attribution to the author(s) and the title of the work, journal citation and DOI.
Published under licence by IOP Publishing Ltd

Electrospinning is a technique using electrostatic force to produce fine fibers from a polymer solution, fibers produced from this electrospinning have a small diameter (micrometer or nanometers) and a surface area higher than that obtained from conventional spinning process [5, 6]. Electrospun nanofiber has excellent characteristics such as the vast surface area to volume ratio and high porosity with small pore sizes [7]. Other reports, TiO_2 fibers showed high surface-to-volume ratios and a high number of active surface sites for effective photocatalytic [8, 9]. Chitosan nanofibers could become a new material for virus removal in drinking water applications. Some other applications of chitosan-based chitosan nanofibers including; PVA/chitosan nanofibers were used as a membrane to remove metal ions in a fixed-bed reactor, in particular for the adsorption of metals nickel and cobalt [10]. Chitosan nanofibers combined with TiO_2 is used for adsorption of heavy metal ions Pb (II) and Cu (II) [11].

Titanium dioxide (TiO_2) has been successfully studied for photovoltaic application [12]. Moreover, TiO_2 has also been used as photocatalytic due to its chemical stability and availability in commercial [13]. The size of TiO_2 could be reduced to increase the reactive surface area for some applications [14]. In particular, wastewater treatment for environmental preservation based on TiO_2 powder has been applied in photocatalytic for reducing many organic pollutants effectively. However, this still leaves some problems when applied in environments such as rivers, lakes, and sea [15]. The main problem was due to the difficulty of recovery after the photocatalytic treatment process. In this case, the TiO_2 particles were not easy to separate from the waste [16]. To overcome this problem, we developed photocatalytic based nanofibers synthesized using electrospinning methods [17].

Many applications of PVA/TiO_2/chitosan nanofibers, such as a membrane for filtration, membrane adsorption, for and membrane for photocatalytic application. For this application nanofiber membrane must provide high stability especially in water. The stabilization method can use chemical cross-linking agents or physical treatment methods to improve the water stability of nanofibers membrane [18 – 20]. In this paper, we studied the various stabilization method to increase the stability of the nanofiber membrane in water. This paper also discusses the effect of stabilization method and on the morphology of the nanofibers membrane especially diameter and porosity of nanofibers membrane.

2. Materials and methods

All material was used without further purification: PVA (Mw: 89.000 $g \cdot mol^{-1}$ to 98.000 $g \cdot mol^{-1}$, 99 % hydrolyzed), Chitosan (medium molecule weight), Titanium (IV) oxide, anatase, nanopowder < 25 nm particle size, 99.7 % trace metal basis, Glutaraldehyde solution 50 wt% in H_2O, and tetramethylammonium hydroxide (TMAH). All of them purchased from Sigma-Aldrich. PVA/TiO_2 electrospun solution was prepared by dissolving TiO_2 nanoparticle in deionized water at 90 °C for 1 h in magnetic stirrer and adding TMAH solution with ratio TiO_2 to TMAH of 10:1. TiO_2 dispersion adding with PVA powder and then blending the PVA and TiO_2 at a ratio of 4:1. And then stirring them for 1 h at 90 °C. Chitosan (1 wt%) was dissolved in 2 % acetic acid solution at 45 °C for 1 h in magnetic stirrer. PVA/TiO_2/Chitosan solution was obtained by blending the PVA/TiO_2 and Chitosan solution at a ratio of 85:15 and then stirring them for 1 h at 45 °C and then sonicated them for 3 h. The prepared PVA/TiO_2/Chitosan solution was then loaded into a 12-mL plastic syringe equipped with a 0.4-mm needle diameter. A high voltage of 15 kV, with a tip-collector distance of 14 cm, was applied to the solution and fiber collected on the plane collector [21]. Electrospun nanofibers were stabilized with various stabilization method, i.e., Heat treatment and GA crosslinked. Heat treatment was done by heating the fiber membrane in a furnace at 135 °C for 4 h. GA crosslinked were done by immersing in a solution of 2.5 wt% Glutaraldehyde solution (volume ratios of Glu (aq): acetone/ethanol: HCl (aq) = 1:2:0.01) at 40 °C for 4 h. Characterizations of PVA/TiO_2 nanofibers using FTIR and SEM-EDAX. FTIR (Shimadzu, Type: IR Prestige21) were used to determine the functional group of PVA/Chitosan/TiO_2 nanofiber. The nanofibers surface morphology was studied using SEM (FEI, Type: Inspect-S50). Image processing software was used for measuring the diameter of the nanofibers captured in the SEM micrograph. The EDAX used to investigate the content elements of electrospun fibers, especially presence of TiO_2 nanoparticle on the fibers.

3. Results and discussion

The FTIR spectra of PVA/TiO_2/Chitosan nanofiber with various stabilization method are shown in Figure 1. The FTIR spectra for all sample is same. The Broadband at 3200-3550 cm^{-1} indicates the O-H stretching. The peaks at 2931.8 cm^{-1} indicate C-H- stretching. The peaks at 1342.46 cm^{-1} indicate the CH banding. The broad absorption bands at 1095.57 cm^{-1} indicate the C-O stretching [10]. The peaks at 856.39 cm^{-1} were assigned C-C vibrations of PVA [22]. The chitosan vibration occurs in the broadband at 3425.58 cm^{-1} which is due to the OH stretching. The band at 1573.91 cm^{-1} is assigned for the NH bending (amide II) (NH2) while the small peak at 1658.78 cm^{-1} indicates the C=O stretching (amide I) O=C-NHR. The bands at 2924.09 cm^{-1} were assigned C-H stretching [23]. The band at 1381.03 cm-1 is due to CH3 wagging. The vibrational band observed at 2924.09 cm^{-1} is associated with the C-H stretching from alkyl groups. The absorption corresponding to the C-O stretching occurs at 1080.14 cm^{-1}, and broadband around (550 to 800) cm^{-1} is attributed to the Ti-O-Ti band [24].

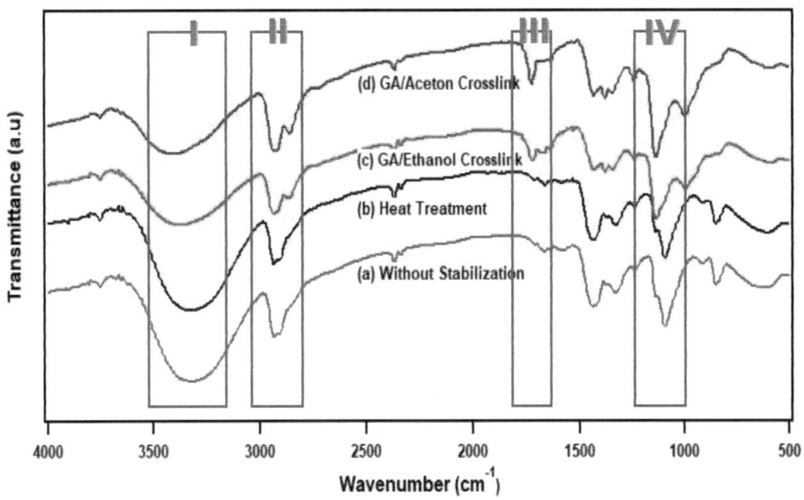

Figure 1. FTIR spectra of PVA/TiO_2/Chi nanofibers (a) without stabilization (b) heat treatment, (c) GA crosslink with ethanol, (d) GA crosslink with acetone.,

Effect of stabilization method regarding FTIR results seen at wavelengths around 1720 cm^{-1} and 1095 cm^{-1} (area III and IV). The PVA/TiO_2/Chi nanofibers with the GA crosslink show a new peak at the wavelength of 1720 cm^{-1}, this peak indicates the presence of C = O stretching of GA and 2870 cm^{-1} (area II) C-H stretching of GA. While The PVA/TiO_2/Chi nanofibers without stabilization and heat treatment the fibers do not show up. In The PVA/TiO_2/Chi nanofibers with crosslink GA, there was a shift from 1095 cm^{-1} to 1141 cm^{-1}. It indicates the interaction between C-O from PVA and C-O from GA [25].

PVA is readily soluble in water, so the PVA/TiO_2/Chitosan nanofiber is also readily soluble in water and will be damaged if it is used for applications in water. Therefore, this research is preserved/stabilized so that the resulting nanofiber will be stable/insoluble when applied to the liquid, especially for photocatalytic application. In this study there are several stabilization methods used are heat treatment and GA crosslink. The SEM image of PVA/TiO_2/Chitosan nanofibers with various stabilization method is shown in Figure 2. Figure 2 shows that the PVA/TiO_2/Chitosan nanofiber with heat treatment is compacted and has a smaller diameter, while the crosslink GA nanofiber results in wrapping so that the resulting fiber has a larger diameter. Nanofiber results of crosslink GA by using acetone solvent have a cleaner surface than those using ethanol solvent.

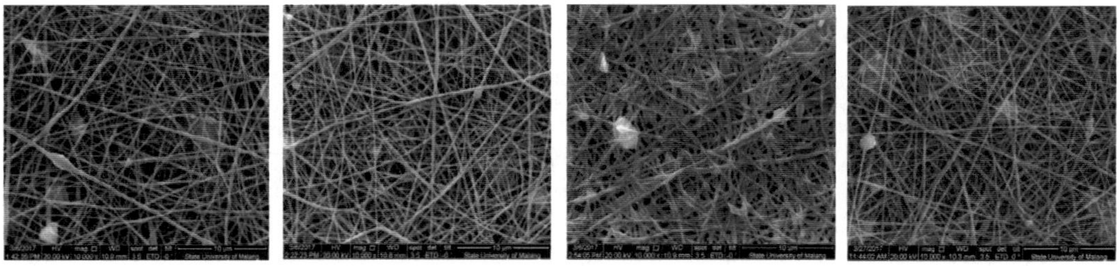

(a) (b) (c) (d)

Figure 2. SEM image of PVA/TiO_2/Chi nanofibers before immersing in water (a) without stabilization (b) heat treatment, (c) GA crosslink with ethanol, (d) GA crosslink with acetone.

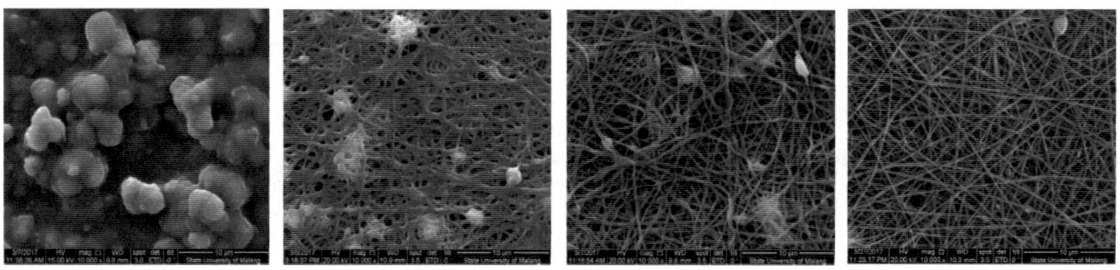

(a) (b) (c) (d)

Figure 3. SEM image of PVA/TiO_2/Chi nanofibers after immersing in water for 20 days (a) without stabilization (b) heat treatment, (c) GA crosslink with ethanol, (d) GA crosslink with acetone.

The stabilized nanofiber was tested for water resistance by immersing in water for 24 h. Figure 3a showed that the nanofiber without stabilization dissolves in the water so that none of the fibers are visible and only particles of TiO_2 in the gel solution are visible. Nanofiber with heat treatment has low stability, as shown in Figure 3b the fibers still exist but it starts to deteriorate and swell due to water uptake. Nanofibers with heat treatment do not dissolve after soaking but present a certain degree of aggregation and swelling, which indicates that heat treatment can improve the water stability of fibers as well as the degree of physical stabilization and crystallization [1]. Nanofiber with crosslink GA has high stability as shown in Figure 3c and Figure 3d. The fiber is still intact with the same morphological form as before soaked in water, no damage at all in fiber. Stabilization with crosslink GA has high stability. Chemical crosslinking of PVA nanofibers treated in GA/methanol shown in this work is apparently more useful in stabilizing them against disintegration in water [26]. It is considered to be applied in the fiber because fiber will be used for photocatalyst application which directly contacts to water.

IOP Conf. Series: Materials Science and Engineering **367** (2018) 012004 doi:10.1088/1757-899X/367/1/012004

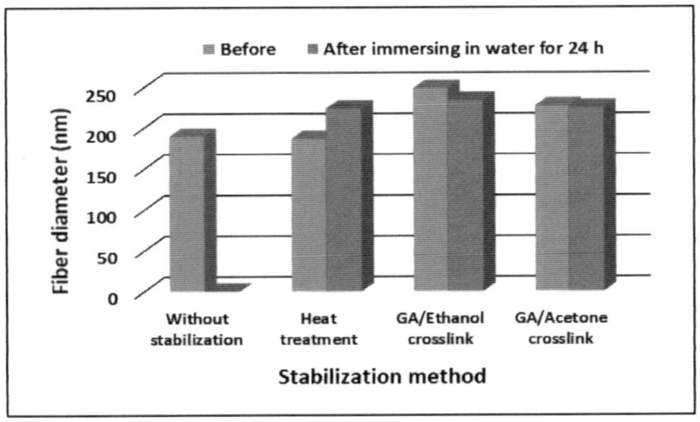

Figure 4. Fibers diameter of PVA/TiO_2/Chi nanofibers before and After immersing in water for 20 days with various stabilization method.

The result of the diameter analysis is shown in Figure 4. The nanofiber without stabilization has a diameter of 190 nm, and after being soaked for 20 days, one of fiber diameter is detectable. Nanofiber with heat treatment has a diameter of 187 nm, and after soaked for 20 days, the diameter of the fiber enlarged about of 224 nm because of fiber have swelling due to immersion in the water. Nanofiber with a crosslinked acid with ethanol as a solvent having a diameter of 249 nm, and after being soaked for 20 days, the fiber diameter decreased to 234 nm, this is because most of the GA did not react in the crosslinking process dissolved in the water during immersion. Nanofiber with a crosslinked acid with ethanol as a solvent having a diameter of 227 nm, and after soaking for 20 days, the fiber diameter has not changed in size. It shows the high stability of this nanofiber.

The EDAX was used to investigate the content elements of electrospun fibers. EDAX result of the PVA/TiO_2/Chi composite nanofibers before water-immersion tested is shown in Figure 5 and after water-immersion tested is shown in Figure 6. The summary results are listed in Table 1.

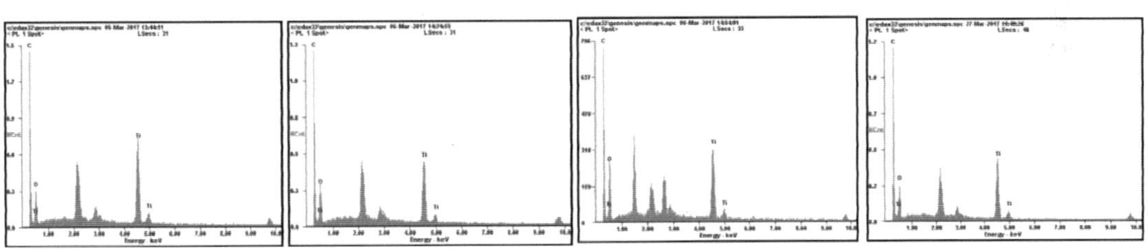

(a) (b) (c) (d)

Figure 5. SEM-EDAX of PVA/TiO_2/Chi nanofibers before immersing in water (a) without stabilization (b) heat treatment, (c) GA crosslink with ethanol, and (d) GA crosslink with acetone.

IOP Conf. Series: Materials Science and Engineering **367** (2018) 012004 doi:10.1088/1757-899X/367/1/012004

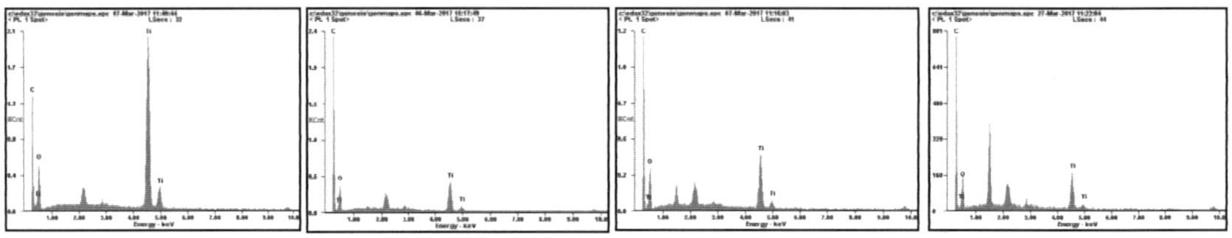

(a) (b) (c) (d)

Figure 6. SEM-EDAX of PVA/TiO_2/Chi nanofibers after immersing in water (a) without stabilization (b) heat treatment, (c) GA crosslink with ethanol, and (d) GA crosslink with acetone

Table 1. Summary of EDAX of PVA/TiO_2/Chi nanofibers before and after immersing in water.

SEM-EDAX	without stabilization	heat treatment	GA crosslink with ethanol	GA crosslink with acetone
Before immersing in H2O				
C	49.23 %	50.47 %	44.51 %	58.95 %
O	26.86 %	29.87 %	36.30 %	28.28 %
Ti	23.90 %	19.66 %	19.19 %	12.77 %
After immersing in H2O				
C	30.90 %	62.83 %	53.65 %	58.55 %
O	29.94 %	25.06 %	30.49 %	28.68 %
Ti	39.16 %	12.12 %	15.86 %	12.77 %

The elemental analysis reveals the presence of TiO2 nanoparticle on the fibers, the element ratio of C, O, and Ti for PVA/TiO_2/Chi nanofibers before stabilization are 49.23 %, 26.86 %, and 23.90 % respectively in the entire area of PVA/TiO_2/Chi nanofibers. It proves that the nanofiber is containing TiO_2. Results of EDAX nanofiber PVA/TiO_2/Chi without stabilization showed that the Ti composition increased from 23 % to 39 % while the composition C decreased after immersion. It suggests that after immersion some of the polymer (PVA) dissolves in water and the fibers become damaged so that the remaining ones are TiO2 which is enveloped in a gel solution of PVA. The results of EDAX nanofiber PVA/TiO_2/Chi with heat treatment stabilization method showed that the Ti composition decreased from 19 % to 12 % while the composition C increase from 50 % to 62 % after immersion. This suggests that after partial immersion of the polymer (PVA) absorbs water so that it is swelling and the diameter of the fiber enlarges and the fiber swells. Results of EDAX nanofiber PVA/TiO_2/Chi with GA crosslink stabilization method show that the composition of Ti and C and O is relatively fixed. It shows that before and after immersion of fiber condition still no change of composition nor change of surface morphology of fiber. It shows that nanofiber PVA/TiO_2/Chi with GA crosslink stabilization method has high stability when applied in water.

4. Conclusions

The PVA/Chitosan/TiO_2 nanofibers were successfully stabilized in water with various stabilization method. The SEM results show that the nanofiber formed with a smooth surface with a diameter of 190 nm and has grains of TiO2 particles in it. These results are supported by EDAX results which show a high TiO_2 content in fiber reaching 23 %. The SEM result shows that after immersing in water for 24 h, PVA/Chitosan/TiO_2 nanofibers without stabilization rapidly shrink and dissolve after dipped in water and finally form transparent gel shape material with a grain of TiO_2 nanoparticle. PVA/Chitosan/TiO_2 nanofibers with heat treatment don't dissolve after immersing but present a certain degree of aggregation and swelling, which indicates that heat treatment can improve the stability of the fiber but still low. While PVA/Chitosan/TiO_2 nanofibers with GA crosslink don't dissolve and no change form of the fiber after immersing 24 h and same as before immersing. It indicates that PVA/Chitosan/TiO_2 nanofibers with GA crosslink have excellent stability in water. Nanofiber with GA crosslinks has better morphology when using acetone as a solvent than ethanol. The nanofiber stabilization results indicate that the stabilization method with GA crosslink is proven to have high resilience or stability in the water, which will be very useful for maintaining fiber durability when used for photocatalytic applications.

Acknowledgments
This research was supported by the Directorate General of Higher Education (DIKTI), the ministry of research and higher education, the Republic of Indonesia Contract No PUPT Lanjutan Fiscal Year 2017: 2383/UN1.PIII/DIT-LIT/LT/2017. The authors would also thank Mr. Kunto Wandono for maintaining the electrospinning machine.

References

[1] Liu R, Liu T and Zhao N 2012 Study on Stabilized of Poly (Vinyl Alcohol) Nanofibers Based Sandwich Structure Purification Material **537** 473–6

[2] Jia Y, Gong J, Gu X, Kim H and Dong J 2007 Fabrication and characterization of poly (vinyl alcohol)/ chitosan blend nano W bers produced by electrospinning method *Carbohydr. Polym.* **67** 403–9

[3] Mi X, Vijayaragavan K S and Heldt C L 2014 Virus adsorption of water-stable quaternized chitosan nanofibers *Carbohydr. Res.* **387** 24–9

[4] Liu R, Xu X, Zhuang X and Cheng B 2014 Solution blowing of chitosan/PVA hydrogel nanofiber mats *Carbohydr. Polym.* **101** 1116–21

[5] Bhardwaj N and Kundu S C 2010 Electrospinning : A fascinating fi ber fabrication technique **28** 325–47

[6] Garg K and Bowlin G L 2011 Electrospinning jets and nanofibrous structures *Biomicrofluidics* **5** 1–19

[7] Elsabee M Z, Naguib H F and Morsi R E 2012 Chitosan based nanofibers, review *Mater. Sci. Eng. C* **32** 1711–26

[8] Choi D, Hwang C, Lee J, Lee I, Oh I and Park J 2013 Characterization of TiO_2 fibers by combined electrospinning method and hydrothermal reaction **106** 41–4

[9] Prahsarn C, Klinsukhon W and Roungpaisan N 2011 Electrospinning of PAN / DMF / H_2O containing TiO 2 and photocatalytic activity of their webs *Mater. Lett.* **65** 2498–501

[10] Esmaeili A and Beni A A 2014 A novel fixed-bed reactor design incorporating an electrospun PVA/chitosan nanofiber membrane *J. Hazard. Mater.* **280** 788–96

[11] Razzaz A, Ghorban S, Hosayni L, Irani M and Aliabadi M 2015 Chitosan nanofibers functionalized by TiO_2 nanoparticles for the removal of heavy metal ions *J. Taiwan Inst. Chem. Eng.* **0** 1–11

[12] Gupta S M and Tripathi M 2011 A review of TiO2 nanoparticles *Chinese Sci. Bull.* **56** 1639–57

[13] Bo Y I N, Ji-tong W, Wei X U, Dong-hui L and Wen-ming Q 2013 Preparation of TiO_2/ mesoporous carbon composites and their photocatalytic performance for methyl orange degradation *New Carbon Mater.* **28** 47–54

[14] Yun J, Jin D, Lee Y and Kim H 2010 Photocatalytic treatment of acidic waste water by electrospun composite nano fi bers of pH-sensitive hydrogel and TiO_2 *Mater. Lett.* **64** 2431–4

[15] Im J S, Kim M I I and Lee Y-S 2008 Preparation of PAN-based electrospun nanofiber webs containing TiO_2 for photocatalytic degradation *Mater. Lett.* **62** 3652–5

[16] Qu X, Alvarez P J J and Li Q 2013 Applications of nanotechnology in water and wastewater treatment *Water Res.* **47** 3931–46

[17] Nasikhudin, E P Ismaya, M Diantoro, A Kusumaatmaja and K T Preparation of PVA / TiO 2 Composites Nanofibers by using Electrospinning Method for Photocatalytic Degradation **12011**

[18] Kar K and Wong H 2010 Effect of annealing on aqueous stability and elastic modulus of electrospun poly (vinyl alcohol) fibers 2456–65

[19] Sencadas V, Correia D M, Ribeiro C, Moreira S, Botelho G, Gómez Ribelles J L and Lanceros-Mendez S 2012 Physical-chemical properties of cross-linked chitosan electrospun fiber mats *Polym. Test.* **31** 1062–9

[20] Cates R S 2010 Influence of crosslink density on swelling and conformation of surface-constrained Poly (N- isopropylacrylamide) hydrogels

[21] Nasikhudin, Puspitasari I, Diantoro M, Kusumaatmaja A and Triyana K 2017 *Effect of blend ratio on morphology and swelling properties of PVA/chitosan nanofibers* vol 901 MSF

[22] Alhosseini S N, Moztarzadeh F, Mozafari M, Asgari S, Dodel M, Samadikuchaksaraei A, Kargozar S and Jalali N 2012 Synthesis and characterization of electrospun polyvinyl alcohol nanofibrous scaffolds modified by blending with chitosan for neural tissue engineering *Int. J. Nanomedicine* **7** 25–34

[23] El-hefian E A, Nasef M M and Yahaya A H 2010 The Preparation and Characterization of Chitosan / Poly (Vinyl Alcohol) Blended Films *E-Journal Chem.* **7** 1212–9

[24] Abbasizadeh S, Keshtkar A R and Mousavian M A 2013 Preparation of a novel electrospun polyvinyl alcohol/titanium oxide nanofiber adsorbent modified with mercapto groups for uranium(VI) and thorium(IV) removal from aqueous solution *Chem. Eng. J.* **220** 161–71

[25] Mansur H S, Sadahira C M, Souza A N and Mansur A A P 2008 FTIR spectroscopy characterization of poly (vinyl alcohol) hydrogel with different hydrolysis degree and chemically crosslinked with glutaraldehyde *Mater. Sci. Eng.* **28** 539–48

[26] Cao S, Liu Z, Hu B and Liu H 2010 Stabilization of electrospun poly (vinyl alcohol) *Nanofibrous* **28** 781–8

IOP Conf. Series: Materials Science and Engineering **367** (2018) 012005 doi:10.1088/1757-899X/367/1/012005

Preparation of Calcium Carbonate (from Shellfish)/Magnesium Oxide Composites as an Antibacterial Agent

Z. Jannah[1], H. Mubarok[1], F. Syamsiyah[2], A. A. H Putri[3], and L. Rohmawati[4]

[1]Department of Physics, Universitas Negeri Surabaya, Jl. Ketintang, Surabaya 60231, Indonesia
[2]Department of Biology, Universitas Negeri Surabaya, Jl. Ketintang, Surabaya 60231, Indonesia
[3]Department of Chemistry, Universitas Negeri Surabaya, Jl. Ketintang, Surabaya 60231, Indonesia
[4]Center for Materials Science Studies, Department of Physics, Universitas Negeri Surabaya, Jl. Ketintang, Surabaya 60231, Indonesia

E-mail: zjannah14@gmail.com

Abstract. We have performed research on antibacterial substance from a natural substance, one of them is calcite from shellfish (Anadara granosa) in Kenjeran Beach Surabaya. This calcite is composed of magnesium oxide using PEG 4000 (Polyethylene glycol) as a solvent and then heated at 800 °C for 30 minutes. Weight variety of calcite used was 80% wt, 85% wt, and 90 wt%. Subsequently, that composites characterized using XRD, antibacterial activity test (Escherichia coli and Staphylococcus aureus), and SEM. The result of antibacterial assay shows that composite of $CaCO_3$/MgO with 80% wt composition have the best activity inhibitory of 31,96 mm for Escherichia coli bacteria and 32.26 mm for Staphylococcus aureus bacteria.

Keyword: calcium carbonate, shellfish, magnesium oxide, composite, antibacterial.

1. Introduction

Food entered into the human body through the mouth and teeth. Both of these organs have a very vital role which is vulnerable to the entry of pathogenic bacteria and viruses which may cause various types of diseases. There are more than 500 species of bacteria living in the oral cavity of humans [1], [2]. The bacteria live in the oral cavity, which can form plaque on the teeth, and cause inflammation of the oral cavity [3]. One type of bacteria in dental plaque is *Staphylococcus aureus* [4]. And other bacteria in dental plaque is *Escherichia coli* [5].

The antibacterial agent is required which can inhibit bacterial activity in the oral cavity to reduce the disease in the oral cavity and tooth caused by bacterial buildup. Generally, to control the activity of bacteria in the oral cavity used abrasive substances such as calcite and hydroxyapatite contained in toothpaste to control bacterial activity in the oral cavity. The abrasive material can be obtained from calcium carbonate contained in the shellfish which content of 98% [6]. Calcium carbonate is known to use as a material to clean up the remains of food, stains on the teeth and kill bacteria [7]. However,

Content from this work may be used under the terms of the Creative Commons Attribution 3.0 licence. Any further distribution of this work must maintain attribution to the author(s) and the title of the work, journal citation and DOI.
Published under licence by IOP Publishing Ltd

abrasive is not enough to remove bacteria because it has no antibacterial activity [8]. Therefore, it needs materials that have two functions such as an abrasive as well as shows an antibacterial properties.

Antibacterial compounds can be made from organic and inorganic materials. Inorganic materials are more in demand because they have good stability compared to organic one [9], [10]. Magnesium oxide is one of the inorganic materials that have good antibacterial activity [11] and as safe for humans [12], [13]. Thus the composite of $CaCO_3$/MgO fit to two functions as abrasive and antibacterial.

Yamamoto et al.[14] reported his study of antibacterial characteristics of carbon-coated $CaCO_3$/MgO powder which led to the pyrolysis of poly(vinyl alcohol)-dolomite mixture showed strong antibacterial properties in *Staphylococcus aureus* and *Escherichia coli*. The antibacterial action on *Staphylococcus aureus* was greater than towards *Escherichia coli* [8]. They used the in-situ method to obtain the $CaCO_3$/MgO composite. So more expensive tools and materials are needed for high sample preparation and heating with long holding time. The further reported from XRD analysis that impurities such as Al, Si, and Fe [8] existed. The $CaCO_3$/MgO composite can be prepared by an ex-situ method that combines pure $CaCO_3$ and MgO materials.

In this work, we report the study to determine the inhibitory power of composite material made by $CaCO_3$/MgO to the activity of *Escherichia coli* and *Staphylococcus aureus* bacteria. We also report the preparation of $CaCO_3$/MgO composite material by the optimum clear zone and able to utilize shell waste shells to be extracted into calcium carbonate and composed with magnesium oxide as one of the antibacterial ingredients of natural materials.

2. Materials and methods

2.1 Preparation of $CaCO_3$ Extract

The stage of extracting calcium carbonate from shellfish (*Anadara granosa*) begins by cleansing the fur shells from feathers using 2 M HCl. Then dried under the sun for 24 hours and calcined at 900 °C for 5 hours. Furthermore, the shellfish was smoothed and flowed CO_2 gas with a speed of 2.8 L/min and then precipitated for 24 hours. The precipitate was filtered and heated at 90 °C for 24 hours to obtain calcite phase on calcium carbonate.

2.2 Preparation of $CaCO_3$/ MgO Composites

This composite was prepared by using the ex-situ method of combining pure materials. The calcium carbonate and magnesium oxide powder were dissolved in PEG 4000 solution then stirred using magnetic stirrer at 80 °C for 30 minutes at 6 rpm. The solution was then filtered and calcined at 800 °C for 30 minutes. The calcined product pounded to get $CaCO_3$/MgO composites powder. The calcite mas was varied as 80% wt, 85% wt, and 90 wt%.

2.3 Antibacterial assay

Referring to [14] the research using *Escherichia coli* (hereafter, *E. coli*) bacteria as gram-negative bacteria and *Staphylococcus aureus* (hereafter, *S. aureus*) as gram-positive bacteria. Tests were performed by isolation of $CaCO_3$/MgO composite and bacterial isolation. The bacteria then incubated the bacteria at a temperature of 28-30 °C for 24 hours to observe the bacterial reactivity of the $CaCO_3$/MgO alloy material. A clear zone of inhibiting zone indicated the antibacterial activity test. The diameter expresses this activity. Based on the dimension of the inhibition zone, one can obtain bacterial activity against $CaCO_3$/MgO composite material. The inhibitory power was calculated by measuring the diameter of the clear visible region. The width of the clear zone was then reduced by 6 mm from the diameter of the well [15]. The measurement results are presented as an average for each treatment.

IOP Conf. Series: Materials Science and Engineering **367** (2018) 012005 doi:10.1088/1757-899X/367/1/012005

3. Results and discussions

3.1 XRD Characterization

The X-ray data collection specification uses Philips X'Pert MPD (Multi Purpose Diffractometer) type system, with anodic radiation source Cu, 40 kV, 30 mA, CuKα 1.54056 wavelength and Bragg-Brentano optical light. XRD patterns of the CaCO3 powder sample from shellfish analyzed quantitatively with Match Software. The technique used searches and match database on application to identify the phase formed. This technique also applies to the analysis of XRD composite CaCO3/MgO. Based on Fig. 1, it shows that the extraction results in this study were 100% CaCO3, with miller index [012], [104], [110], [113], [202], [018], [116], [1010].

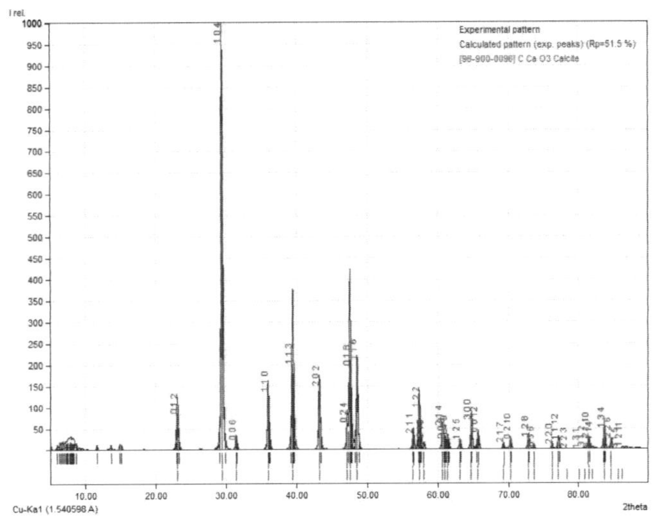

Figure 1. XRD patterns of the calcium carbonate extract in the analysis using Match! software.

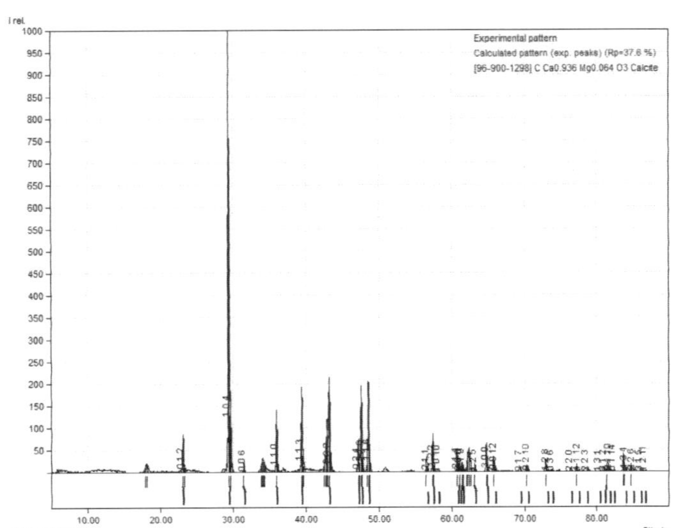

Figure 2. XRD patterns of $CaCO_3$/MgO composite in the analysis using Match! software.

(a)

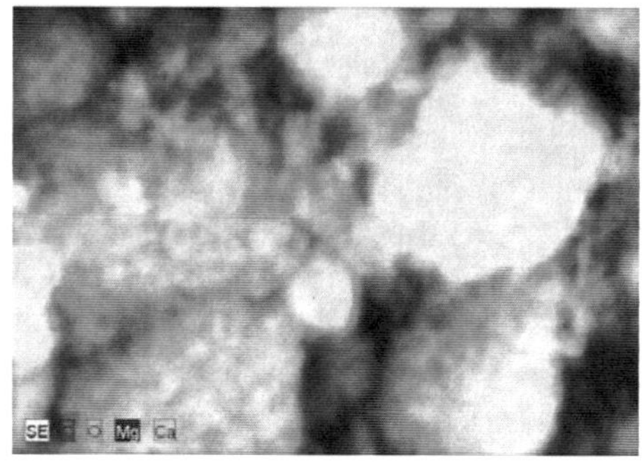

(b)

Figure 3. SEM-EDX image of $CaCO_3$/MgO composite (80% wt $CaCO_3$ and 20% wt MgO) magnification x15k (a) SEM, (b) SEM-EDX $CaCO_3$/MgO.

The results of the composite $CaCO_3$/MgO XRD patterns are shown in Fig. 2 which shows the presence of peak at 2θ $CaCO_3$ or MgO diffraction angles corresponding to the available databases in the Match, *e.g.* at 42.8°, and 62.17° for MgO. The results of the Match analysis indicates that the phase formed in Fig. 2 is composed by 64.2% calcite $CaCO_3$ and 35.8% MgO.

3.2 SEM-EDX Analysis

Based on Fig.3 SEM-EDX results for composites having the greatest inhibitory power in composites 80% wt $CaCO_3$ and 20% wt MgO formed composite particle bonds. Fig.3 (a) Composite morphology is shaped like a solid cube which indicating the calcite phase $CaCO_3$ [16] is supported with Fig.3 (b) light blue. Calcite is the most stable phase when compared with other phases such as vaterite and aragonite.

The composite is composed of C, O, Mg and Ca atoms with weight percentage and atoms symbolized by different colors (Figure 3). In the picture there is a dark and bright side, the dark side shows the location of Mg atoms while the light-colored side shows the location of Ca atoms. The red spots on the image indicate the presence of C atoms, which are not very dominant. This sample is dominated by O atoms that can be seen from the point of green color that is quite large and evenly distributed. It is very reasonable because the O atoms in this composite have four molar fraction of 3 atoms in $CaCO_3$ and one atom from MgO.

3.3 Clear Zone of Antibacterial Activity

The results of the antibacterial assay against *E. coli* and *S. aureus* bacteria showed that the highest clear zone in the composite was 80% wt $CaCO_3$, 20% wt MgO for each bacteria (Table 2). $CaCO_3$/MgO composites obtained at 800°C have clear zone diameter greater than MgO such as research conducted by the data shown in Table 3 [13].

Table 1. Percentage by weight of composite atoms of $CaCO_3$/MgO

Element	% wt	% Atoms	Color
O	60.35	67.64	Green
Mg	16.92	12.48	Dark blue
Ca	13.45	6.02	Light blue
C	9.29	13.86	Red

Table 2. Clear Zone Diameter of $CaCO_3/MgO$ obtained at 800°C.

Extract Composition	**Clear Zone Diameter (mm)**	
	***S. aureus* (Gram positive)**	***E. coli* (Gram negative)**
90% wt $CaCO_3$, 10% wt MgO	18.30	15.13
85% wt $CaCO_3$, 15% wt MgO	23.10	23.06
80% wt $CaCO_3$, 20% wt MgO	32.26	31.96

Table 3. Clear Zone Diameter of MgO.

Annealing temperature (°C)	**Clear Zone Diameter (mm)**	
	***S. aureus* (Gram positive)**	***E. coli* (Gram negative)**
300	23	21
500	21	19
700	19	17

The existence of differences in structure and chemical composition in both types of bacteria cause antibacterial inhibition in *S. aureus* bacteria greater than *E. coli*. The structure and chemical composition of the *S. aureus* cell surface consist only of the peptidoglycan layer, whereas the surface of *E. coli* cells includes of a thin layer of lipid, lipopolysaccharide, and peptidoglycan [8]. This research relevant to previous studies suggesting that the antibacterial inhibition of *S. aureus* bacteria is more significant than that of *E. coli* [8], [11], [14].

The composite with concentration 80% wt $CaCO_3$, 20% wt MgO is able to inhibit the growth of *Escherichia coli* and *Staphylococcus aureus* bacteria. $CaCO_3/MgO$ is the inorganic material of metal oxide. The presence of metal elements causes the increase of pH [14] in bacteria when the membrane cell contact with the material so that the metabolism of bacteria is disrupted. MgO has metal elements less than $CaCO_3/MgO$ composite. It is more effectively used as an antibacterial agent.

Antibacterial results of the study can be classified as antibacterial in the medium category. This category is based on observations of the bright zone formed by each composite. The results showed that each concentration has different inhibitory power based on the clear zone formed. Clear zones visible around the well characterized antibacterial activity. The mechanism of action of microbial control with antimicrobial compounds is to cause barriers to the formation of cell walls. Failure of this cell wall synthesis process will cause the cell to die due to lysis. The inhibition of cell wall synthesis and protein is due to these antimicrobial compounds by joining the 50S ribosomal subgroup so that all processes are inhibited [7].

4. Conclusion

The $CaCO_3/MgO$ composite prepared has a very variable inhibitory of bacterial activity (*E. coli* and *S. aureus*). The best inhibitory effect on bacterial activity in this study was 80% wt $CaCO_3$ and 20% wt MgO composite, ie 31.96 mm for *E. coli* bacteria and 32.26 mm for *S. aureus* bacteria. This study is stating antibacterial composite $CaCO_3/MgO$ activity on *S. aureus* is stronger than on *E. coli*.

Acknowledgments
Acknowledgment of the researcher submitted to Kemenristek DIKTI who has funded this research through Student Creativity Programs in 2017 and LPPM State University of Surabaya which facilitate registration of method patent claimed in this research.

References

[1] I. Kroes, P. W. Lepp, and D. A. Relman, "Bacterial diversity within the human subgingival crevice.," *Proc. Natl. Acad. Sci. U. S. A.*, vol. 96, no. 25, pp. 14547–52, 1999.

[2] B. J. Paster *et al.*, "Bacterial Diversity in Human Subgingival Plaque," *J. Bacteriol.*, vol. 183, no. 12, pp. 3770–3783, 2001.

[3] K. Okuda and Y. Ebiraha, "Relationships between chronic oral infectious diseases and systemic diseases.," *Bull Tokyo Dent Coll.*, vol. 39, no. 3, pp. 165–74, 1998.

[4] A. Tada, H. Senpuku, Y. Motozawa, A. Yoshihara, N. Hanada, and H. Tanzawa, "Association between commensal bacteria and opportunistic pathogens in the dental plaque of elderly individuals," *Clin. Microbiol. Infect.*, vol. 12, no. 8, pp. 776–781, 2006.

[5] M. R. Islam *et al.*, "A sensitive and simple plaque formation method for the Stx2 phage of Escherichia coli O157: H7, which does not form plaques in the standard plating procedure," *Plasmid*, vol. 67, no. 3, pp. 227–235, 2012.

[6] N. A. Junaidi, A. Hazmi, A. B. Zakaria, Md. Zuki, M. Mustapha, Noordin, Abu, Jalila and Yusof, "Mineral composition of the cockle (Anadara granosa) shells of West Coast of Peninsular Malaysia and it's potential as biomaterial for use in bone repair," *UPM institutional repository*, vol. 6, no. 5. pp. 591–594, 2007.

[7] A. El-Ishaq, L. Bello, and A. A. A. Aidi, "The Role of various toothpaste in reduction of bacteria load in the mouth", *Extensive Journal of Applied Sciences* -2015-3-2, 49-54.

[8] O. Yamamoto, T. Ohira, K. Alvarez, and M. Fukuda, "Antibacterial characteristics of $CaCO_3$-MgO composites," *Mater. Sci. Eng. B Solid-State Mater. Adv. Technol.*, vol. 173, no. 1–3, pp. 208–212, 2010.

[9] C. J. Hewitt, S. T. Bellara, A. Andreani, G. Nebe-von-Caron, and C. M. Mcfarlane, "An evaluation of the antibacterial action of ceramic powder slurries using multiparameter flow cytometry.," *Biotechnol. Lett.*, vol. 23, pp. 667–675, 2001.

[10] S. Makhluf, R. Dror, Y. Nitzan, Y. Abramovich, R. Jelinek, and A. Gedanken, "Microwave-assisted synthesis of nanocrystalline MgO and its use as a bacteriocide," *Adv. Funct. Mater.*, vol. 15, pp. 1708–1715, 2015.

[11] Z. X. Tang and B. F. Lv, "MgO nanoparticles as antibacterial agent: Preparation and activity," *Brazilian J. Chem. Eng.*, vol. 31, no. 3, pp. 591–601, 2014.

[12] P. K. Stoimenov, R. L. Klinger, G. L. Marchin, and K. J. Klabunde, "Metal oxide nanoparticles as bactericidal agents.," *Langmuir*, vol. 18, pp. 6679–6686, 2002.

[13] M. Sundrarajan, J. Suresh, and R. R. Gandhi, "A comparative study on antibacterial properties of MgO nanoparticles prepared under different calcination temperature," *Dig. J. Nanomater. Biostructures,* vol. 7, pp. 983–989, 2012.

[14] O. Yamamoto *et al.*, "Antibacterial characteristics of carbon-coated $CaCO_3$/MgO powder led by the pyrolysis of poly (vinyl alcohol)-dolomite mixture," *Carbon N. Y.*, vol. 46, no. 7, p. 1111, 2008.

[15] R. Jagessar and G. Gomes, "Selective Antimicrobial properties of Phyllanthus acidus leaf extract against Candida albicans, Escherichia coli and Staphylococcus aureus using Stokes Disc diffusion, Well diffusion, Streak plate and a dilution method," *Nat. Sci.*, vol. 6, no. 2, pp. 1545–740, 2008.

[16] R. P. Allaker (2012), "The Use of Antimicrobial Nanoparticles to Control Oral Infections". In: Cioffi N., Rai M. (eds) *Nano-Antimicrobials*. Springer, Berlin, Heidelberg

Effect of Synthesis Parameters on Crystals Structures and Magnetic Properties of Cobalt Nickel Ferrite Nanoparticles

A. K. Agustina[1], J. Utomo[1], E. Suharyadi[1], T. Kato[2] and S. Iwata[3]

[1]Departement of Physics, Gadjah Mada University, Yogyakarta, Indonesia
[2]Departement of Electronics, Nagoya University, Japan
[3]Institutes of Materials and System for Sustainability, Nagoya University, Japan

E-mail : esuharyadi@ugm.ac.id

Abstract. $Co_{0.5}Ni_{0.5}Fe_2O_4$ magnetic nanoparticles have been successfully synthesized by co-precipitation method with varying synthesis temperature and NaOH concentration. X-ray diffraction (XRD) analysis confirmed that $Co_{0.5}Ni_{0.5}Fe_2O_4$ nanoparticles have spinel cubic crystal structure. The particle size of the sample synthesized at 3 M NaOH was 7.3 nm and then increased with the increase of NaOH concentration. It is caused by the increase of particle growth rate. The particle size of the sample synthesized at 30 °C was 10.2 nm and then increased with the increase of temperature. The lattice parameter of the sample was constant at 8.5 Å with varying synthesis temperature and NaOH concentration. The maximum magnetization and the coercivity of the sample synthesized at 3 M NaOH was 11.3 emu·g^{-1} and 0.1 kOe, respectively, and then increased with the increase of NaOH concentration. It is caused by the increase of crystallite size in a single domain. The maximum magnetization and the coercivity of the sample synthesized at 30 °C was 9.9 emu·g^{-1} and 0.1 kOe, respectively, and then increased with the increase of temperature. Transmission Electron Microscopy (TEM) image showed that the sample was agglomerated. Selective Area Electron Diffraction (SAED) image shows that sample is confirmed peak is appearing in XRD. The Fourier Transform Infra-Red (FTIR) spectra confirmed that vibration of octahedral and tetrahedral metal-oxide bond is 601.8 cm^{-1} and 370.3 cm^{-1}. The purpose of this work is to investigate the effect of synthesis temperature and NaOH concentration on the relation between crystal structure and magnetic properties.

Keywords. Cobalt Nickle Ferrite, Magnetic Properties, Nanoparticles, Synthesis Parameter

1. Introduction

Recently, magnetic nanoparticles have been interested subject, because of unusual fundamental characteristic as well as their application such as magnetic drugs delivery and recording storage. MFe_2O_4 Spinel ferrite (where M = Co^{2+}, Ni^{2+}, Fe^{2+}, Mg^{2+}) is an important group of magnetic materials because of their magnetic properties and wide practical applications. Due to the unique structure, morphology, chemical, mechanical, optical, electrical and magnetic properties, the spinel ferrite magnetic nanoparticles can be studied for the next generation nanodevices [1]. Among spinel ferrites, $CoFe_2O_4$ is especially interesting because of their characters as inverse spinel, hard magnetic materials, high coercivities and magnetic saturation [2, 3]. High physical and chemical stability make cobalt ferrite suitable for high-density magnetic recording disk [4]. On the other hand, $NiFe_2O_4$ is a

Content from this work may be used under the terms of the Creative Commons Attribution 3.0 licence. Any further distribution of this work must maintain attribution to the author(s) and the title of the work, journal citation and DOI.
Published under licence by IOP Publishing Ltd

well-known inverse spinel, a soft magnetic material with low coercivity and magnetic saturation that can be applied to electronic equipment such as transformation and inductor [5, 6]. If cobalt ferrite and nickel ferrite combined into $Co_{0.5}Ni_{0.5}Fe_2O_4$ will present the new characteristics of crystal structure and magnetism properties [7]. The $Co_{0.5}Ni_{0.5}Fe_2O_4$ characteristic can be applied to a circulator insulator device as a signal rectifier [8, 9].

Coprecipitation method is widely used for the preparation of ferrites due to the overriding advantages such as easily synthesis parameters control, homogeny particles size distribution, effectively and relatively simple to control nucleation growth [1]. Maaz *et al.* successfully fabricated $Co_{0.5}Ni_{0.5}Fe_2O_4$ by coprecipitation method [9]. However, the relation among preparation condition, structural and magnetic has not been investigated in detail. Therefore, it is necessary to investigate the relation between preparation condition, structural and magnetic properties. Moreover, study about a synthesis of cobalt-nickel ferrite by varying NaOH concentration and synthesis temperature by coprecipitation method on crystal structure and magnetic properties is very few. The purpose of this work is to investigate the effect of synthesis temperature and NaOH concentration on the crystal structure. The relation between crystal structure and magnetic properties of $Co_{0.5}Ni_{0.5}Fe_2O_4$nanoparticles is also investigated in detail to give more comprehensive information.

2. Materials and methods

Nanoparticles of$Co_{0.5}Ni_{0.5}Fe_2O_4$were synthesized by coprecipitation method with synthesis parameters variations (NaOH concentrations and temperatures). Precursors as starting material were $CoCl_2 \cdot 6H_2O$, $NiCl_2 \cdot 6H_2O$ and $FeCl_3 \cdot 6H_2O$ (all from Merck, Germany) with dissolving in distilled water and stirring to obtain homogeneous solutions. All precursors were mixed in the calculated molar ratio 1:1:4. The solution was mixed to HCl as a catalyst to solve the ionic bond between metal cation and a chloride anion. The mix solution was added wisely drop by drop at NaOH solution (3 M, 6 M, 10 M and 12 M) for 60 min in varying temperatures (30 °C, 60 °C, 90 °C and 120 °C). The precipitate was washed several times to remove salt contents and other impurities by distilled water. The precipitate of powders was dried at 90 °C in furnace 5 h. The powders of $Co_{0.5}Ni_{0.5}Fe_2O_4$ nanoparticles were characterized by X-ray diffractometer (XRD) Shimadzu XD with Cu-Kα radiation (1.56 Å) to confirm the crystal structure. The morphology and diffraction pattern were studied by Transmission Electron Microscope (TEM) Jeol JEM-1400. The magnetic properties were studied by Vibrating Sample Magnetometer (VSM) Riken Denshi Co Ltd. The functional groups were identified by Fourier Transform Infrared spectroscopy (FTIR) Shimadzu-21 in the wave number 4000 cm^{-1} to 400 cm^{-1}.

3. Results and discussion

All samples exhibit a characteristic peak of crystallite plane of (311), which explain the cubic spinel structure of nanoparticles $Co_{0.5}Ni_{0.5}Fe_2O_4$, as shown in Figure 1 and Figure 2. The ferrite phase diffraction peaks were adjusted through the JCPDS $CoFe_2O_4$ (No. 79-1744) and $NiFe_2O_4$ (No. 74-2081) database as standard [9]. The several peaks of other crystallite planes, (220), (400), (511) and (440) were detected but not clearly seen in all samples. The XRD peaks become narrower and sharper with the increase of NaOH concentrations. It indicated that there is an enhancement of crystallinity. Precipitation process is accelerated with the increase of NaOH concentration. The faster precipitation process is used to increased crystallite growth. The XRD peaks become narrower and sharper with the increase of synthesis temperatures. It indicated that there is an enhancement of crystallinity [10]. The increasing synthesis temperatures attributed to increasing thermal energy and crystal growth rate.

Table. 1 shows that the crystallite size increased with the increase of NaOH concentrations (3 M, 6 M and 10 M). The increase of NaOH attributed to accelerating the rate of particles growth. Crystallite size of $Co_{0.5}Ni_{0.5}Fe_2O_4$nanoparticlesdecreased by 12 M NaOH. The solution has been saturated with the increase of nucleation rate. $Co_{0.5}Ni_{0.5}Fe_2O_4$ nanoparticles became small and stable. Particle size and crystallinity were controlled by the OH^- ions in coprecipitation. OH^- ions determined the saturation level at the nucleation-growth process and crystallization [11]. The concentration of 12 M NaOH affects the stable reaction for the growth of crystallite size. The nucleation process at a lower temperature (30 °C and 60 °C) wasn't stable. $Co_{0.5}Ni_{0.5}Fe_2O_4$ nanoparticles at a higher temperature (90 °C) attributed to increasing of nucleation process. The increase of nucleation process causes the particles became smaller and stable. The crystallite size increased with the increase of temperature and maximum at 120 °C. It is due to the increase of thermal energy. Higher thermal energy attributed to accelerating the atomic diffusion. The particle growth rate increased by the increase of atomic motion. The changing synthesis parameters do not affect the atomic density. Because of the lattice parameters did not change with varying synthesis parameters.

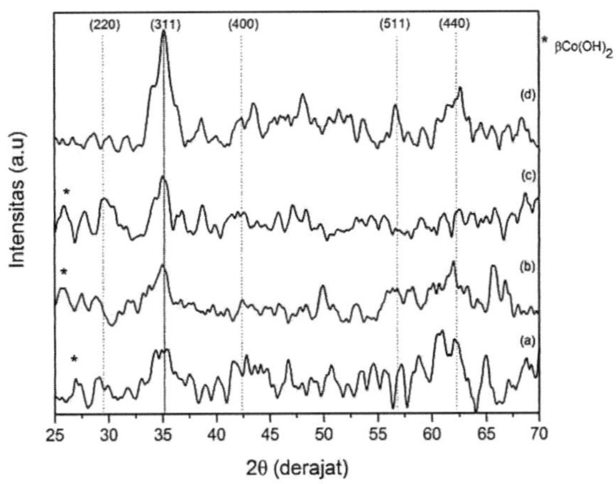

Figure 1. XRD pattern of Co0,5Ni0,5Fe2O4 nanoparticles at (a) 3, (b) 6, (c) 10 and (d) 12 M NaOH (with synthesis temperature constant at 90 °C)

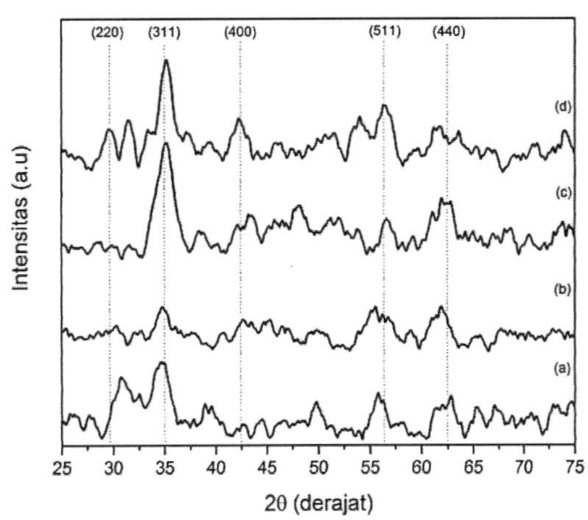

Figure 2. XRD pattern of $Co_{0.5}Ni_{0.5}Fe_2O_4$ nanoparticles at (a) 30 °C, (b) 60 °C, (c) 90 °C, and 120 °C (with NaOH concentration constant at 12 M)

From TEM images shows that nanoparticles of $Co_{0.5}Ni_{0.5}Fe_2O_4$ are agglomerated, as shown in Figure 3 and Figure 4. Agglomeration due to the higher magnetic and interaction electrostatic forces of nanoparticles with small size and large surface [12, 13]. SAED pattern shows that nanoparticles of the $Co_{0.5}Ni_{0.5}Fe_2O_4$ ring more prominent with the increase of NaOH concentration. Because of the higher NaOH concentration have higher crystallinity.

Table 1. Lattice parameter (*a*), crystallite size, and density of $Co_{0.5}Ni_{0.5}Fe_2O_4$ nanoparticles

NaOH concentration (M)	*a* (Å)	Crystalite size (nm)	Density (g·cm^{-3})	Synthesis temperature (°C)	*a* (Å)	Crystalite size (nm)	Density (g·cm^{-3})
3	8.5	7.3	5.1	30	8.7	10.2	4.8
6	8.5	7.7	5.1	60	8.5	11.7	5.0
10	8.5	10.9	5.1	90	8.4	8.5	5.2
12	8.4	8.5	5.2	120	8.5	10.9	5.2

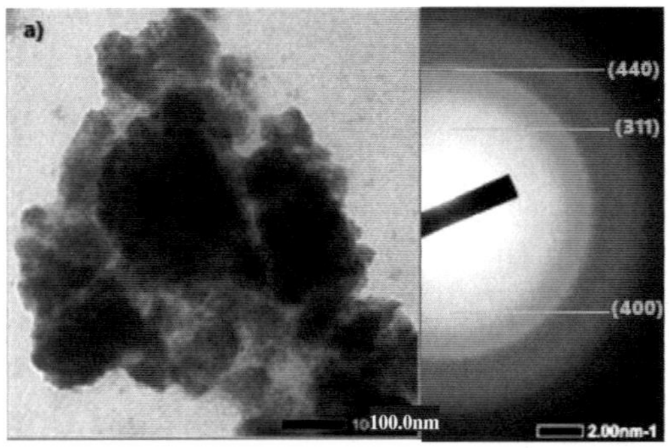

Figure 3. Morphology and SAED at 6 M NaOH

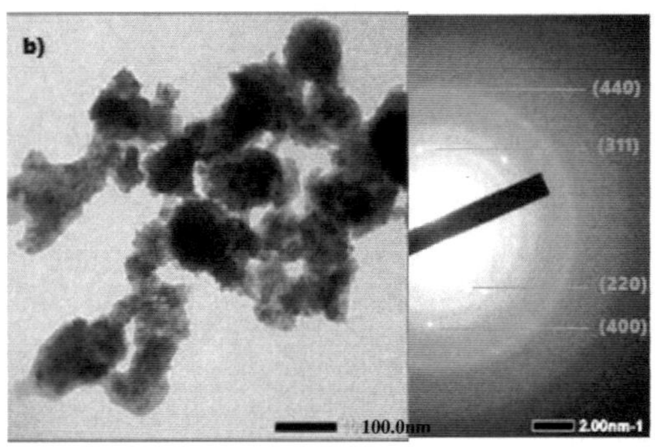

Figure 4. Morphology and SAED at 12 M NaOH

FTIR result shows that the vibrational spectra of $Co_{0.5}Ni_{0.5}Fe_2O_4$ nanoparticles are present in the wave number 370.3 cm^{-1}, 601.8 cm^{-1}, 10654,7 cm^{-1}, 1350.2 cm^{-1}, 1481.3 cm^{-1}, 1635.6 cm^{-1}, 2337.7 cm^{-1} and 3425.6 cm^{-1}, as shown in Figure 5. The band of 370.33 cm^{-1} and 601, 79 cm^{-1} are indicated the stretching of M-O in octahedral and tetrahedral site [14]. The band of 1064.71 cm^{-1} are showed the absorption of Fe-O-H functional group. The banding in 1350.17 cm^{-1} to 1635.64 cm^{-1} have detected the presence of O-H banding [15]. Stretching of H-O-H appears at 3425.58 cm^{-1}. Hydroxyl functional groups are appeared by a not perfectly drying process of the slurry at the furnace. $Co_{0.5}Ni_{0.5}Fe_2O_4$ nanoparticles contain a water solvent. C-H carboxyl stretching was detected in 2337.72 cm^{-1} [14]. The synthesis by coprecipitation base method was allowed the $Co_{0.5}Ni_{0.5}Fe_2O_4$ nanoparticles by the outside air contact.

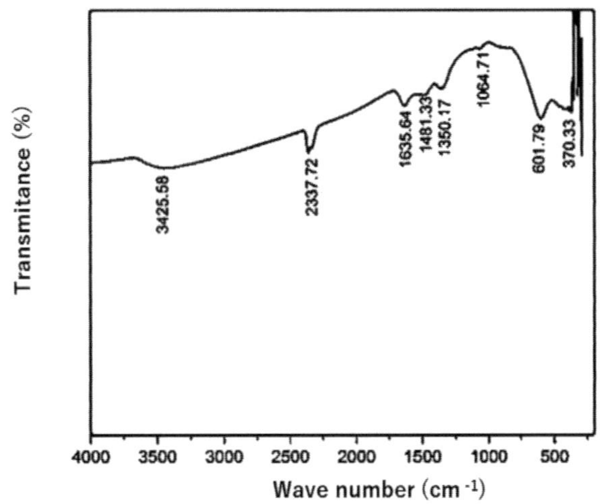

Figure 5. FTIR spectrum of $Co_{0.5}Ni_{0.5}Fe_2O_4$ at 12 M NaOH, 90 °C

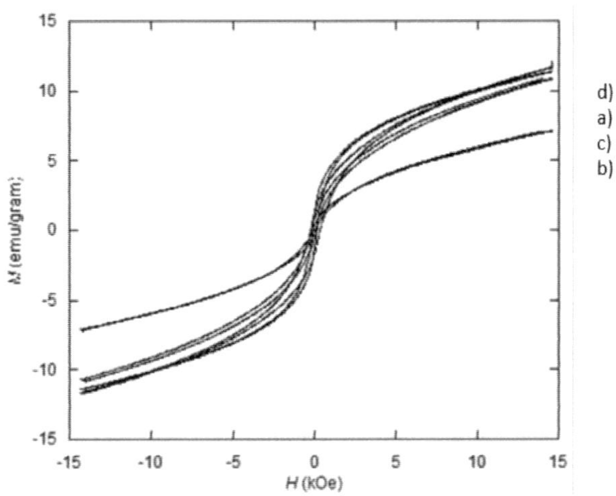

Figure 6. VSM hysteresis loops of $Co_{0.5}Ni_{0.5}Fe_2O_4$ nanoparticle at a) 3, b) 6, c) 10, and d) 12 M NaOH (with synthesis temperature constant at 90 °C)

Hysteresis loops confirm that all samples do not fully saturate at a maximum external field of 15kOe, as shown in Figure 6 and Figure 7. However, all the samples have ferromagnetic behavior. Table. 2 shows the coercivity increased with the increase of crystallite size. It is due to the increase of magneto crystalline energy on the surface of $Co_{0.5}Ni_{0.5}Fe_2O_4$nanoparticlesby the increase of surface area and crystallite size. Small coercivity in the small crystallite size can be related to the single domain. Meanwhile, coercivity decreased with the increase of crystallite size for Synthesis temperature. This phenomenon indicated $Co_{0.5}Ni_{0.5}Fe_2O_4$nanoparticles in the multidomain structure. The maximum magnetization increased with the increase of $Co_{0.5}Ni_{0.5}Fe_2O_4$nanoparticles crystallinity. The crystallinity of$Co_{0.5}Ni_{0.5}Fe_2O_4$nanoparticles depends on NaOH concentration and temperature synthesis. The higher crystallinity changes magnetic moment direction easily [16].

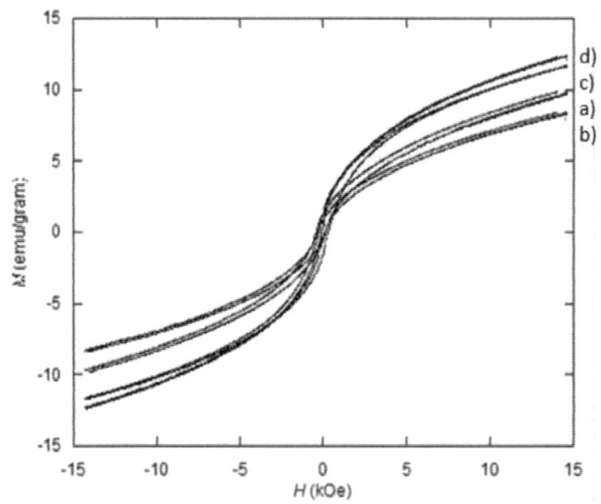

Figure 7. VSM hysteresis loops of $Co_{0.5}Ni_{0.5}Fe_2O$ nanoparticle at a) 30 °C b) 60 °C c) 90 °C and d) 120 °C (with NaOH concentration constant at 12 M)

Table 2. Crystallite size, coercivity, and maximum magnetization of $Co_{0.5}Ni_{0.5}Fe_2O_4$ nanoparticles

NaOH concentration (M)	Crystalite size (nm)	Coercivity (kOe)	Maximum Magnetization (emu·g^{-1})	Synthesis temperature (°C)	Crystalite size (nm)	Coercivitay (kOe)	Maximum Magnetization (emu·g^{-1})
3	7.3	0.1	11.3	30	10.2	0.1	9.9
6	7.7	0.2	7.1	60	11.7	0.1	8.4
10	10.9	0.2	10.8	90	8.5	0.2	12.0
12	8.5	0.2	12.0	120	10.9	0.2	12.3

4. Conclusions

$Co_{0.5}Ni_{0.5}Fe_2O_4$ magnetic nanoparticles have been successfully synthesized by co-precipitation method with varying synthesis temperature and NaOH concentration. Lattice parameters of $Co_{0.5}Ni_{0.5}Fe_2O_4$nanoparticles are constant by varying synthesis temperatures and NaOH concentrations. Crystallinity increased by the increase of NaOH concentration and synthesis temperature. The average crystallite size of $Co_{0.5}Ni_{0.5}Fe_2O_4$ nanoparticles is in the ranges 7.3 nm - 11.7 nm. The density has not changed by the variation of NaOH concentrations and synthesis temperatures. The magnetization of $Co_{0.5}Ni_{0.5}Fe_2O_4$Nanoparticles was influenced by crystallinity and crystallite size with ferromagnetic behavior. NaOH variations result shown that the $Co_{0.5}Ni_{0.5}Fe_2O_4$ nanoparticles have single domain because of crystallite size increased by the increase of coercivity. The low-temperature variations (30 °C to 60 °C) results show that the sample has multiple domains due to small coercivity and large crystallite size. The maximum magnetization increased by the increase of crystallinity.

Acknowledgments

Research Grant of Nanofabrication Platform Consortium Project of Nagoya University, Ministry of Culture, Sports, Science and Technology (MEXT) Nano-Project Platform, Japan and Grant of "*Penelitian Berbasis* Kompetensi"Ministry of Research, Technology and Higher Education of the Republic Indonesia (*Kementerian Ristek Dikti*).

References

[1] Kumar, A., Yadav, N., Rana, D. S., Kumar, P., Arora, M., & Pant, R. P. (2015). Structural and magnetic studies of the nickel doped CoFe2O4 ferrite nanoparticles synthesized by the chemical co-precipitation method. *Journal of Magnetism and Magnetic Materials*, *394*, 379-384.

[2] Ahmad, M. R., Jamil, Y., Tabasuum, A., & Hussain, T. (2015). Refinement in the structural and magnetic properties of Co0. 5Ni0. 5Fe2O4 and its application as laser micro-propellant using ablation confinement. *Journal of Magnetism and Magnetic Materials*, *384*, 302-307.

[3] Kim, Y. I., Kim, D., & Lee, C. S. (2003). Synthesis and characterization of CoFe2O4 magnetic nanoparticles prepared by temperature-controlled coprecipitation method. *Physica B: Condensed Matter*, *337*(1-4), 42-51.

[4] Houshiar, M., Zebhi, F., Razi, Z. J., Alidoust, A., & Askari, Z. (2014). Synthesis of cobalt ferrite (CoFe2O4) nanoparticles using combustion, coprecipitation, and precipitation methods: A comparison study of size, structural, and magnetic properties. *Journal of Magnetism and Magnetic Materials*, *371*, 43-48.

[5] Deraz N M and Abd-Alkader O H 2014 *J. Pure Appl. Microbio.* **7** pp 333–339

[6] Shofiah S, Muflihatun and Suharyadi E 2015 Crystal Structures and Magnetic Properties of Polyethylene Glycol (PEG-4000) and Silica-Encapsulated Nickel Ferrite ($NiFe_2O_4$) Nanoparticles. *AIP Conference Proceedings* **1725** pp 020079-1–020079-6

[7] Maqsood, A., Khan, K., Anis-ur-Rehman, M., & Malik, M. A. (2011). Spectroscopic and magnetic investigation of NiCo nanoferrites. *Journal of Alloys and Compounds*, *509*(27), 7493-7497.

[8] Xiang, J., Chu, Y., Zhang, X., & Shen, X. (2012). Magnetic and microwave absorption properties of electrospun $Co_{0.5}Ni_{0.5}Fe_2O_4$ nanofibers. *Applied Surface Science*, *263*, 320-325.

[9] Maaz, K., Karim, S., Lee, K. J., Jung, M. H., & Kim, G. H. (2012). Effect of temperature on the magnetic characteristics of $Ni_{0.5}Co_{0.5}Fe_2O_4$ nanoparticles. *Materials Chemistry and Physics*, *133*(2-3), 1006-1010.

[10] Hankare, P. P., Jadhav, S. D., Sankpal, U. B., Chavan, S. S., Waghmare, K. J., & Chougule, B. K. (2009). Synthesis, characterization and effect of sintering temperature on magnetic properties of MgNi ferrite prepared by co-precipitation method. *Journal of Alloys and Compounds*, *475*(1-2), 926-929.

[11] Safi, R., Ghasemi, A., Shoja-Razavi, R., & Tavousi, M. (2015). The role of pH on the particle size and magnetic consequence of cobalt ferrite. *Journal of Magnetism and Magnetic Materials*, *396*, 288-294.

[12] Ati M A *et al* 2014 *Int. J. Sci. Eng. R.* **5** pp 927–930

[13] Patil, R. P., Hankare, P. P., Garadkar, K. M., & Sasikala, R. (2012). Effect of sintering temperature on structural, magnetic properties of lithium chromium ferrite. *Journal of Alloys and Compounds*, *523*, 66-71.

[14] Torkian, S., Ghasemi, A., & Razavi, R. S. (2017). Cation distribution and magnetic analysis of wideband microwave absorptive $Co_xNi_{1-x}Fe_2O_4$ ferrites. *Ceramics International*, *43*(9), 6987-6995.

[15] Srinivas, C., Tirupanyam, B. V., Meena, S. S., Yusuf, S. M., Babu, C. S., Ramakrishna, K. S., ... & Sastry, D. L. (2016). Structural and magnetic characterization of co-precipitated $Ni_xZn_{1-x}Fe_2O_4$ ferrite nanoparticles. *Journal of Magnetism and Magnetic Materials*, *407*, 135-141.

[16] Gharagozlou, M. (2009). Synthesis, characterization and influence of calcination temperature on magnetic properties of nanocrystalline spinel Co-ferrite prepared by polymeric precursor method. Journal of Alloys and Compounds, 486(1-2), 660-665.

Heating Effect on Manufacturing $Li_4Ti_5O_{12}$ Electrode Sheet with PTFE Binder on Battery Cell Performance

S. Priyono[1], B. M. Lubis[2], S. Humaidi[2], and B. Prihandoko[1]

[1]Research Center for Physics – Indonesian Institute of Sciences Kawasan PUSPIPTEK Gedung 442 Serpong, Tangerang Selatan 15314, Banten, Indonesia Telp. 021-7560570, Fax 021- 7560554
[2]Department of Physics- Faculty of Mathematics and Natural Sciences-University of North Sumatera Jl. Bioteknologi I Kampus USU Medan, 20155, Indonesia.

E-mail : slam013@lipi.go.id

Abstract. The synthesis of $Li_4Ti_5O_{12}$ (LTO) and study of the heating effect on the manufacturing process of LTO sheet on the electrochemical performance have been investigated. LTO anode material composed with $LiOH.H_2O$, TiO_2 as raw materials were synthesized by the solid-state process. All raw materials were stoichiometrically mixed and milled with a planetary ball mill for 4 h to become the precursor of LTO. The precursor was characterized by Simultaneous Thermal Analyzer (STA) to determine sintering temperature. The STA analysis revealed that the minimum temperature to sinter the precursor was 600 °C. The precursor was sintered by using high-temperature furnace at 900 °C for 2 h in air atmosphere. The final product was ground and sieved with a screen to get finer and more homogenous particles. The final product was characterized by X-ray Diffraction (XRD) to determined crystal structure and phases. LTO sheet was prepared by mixing LTO powders with PTFE and AB in ratio 85:10:5 wt% by varrying heating process with 40 °C, 50 °C and 70 °C to become slurry. The slurry was coated on Cu foil with doctor blade method and dried at 80 °C for 1 h. LTO sheet was characterized by FTIR to analyze functional groups. LTO sheet was cut into circular discs with 16 mm in diameter. LTO sheet was arranged with a separator, metallic lithium and electrolyte become coin cell in a glove box. Automatic battery cycler was used to measure electrochemical performance and specific capacity of the cell. From the XRD analysis showed that single phase of LTO phase with a cubic crystal structure is formed. FTIR testing showed that there are stretching vibrations of Ti-O and H-F from tetrahedral TiO_6 and PTFE respectively. Increasing temperature on LTO sheet manufacturing doesn't change the structure of LTO. Cyclic voltammetry analysis showed that sample with heating of 40 °C showed better redox process than others. Charge-discharge test also showed that sample with heating of 40 °C has higher specific capacity than other samples with 53 $mAh \cdot g^{-1}$.

Keywords. Cyclic voltammetry, Electrode Sheet, LTO, PTFE, TiO_2.

Content from this work may be used under the terms of the Creative Commons Attribution 3.0 licence. Any further distribution of this work must maintain attribution to the author(s) and the title of the work, journal citation and DOI.
Published under licence by IOP Publishing Ltd

1. Introduction

Energy consumption relying on fossil fuel is predicted having a bad future impact on the economy and ecology of the world [1] because fossil fuel is not sustainable and put carbon gas pollution on air that causing global warming. One of the enormous sector emitting carbon gas pollution into the world is transportation. This problem can be solved by replacing the conventional engine with the battery as energy sources. The battery is a device to store energy for supplying electrical power [1]. Recently, lithium-ion batteries have been emerged as the most promising energy storage technology and widely used in some applications such as portable devices and electric vehicles [2]. This is due to its advantages such as high energy density, lighter weight, no memory effect, and low self-discharge rate when compared to other rechargeable batteries [3].

Battery components include anode, cathode, electrolyte, separator and currents collector. They are assembled to be coin cells for battery performance testing. Before assembling process, the anode and cathode must be made in sheets form. Forming of anode and cathode sheets needs a binder.

Nowadays, Lithium Titanium Oxide ($Li_4Ti_5O_{12}$) called LTO and used as the anode is considered to be a better choice than conventional graphite. It is caused by its zero-strain structure, small particle size, no Solid Electrolyte Interphase (SEI) formation, etc. Therefore, lithium ion battery with LTO anode shows a very long-life cycle [4].

Binder is one of the main raw materials for making anode and cathode sheets in battery [5]. A good binder has the following requirements; right material composition, good adhesion, high conductivity, proper coating thickness, and stable or not blend with other mixed materials. The conventional binder used is PVDF (Polyvinylidene fluoride). PVDF has a drawback. It can only be deluded with N N-Dimethyl-acetamide (DMAC) solvent. However, DMAC is toxic, relatively high price, and difficult to decompose.

Water is a saver and more secure that can be used in making a slurry. Nevertheless, PVDF is not able to dissolve into it. Therefore, it is needed another binder to replace PVDF, that is PTFE. PTFE (Polytetrafluoroethylene) is environmentally friendly. It uses water solvent that is not harmful to health. PTFE has excellent chemical resistance, high heat, and cold resistance, insulating, resistant to UV rays (ultraviolet), good weather resistance, low friction coefficient, anti-fracture, flexural, high resistance to fire and low water absorption.

In this research, anode powder, $Li_4Ti_5O_{12}$ (LTO), had been made with raw materials of $LiOH\cdot H_2O$ and TiO_2. Following that, anode powder was used to make electrode sheet. In this study is focused on using PTFE binder with varying temperature on the heating process to know the characteristic of battery performances.

2. Materials and methods

The $Li_4Ti_5O_{12}$ powder was synthesized via Solid state reaction with using raw materials such as $LiOH\cdot H_2O$ (technical) as Lithium (Li) source, and TiO_2 (technical) as Titanium (Ti) source. The synthesized material follows stoichiometry formula as stated in Equation 1:

$$4LiOH\cdot H_2O + 5TiO_2 \rightarrow Li_4Ti_5O_{12} + H_2O + O_2 \qquad (1)$$

All raw materials were mixed, mashed and screened to get precursor of $Li_4Ti_5O_{12}$. The precursor was milled with Planetary Ball Milling (PBM) with speed 400 rpm for 4h. Following that, the precursor was sintered at 900 °C for 2 h with 2 °C heat rate in an air atmosphere. The final product was screened 200#. The synthesized materials were characterized by using X-ray diffractometer (XRD, Rigaku) to know the phases.

The slurry was prepared by mixing active powder ($Li_4Ti_5O_{12}$) with a binder (PTFE) and conductive agent (acetylene black) at weight ratio 85:10:5. Aquadest was used as a solvent to soluble the binders. All materials were mixed, stirred, and heated by varying temperature heating at 30 °C, 50 °C and 70 °C in a solvent to become slurry and then coated with Cu foil.

The anode sheet was cut into a circular form with a diameter of 16 mm. The circle sheet was tested by FTIR (Thermo Scientific Nicolet iS10) to study the functional group of the sheet. The testing was conducted at a range of 500 cm^{-1} to 4000 cm^{-1} wavenumbers.

Before battery performance testing, the sheets were assembled in cell coin in the glove box. Cyclic Voltammetry (CV) and Charge-Discharge (CD) testing by WBCS instrument (WBCS 3000, Automatic Battery Cycler Ver.3.2) were performed on the coin cells to observe the battery performance. In CV testing, battery voltage used was starting from 2.8 V to 0.5 V at a scan rate of 0.1 $mV \cdot s^{-1}$ in 1 cycle. Meanwhile, in CD testing, it was starting from 1.0-2.8 V with a scan rate of 0.1 C for 1 cycle.

3. Results and discussion

The XRD result displayed in Figure 1 showed that LTO crystal phase is formed without any impurities. All the peaks are high and sharp, and those indicate high crystallinity of the sample. The diffraction peaks of LTO peaks are at 2θ of 18.33°, 35.57°, 37.21°, 43.24°, 47.35°, 57.21°, 62.83°, 66.07°, 74.34°, 75.36°, and 79.34°. The crystal planes (hkl) are 111, 311, 222, 400, 331, 333, 440, 531, 533, 622, and 444. The peaks pattern is corresponding to LTO ICDD database No #00-049-0207. The crystal structure of LTO sample is cubic with a space group of Fd-3m and lattice parameter of 8.373 Å.

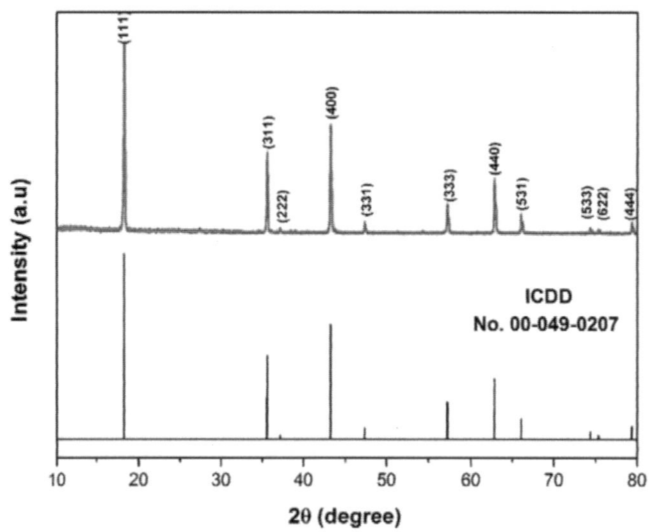

Figure 1. XRD pattern of LTO powder synthesized via solid-state reaction

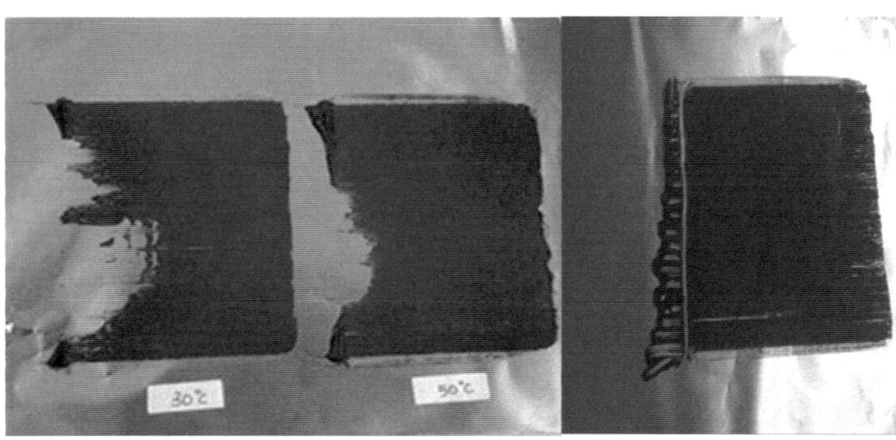

Figure 2. LTO sheets with PTFE binder with the different heating process.

Figure 2 shows electrode sheets of LTO prepared by PTFE binder via doctor blade method. All samples physically show good sample, smooth, and homogenous. Samples also have well adhered and can't be brittle.

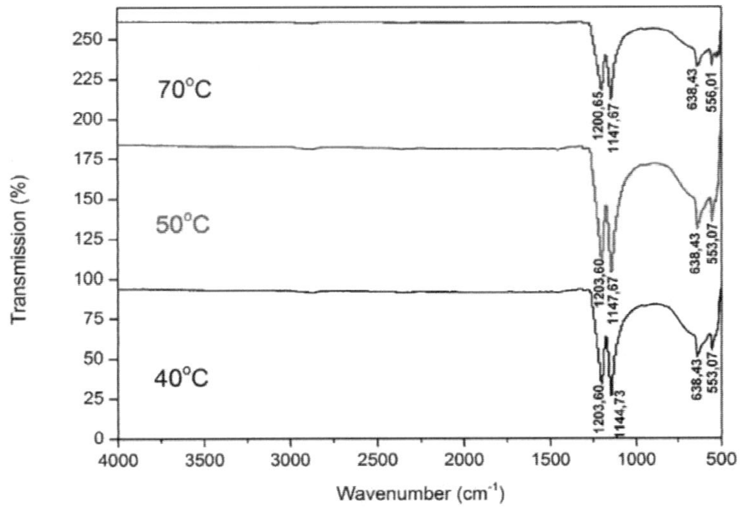

Figure 3. The infrared spectrum of the sample at a temperature of 40 °C, 50 °C and 70 °C.

Figure 3 shows FTIR pattern of LTO sheet using PTFE as a binder with varying temperature. All samples indicate identical spectrum in fingerprint region. The peaks observed at a wave number of 553 cm^{-1} and 638 cm^{-1} denoted the symmetric stretching vibration of TiO_6 octahedral of LTO [6]. Also in the spectrum, the band at wave number 1144.73 cm^{-1} to 1200.65 cm^{-1} represent H-F bonds derived from PTFE binder [7]. Sample with heating 50 °C shows a little bit higher, wider and sharper than others in H-F bond, and it indicates that this sample has a stronger bond. There are no peaks at Hydroxyl bond (at 3000 cm^{-1} to 3500 cm^{-1}) it indicates that sample is dry enough and ready to be used as a negative electrode of the battery.

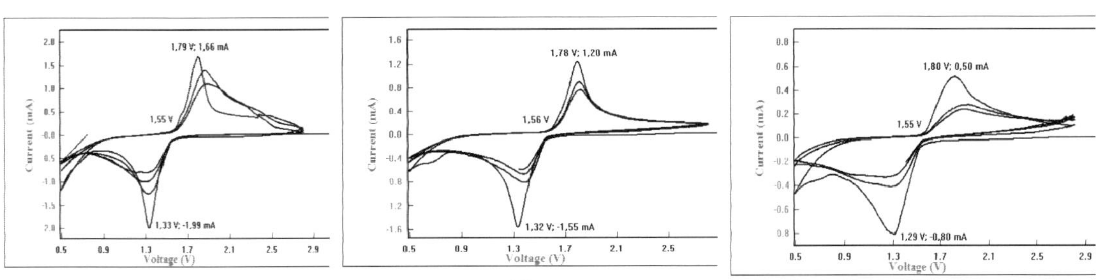

A B C

Figure 4. Cyclic voltammograms of LTO using PTFE binders with varying heating temperature (A). 30 °C, (B) 50 °C, and 70 °C

The cyclic voltammetry measurement was tested in the voltage range from 0.5 V to 2.8 V and vice versa for 3 cycles at a scan rate 0.1 $mV \cdot s^{-1}$ in order to well understand the Li^{+} intercalation/deintercalation kinetics of the electrode material. Figure 4 shows cyclic voltammograms of LTO coated onto Cu foil with PTFE binder at different heating temperature. All samples show a clear pair of redox peak observed below 2 V those indicated that samples could be rechargeable and show properties of secondary batteries. Redox peaks are associated with reduction and oxidation of Ti^{3+}/Ti^{4+} [10].

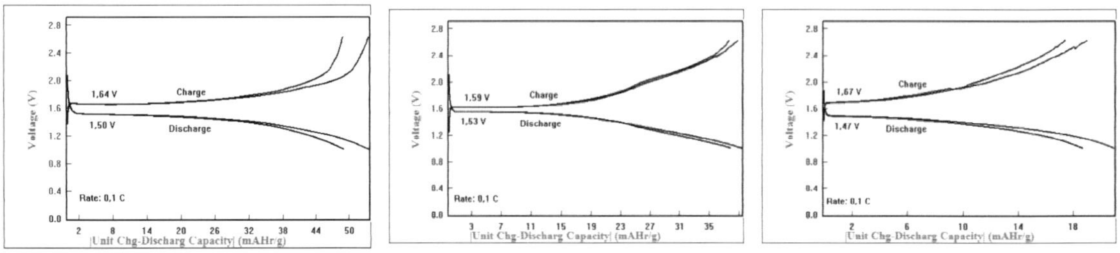

A B C

Figure 5. Galvanostatic charge and discharge curve of $Li_4Ti_5O_{12}$.

Figure 5 shows the charge-discharge curve of a cell for the first two cycles between 1.0 V and 2.8 V at a rate of 0.1 C. It can be seen from Fig 5 that the plateau is clearly in the charge-discharge curve at the potential of 1.55 V (versus Li+/Li). This behavior corresponds to the solid-state redox of Li in $LiTi_4O_{12}$ involved with Li^+ ion extraction and insertion. Sample A shows higher specific capacity than sample B and C with 54 $mAh{\cdot}g^{-1}$. Increasing in temperature slurry will decrease specific capacity and increase the polarization of samples. Higher temperature will evaporate more water content in slurry so make the bond between the active material and current collector become weak, so it will make electron difficult to across through electrode.

4. Conclusions

$Li_4Ti_5O_{12}$ powder and its electrode sheet were successfully prepared by solid-state reaction and doctor blade method respectively. The structural analysis revealed that the prepared sample was highly crystalline. Sample heated 50 °C had stronger and better bond than others, and also showed good intercalation and de-intercalation process. But sample heated 40 °C has the higher specific capacity with 55 °C, and it's greater than the previous experiment.

Acknowledgments

The author would like to acknowledge for financial and experimental tools support from Minister Research, Technology and Higher Education (Menristek-Dikti); and Research Center for Physics – Indonesian Institute of Sciences (LIPI). The author also would like to thank the Battery Research Group in.

References

[1] Vikström H, Davidsson S. and Höök M 2013 Lithium availability and future production outlooks *Applied Energy* **110** pp 252–266
[2] Anuphappharadorn S *et al* 2014 Comparison the economic analysis of the battery between lithium-ion and lead-acid in PV stand-alone application *Energy Procedia* **56** pp 352–358
[3] Wang, X., Adelmann, P., & Reindl, T. 2012. Use of $LiFePO_4$ batteries in stand-alone solar system. *Energy Procedia*, 25, 135-140.
[4] Zhao R, Liu J, and Gu J 2017 A comprehensive study on Li-ion battery nail penetrations and the possible solutions *Energy* **123** pp 392–401
[5] Han, X., Ouyang, M., Lu, L., & Li, J. 2014. Cycle life of commercial lithium-ion batteries with lithium titanium oxide anodes in electric vehicles. *Energies*, 7(8), 4895-4909.
[6] Zhong, C *et al* 2016 *Electrolytes for electrochemical supercapacitors* CRC Press.
[7] Shu J 2008 Study of the interface between $Li_4Ti_5O_{12}$ electrodes and standard electrolyte solutions in 0.0–5.0 V *Electrochemical and Solid-State Letters* **11** pp A238–A240

[8] Zhao Y *et al* 2013 High-performance alkaline fuel cells using crosslinked composite anion exchange membrane *Journal of Power Sources* **221** pp 247–251
[9] Nordh T 2013 $Li_4Ti_5O_{12}$ as an anode material for Li ion batteries in situ XRD and XPS studies (Thesis) *University Uppsala*
[10] Yuan T *et al* 2010 Synthesis of pristine and carbon-coated $Li_4Ti_5O_{12}$ and their low-temperature electrochemical performance *Journal of Power Sources* **195** pp 4997–5004.

IOP Conf. Series: Materials Science and Engineering **367** (2018) 012008 doi:10.1088/1757-899X/367/1/012008

Characteristics of silica rice husk ash from Mojogedang Karanganyar Indonesia

R. Suryana[1], Y. Iriani[1], F. Nurosyid[1], and D. Fasquelle[2]

[1]Department of Physics, Faculty of Mathematics and Natural Sciences, Universitas Sebelas Maret, Indonesia
[2]Unité de Dynamique et Structure des Matériaux Moléculaires, Universite du Littoral Cote d'Opale, Calais, France

E-mail: rsuryana@staff.uns.ac.id

Abstract. Indonesia is one of the countries in the world as the most abundant rice producer. Many researchers have demonstrated that the highest composition in the rice husk ash (RHA) is silica. Some of the advantages in utilizing silica as the raw material is the manufacture of ceramics, zeolite synthesis, fabrication of glass, electronic insulator materials, and as a catalyst. The amount of silica from rice husk ash is different for each region. Therefore, the study of silica from RHA is still promising, especially rice organic fertilizers. In this study, the rice came from Mojogedang Karanganyar Indonesia. Rice husk was dried under the solar radiation. Then the rice husk was heated in two steps: the first step at a temperature of 300°C and the second step at a temperature of 1200°C with a holding time at 2 h and 1 h, respectively. Furthermore, the temperature of the second step was varied at 1400 °C and 1600 °C. This heating process produced RHA. The content of RHA was observed on the EDAX spectrums while the morphology was observed from SEM images. The crystal structure of RHA was determined from XRD spectrums. The EDAX spectrums showed that RHA composition was dominated by elements Si and O for all the heating temperature. SEM images showed an agglomeration towards larger domains as heating temperatures increase. Analysis of XRD spectra is polycrystalline silica formed with the significant crystal orientation at 101, 102 and 200. The intensity of 101 increases significantly with increasing temperature. It is concluded that the crystal growth in the direction of 101 is preferred.

1. Introduction

Indonesia is one of the biggest rice-producer countries in the world. Thus, there is a huge of rice husk (RH) waste. It still widely used as a fuel by burning it. Approximately 20% of ash can be obtained from burning RH [1-3]. Many researchers report that the rice husk ash (RHA) contains about 87%-98% of silica (SiO_2) [4-6]. The rest is metallic impurities such as K_2O, Al_2O_3, CaO, MgO, Na_2O, Fe_2O_3 with the percentages of each compound are less than 1% [3, 7, 8].

In industrial application, silica is the raw material in the manufacture of ceramics, zeolite synthesis, glass, electronic insulator materials, and as a catalyst [4, 8]. Recently, silica from RHA because of the high silica content is found to be the source of silicon (Si) through common fabrication methods such as submerged electric arc method and magnesium (Mg) reduction [3, 5, 9, 10]. It is applied in many electronic devices. One is semiconductor grade silicon (SGS) which is the material for a solar cell that it becomes one of the most attractive studies to overcome the energy crisis issues nowadays [3]. Silica

Content from this work may be used under the terms of the Creative Commons Attribution 3.0 licence. Any further distribution of this work must maintain attribution to the author(s) and the title of the work, journal citation and DOI.
Published under licence by IOP Publishing Ltd

must have high purity and crystallinity to produce semiconductor grade silicon [11]. Therefore, we have been conducting a study to gain a high purity (>99 %~100%) and crystallinity of silica in the RHA.

Table 1. The results of RHA fabrication in the previous research [12]

Sintering temperatures (°C)	Percentage of silica in RHA (%)	Products and crystallinity (%)
1000	95.11	Crystalline and 97.43
900	94.72	Amorphous
800	94.96	Amorphous

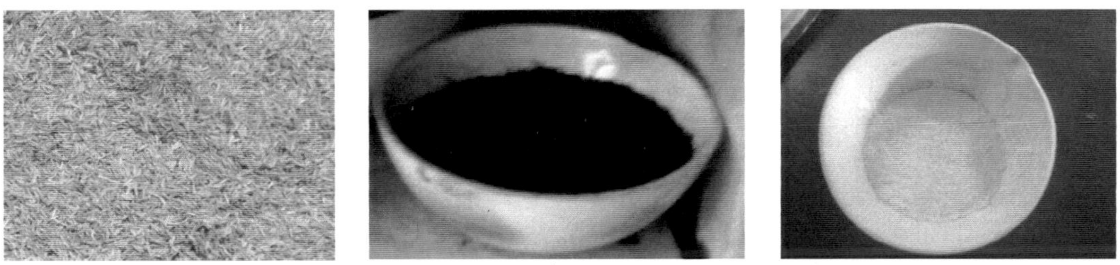
a b c

Figure 1. Images of (a) RH (b) RH charcoal and (c) RHA

The silica content in RHA is affected by the crop variety, soil chemistry, climate condition, fertilizer condition, and even the geographic location [8, 9]. Additionally, other factors such as the temperature, holding time, the rate and heating technique also have their impact [8]. Several studies have reported that silica is an amorphous phase when it is heated up to low temperature (500°C-700°C). It forms high purity and crystalline at a high sintering temperature (>800°C) depended on the time combustion [3, 6]. Thus, the study to produce high purity silica in RHA (>99 %~100%) is still challenging.

In our previous study [12], RHA has been synthesized by two steps heating; the first heating called carbonization at a temperature of 300°C and the second heating called sintering at different temperatures of 800°C, 900°C, and 1000°C. The results in Table 1 showed that silica RHA was still amorphous at the temperatures of 800°C and 900°C and changed to the crystalline phase at the temperature of 1000°C. It yielded silica with more than 90% for each sintering temperature. However, the metallic impurities such as CaO, P_2O_5, Al_2O_3, K_2O, Cl, SO_3, Fe_2O_3, MnO, Cr_2O_3, NiO, SnO_2, and ZnO remained. Consequently, the heating process has been done 3 times and the sintering temperatures have been increased in this study.

2. Materials and methods

The synthesis of silica included washing the RH, drying, carbonization, and sintering. The RH was prepared from Mojogedang Karangayar, Indonesia displayed by Fig. 1a. It was cleaned to remove the dust and sand attached to the rice husk. Drying was directly exposed under the sun to eliminate water content in rice husk after cleaning process. The carbonization was carried out using furnace at a temperature of 300 °C for 1 h in the air ambient. It was to decompose the volatile organic substances (cellulose, hemicellulose, and lignin) contained in the RH. It produced RH in the black color shown in Fig. 1b. This black RH composed much of carbon evaporated on heating > 400 °C [3]. It was then sintered in two sintering processes; first sintered at a temperature of 900 °C for 2 h producing RHA shown by Figure 1c and second sintered at varied temperatures of 1200 °C, 1400 °C, and 1600 °C with the holding time of 6 h for each variation. The second sintering was performed to remove the metallic impurities left after the first sintering. The RHA was then characterized by X-ray Diffraction (XRD) and Scanning Electron Microscope-Electron Dispersive X-Ray (SEM-EDX) equipment.

3. Results and discussion

Figure 2 presents XRD patterns of the samples formed at different temperatures. All the diffraction peaks are in good agreement with literature data (ICDD #82121410). They are defined as SiO_2 with a tetragonal structure and space group of $P4_12_12$. Additionally, the highest peaks were investigated as the increased temperature displayed by the (101) plane in Fig. 2. Based on the Scherrer formula as performed in ref. [13], it implies that the crystallite size of silica grew with increasing temperature presented in Table 2. It also indicates improving crystallization of the material which means high crystallinity achieved [6]. The most substantial peak showing by Bragg's plane of (101) is one of the fingerprints of silica cristobalite phase. It leads to being the key of the fabricating single crystal SGS for a solar cell in our next study. The single crystal is more easily grown on a substrate than the poly-crystal.

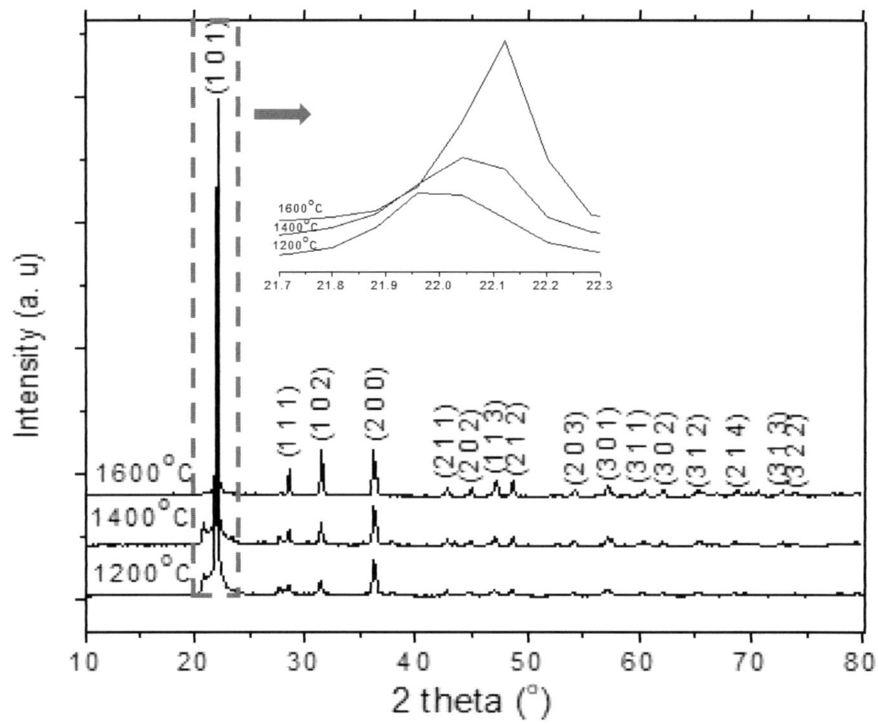

Figure 2. XRD patterns of RHA at different temperatures

Table 2. Crystallite size of RHA at different temperatures

Temperatures (°C)	Crystallite size (nm)		
	(101)	(102)	(200)
1200	25.39	26.25	24.46
1400	31.18	26.42	32.59
1600	56.20	42.16	37.49

IOP Conf. Series: Materials Science and Engineering **367** (2018) 012008 doi:10.1088/1757-899X/367/1/012008

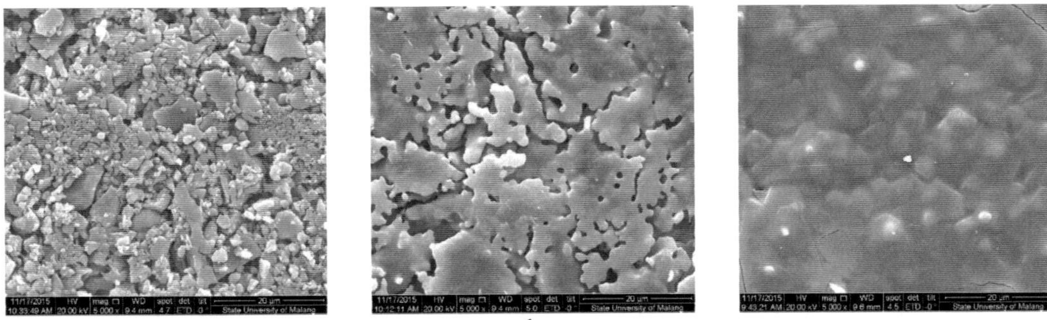

a b c

Figure 3. SEM images of RHA at temperatures of (a) 1200 °C (b) 1400 °C and (c) 1600 °C

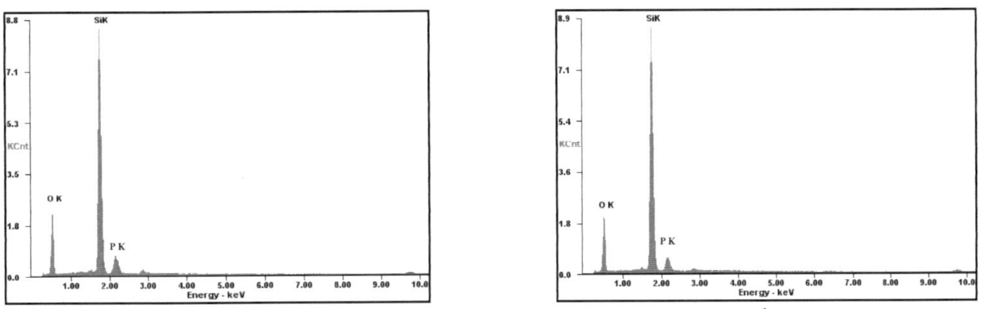

a b

Figure 4. EDX spectrum of RHA at temperatures of (a) 1200 °C and (b) 1400 °C

Figure 3 shows the morphologies of the RHA at different temperatures. It is clear that the samples agglomerated with increasing temperature, which represents the crystal growth in the samples. Therefore, the SEM results are corresponding to the XRD patterns of the samples. This phenomenon excludes whose silica powder was melted at the sintering temperature of 1600 °C. It is due to the melting point of silica is around 1600 °C [14].

In order confirm the compounds in RHA, the samples with sintering temperatures of 1200 °C and 1400 °C were further examined by EDX. This equipment can be used to identify the chemical compositions of material. As revealed in Fig. 4, Si and O are the main components in RHA. The others elements with low intensity are assigned to phosphor (P). The elements determination of EDX result in RHA refers to the study reported by Hadipramana et al [2] and Yavakkumar et al [4].

The percentage of silica in RHA has not been reported in this study. However, lost in many type metallic contaminants and only leaving the little amount of phosphor indicates the higher content of silica in RHA obtained in this study compared to the previous study [12]. As a result, it can be considered as the raw material for producing silicon.

4. Conclusions

High content and crystallinity of silica from RHA can be extracted by two steps of sintering. Thus, it can be considered as the source of silicon. The crystal growth of silica at (101) orientation is indicating a silica crystobalite phase which able to be a basis for fabricating single crystal silicon which is easily grown on a substrate.

Acknowledgment
Authors wish to thank Kementerian Riset, Teknologi dan Pendidikan Tinggi for financial support through Program Research and Innovation in Science and Technology Project (RISET-Pro) 2016 No. 1595. 1/D3/PG/2016.

References

[1] Hieu, N. M., Korobochkin, V. V., & Tu, N. V. (2015). A study of silica separation in the production of activated carbon from rice husk in Viet Nam. *procedia Chemistry*, **15** 308-312.

[2] J. Hadipramana, F.V. Riza, I.A. Rahman, L.Y. Loon, S.H. Adnan, A.M.A. Zaidi, 2016 Pozzolanic Characterization Of Waste Rice Husk Ash (RHA) From Muar, Malaysia. *IOP Conference Series: Materials Science and Engineering*, **160** 012066.

[3] B.O. Ayomanor, K. Vernon-Parry, 2016 Potential Synthesis of Solar-Grade Silicon from Rice Husk Ash, *Solid State Phenomena*, **242** 41-47.

[4] R. Yuvakkumar, V. Elango, V. Rajendran, N. Kannan, 2012 High-purity nano silica powder from rice husk using a simple chemical method, *Journal of Experimental Nanoscience*, **9**(3) 272-281.

[5] A. A. França, J. Schultz, R. Borges, F. Wypych, A. S. Mangrich,(2017). Rice Husk Ash as Raw Material for the Synthesis of Silicon and Potassium Slow-Release Fertilizer. *Journal of the Brazilian Chemical Society*, **28**(11) 2211-2217.

[6] R.A. Bakar, R. Yahya, S.N. Gan, (2016) Production of high purity amorphous silica from rice husk, *Procedia Chemistry*, **19** 189-195.

[7] R. Prasad, M. Pandey, (2012) Rice husk ash as a renewable source for the production of value added silica gel and its application: an overview, *Bulletin of Chemical Reaction Engineering & Catalysis*, **7**(1) 1-25.

[8] A. Kumar, K. Mohanta, D. Kumar, Om Parkash, (2012) Properties and industrial applications of rice husk: a review, *International Journal of Emerging Technology and Advanced Engineering*, **2**(10) 86-90.

[9] A. Onojah, A. N. Amah, B. O. Ayomanor, (2012) Comparative studies of silicon from rice husk ash and natural quartz, *American Journal of Scientific and Industrial Research*, **3**(3) 146-149.

[10] T. Okutani, (2009) Utilization of silica in rice hulls as raw materials for silicon semiconductors, *Journal of Metals Materials and Minerals*, **19** 51-59.

[11] C. P. Lund, W. Zhang, P. J. Jennings, P. Singh, (2000), in Solar 2000: Renewable Energy Transforming Business, 38th Annual Conference of the Australian and New Zealand Solar Energy Society, 421-429.

[12] D. K. Sandi, Program Kreativitas Mahasiswa-Penelitian, (2014) (unpublished).

[13] R. Suryana, O. Nakatsuka, S. Zaima, (2011) Formation of Palladium Silicide Thin Layers on Si (110) Substrates, *Japanese Journal of Applied Physics*, **50** 05EA09.

[14] National Center for Biotechnology Information. PubChem Compound Database; CID=24261, https://pubchem.ncbi.nlm.nih.gov/compound/24261 (accessed Mar. 28, 2017).

Nanostructure Analysis for Microwave Absorption Properties of Fe_3O_4 Particles by Symmetry Top Rotational Molecular Model

Y. P. Hardianto[1], A. Taufiq[1], A. Hidayat[1], S. Sunaryono[1], A. Listanti[1] and H. Wisodo[1]

[1]Department of Physics, Faculty of Mathematics and Natural Sciences, Universitas Negeri Malang, Jalan Semarang No 5, Malang 65145, Indonesia

Email: ahmad.taufiq.fmipa@um.ac.id

Abstract: In the last years, magnetite (Fe_3O_4) nanoparticles have been prepared with various methods to produce the Fe_3O_4 with high performance especially in microwave absorption properties. In general, the characterization of reflection loss using microwave of the Fe_3O_4 particles has been carried out in the range of 2-18 GHz. In this work, the rotational molecular model was developed to analyze the reflection loss peaks in the range of the frequency. The primary technique of simulation was an interpolation of peak frequencies originating from the molecular rotational transition of the Fe_3O_4 particles. The data analysis presented that the mechanism occurs at the electromagnetic absorption of the Fe_3O_4 particles is rotational molecular. Interestingly, it is presented that the result of the simulation is in line with the results of the characterizations of several materials that have been reported in some literature.

Keywords: Magnetite, reflection loss, microwave absorption, rotational molecular model.

1. Introduction

In the last decades, the research by focusing on the microwave absorption properties of materials has been developed significantly [1]. One of the bases that trigger such development is related to the various applications such as defense technology, especially for radar absorbing material [2]. Nowadays, one of the magnetic material groups that has excellent performance as the radar absorbing material is magnetite (Fe_3O_4) nanoparticles [3], especially in the microwave range.

So far, the Fe_3O_4 nanoparticles have been synthesized in many morphologies, shapes, and sizes to produce a better performance in absorbing microwave. Other strategies have also been conducted to produce the Fe_3O_4 nanoparticles in nanocomposites [4–8], core shells [9-11], or coated with other materials [12,13]. Unfortunately, studies that exploring microscopic phenomena related to the microwave absorption properties of the Fe_3O_4 nanoparticles are rarely reported. Therefore, the study by focusing on the microscopic view of the microwave absorption of the Fe_3O_4 nanoparticles is a great importance, both through the development of theoretical, computational, and experimental research.

Based on the previous studies, the microwave absorption values of the various materials using Fe_3O_4 as the main compound have maximum were in the range of 2-12 GHz [14–16]. From such literature, it is known that the reflection loss tends to increase by decreasing the thickness of the

Content from this work may be used under the terms of the Creative Commons Attribution 3.0 licence. Any further distribution of this work must maintain attribution to the author(s) and the title of the work, journal citation and DOI.
Published under licence by IOP Publishing Ltd

samples and by increasing frequency. Therefore, it is essential to investigate the microscopic phenomenon regarding microwave absorption of Fe_3O_4 nanoparticles. The importance of this study is not only to reveal fundamental physical phenomena in the material but also to predict and improve microwave absorption performance for further application development. In this paper, we report the symmetry top rotational molecular model and microwave absorption calculation method for the Fe_3O_4 nanoparticles.

2. Materials and methods

In this work, the calculation method was employed by symmetry top rotational molecular model. Based on the crystal structure, Fe_3O_4 is a material consisting of two substructures, i.e., tetrahedral and octahedral structures. The crystal structure of the Fe_3O_4 is visualized in Figure 1. The two substructures have different characters in absorbing electromagnetic wave due to the different inertial moment resulting in the different energy levels of them. The inertial moment for the substructures are represented in Equation 1-3.

$$I_{oct} = 4m_{oxy}\,r_A^{\,2} \quad (1)$$

$$I_{tet//} = 2m_{oxy}(1-\cos\theta)r_B^{\,2} \quad (2)$$

$$I_{tet\perp} = m_{oxy}(1-\cos\theta)r_B^{\,2} + \frac{m_{oxy}}{m_{Fe}+4m_{oxy}}(m_{Fe}+m_{oxy})(1+2\cos\theta)r_B^{\,2} + ((m_{Fe}+3m_{oxy})r_B + 6m_{oxy}r_B\left(\frac{1}{3}(1+2\cos\theta)\right)^{1/2})r_B \quad (3)$$

where I_{oct} is the inertia moment for octahedral phase of the Fe_3O_4 particle, $I_{tet//}$ and $I_{tet\perp}$ are the inertia moments for tetragonal phase of the Fe_3O_4 particle, m_{oxy} is the mass of oxygen atom (16 amu), m_{Fe} is the mass of an iron atom (55.847 amu), r_A is the distance of oxygen atom to Fe (+3) atom (2.066 Å) in the crystal balance condition [17]. Based on the data, we propose the rotational Hamiltonian of tetrahedral and octahedral for Fe_3O_4 particles as presented in Equation 4-7.

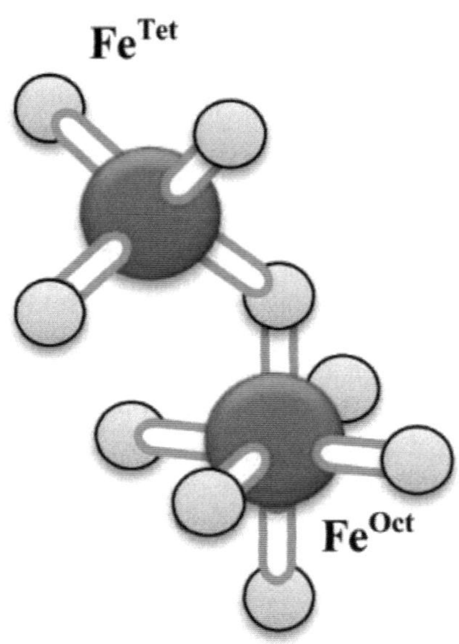

Figure 1. Tetrahedral and octahedral structure of Fe_3O_4 particles

$$H_{okt} = \frac{J^2}{2I_{okt}} \tag{4}$$

$$H_{tet} = \frac{J^2}{2I_{tet\perp}} + (\frac{1}{2I_{tet//}} - \frac{1}{2I_{tet\perp}})J_z^{\ 2} \tag{5}$$

where H_{okt} is Hamiltonian for the octahedral phase of the Fe_3O_4 particle and H_{tet} is Hamiltonian for the tetragonal phase of the Fe_3O_4 particle.

If the two Hamiltonian are subjected to Eigenfunction of the Fe_3O_4 particles, the Eigenvalue will be presented in Equation 6-7.

$$H_{okt}\left|J.K\right\rangle = \frac{J^2}{2I_{okt}}\left|J.K\right\rangle = \frac{J(J+1)\hbar^2}{2I_{okt}}\left|J.K\right\rangle \tag{6}$$

$$H_{tet}\left|J.K\right\rangle = \frac{J^2}{2I_{tet\perp}}\left|J.K\right\rangle + (\frac{1}{2I_{tet//}} - \frac{1}{2I_{tet\perp}})J_z^{\ 2}\left|J.K\right\rangle = \left(\frac{J(J+1)\hbar^2}{2I_{tet\perp}} + (\frac{1}{2I_{tet//}} - \frac{1}{2I_{tet\perp}})K^2\hbar^2\right)\left|J.K\right\rangle \tag{7}$$

where J is quantum number of molecular orbit with the value of 0 to ∞, while $M = J, J\text{-}1, \ldots, \text{-}J$.

Based on the *Raman selection rules* [18], the rotational transition can be found for $\Delta J = \pm1, \pm2$. Therefore, it will imply on the change of allowed energy level as shown in Equation 8.

$$\Delta E_{okt} = \frac{\hbar^2}{2I_{okt}}(2J+1) \text{ or } \Delta E_{okt} = \frac{\hbar^2}{2I_{okt}}(4J+6) \tag{8}$$

The above requirement can also apply to the change of energy level on tetrahedral and octahedral structures of Fe_3O_4 particles. Besides that, in this work, we also apply the rotational energy of Fe_3O_4 particles. Therefore, the visualization of the absorption intensity of the Fe_3O_4 particles depends on the ratio of the number of the particles in the certain level and ground state [19].

$$\frac{N_J}{N_0} = (2J+1)\exp\left(-\frac{E_J}{k_B T}\right) \tag{9}$$

where J is total orbital quantum number, N_J is molecular number on J-level, N_0 is molecular number on ground state, E_J is rotational energy level, and T is temperature when the photons directed to the samples.

Using Raman selection rule, we calculate and visualize the absorption of the electromagnetic intensity of Fe_3O_4 particles as a function of frequency using Borland Delphi 7. The primary technique for the calculation is by determining the change of energy level of the Fe_3O_4 particles using Equation 8 and conducting similarity with the Einstein's photon energy ($E = hf$). Based on the calculated frequency, we interpolate for all frequencies of the electromagnetic wave that suitable with the absorbed photon intensity as shown in Equation 9.

3. Results and discussion

Based on the theory of quantum theory, the intensity of the absorbed photon is proportional to the square of the transition moment represented by Equation 10.

$$R_r = \int \psi_r^{\ *} \mu \psi_r^{\ ''} d\tau \tag{10}$$

Where $\psi_r^{\ ''}$ and $\psi_r^{\ *}$ are two functions of the state of the system in which the molecular transition occurs [20]. For the rotational system, the transition moment of the system is proportional to the number of molecules at that level [19], where the formulation is given by Equation 9.

According to the above calculations, the visualization of microwave absorption by Fe_3O_4 particles is obtained as the curve in Figure 2. Based on the curve, it is seen that the intensity of the photon

IOP Conf. Series: Materials Science and Engineering **367** (2018) 012009 doi:10.1088/1757-899X/367/1/012009

absorbed by the Fe_3O_4 particles increases with the increase of the quantum number of molecular orbitals J. The frequency characteristics of the photons absorbed by Fe_3O_4 particles ranging in the range of 2-12 GHz. It is intuitively understood that the above visualization of the theoretical results is in accordance with the result of microwave absorption characters for the samples with the main materials of Fe_3O_4 particles. It is due to the suitability of the absorption curve obtained from the characterization and theoretical visualization. In addition, the microwave absorption characters for the sample with the thickness variation obtained a curve that is not different from the expectation, i.e., the maximum absorption occurs at a frequency of about 10 GHz as shown in Figure 3 [21].

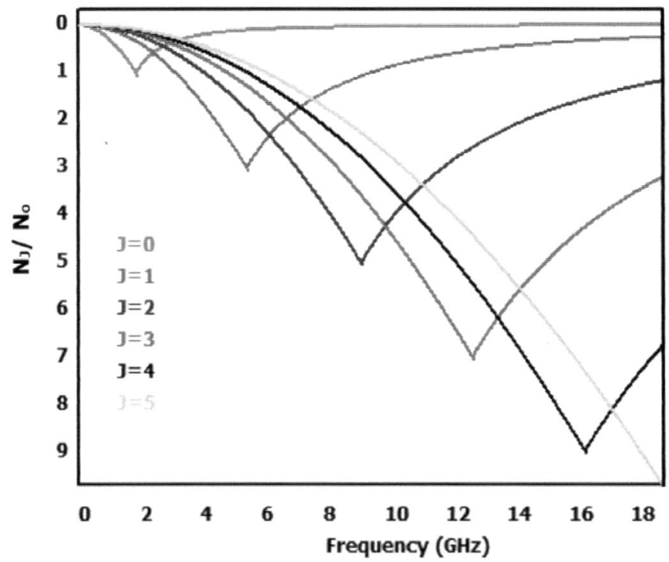

Figure 2. Visualization of ratio of molecule number and frequency of absorbed photon

Physically, the absorption of electromagnetic waves of the materials including Fe_3O_4 particles has a close relationship with the optical quantum phenomenon that is the phenomenon of stimulation absorption. If a photon comes to a molecule at a particular energy level, a jumping of the molecular energy level from low energy to higher energy can occur as long as its frequency is proportional to the difference between the two energy levels of the molecule. This jumping energy must be accompanied by photon absorption events [20]. Schematically, the absorption event is illustrated in Figure 4.

Drude Lorentz's theory has described the complex permittivity as one of the causes of electromagnetic wave absorption [22]. Thus, it can be concluded that the reflection loss value obtained from the permittivity complex and permeability complex [23], is proportional to the intensity of photons absorbed by the material. Therefore, the visualization curve of the rotational transition, which is a theoretical model, the intensity of photons absorbed by Fe_3O_4 particles can be compared to the reflection loss curve of the experiment.

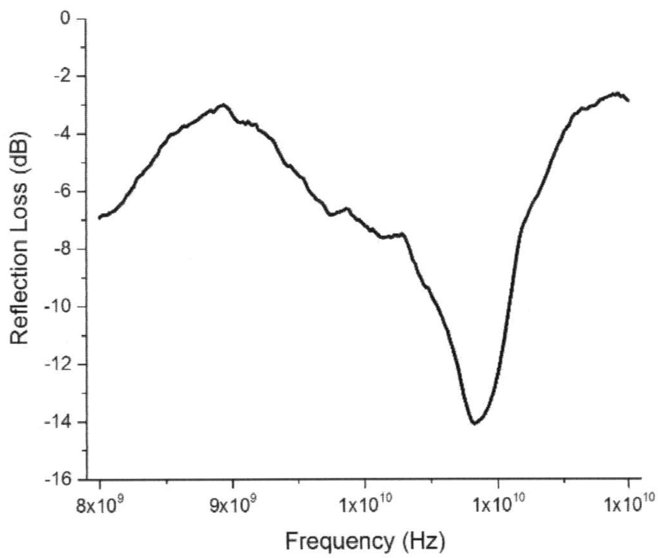

Figure 3. The reflection loss of the prepared Fe_3O_4 nanoparticles

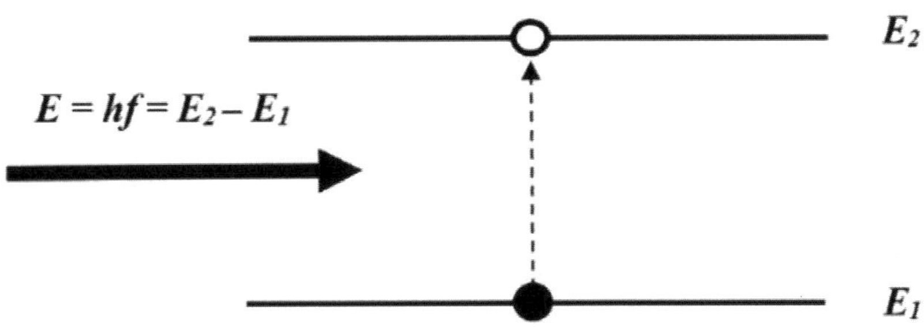

Figure 4. Schematic of two energy levels of particle population stimulated by photon

From the characterization of microwave absorption properties of the material, it can be seen that the thickness of the sample causes a change in the level of frequency of absorbed electromagnetic waves. The thicker the sample layer, the lower the absorbed electromagnetic wave frequency. This phenomenon is based on the physics concept for single-layer absorber as shown in Equation 11-12 [23]:

$$RL(dB) = 20Log\left|\frac{Z_{in}-1}{Z_{in}+1}\right| \quad (11)$$

Where RL and Z_{in} are the reflection loss and input impedance, respectively.

$$Z_{in} = \left(\frac{\mu_r}{\varepsilon_r}\right)^{\frac{1}{2}} \tanh\left[j\left(\frac{2\pi fd}{c}\right)(\mu_r\varepsilon_r)^{\frac{1}{2}}\right] \quad (12)$$

Where ε_r is the complex permittivity of the sample, f is the frequency, d is the thickness of the sample, and c is the speed of light in vacuum.

Based on the theoretical visualization of the rotational model of the molecule, the frequency and intensity of electromagnetic waves absorbed by the Fe_3O_4 particles depend on the quantum number of molecular orbitals *J*, where the higher the quantum number of the molecular orbital, the higher the absorbed intensity [24]. It can be understood by referring the equations that become the basis of the visualization, Equation 8-9. From the experimental data and theoretical model, there is a similarity of characteristics related to the shape of the curve. This phenomenon related to the quantum number of molecular orbitals J in quantum mechanics is a quantum number that describes the shape of a molecular orbit. When samples have different thicknesses, the attractive forces experienced by atoms in the molecule are also different. Thus, the shape of the orbit of atoms in the molecule changes implying a change in the quantum number of molecular orbitals *J*. The reasoning that causes the difference in sample thickness affects the peak frequency shift for reflection loss.

4. Conclusions

Based on the calculation and visualization of the absorbed electromagnetic intensity by Fe_3O_4 particles via rotational molecular transition approach, it is identified that the absorption pattern of the Fe_3O_4 particles has a similar pattern with the experimental data. It also gives information that the mechanism occurs at the electromagnetic absorption of the Fe_3O_4 particles is the rotational molecular mechanism. Therefore, for further experiment, the rotational molecular mechanism can be taken into account for investigating microwave absorption, especially for the Fe_3O_4 particles.

Acknowledgments
The authors would like to thank KEMENRISTEKDIKTI RI for serving "*Hibah PDUPT*" (for AT) in 2017-2018.

References

[1] Li Y, Chen G, Li Q, Qiu G and Liu X 2011 Facile synthesis, magnetic and microwave absorption properties of Fe_3O_4/polypyrrole core/shell nanocomposite *J. Alloys Compd.* **509** 4104–7.
[2] Wang Y, Li T, Zhao L, Hu Z and Gu Y 2011 Research Progress on Nanostructured Radar Absorbing Materials *Energy Power Eng.* **03** 580–4.
[3] Shi L, Zhao Y, Li Y, Han X and Zhang T 2017 Octahedron Fe_3O_4 particles supported on 3D MWCNT/graphene foam: In-situ method and application as a comprehensive microwave absorption material *Appl. Surf. Sci.* **416** 329–37.
[4] Hou C, Li T, Zhao T, Liu H, Liu L and Zhang W 2013 Electromagnetic wave absorbing properties of multi-wall carbon nanotube/ Fe_3O_4 hybrid materials *New Carbon Mater.* **28** 184–90.
[5] Cao J, Fu W, Yang H, Yu Q, Zhang Y, Wang S, Zhao H, Sui Y, Zhou X, Zhao W, Leng Y, Zhao H, Chen H and Qi X 2010 Fabrication, characterization and application in electromagnetic wave absorption of flower-like ZnO/ Fe_3O_4 nanocomposites *Mater. Sci. Eng. B* **175** 56–9.
[6] Amir M, Güner S, Yıldız A and Baykal A 2017 Magneto-optical and catalytic properties of Fe_3O_4@HA@Ag magnetic nanocomposite *J. Magn. Magn. Mater.* **421** 462–71.
[7] Bera R, Das A K, Maitra A, Paria S, Karan S K and Khatua B B 2017 Salt leached viable porous Fe_3O_4 decorated polyaniline–SWCNH/PVDF composite spectacles as an admirable electromagnetic shielding efficiency in extended Ku-band region *Compos. Part B Eng.* **129** 210–20.
[8] Li S, Zhou Z, Zhang T, Jiang G and Su R 2014 Synthesis and characterization of Ag/Fe_3O_4 electromagnetic shielding particles *J. Magn. Magn. Mater.* **358–359** 27–31.
[9] Khani O, Shoushtari M Z, Jazirehpour M and Shams M H 2016 Effect of carbon shell thickness on the microwave absorption of magnetite-carbon core-shell nanoparticles *Ceram. Int.* **42** 14548–56.

[10] Chen Y-J, Xiao G, Wang T-S, Ouyang Q-Y, Qi L-H, Ma Y, Gao P, Zhu C-L, Cao M-S and Jin H-B 2011 Porous Fe_3O_4/Carbon Core/Shell Nanorods: Synthesis and Electromagnetic Properties *J. Phys. Chem. C* **115** 13603–8.
[11] Chen Y-J, Zhang F, Zhao G, Fang X, Jin H-B, Gao P, Zhu C-L, Cao M-S and Xiao G 2010 Synthesis, Multi-Nonlinear Dielectric Resonance, and Excellent Electromagnetic Absorption Characteristics of Fe_3O_4/ZnO Core/Shell Nanorods *J. Phys. Chem. C* **114** 9239–44.
[12] Qiang C, Xu J, Zhang Z, Tian L, Xiao S, Liu Y and Xu P 2010 Magnetic properties and microwave absorption properties of carbon fibers coated by Fe_3O_4 nanoparticles *J. Alloys Compd.* **506** 93–7
[13] Meng X, Wan Y, Li Q, Wang J and Luo H 2011 The electrochemical preparation and microwave absorption properties of magnetic carbon fibers coated with Fe_3O_4 films *Appl. Surf. Sci.* **257** 10808–14.
[14] Chu H, Zeng Q, Chen P, Yu Q, Xu D, Xiong X and Wang Q 2017 Synthesis and electromagnetic wave absorption properties of matrimony vine-like iron oxide/reduced graphene oxide prepared by a facile method *J. Alloys Compd.* **719** 296–307.
[15] Liu X, Cao K, Chen Y, Ma Y, Zhang Q, Zeng D, Liu X, Wang L-S and Peng D-L 2017 Shape-dependent magnetic and microwave absorption properties of iron oxide nanocrystals *Mater. Chem. Phys.* **192** 339–48.
[16] Zheng Y, Wang X, Wei S, Zhang B, Yu M, Zhao W and Liu J 2017 Fabrication of porous graphene- Fe_3O_4 hybrid composites with outstanding microwave absorption performance *Compos. Part Appl. Sci. Manuf.* **95** 237–47.
[17] Friák M, Schindlmayr A and Scheffler M 2007 *Ab initio* study of the half-metal to metal transition in strained magnetite *New J. Phys.* **9** 5–5.
[18] Atkins P W and Friedman R 2011 *Molecular quantum mechanics* (Oxford ; New York: Oxford University Press).
[19] Hollas J M 2004 *Modern spectroscopy* (Chichester ; Hoboken, NJ: J. Wiley).
[20] Fox M 2006 *Quantum optics: an introduction* (Oxford ; New York: Oxford University Press).
[21] Taufiq A, Bahtiar S, Sunaryono, Hidayat N, Hidayat A, Mufti N, Diantoro M, Fuad A, Munasir, Rahmawati R, Adi W A, Pratapa S and Darminto 2017 Preparation of Superparamagnetic $Zn_{0.5}Mn_{0.5}Fe_2O_4$ Particle by Coprecipitation-Sonochemical Method for Radar Absorbing Material *IOP Conf. Ser. Mater. Sci. Eng.* **202** 012024.
[22] Moliton A 2007 *Basic electromagnetism and materials* (New York, NY: Springer).
[23] Che R C, Peng L-M, Duan X F, Chen Q and Liang X L 2004 Microwave Absorption Enhancement and Complex Permittivity and Permeability of Fe Encapsulated within Carbon Nanotubes *Adv. Mater.* **16** 401–5.
[24] Alexander A J, Kroto H W and Walton D R M 1976 The microwave spectrum, substitution structure and dipole moment of cyanobutadiyne, HCCCCCN *J. Mol. Spectrosc.* **62** 175–80.

Structure Analysis of Fe_3O_4@SiO_2 Core Shells Prepared from Amorphous and Crystalline SiO_2 Particles

Munasir[1,2], A. S. Dewanto[1], D. H. Kusumawati[1,2], N. P. Putri[1,2], A. Yulianingsih[1], I. K. F. Sa'adah[1], A. Taufiq[3], N. Hidayat[3], S. Sunaryono[3], Z. A. I. Supardi[1,2]

[1]Department of Physics, Faculty of Mathematics and Natural Science, Universitas Negeri Surabaya, Jl. Ketintang, Surabaya, 60231, Indonesia
[2]Research Center for Advanced Materials, Universitas Negeri Surabaya, Jl. Ketintang, Surabaya 60231, Indonesia
[3]Department of Physics, Faculty of Mathematics and Natural Science, Universitas Negeri Malang, Jl. Semarang 5, Malang 64145, Indonesia

E-mail: munasir_physics@unesa.ac.id

Abstract: This article reports the results synthesis of crystalline (Fe_3O_4@c-SiO_2) and amorphous (Fe_3O_4@a-SiO_2) nanoparticles from natural resources (iron sand and silica sand). The synthesis of Fe_3O_4 and SiO_2 nanoparticles used co-precipitation and hydrothermal-coprecipitation methods with polyethylene glycol (PEG) 4000 as a template. The XRD data analysis presented that the amorphous SiO_2 particles were successfully produced using hydrothermal and co-precipitation methods. The XRD data analysis also presented that the crystalline phases were formed in quartz and tridymite phases after calcination process of the amorphous phase. SEM images exhibited that the amorphous phase had different particle size and morphology from the crystalline phase. FTIR spectra presented some absorption peaks of new functional groups indicating the existence of Si-O-Si (silanol), Fe-O, C-N, and Fe-O-Si as new functional groups.

Keywords. Core-shell, crystalline and amorphous structures, Fe_3O_4@SiO_2, iron sand, nanoparticle, and silica sand.

1. Introduction

Iron oxide nanoparticle particle has special superior characteristics comparing to other mineral compounds. Naturally, the iron oxide is easily found in iron sand and andesite [1], [2]. The iron oxide has various applications of technologies such as for data saving [3], sensor, spintronic [4], drug delivery system, medical diagnosis [3]–[6], immobilization of protein and enzyme separations [5], [7], [8], and water absorbent [9]–[14]. Related to these applications, particle size strongly determines the electrical, optical, and magnetic properties of the magnetic nanoparticles. Magnetite (Fe_3O_4), besides maghemite (γ-Fe_2O_3) and hematite (α-Fe_2O_3), is an iron oxide that possesses a strong magnetic property. Fe_3O_4 is commonly known as black iron oxide, magnetic iron ore, loadstone, ferrous ferrite, or a Hercules stone that shows the strongest magnetic characteristic [15]–[17].

Despite Fe_3O_4, silica (SiO_2) also has special characteristics that superior to be applied in various applications. In general, the SiO_2 can be found in rocks [1], [18] of about 60 % for example in source

Content from this work may be used under the terms of the Creative Commons Attribution 3.0 licence. Any further distribution of this work must maintain attribution to the author(s) and the title of the work, journal citation and DOI.
Published under licence by IOP Publishing Ltd

rocks (diorite, gabbro), silica sand, clay or mud, organic material like rice husk ash, and bagasse ash [19]–[21]. The common forms of the SiO_2 crystal structure are quartz, cristobalite, and tridymite [18]. Typically, the SiO_2 nanoparticle has an amorphous phase, and it can be commercially produced as silica gel, fume-silica, and so forth. Several applications of SiO_2 nanoparticles for examples such as medical purposes [22–24], as additives for rubber and plastics [25–27], as fillers for composite construction concrete [28–30], as stabilizers and agents drug delivery and theranostics [23, 31], and as a heavy metal absorbent material in a water filter [14], [22].

In order to increase the application performance of the Fe_3O_4/SiO_2 particles effectively, it is essential to prepare the particles using the inexpensive method from natural resources as raw materials. Based on the previous reports, various methods have been employed such as hydrolysis, microemulsions, and co-precipitation. Meanwhile, the Fe_3O_4/SiO_2 nanocomposites have been fabricated via coprecipitation method [1], [15], [23], [24]. In this work, we exploited a simple co-precipitation method at room temperature which easier to control the particle size of the samples. The main purpose of this research was the synthesis of Fe_3O_4 @ SiO_2 core shells based on natural materials using polyethylene glycol (PEG) 4000 as a template, and analyzing their structure both in amorphous phase (a-SiO_2) and crystalline phase (c-SO_2).

2. Materials and methods

The materials used in this research were silica sand, iron sand, sodium hydroxide (NaOH, 7 M, PA), chloride acid (HCl, 2 M, PA), distilled water, ammonium hydroxide (NH_4OH, 6.5 M, PA), and polyethylene glycol (PEG-4000, PA). The SiO_2 particle formation from silica sand followed equations (1-3), while the fabrication of the Fe_3O_4 particle from iron sand followed equation 4. The preparation process of the SiO_2 particle was initiated by an extraction process to form $Na_2O.xSiO_3$ and $Si(OH)_4$ silica gel with NaCl as a compound of the remaining product.

$$xSiO_{2(s)} + 2NaOH_{(l)} \rightarrow Na_2OxSiO_{3(aq)} + H_2O_{(l)} \quad (1)$$

$$Na_2OxSiO_{3(aq)} + H_2O_{(l)} + 2HCl_{(aq)} \rightarrow Si(OH)_{4(aq)} + 2NaCl_{(aq)} \quad (2)$$

$$Si(OH)_{4(aq)}\ (Drying) \rightarrow SiO_{2(s)} + 2H_2O_{(g)} \quad (3)$$

$$2FeCl_{3(l)} + FeCl_{2(l)} + H_2O_{(l)} + 8NH_4OH_{(l)} \rightarrow Fe_3O_{4(s)} + 8NH_4Cl_{(l)} + 5H_2O_{(l)} \quad (4)$$

The Fe_3O_4 powder from iron sand was reacted with HCl and followed by NH_4OH to form a Fe_3O_4 particle. Furthermore, the preparation of the Fe_3O_4@SiO_2 core shells was conducted using PEG-4000 as a template, and then it will be removed after calcination at (50 to 60) °C. The Fe_3O_4, SiO_2, and PEG were set based on their respective mass composition by comparison of 2:3:3 before draining process at below 60 °C. The crystal structure characterization of the samples was performed using XRD (based on Cu-Kα), while identification of a special functional group of SiO_2 (amorphous and crystalline phases), Fe_3O_4 and Fe_3O_4@SiO_2 employed FTIR. The profile analysis of the samples (morphology and particle size) was done using SEM-EDX.

3. Results and discussion

The diffraction patterns are presented in Figure 1, while the data analysis is presented in Table 1. The figure reveals the existence of a Fe_3O_4 phase with the characteristic peaks (at 2θ): 30.20°, 35.51°, 37.23, 43.16°, 53.51°, 57.01°, 62.60° and 74.18°. These peaks are by their respective crystalline indices of (220), (311), (222), (400), (422), (511), (440), and (533) originating from an inverse cubic spinel [17]. In Table 1, the crystal field or peaks (*hkl*) of the experimental sample characteristics are not significantly different from the previous research and shows a similar pattern to the PDF database 00-019-0629 (shown in Table 1). From the results of quantitative analysis using Reitica software with Rietveld method, and modeling referring to suitable the crystallographic database. Moreover, the

IOP Conf. Series: Materials Science and Engineering **367** (2018) 012010 doi:10.1088/1757-899X/367/1/012010

sample had a particle diameter of about ~ (25 ± 3.5) nm (d < 100 nm), and was identified as a cubic structure.

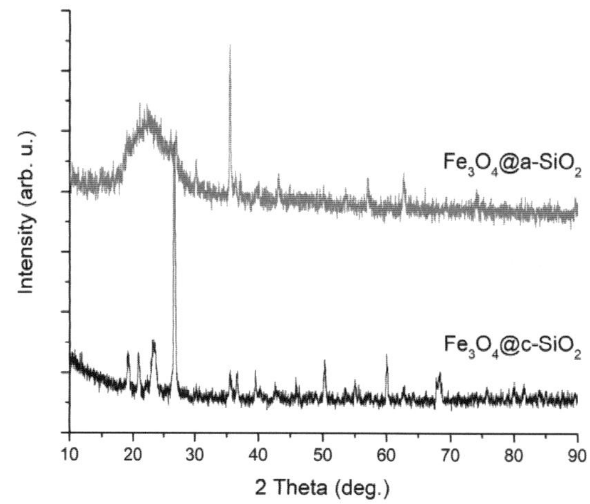

Figure 1. XRD patterns of Fe_3O_4@c-SiO_2 and Fe_3O_4@a-SiO_2 particles: c and a represent the crystalline and amorphous phases.

From Figure 1, it is also found a broad peak at $2\theta \approx 20^o$ to 24° presenting a peak of the amorphous silica. Therefore, the figure proves a combination of the Fe_3O_4 particle and the c-SiO_2 particle. Fe_3O_4 particles that added to the SiO_2 particle as a matrix for crystal or amorphous phase affected the size of the Fe_3O_4@ SiO_2. This phenomenon can be predicted by the growth of finite particles by a large number of Si-O-Si chains on amorphous phase. The diffraction patterns among the Fe_3O_4/c-SiO_2 and Fe_3O_4/a-SiO_2 appears to significantly differ except for a Fe_3O_4 crystal plane identified on the Fe_3O_4/a-SiO_2 diffraction pattern. The results of the phase analysis of the c-SiO_2 sample presented characteristic peaks at 2θ and crystal plane (*hkl*) as shown in Table 1. Hence, the characteristic peaks of the Fe_3O_4/c-SiO_2 composite were identified at 2θ = 30.11°, 35.54°, 43.1°, 56.9° and 62.5° corresponding to their respective crystal plane, i.e. (220), (311), (422), (511), (440).

Table 1. Data analysis of the XRD patterns of the samples

Oxide	*Peak*	*1*	*2*	*3*	*4*	*5*	*6*	*7*	*8*	*9*	*Description*
Fe_3O_4	2θ (deg)	30.07	35.46	43.17	53.55	56.97	62.77				Experimental data
		30.10	35.40	43.10	53.40	56.90	62.60				PDF 00-019-0629
	(hkl)	(220)	(311)	(400)	(422)	(511)	(440)				
SiO_2	2θ (deg)	20.8	26.6	36.5	39.4	45.7	50.1	54.8	59.9	68.1	Experimental data
		20.6	26.6	36.5	39.4	45.7	50.1	54.8	59.9	68.1	PDF 96-901-26101
	(hkl)	(100)	(011)	(110)	(102)	(201)	(112)	(022)	(121)	(203)	

The absorption peaks of the synthesized Fe_3O_4@SiO_2 particles were adjusted to the character peaks of the Fe_3O_4, and SiO_2 resulted from previous research. The analysis was done to identify the functional groups of the samples as shown in Figure 2 and as explained in Table 2. It appears a wavenumber of 543 cm^{-1} that confirmed a vibration on the Fe-O bond, which is a characteristic of the Fe_3O_4. Besides that, the peaks among the Fe_3O_4 and Fe_3O_4@c-SiO_2 or Fe_3O_4@a-SiO_2 particles could also be seen. On the wave number of 1130 cm^{-1} represents the asymmetric vibration of the Si-O-Si and

SiO_2 bonds and a Si-OH bond is found at 900 cm^{-1}. The 3014-3498 cm^{-1} wavenumbers show O-H (water molecule) bond stretching in SiO_2. Such functional groups have similarities with the characteristics of the Fe_3O_4@c-SiO_2 particles obtained from the previous research [25], [26].

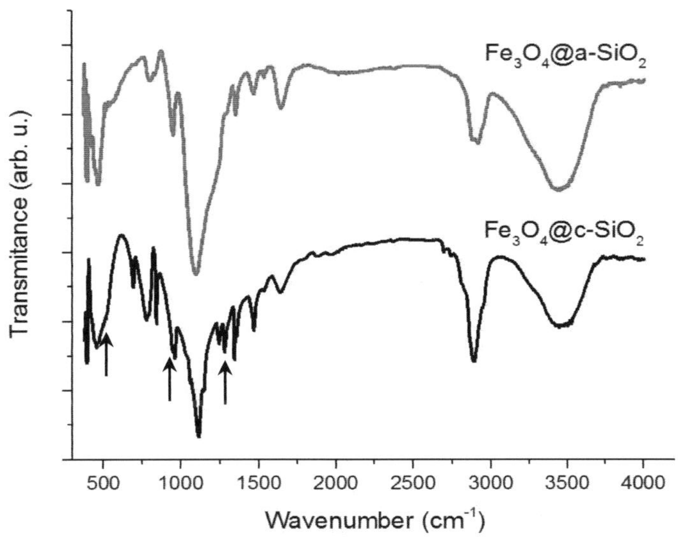

Figure 2. FTIR spectra of Fe_3O_4@c-SiO_2 and Fe_3O_4@a-SiO_2 particles

Table 2. Data analysis results of FITR spectra of the samples

No	*Wave Number (cm^{-1})*	*Functional group*	*Wave Number literature (cm^{-1})*	*Reference*
1	3465	O-H stretching	3650-3200	[27]
2	2870	O-H	1640-3483	[28]
3	1386	Fe-O	1390	[27]
4	1097	Si-O-Si stretching vibration Fe-O-Si stretching	1070-1080	[27], [29]
5	950	Si-O	964	[18], [30]
6	800	Si-O-Si	798	[18], [31]
7	569	Si-O-Fe	570	[1], [32]
8	468	Si-O	461	[18], [31]

According to the results of functional groups adjustment of the Fe_3O_4@a-SiO_2 particles as shown in Table 2, there is a bending vibration on the Si-O bond at the wave number of 468 cm^{-1}. The characteristic peaks between Fe_3O_4 and SiO_2 molecules can be observed as follows: at 569 cm^{-1} and 1097 cm^{-1} wavenumbers identified as stereo types of Fe-O-Si bonds, and wave number 1386 cm^{-1} that identify as the Fe-O group vibration. Thus the process of formation of the Fe_3O_4@SiO_2 composite has successfully produced. At the same time, the presence of Fe-O group indicates Fe_3O_4 particles have an interface bond with the SiO_2 matrix. As shown in Figure 2, there are three arrows showing vibration waves on Fe-O (1386 cm^{-1}) and Fe-O-Si (569 cm^{-1} and 1097 cm^{-1}). On the other hand, the peaks at 468, 800, 950, and 1097 cm^{-1} present asymmetrical vibrations and symmetrical stretching of Si-O-Si siloxane and silanol bonds on silica. Therefore, it indicates that the Fe_3O_4 particles can succeed to form a core formation shell with SiO_2 as a layer [1, 18, 40, 41].The stretching and bending vibration of the H-O-H bond are found at wave numbers of 1641-3465 cm^{-1}. Such functional groups have almost similar characteristics with the Fe_3O_4@a-SiO_2 particles obtained from the references [27], [28].

IOP Conf. Series: Materials Science and Engineering **367** (2018) 012010 doi:10.1088/1757-899X/367/1/012010

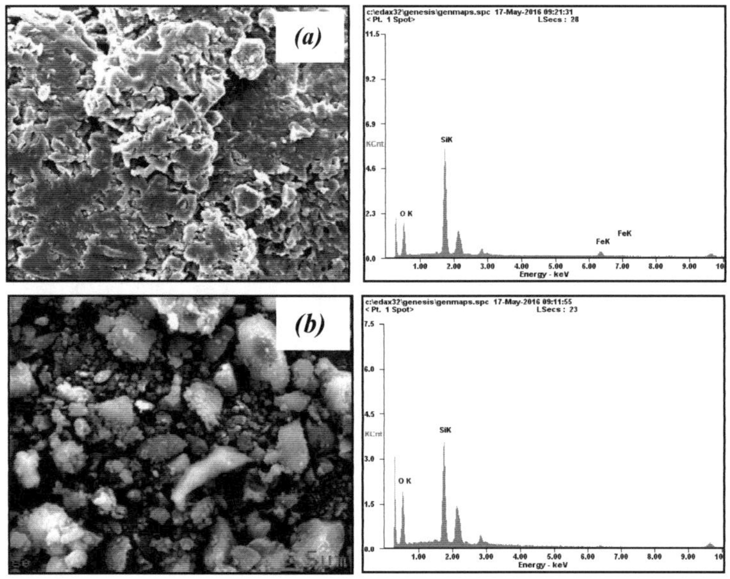

Figure 3. SEM-EDX profiles of $Fe_3O_4@SiO_2$, with: (a) a-SiO_2 and (b) c- SiO_2 particles

Table 3. Atomic elements of $Fe_3O_4@SiO_2$ particles

Element	*Fe_3O_4@a-SiO_2*		*Fe_3O_4@c-SiO_2*	
	Wt%	*At%*	*Wt%*	*At%*
O	40.75	57.15	46.15	56.21
Si	47.94	38.30	42.50	39.22
Fe	11.31	04.55	11.35	04.57

Figure 3a Fe_3O_4@a-SiO_2 (for SiO_2 amorphous phase) and Figure 3b Fe_3O_4@ c-SiO_2 (for SiO_2 crystalline phase) are the results of the elective analysis (EDX) for different samples. It appears that the dominant elements of the constituent atom are Si, Fe, and O. The detailed quantitative analysis of the atomic element is presented in Table 3. The elemental atomic content: (i) O Oxygen is 57.15 (At%) and 56.21 (At%); (ii) Si is 38.30 (At%) and 39.22 (At%); (iii) Fe is 4.55 (At%) and 4.57 (At%) respectively. The Fe_3O_4@a-SiO_2 and Fe_3O_4@c-SiO_2 samples show the composition of the constituent elements of the same atom. Morphologically, however, the two samples are different. The Fe_3O_4@ a-SiO_2 sample tends to form amorphous formations as shown in Figure 3a comparing to the crystal formation for Fe_3O_4@c-SiO_2 as shown in Figure 3b. The Fe_3O_4@a-SiO_2 sample tends to form an amorphous phase whereas for Fe_3O_4@c-SiO_2 samples tend to have a crystal phase. Furthermore, Figure 4 represented SEM images with different magnification for all samples show that the crystalline sample has different shape and size of the amorphous sample. The amorphous sample tends to form irregular shape comparing to the crystalline phase. Interestingly, using TEM experiment as shown in Figure 5, the Fe_3O_4@a-SiO_2 sample has successfully produced core-shell structure. Based on the data analysis, it was found that Fe_3O_4 particles presented as core and the a-SiO_2 particles presented shell.

(a) (b)

Figure 4. SEM profiles of (a) by a-SiO_2 and (b) c- SiO_2 particles in higher magnification than Figure 3

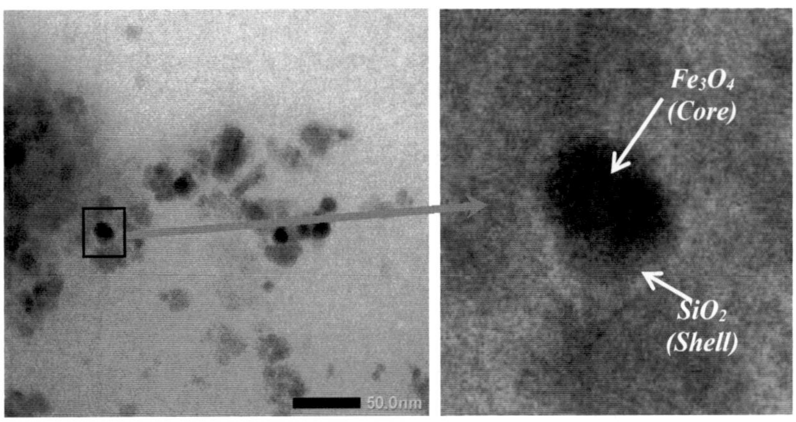

Figure 5. TEM profiles of Fe_3O_4@a-SiO_2 Core-Shell

4. Conclusions

The Fe_3O_4@c-SiO_2 and Fe_3O_4@a-SiO_2 particles have successfully prepared in crystalline and amorphous phases, respectively. The XRD and SEM data analysis showed that the Fe_3O_4 particles had a particle size below 100 nm and served as core covered by SiO_2 particles. The functional groups of the samples exhibited the formation of the SiO_2 and Fe_3O_4 particles.

Acknowledgments

This research was supported by the Ministry of Research, Technology, and Higher Education of Indonesia through the Research Grant.

References

[1] Munasir M *et al* 2017 Composites of Fe_3O_4/SiO_2 from Natural Material Synthesized by Co-Precipitation Method *IOP Conf. Mater. Sci. Eng.* **202** p 012057

[2] Taufiq A *et al* 2015 Nano-Structural Studies on Fe3O4 Particles Dispersing in a Magnetic Fluid Using X-Ray Diffractometry and Small-Angle Neutron Scattering *Mater. Sci. Forum* **827** 213–218

[3] Weller D and Moser A 1999 Thermal Effect Limits in Ultrahigh-Density Magnetic Recording *IEEE Trans. Magn.* **35** 4423–4439

[4] Beg MS *et al* 2017 Porous Fe_3O_4-SiO_2 core-shell nanorods as high-performance MRI contrast agent and drug delivery vehicle *J. Magn. Magn. Mater.* **428** 340–347

[5] Tang L and Cheng J 2013 Nonporous silica nanoparticles for nanomedicine application *Nano Today* **8** 290–312

[6] Lungu I I, Rădulescu M, Mogoşanu G D, and Grumezescu A M 2016 pH sensitive core-shell magnetic nanoparticles for targeted drug delivery in cancer therapy *Rom. J. Morphol. Embryol.* **57** 23–32

[7] Yang S, Zhang X, Zhao W, Sun L, and Luo A 2016 Preparation and evaluation of Fe_3O_4 nanoparticles incorporated molecularly imprinted polymers for protein separation *J. Mater. Sci.* **51** 937–949

[8] Hilczer A, Nowicki M, Czajka R, and Łęska B 2013 Morphology and magnetic properties of Fe_3O_4-alginic acid nanocomposite *Mater. Tehnol.* **1309.7216**

[9] Zhang *Y et al* 2016 Nanomaterials-enabled water and wastewater treatment *NanoImpact.* **3–4** 22–39

[10] Singh S, Barick K C, and Bahadur D 2013 Functional Oxide Nanomaterials and Nanocomposites for the Removal of Heavy Metals and Dyes *Nanomater Nanotechnol* **3** p 20

[11] Olvera R C, Silva S L, Robles-Belmont E, and Lau E Z 2017 Review of nanotechnology value chain for water treatment applications in Mexico *Resour-Effic. Technol.* **3** 1–11

[12] Montero-Campos V and Puente-Urbina A 2016 Continuous-Flow Removal of Arsenic in Drinking Water by Filtering down through Fe_3O_4@SiO_2 Magnetic Composite *J. Water Resour. Prot.* **08** 619–630

[13] Tiwari D K, Behari J, and Sen P 2018 Application of Nanoparticles in Waste Water Treatment *World Appl. Sci. J.* **3** 417–433

[14] Ahmaruzzaman M and Gupta V K 2011 Rice Husk and Its Ash as Low-Cost Adsorbents in Water and Wastewater Treatment *Ind. Eng. Chem. Res.* **50** 13589–13613

[15] Sonmez M *et al* 2015 Synthesis and Applications Of Fe3O4/SiO2 Core-Shell Materials *Curr. Pharm. Des.* **21** 5324–5335

[16] Bahtiar S *et al* 2017 Synthesis Investigation on Structural and Magnetic Behaviors of Spinel M-Ferrite [M = Fe; Zn; Mn] Nanoparticles from Iron Sand *IOP Conf. Ser. Mater. Sci. Eng.* **202** p 012052

[17] Abbas M, Takahashi M, and Kim C 2013 Facile sonochemical synthesis of high-moment magnetite Fe_3O_4 nanocube *J. Nanoparticle Res.* **15**

[18] Munasir, Triwikantoro, Zainuri M, and Darminto 2015 Synthesis of SiO_2 nanopowders containing quartz and cristobalite phases from silica sands *Mater. Sci-Pol.* **33** 47–55

[19] Affandi S *et al* 2009 A facile method for production of high-purity silica xerogels from bagasse ash *Adv. Powder Technol.* **20** 468–472

[20] Nittaya A N and Thuadaij 2008 Preparation of Nanosilica Powder from Rice Husk Ash by Precipitation Method *Chiang. Mai. J. Sci* **.35** 206–211

[21] Trabelsi W Benzina M and Bouaziz S 2009Physico-chemical characterisation of the Douiret sand Southern Tunisia: Valorisation for the production of Silica Gel *Phys. Procedia* **2** 1461–1467

[22] Barbé C *et al* 2013 Silica Particles: A Novel Drug-Delivery System *Adv. Mater.* **16** 1959–1966

[23] Kwon S *et al* 2013 Silica-based mesoporous nanoparticles for controlled drug delivery *J Tissue Eng* **4** p 2041731413503357

[24] Karimi Pasandideh E *et al* 2016 Silica-coated magnetite nanoparticles core-shell spheres Fe_3O_4@SiO_2 for natural organic matter removal *J. Environ. Health. Sci. Eng.* **14** p 21

[25] Conradi M 2013 Nanosilica-Reinforced Polymer Composite *Mater. Technol.* **47** 285–293

[26] Marquis D M, Guillaume É, and Chivas-Joly C 2011 Properties of Nanofillers in Polymer *Nanocomposites. Polym. Anal. Methods.* **InTech**

[27] Chen Y *et al* 2008 Natural rubber nanocomposite reinforced with nano silica *Polym. Eng. Sci.* **48** 1674–1677

[28] Said AM Zeidan MS Bassuoni MT and Tian Y 2012 Properties of concrete incorporating nano-silica *Constr. Build Mater.* **36** 838–844

[29] Park J T, Lee K J, Kang M-S, Kang Y S, and Kim J H 2007 Nanocomposite polymer

electrolytes containing silica nanoparticles: Comparison between polyethylene glycol and polyethylene oxide dimethyl ether *J. Appl. Polym. Sci.* **106** 4083–4090

[30] Mizutani T, Arai K, Miyamoto M, and Kimura Y 2006 Application of silica-containing nano-composite emulsion to wall paint: A new environmentally safe paint of high performance *Prog. Org. Coat.* **55** 276–283

[31] Luo Z, Xu Y, and Ye Q 2015 Effect of nano-SiO_2-LDPE packaging on biochemical sensory and microbiological quality of Pacific white shrimp Penaeus vannamei during chilled storage *Fish Sci.* **81** 983–993

[32] Anjum M, Miandad R, Waqas M, Gehany F, and Barakat M A 2016 Remediation of wastewater using various nano-materials *Arab. J. Chem.*

[33] Peternele W S *et al* 2014 Experimental Investigation of the Coprecipitation Method: An Approach to Obtain Magnetite and Maghemite Nanoparticles with Improved Properties *J. Nanomater* **2014** 1–10

[34] Majeed J, Ramkumar J, Chandramouleeswaran S, and Tyagi A K 2014 Fe_3O_4@SiO_2 core-shell nanoparticles: Synthesis characterization and application in environmental remediation *AIP Conference Proceedings* **1591** pp 605–607

[35] Bini R A *et al* 2012 Synthesis and functionalization of magnetite nanoparticles with different amino-functional alkoxysilanes *J. Magn. Magn. Mater.* **324** 534–539

[36] Silva V A J *et al* 2013 Synthesis and characterization of Fe_3O_4 nanoparticles coated with fucan polysaccharides *J. Magn. Magn. Mater.* **343** 138–143

[37] Quy D V *et al* 2013 Synthesis of Silica-Coated Magnetic Nanoparticles and Application in the Detection of Pathogenic Viruses *J. Nanomater* **2013** 1–6

[38] Munasir, Sulton A, Triwikantoro, Zainuri M, and Darminto 2013 Synthesis of silica nanopowder produced from Indonesian natural sand via alkalifussion route *AIP. Conf. Proc. AIP Publishing* **1555** pp 28–31

[39] Hui C *et al* 2011 Core-shell Fe_3O_4@SiO_2 nanoparticles synthesized with well-dispersed hydrophilic Fe3O4 *Nanoscale* **3** 701–705

[40] Alizadeh A *et al* 2012 Biguanide-Functionalized Fe_3O_4/SiO_2 Magnetic Nanoparticles: An Efficient Heterogeneous Organosuperbase Catalyst for Various Organic Transformations in Aqueous Media Bull *Korean Chem. Soc.* **33** 2546–2552

[41] Jal P K, Sudarshan M, Saha A, Patel S, and Mishra B K 2004 Synthesis and characterization of nanosilica prepared by precipitation method *Colloids Surf. Physicochem. Eng. Asp.* **240** 173–178

[42] Ahangaran F, Hassanzadeh A, and Nouri S 2013 Surface modification of Fe_3O_4@SiO_2 microsphere by silane coupling agent *Int. Nano. Lett.* **3** p 23

[43] Hui C *et al* 2011 Core-shell Fe_3O_4@SiO_2 nanoparticles synthesized with well-dispersed hydrophilic Fe3O4 seeds *Nanoscale* **3** 701–705

Preparation of SiC Fibers from Indonesian Natural Resource Prepared by Electrospinning Method

A. Fuad[1,2], N. Fatriani[1], U. Kultsum[1], A. Taufiq[1,2], C.I. Yogihati[1] and M. Diantoro[1,2]

[1]Department of Physics, Faculty of Mathematics and Natural Sciences, Universitas Negeri Malang, Jl. Semarang No. 5, Malang 65145, Indonesia
[2]Central Laboratory for Mineral and Advanced Materials, Faculty of Mathematics and Natural Sciences, Universitas Negeri Malang, Jl. Semarang No. 5, Malang 65145, Indonesia

E-mail : abdulloh.fuad.fmipa@um.ac.id

Abstract. SiC fibers were prepared via electrospinning method. The starting materials were polyvinyl alcohol, sucrose, and SiO2. The SiO2was extracted from natural pyrophyllite. In this work, we report the effect of polyvinyl alcohol and applied voltage on structural SiC fibers. The samples were then characterized using XRF, FTIR, and SEM-EDX. Based on the results of data analysis, it was shown that the optimum composition of polyvinyl alcohol in the SiC fibers was of 13.75% (w/v). It was also found that the SiC fibers had a smooth surface with a uniform diameter. Furthermore, the optimum applied voltage was set up at about 24 kV.

Keyword. Electrospinning, poly, polyvinyl alcohol, SiC Fiber, and sucrose.

1. Introduction

Recently, silicon carbide (SiC) fibers become essential materials due to their potential applications in many areas such as aerospace, aviation, and nuclear energy. In general, such claim is related to the particular characteristics of SiC fibers, for example, high tensile strength, high elastic modulus, high-temperature resistivity, excellent corrosion and wear resistance [1–4]

Generally, various preparation methods in fabricating SiC fibers could be conducted by using pyrolysis and electrospinning [5–7]. Interestingly, the electro spinning method becomes effective and efficient method because it can be performed with easy way, inexpensive, simple experimental setup, and easy to control the fibers [8]. SiC fibers generally prepared using commercial Si powders or polycarbo methylsilaneas Si elemental source [6, 9]. On the other hand, in Indonesia, Si is available abundantly in nature almost in all area. Practically, natural Si can be obtained from SiO2containing several impurities such as TiO2, CaO, and Al2O3. Such contaminants could be trough out by leaching process to captureSiO2particles with high purity.

In this work, SiC fibers were prepared using SiO2particles as the main precursor extracted from natural Indonesian pyrophyllite. The extractedSiO2particles were then modified to form SiC nanoparticles using the sonochemical method. The SiC fibers were obtained using electrospinning method. Furthermore, in this paper, we report the effect of polyvinyl alcohol and applied voltage on structural SiC fibers.

Content from this work may be used under the terms of the Creative Commons Attribution 3.0 licence. Any further distribution of this work must maintain attribution to the author(s) and the title of the work, journal citation and DOI.
Published under licence by IOP Publishing Ltd

IOP Conf. Series: Materials Science and Engineering **367** (2018) 012011 doi:10.1088/1757-899X/367/1/012011

2. Materials and methods

The SiO2 used in this study was obtained from local pyrophyllite ore. Based on the XRF characterization, the chemical composition of the pyrophyllite was Si (64.6 %), Al (16.0 %), Fe (9.0 %), Ti (7.2 %), and other elements. The pyrophyllite was ground using a mortar and a pestle. 20 grams of the ground sample were then leached using 5 M NaOH solution for 3 h and at a boiling temperature. After leaching process, the solution was diluted and filtered to separate undissolved materials. The reaction expected during this process was like in the Equation 1.

$$SiO_2 + NaOH \rightarrow Na_2SiO_3 + H_2O \quad (1)$$

Furthermore, HCl was added dropwise to the obtained solution until the SiO_2 gel was formed. Finally, theSiO_2 gel was rinsed using distilled water and dried to produce white-colored SiO_2powder.The reaction expected during the sintering process using magnesium was like the Equation 2.

$$Na_2SiO_3 + HCl \rightarrow SiO_2 + NaOH + H_2O \quad (2)$$

The SiC nanoparticles were synthesized using a sonochemical method. 3 g SiO_2 and 9 g sucrose were added to a 240 mL distilled water and ethanol mixture. Next, the mixture was stirred and continued by sonication process for one hour at 70 °C. The result was then put into a furnace for the following preheating process for 1 h at 180 °C. After that, the process continued by annealing at 950 °C in argon condition until SiC nanoparticles were produced.

To produce SiC nao fibers, 1 g SiC particles were mixed with PVA solution with concentration variation of 12.5 %, 13.75 %, and 15 % *w/v*. Eachof the solutionswas then stirred and sonicated for 1 h to disperse the SiC nanoparticles. The sonicated solution was then put into a syringe with 24 gauge needle. In this electrospinning process, the parameters were maintainedat18 kV, 20 kV, 22 kV, and 24 kV.

The percentages of elements, size, and morphology of the samples were identified through x-ray fluorescence (XRF, PANalytical) and scanning electron microscopy and energy-dispersive x-ray spectroscopy (SEM-EDX, FEI INSPECT-S50) characterizations.

3. Results and discussion

The prepared SiO2 was then extracted to form SiC nanoparticles. The EDX test results showed that the SiC nanoparticles had a high purity with the Si, C, and O contents (Table 1). Figure 1 shows the FTIR test result on the SiC nanoparticles. The sample showed the existence of a Si-C functional group at the wavenumber of 798.5 cm^{-1}, a Si-O-C functional group at the wavenumbers of 999.13 cm^{-1}, and a Si-O functional group at 1200 cm^{-1}. Such findings are in good agreement with the database of experimental results of the researchers, in which the Si-C functional group formed at a wave number of around 754 cm^{-1} until 840 cm^{-1}[6,10–13].

Table 1. Element content of SiC nanoparticles

No.	Element	Content of element (%)
1.	Si	50.94
2.	C	44.05
3.	O	5.01

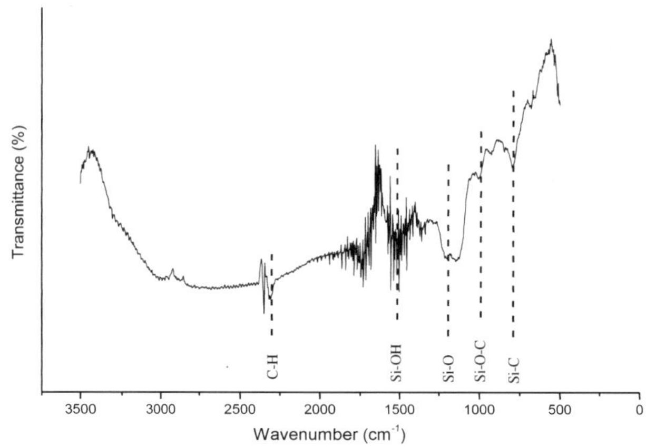

Figure 1. Result of FTIR Characterization of SiC Sample

The SEM results of SiC nanofibers with PVA concentrations of 12.5 %, 13.75 %, and 15 % w/v.10 % *w/v*is shown in Figure 2. It can be seen that the produced SiC nanofibers were formed in beaded-structures, which indicated the low concentration of PVA solution. Generally, in electrospinning, the polymer solution has critical value in order to obtain the fibers. Kim *et al.* reported that beaded-structure SiC nanofibers can be overcome by increasing volume concentration of PVA [14]. Figure 2 expresses the SEM results of the SiC nanofibers with the 12.5 %, 13.75%, and 15 % *w/v* variations of PVA solution. The produced SiC nanofibers tended to be straight and to possess a smooth and bead-free surface. An increase in the PVA solution concentration increased the SiC nanofiber diameter. The SiC nanofibers with 12.5 % and 15 % *w/v* PVA concentrations have different diameters. The percentages of SiC nanofiber element contents identified via EDX are shown in Table 2. The existence of the O element and a high proportion of C element resulted from the carbon chain of the PVA.

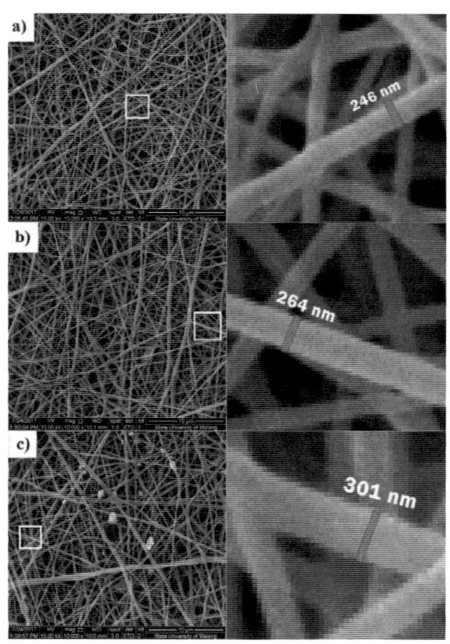

Figure 2. Morphology of SiC nanofibers synthesized by electrospinning method using a PVA solution of (a) 12.5% w/v; (b) 13.75% w/v; and (c) 15% w/v.

Table 2. Elemental contents of SiCNanofibers

No.	PVA Concentration (% w/v)	Element Contents (%)		
		Si	C	O
1.	12.50	1.64	63.20	35.16
2.	13.75	1.58	62.83	35.59
3.	15.00	1.89	64.23	33.88

In this research, we also report the effect of voltage variations on the structures of SiC nanofiber as presented in Figure 3. Theoretically, the voltage plays an important role in producing SiC nanofibers. The figure shows that the voltage gives an effect on the structural change of the SiC nanofibers. From the figure, the higher the voltage, the lower the diameter of the SiC nanofibers. Physically, an increasing voltage causes the increasing strain of the polymer originating from increasing Coulomb force and electric field and finally reducing the diameter of the fibers [8].

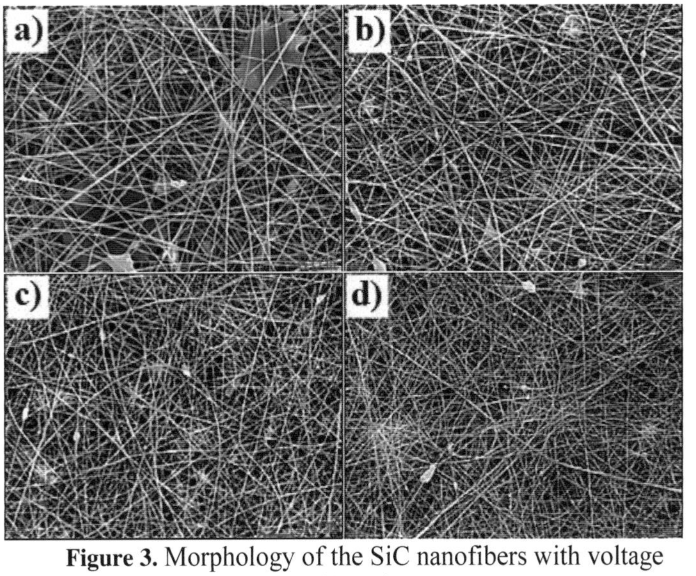

Figure 3. Morphology of the SiC nanofibers with voltage variations of a) 18 kV, b) 20 kV, c) 22 kV; and d) 24 kV.

4. Conclusions

In this research, SiC nanofibers were prepared from a natural resource using electrospinning method. The SiC nanoparticles were dispersed in a polymer solution by the polymer concentration and applied voltage affected the structure and diameter of the SiC nanofibers. The optimum level of the PVA to obtain the proper SiC nanofibers was 13.75 %, and the applied voltage was set up at about 24 kV, which had straight structured SiC fibers with a smooth surface.

References

[1] Cao S, Wang J, and Wang H 2016 High-temperature behavior and degradation mechanism of SiC fibers annealed in Ar and N2 atmospheres *J. Mater. Sci.* **51** pp 4650–4659
[2] Dong, S.M *et al* 2001 Characterization of nearly stoichiometric SiC ceramic fibres *J. Mater. Sci.* **36** pp 2371–2381
[3] Lee J S and Yano T, 2004 Fabrication of short-fiber-reinforced SiC composites by polycarbosilane infiltration. *J. Eur. Ceram. Soc.* **24** pp 25–31.
[4] Xiong L, Xu Y, Li Y, and Xia X 2008 Synthesis of SiC ceramic fibers fromnuclear reactor irradiated polycarbosilane ceramic precursor fibers *J. Mater. Sci.* **43** 4849–4855
[5] Elyassi B *et al* 2009 Effect of polystyrene on the morphology and physical properties of silicon

carbide nanofibers *Mater. Chem. Phys.* **118** pp 259–263
[6] Liu H A and Balkus K J 2009 Electrospinning of beta silicon carbide nanofibers. *Mater. Lett.* **63** pp 2361–2364
[7] Zhang J *et al* 2016 Novel synthesis of ultra-long single crystalline β-SiC nanofibers with strong blue/green luminescent properties *Ceram. Int.* **42** pp 4600–4606.
[8] Bhardwaj N and Kundu S C 2010 Electrospinning: a fascinating fiber fabrication technique. *Biotechnol. Adv.* **28** pp 325–347
[9] Salinas A *et al* 2016 Production of b-Silicon Carbide Nanofibers using the Forcespinning® Method.
[10] Abro D M *et al* 2016 Characterization of Surface State of Inert Particles: Case of Si and SiC *J. Miner. Mater. Charact. Eng.* **4** pp 62–72.
[11] Moshtaghioun B M *et al* 2012 Rapid carbothermic synthesis of silicon carbide nano powders by using microwave heating *J. Eur. Ceram. Soc.* **32** pp 1787–1794
[12] Raman V, Bahl O P, and Dhawan U 1995 Synthesis of silicon carbide through the sol-gel process from different precursors *J. Mater. Sci.* **30** pp 2686–2693
[13] Sivasankaran S and Kishor Kumar M J 2015 a novel sonochemical synthesis of nano-crystalline silicon carbide ceramic powder and its characterization *Int. J. Recent Sci. Res* pp 2630–2633
[14] Kim B *et al* 2005 Poly (acrylic acid) nanofibers by electrospinning *Mater. Lett.* **59** pp 829–832.

Control of Dielectric Constant and Anti-Bacterial Activity of PVA-PEG/*x*-SnO_2 Nanofiber

M. Diantoro[1,2], L. A. Sari[1], T. Istirohah[1], A. D. Kusumawati[1], Nasikhudin[1,2], and Sunaryono[1,2]

[1]Department of Physics, Faculty of Mathematics and Natural Sciences, Universitas Negeri Malang, Jl. Semarang 5 Malang 65146, Indonesia
[2]Central Laboratory of Minerals and Advanced Materials, Faculty of Mathematics and Natural Sciences, Universitas Negeri Malang, Jl. Semarang 5 Malang 65146, Indonesia

E-mail: markus.diantoro.fmipa@um.ac.id

Abstract. Research in the utilization of organic natural materials for electronic devices and for the biological application becoming extensively studied. We report a comprehensive review of the role of SnO_2 nanoparticle and the effect of light intensity on toxicity properties, anti-bacterial activity, microstructure and electrical properties of PVA-PEG nanofiber films. The PVA-PEG/SnO_2 nanofiber structure has been successfully fabricated on the ITO-glass substrate. Characterization was performed on samples using FTIR, XRD, SEM, toxicity and antibacterial tests, as well as LCR measurement. The presence of various light intensities has also measured the dielectric constant. The addition of SnO_2 nanoparticle influenced the structure of the PVA-PEG/SnO2 nanofiber bonding functional group indicated by the appearance of Sn-O-Sn peaks at 648.08 cm^{-1} and 958 cm^{-1} wavenumbers. The addition of SnO_2 nanoparticles affects the grain size of SnO_2. Addition of SnO_2 nanoparticles increases the detected toxicity voltage but is still below the threshold. It means the compound is not toxic, or safe to use in the body. The film lacks the antibacterial power of S. Aurelius. The addition of nanoparticles SnO_2 increases the dielectric constant but decreases with increasing frequency of input voltage and the intensity of light employed to PVA-PEG/SnO_2 nanofiber. The application of the light intensity reduces the dielectric constant of the PVA-PEG/SnO_2 nanofiber in all nanoparticle doping ranges.

Keywords. Dielectric constant, light intensity, nanoparticles, PVA-PEG/SnO_2 nanofibers, SnO_2, and toxicity.

1. Introduction

Nanoparticle materials have shown a unique of physical, chemical, electrical properties, and magnetic properties [1]. Commonly, many researchers interest to nanoparticles materials especially on the metal oxide such as Fe_3O_4, organic material such as cellulose acetate, polyaniline and its composite [2, 3]. One of them is indicated by a well-known semiconductor of Tin (II) Oxide (SnO_2). SnO_2 is an n-type semiconductor that has a non-stoichiometric structure and wide band gap (Eg = 3.6 eV to 3.8 eV) [4]. Also, SnO_2 has excellent optical and electrical properties. This oxide also relatively inexpensive, and not toxic [5]. In the previous research, SnO_2 is synthesized as powder, bulk, thin film. It also has kind

Content from this work may be used under the terms of the Creative Commons Attribution 3.0 licence. Any further distribution of this work must maintain attribution to the author(s) and the title of the work, journal citation and DOI.
Published under licence by IOP Publishing Ltd

of surface morphology such as micro or nano spherical, fiber [6], rod [7], or wire [8]. SnO_2 nanofiber is applicated as active material like anode of Li battery to enhance its specific capacitance [9–11]. On the other hand, SnO_2 as the primary precursor of synthesis nanofiber desire another material as a binder. Polymer regularly is used as binders such as PVDF [11] and PVA [11–13]. The function of the adhesive is giving elasticity on the nanofiber fabrication. Commonly, electrospinning method is used to fabricate fiber.

In the previous research, PVA-SnO2 nanofiber has been synthesized well which obtain SnO_2 grain size about 100 nm to 300 nm and not homogeneous yield. Therefore, in this study will be done homogenization and reduce the size of a grain of PVA-SnO2 nanofiber by adding PEG as a template.

Polyethylene Glycol (PEG) possesses indispensable unique properties in areas such as biological, chemical, biomedical, and pharmaceutical applications. PEG shows excellent features such as non-toxic, immunogenicity, biocompatible, and water-soluble [14]. SnO_2 nanoparticles composite with PVA-PEG polymer using electrospinning technique obtained PVA-PEG /SnO_2 nanofiber. Specific applications of the SnO_2 nanofiber in the health are diagnosing diabetes, and detecting lung cancer [14]. From our knowledge, it is still difficult to find a comprehensive study of the effects of SnO_2 nanoparticles and light intensity on dielectrics of PVA-PEG/SnO_2 nanofiber on which to open the possibility as a bio environmentally compatible and body bioactive sensors. Therefore, a series of studies on nanoparticle SnO_2 to nanofiber especially PEG is needed. In this paper, we proposed the effect of nanoparticles SnO_2 and light intensity on the dielectric constant of PVA-PEG/SnO_2 nanofiber.

2. Materials and methods

The synthesis of nanoparticles SnO_2 was performed by using coprecipitation method from $SnCl_2$. The $SnCl_2$ powder as much as 8 g dissolved in ethanol (C_2H_5OH) as much as 25 mL at 60 °C and stirred about 30 min using magnetic stirrer. Acetyl Acetone was then added as much as five mL and inserted one mL of PEG 6000 then reflux at 80 °C for 5 h. Furthermore, the solution was allowed to stand for 72 h at 30 °C until the sole was formed, then dried at 100 °C for 3 h. After drying, it was then calcined at 400 °C for 1 h to create the nanoparticles. The next step was deposition of the nanofiber of PVA-PEG/SnO_2 using electrospinning technique.

The first step was to insert 4.3 mL of aquades into 10 mL beaker glass and stirred about 350 rpm at 90 °C for 15 min. Afterwards, introducing 3.5% SnO_2 nanoparticles of total volume, 111 μL of PEG 400, and 12.23 μL of TMAH. Subsequently, the mixtures were stirred at 350 rpm with a temperature of 90 °C for 60 min. When a homogeneous has reached, it followed by ultrasonic irradiation for 1 h using an ultrasonic bath. The obtained then incorporated a 0.5 g PVA and stirred 700 rpm at 120 °C for 60 min and sonicated again for 2 h. The solution then ready for electrospinning to fabricate a PVA-PEG/SnO_2 nanofiber

3. Results and discussion

The addition of the SnO_2 nanoparticles to the PVA-PEG nanofiber produces the PVA-PEG/SnO_2 nanofiber spectrum shown in Figure 1. The results of FTIR spectrophotometric characterization of nanoparticles SnO_2, PVA Gohsenol, PEG 400, and PVA-PEG/SnO_2 nanofiber are shown in Figure 1. SnO_2 nanoparticles show transmittance peaks at 3391 cm^{-1}, and 1568 cm^{-1} which are associated to the vibration of stretching OH, while peaks at a wave number of 625, 930, and 1635 cm^{-1}, indicating the appearance of the stretching mode of Sn-O-Sn [14], [15]. The PVA-PEG/SnO_2 nanofiber in Figure 1 has a similar peak shape such as in PEG 400 and PVA compared to SnO_2 nanoparticles. The indistinguishable from the mixture may be caused the number of nanoparticles is less than the amount of PVA, so the transmittance of PVA is stronger appears in the PVA-PEG/SnO_2 nanofiber. The spectra of PEG-400 at (845 to 900) cm^{-1} indicates the presence of O-O stretching [16], whereas the presence of 1109 cm^{-1} peaks indicates symmetry C-O-C on aliphatic ether. Spectral peak at 1352 cm^{-1} shows the presence of deforming of C-H alkyl while 2882 cm^{-1} indicates the presence of str-C-H alkyl stretching.

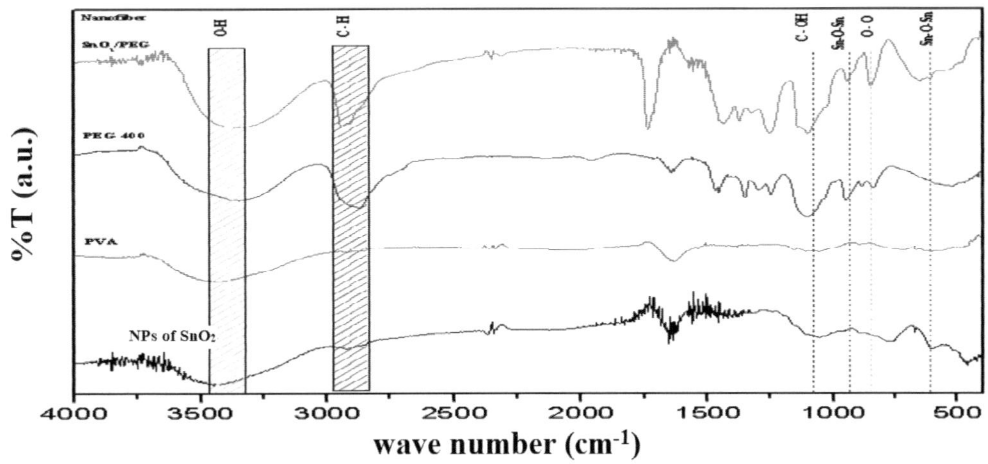

Figure 1. FT-IR Spectra of SnO2 NPs, PVA, PEG, and PVA-PEG/SnO2 Nanofiber.

3.1. Structure of PVA-PEG/SnO_2 Nanofiber Analyses

The diffraction patterns of PVA-PEG/SnO_2 nanofiber films is displayed in Figure 2. The X-ray diffraction pattern of nanoparticle synthesized by the sol-gel method is shown in Figure 2. X-ray diffraction pattern of nanoparticles of SnO_2 showed the result of crystalline according to the previous study [17]. The grain size of SnO_2 nanoparticles of 36.11 nm was obtained by diffusion using Origin program with PearsonVII function obtained FWHM value of 0.25182, then analyzed using Debye Scherer equation. Using the same method, the grain size of SnO_2 nanoparticles in PVA-PEG/SnO_2 nanofiber was 26.94 nm with the FWHM value of 0.33761. The SnO_2 nanoparticles were also analyzed using Rietica program with model data from AMCSD with code 2101853. We obtained that the lattice constant is a = b = 4.73735 Å and c = 3.18640 Å. With these values indicating that the SnO_2 nanoparticles fit with a tetragonal crystal structure under *$P4_2/mnm$* space group. The diffraction peak of the SnO_2 nanoparticles is positioned at a diffraction angle of 26.58°, 33.85°, 37.94°, 51.76°, 54.74° and 65.94°.

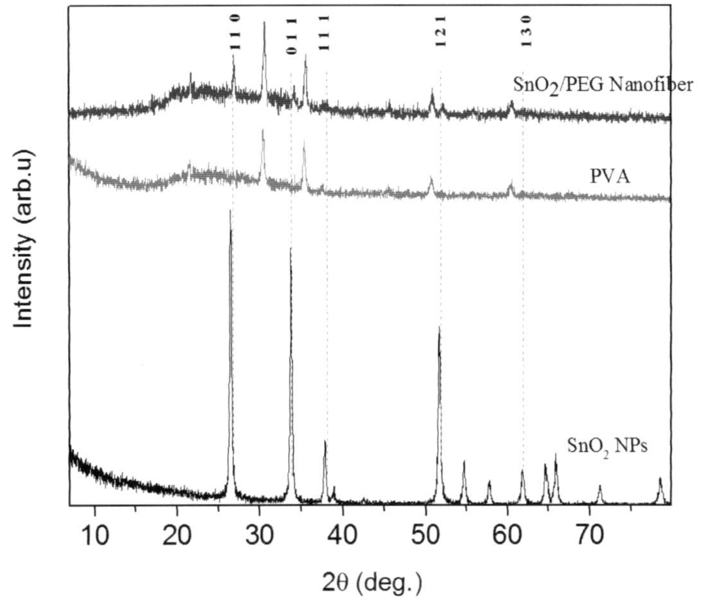

Figure 2. X-Ray Diffraction patterns of Nano SnO_2, PVA, and PVA-PEG/SnO_2 Nanofiber

3.2. Micro structure and morphology of PVA-PEG/SnO_2 nanofiber

A typical EDAX and SEM image of PVA-PEG/SnO_2 Nanofiber is shown in Figure 3. Information from EDAX in Table 3 shows the element of comparison between C, O, and Sn of 11.96 %, 29.46 %, and 03.53 %. Elements C and O are elements of the formation of PEG, while the Sn element comes from the doping given [18]. Based on the information, it shows the formation of PVA-PEG / SnO2 nanofiber in the presence of C, O, and Sn compositions.

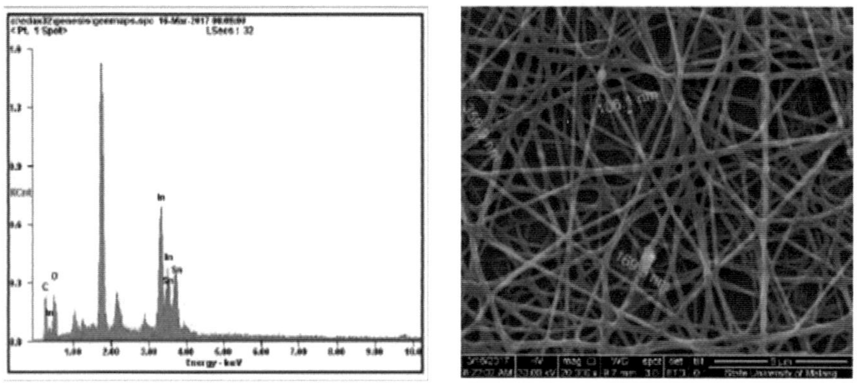

(a) (b)

Figure 3. EDAX analyses (a) and SEM image (b) of PVA-PEG/SnO_2 nanofiber

3.3. Toxicity test of PVA-PEG/SnO_2 Nanofiber

All series of PVA-PEG/SnO_2 Nanofiber have been tested for their toxicity using animal intestinal mus musculus. An example of them is presented in Figure 4. Present a record of intestinal contractions in animals Mus musculus shows the intestine contracts indicated by the wave peaks. In Figure 4 is a record of regular intestinal contractions when compared with intestinal contractions when sampled in Figure 4, the contractions do not change significantly. Based on this, the composition of the PVA-PEG/SnO_2 nanofiber variation sample did not cause the rejection response to the organ.

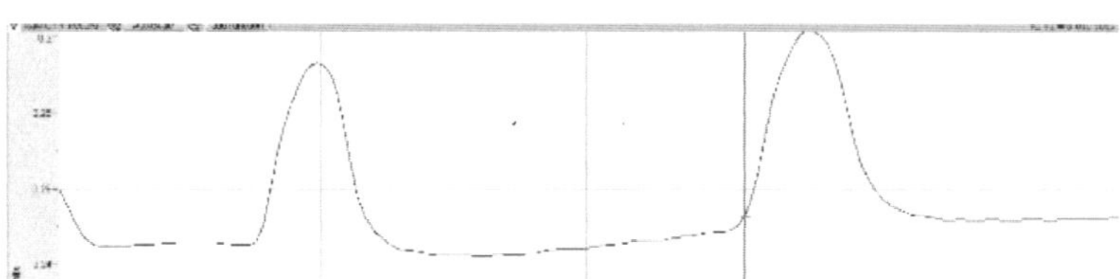

Figure 4. Example of intestinal contraction record after inserting PVA-PEG/SnO_2 Nanofiber

3.4. Antibacterial activity of PVA-PEG/SnO_2 Nanofiber

An example of antibacterial activity for E. Coli of PVA-PEG/SnO_2 Nanofiber is shown in Figure 5. Based on an antibacterial test using S.Aureus bacteria from PVA-PEG/SnO_2 nanofiber did not show the antibacterial power of the material in the absence of clear zone in Figure 5. From the previous study, the antibacterial activity of nanoSnO_2 particles using E.Coli bacteria with different

IOP Conf. Series: Materials Science and Engineering **367** (2018) 012012 doi:10.1088/1757-899X/367/1/012012

concentration of solution indicated the clear presence zone [19, 20]. When compared with antibacterial trials using S. Aureus bacteria there is no clear zone; this suggests the antibacterial ability of SnO2 nanoparticles present in E. coli bacteria.

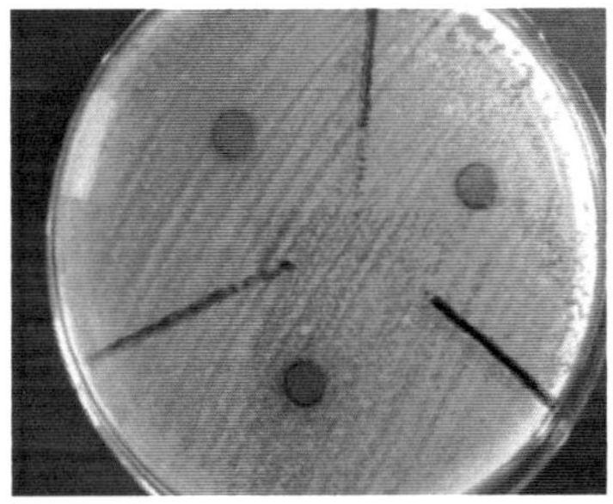

Figure 5. A typical pictogram of Antibacterial PVA-PEG/SnO2 Nanofiber for E.Coli

3.5. Dielectricity

The prepared samples have been characterized using LCR meter for various frequency. The real part of dielectric constant for all samples of PVA-PEG/SnO_2 Nanofiber are presented compactly in Figure 6. Further analyses were performed for dielectric measurement using LCR meter at various light intensity is depicted in 7.

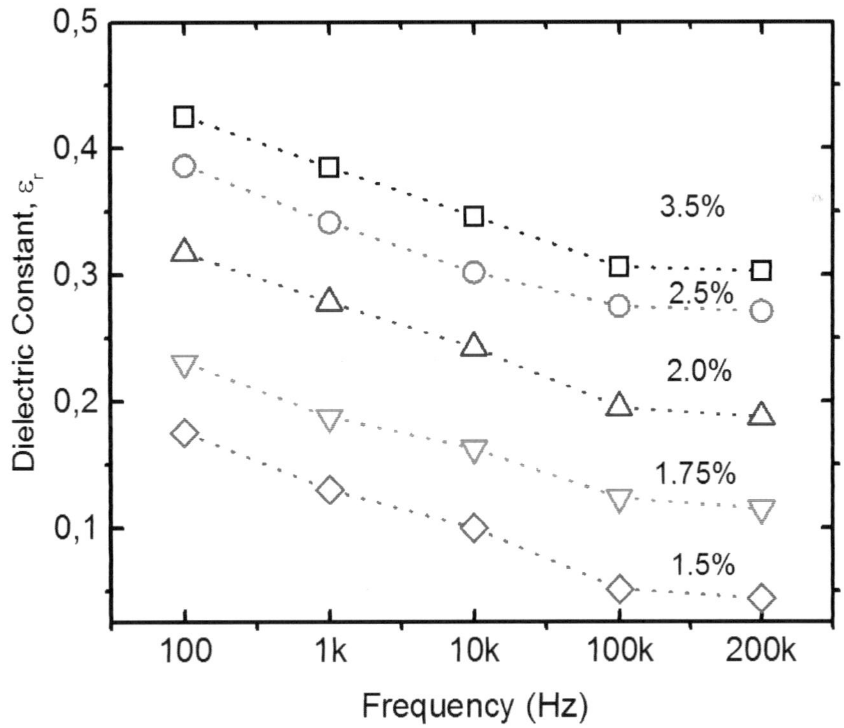

Figure 6. The dielectric constant as a function of Nano SnO and Frequency of PVA-PEG/SnO_2 Nanofiber

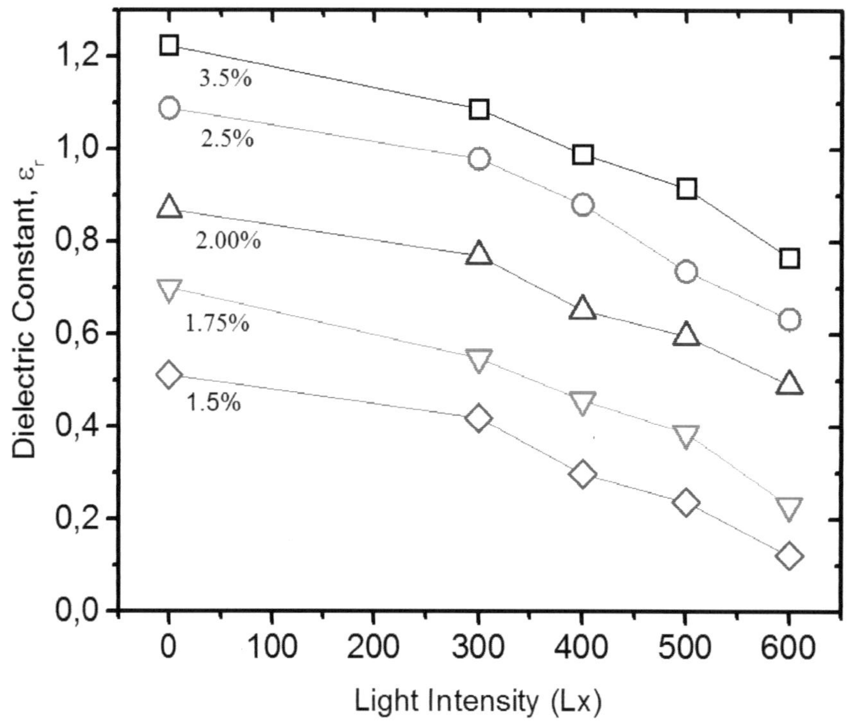

Figure 7. Dielectric constant of different light intensity PVA-PEG/SnO_2 Nanofiber

The higher the intensity of light employed, the dielectric constant decreases. The influence of light on the dielectric constant shows the interaction between light and sample. The light impinges on the sample provides the energy give rise to the excitation of electrons in the atom. This situation produces an atomic swelling that occurs as the radius of the atom gets more prominent. As the ionic radius increases, the electron binding energy becomes weaker, allowing for the electrons to be more easily oriented by the electric screw. In other words, in the view of electronic polarization, to produce the same displacement required a smaller electric field. That is why the dielectric constant decreases.

4. Conclusions

Addition of nanoparticles SnO_2 increases IR spectra, increasing Sn-O-Sn peak in PVA-PEG/SnO_2 nanofiber. The addition of nanoparticles SnO_2 affects the stress of toxicity test in PVA-PEG/SnO_2 nanofiber, but not toxic when applied to the body. The addition of nanoparticles SnO_2 does not change the clear zone of antibacterial assays using S. Aureus microorganisms, so the PVA-PEG/SnO_2 nanofiber has non-antibacterial properties. The addition of nanoparticles SnO_2 the more significant the dielectric constant. The addition of light intensity given the PVA-PEG/SnO_2 nanofiber the smaller the dielectric constant.

Acknowledgements

This research was partially supported by PDUPT Grand of Ministry of Research Technology and Higher Education.

References

[1] Koole R *et al* 2014 *Size effects on semiconductor nanoparticles In Nanoparticles* (Berlin Heidelberg :Springer) pp 13–51

[2] Diantoro M *et al* 2017 Microstructure and dielectric properties of cellulose acetate-ZnO/ITO composite films based on water hyacinth *J. Phys. Conf. Ser.* **853** p 12047

[3] Diantoro M *et al* 2017 Effect of Fe_3O_4 on the Electro-Optic and Magneto-Electric Characteristics of (PANI / Fe_3O_4) -Ag Film *IOP Conf. Ser. Mater. Sci. Eng.* **12062**
[4] Batal M A, Nashed G and Jneed F H 2014 Conductivity and thermoelectric properties of nanostructure tin oxide thin films *Journal of the Association of Arab Universities for Basic and Applied Sciences* **15** pp 15–20
[5] Sahm T *et al* 2007 Sensing of CH4, CO and ethanol with in situ nanoparticle aerosol-fabricated multilayer sensors *Sensors and Actuators B: Chemical* **127** pp 63–68
[6] Shukla G P and Bhatnagar M C 2015 SnO_2 Fiber: Morphology, Optical Property *Journal of Materials Science and Engineering A* **5** pp 377–380
[7] Khuc Q T 2010 The influence of hydrothermal temperature on SnO_2 nanorod formation *Advances in Natural Sciences: Nanoscience and Nanotechnology* **1** p 025010
[8] Lu F C and Wang C Y 2016 April The SnO_2 nanowires Synthesis and Characterization *Meeting Abstracts (The Electrochemical Society)* **24** pp 1259–1259
[9] Fang D *et al* 2014 Hollow SnO_2-ZnO hybrid nanofibers as anode materials for lithium-ion battery *Mater. Res. Express* **1** p 25012
[10] Zhao Y *et al* 2015 Electrospun SnO_2–ZnO nanofibers with improved electrochemical performance as anode materials for lithium-ion batteries *International Journal of Hydrogen Energy* **40** pp 14338–14344
[11] Wei Y *et al* 2017 A Stable Cross-Linked Binder Network for SnO_2 Anode with Enhanced Sodium-Ion Storage Performance *Chemistry Select* **2** pp 11365–11369
[12] Prosanov I Y *et al* 2013 Hybrid material polyvinyl alcohol-stannic acid/stannic oxide *Physics of the Solid State* **55** pp 1519–1524
[13] Hatamie S *et al* 2009 Polymer-embedded stannic oxide nanoparticles as humidity sensors *Materials Science and Engineering: C* **29** pp 847–850
[14] Mostafaei A and Nasirpouri F 2014 Epoxy/polyaniline–ZnO nanorods hybrid nanocomposite coatings: Synthesis, characterization and corrosion protection performance of conducting paints *Progress in Organic coatings* **77** pp 146–159
[15] Thanasanvorakun S *et al* 2008 Characterization of SnO_2 nanowires synthesized from SnO by carbothermal reduction process *Ceramics International* **34** pp 1127–1130
[16] Wang W *et al* 2015 Electrochemistry of myoglobin on graphene–SnO_2 nanocomposite modified electrode and its electrocatalysis *Arabian Journal of Chemistry*
[17] Blessi S *et al* 2014 Preparation and characterization of SnO_2 nanoparticles by hydrothermal method *ChemTech Research* **6** pp 2153–2155
[18] Liu Y *et al* 2008 Improved molten salt synthesis (MSS) for SnO_2 nanorods and nanotwins *Materials Chemistry and Physics* **112** pp 381–386
[19] Bouarroudj T **et al** 2016 Cytotoxic effect of sno2 nanoparticles on alternative cellular model: paramecium tetraurelia *Studia Universitatis Vasile Goldis Seria Stiintele Vietii (Life Sciences Series* **26**
[20] Kumari M M and Philip D 2015 Synthesis of biogenic SnO_2 nanoparticles and evaluation of thermal, rheological, antibacterial and antioxidant activities *Powder Technology* **270** pp 312–319

Band Gap Calculations of Bilayer Graphene and Bilayer Armchair Graphene Nanoribbon

E. Sustini[1], Khairurrijal[1], F. A. Noor[1], and R. Syariati[1]

[1]Physics of Electronics Materials Research Division Faculty of Mathematics and Natural Science, Bandung Institute of Technology

E-mail : euis@fi.itb.ac.id

Abstract. Graphene has a high potential for use in electronic devices because it has a unique characteristic, electrons in massless grasses and has a speed close to the speed of light. Graphene conductivity at temperature space is 40 000 $cm^2 \cdot V^{-1} \cdot s^{-1}$ in the air. In addition, thermal conductivity can reach 2 800 $W \cdot m^{-1} \cdot K^{-1}$. The mechanical properties of graphene are very strong and flexible, and optical properties it has a white light transmittance of 95%. Therefore, graphene is suitable for transparent screen and the flexible electrode. The energy band gap of monolayer graphene can be changed by reducing the size, as for the band gap of bilayer graphene was affected by symmetry between two layers. So, it is interesting to research the band gap characteristics of bilayer armchair graphene nanoribbon (AGNR). In this paper, we have calculated the band gap of bilayer graphene and bilayer AGNR in various parameters such as band width and symmetry between two layers. The method used is tight binding method by only taking into account its nearest neighbors only. The results obtained from this simulation show that the band gap will be more open if asymmetry of the structure bilayer graphene is increasing and the band gap of AGNR will be more open if ribbon width decreased, is not affected by asymmetry.

Keywords. Band gap, graphene, and nanoribbon.

1. Introduction

The unique electronic characteristics of graphene make it the potential for applications on the device both for graphene monolayer and bilayer graphene. Monolayer graphene properties are the same as bilayer graphene. At room temperature in the air both materials have a high electrical conductivity that is up to 40 000 $cm^2 \cdot V^{-1} \cdot s^{-1}$ [1]. While the thermal conductivity of these two materials reaches up to 2 800 $W \cdot m^{-1} \cdot K^{-1}$ [2]. The mechanical properties are strong and flexible [3] while for optical properties are equally transparent with white light transmittance reaching 95 % [4]. Hence both of these materials have very wide applications including transparent screens and flexible electrodes making them suitable for touch screens.

The graphene monolayer with bilayer graphene differs on the possibility of adjusting the width of the band gap. The band gap of monolayer graphene can be changed by reducing its size to the nanoscale. Graphene in this size is one of them named as graphene nanoribbon (GNR) [5]. There are two types of GNR namely, armchair GNR (AGNR) and zigzag GNR (ZGNR). AGNR is known to have semiconductor properties depending on the big of the width while ZGNR has conductor

Content from this work may be used under the terms of the Creative Commons Attribution 3.0 licence. Any further distribution of this work must maintain attribution to the author(s) and the title of the work, journal citation and DOI.
Published under licence by IOP Publishing Ltd

IOP Conf. Series: Materials Science and Engineering **367** (2018) 012013 doi:10.1088/1757-899X/367/1/012013

properties. Another difference, that energy dispersion of monolayer graphene is linear [5]. The bilayer graphene banding of the bandgap may be accomplished by impurity and gate voltage regulation [6]. In addition, the effective mass of bilayer graphene is nonzero (but can be said to be very small) and the parabolic dispersion energy relationship [7] so that the bandgap profile produced by the bilayer will be different from that of the monolayer. In this research, we calculated the band gap of bilayer graphene and bilayer AGNR with parameter asymmetry and width of the ribbon.

2. Theoretical approach

2.1. Dispersion energy of Bilayer Graphene

In one of bilayer graphene lattice, there are four atoms, A_1, B_1, A_2, and B_2 so that the wave function is defined [1].

$$\Phi_{A_m/B_n}(r) = \frac{1}{\sqrt{N}} \sum_{R_{A_m/B_n}}^{N} e^{iKR_{A_m/B_n}} \psi_{A_m/B_n}(\mathrm{r}-\mathrm{R}_{A_m/B_n}) = e^{iKr}\psi_{A_m/B_n}(r) + e^{iK'r}\psi'_{A_m/B_n}(r), \quad (1)$$

$$m = 1,2 \ , \ n = 1,2 \ ,$$

with $R_{A/B}$ representing the atomic position of atoms A and B, N is the number of primitive cells of a lattice, $e^{iKR_{A_m/B_n}}$ representing a periodicity of the lattice, and $\psi_{A_m/B_n}(\mathrm{r}-\mathrm{R}_{A_m/B_n})$ representing the Bloch function. Physically, electrons can move from one atom to another that is the nearest neighbor of the atom. For example, an electron on A_1 atom can move to atoms B_1, A_2, or B_2. From these conditions, the dispersion energy relationship can be determined by using the Schrödinger equation $H\Phi = E\Phi$ so that the matrix forms are

$$\begin{bmatrix} E_{A_1} & -\gamma_0 f(k) & \gamma_4 f(k) & -\gamma_3 f^*(k) \\ -\gamma_0 f^*(k) & E_{B_1} & \gamma_1 & \gamma_4 f(k) \\ \gamma_4^* f(k) & \gamma_1 & E_{A_2} & -\gamma_0 f(k) \\ -\gamma_3 f(k) & \gamma_4 f^*(k) & -\gamma_0 f^*(k) & E_{B_2} \end{bmatrix} \begin{bmatrix} \Phi_{A_1} \\ \Phi_{B_1} \\ \Phi_{A_2} \\ \Phi_{B_2} \end{bmatrix} = \varepsilon_\alpha \begin{bmatrix} 1 & s_0 f(k) & 0 & 0 \\ s_0 f^*(k) & 1 & s_1 & 0 \\ 0 & s_1 & 1 & s_0 f(k) \\ 0 & 0 & s_0 f^*(k) & 1 \end{bmatrix} \begin{bmatrix} \Phi_{A_1} \\ \Phi_{B_1} \\ \Phi_{A_2} \\ \Phi_{B_2} \end{bmatrix} \quad (2)$$

With E_{A_1}, E_{B_1}, E_{A_2}, and E_{B_2} is the initial energy of the electrons in each atom while f (k) describes the geometrical factor of the position of the neighboring atom. As γ_1 is illustrating the parameters coupled between the pair of orbital dimer (identical) electrons of atoms A_2 and B_1. In this parameter, there is no f (k) as it is a vertically coupled pair. γ_3 , explain about the coupling pair of electron orbitals of the non-dimer atoms of A_1 and B_2. In addition, in the second equation, there exists γ_4 a parameter of the interlayer electron orbitals between the dimer and non-dimer pairs of the A_1 and A_2 or B_1 and B_2 atoms. The last parameter for the left-hand side of Equation 2 is that γ_0 it is physically a parameter of an electron jump from one atom to the nearest neighbor atom on one layer. Visually, the role of these parameters can be seen in Figure 1.

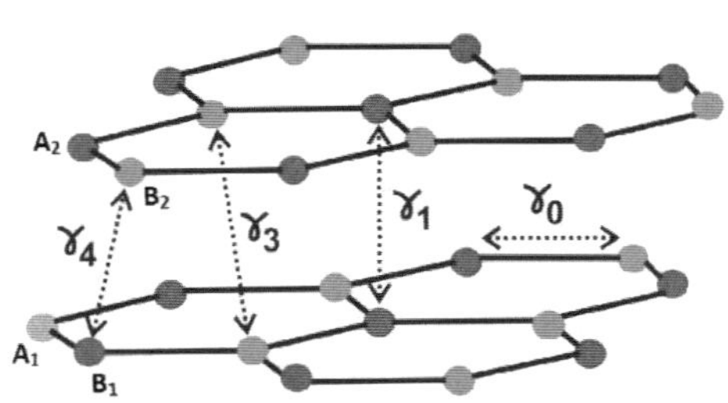

Figure 1. Bilayer graphene structure with parameters of electron leap

The wave number $k = K + p$, K is the point at which the conduction band and the valence band meet, whereas p is the vector shift. Assuming the overlap parameter s_0 and s_1 zero, with $v_0 = {}^{3a\gamma_0}\!/_{2\hbar}$, $v_3 = {}^{3a\gamma_3}\!/_{2\hbar}$, $v_4 = {}^{3a\gamma_4}\!/_{2\hbar}$ and taking into account the condition of the asymmetry bilayer graphene, γ_4 will be zero, the value $E_{A_1} = E_{B1} = -U/2$, and $E_{A_2} = E_{B2} = U/2$ (U potential energy) so will be obtained

$$\varepsilon_\alpha = \pm\sqrt{\frac{\gamma_1^2}{2} + \frac{U^2}{4} + \left(v_0^2 + \frac{v_3^2}{2}\right)p^2 + (-1)^\alpha\sqrt{\frac{1}{4}\left(\gamma_1^2 - v_3^2 p^2\right)^2 + v_0^2 p^2\left(\gamma_1^2 + U^2 + v_3^2 p^2\right) + 2\gamma_1 v_3 v_0^2 p^3 \cos 3\varphi}}.$$

The form of p is the resultant vector of p_x and p_y which can be written mathematically $p = \sqrt{p_x^2 + p_y^2}$, φ , is the angle between p_x and p_y, and the constant α is degeneration of the bilayer graphene gap having a value of 1 and -1. As for the $\pm$ sign, describes the conduction band and the valence band. From these parameters, bandgap can be searched by calculating the difference between the energy contained in the conduction band and the valence band.

2.2. Dispersion energy of Bilayer Graphene

To obtain the dispersion energy relation of the bilayer AGNR, the first thing to do is to assume that one of the lengths of the bilayer graphene is made on a nanometer scale. In this simulation, it is assumed the size in the x-direction is restricted. Physically, the restriction will make the wave number in the x-direction has a unique value. This condition leads to a boundary condition that has a range $r = 0$ and $r = w_{ac} + \sqrt{3}$ is illustrated in Figure 2. With this boundary condition, the wave function in Equation 1 applies

$$\Phi_{A_{1/2}}(\mathrm{r}=0) = \Phi_{B_{1/2}}(\mathrm{r}=0) = \Phi_{A_{1/2}}(\mathrm{r} = w_{ac} + \sqrt{3}\,\mathrm{a}) = \Phi_{B_{1/2}}(\mathrm{r} = w_{ac} + \sqrt{3}\,\mathrm{a}) = 0. \quad (3)$$

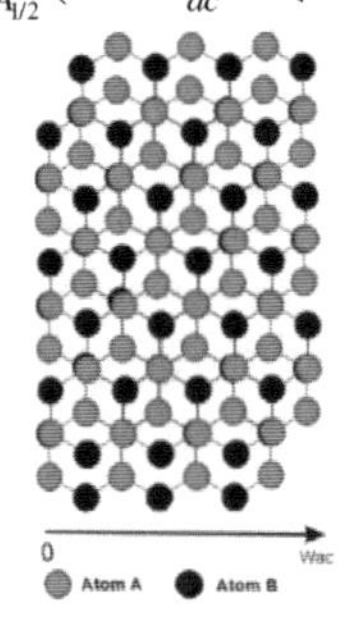

Figure 2. Bilayer AGNR Structure

The solution form for ψ_{A_m/B_n} and ψ'_{A_m/B_n} in Equation 1 by Ansatz is given sinusoidal form which is written as follows [5]

$$\psi_{A_m/B_n} = A e^{ip_x x} + B e^{-ip_x x},$$

$$\psi'_{A_m/B_n} = C e^{ip_x x} + D e^{-ip_x x}.$$

Here A, B, C, and D are constants. By using the boundary conditions in Equation 3 it will be obtained

$$A + B + C + D = 0,$$

$$A e^{i(p_x+K)(w_{ac}+\sqrt{3}a)} + D e^{-i(p_x+K)(w_{ac}+\sqrt{3}a)} + B e^{-i(p_x-K)(w_{ac}+\sqrt{3}a)} + C e^{i(p_x-K)(w_{ac}+\sqrt{3}a)} = 0.$$

The condition of the boundary conditions is met if $B=-C$ and $A=D=0$ and $\sin[(p_x - K)(w_{ac}+\sqrt{3})]=0$ produces. So, the wave number for the x-direction is written to be

$$p_x = \frac{n\pi}{w_{ac}+\sqrt{3}a} + \frac{2\pi}{3\sqrt{3}a}.$$

With w_{ac} is the width of the AGNR bilayer.

$$\varepsilon_\alpha = \pm\sqrt{\frac{\gamma_1^2}{2}+\frac{U^2}{4}+\left(v_0{}^2+\frac{v_3{}^2}{2}\right)p^2+(-1)^\alpha\sqrt{\frac{1}{4}\left(\gamma_1^2-v_3{}^2p^2\right)^2+v_0{}^2p^2\left(\gamma_1^2+U^2+v_3{}^2p^2\right)+2\gamma_1 v_3 v_0^2 p^3\left(\left(\frac{p_x}{p}\right)^3-3\frac{p_y{}^2p_x}{p^3}\right)}}.$$

3. Results and discussion

Figure 3 (a) shows the bandgap of the bilayer graphene under normal conditions and when there is a high asymmetry called the Mexican Hat Effect. From the picture, it is seen that the bandgap of bilayer graphene actually has zero band gap. However, when the arrangement of its high asymmetry structure the band gap was becoming open. This phenomenon occurs because when the bilayer graphene structure in that state the value of Δ parameter equals with γ_1 [7]

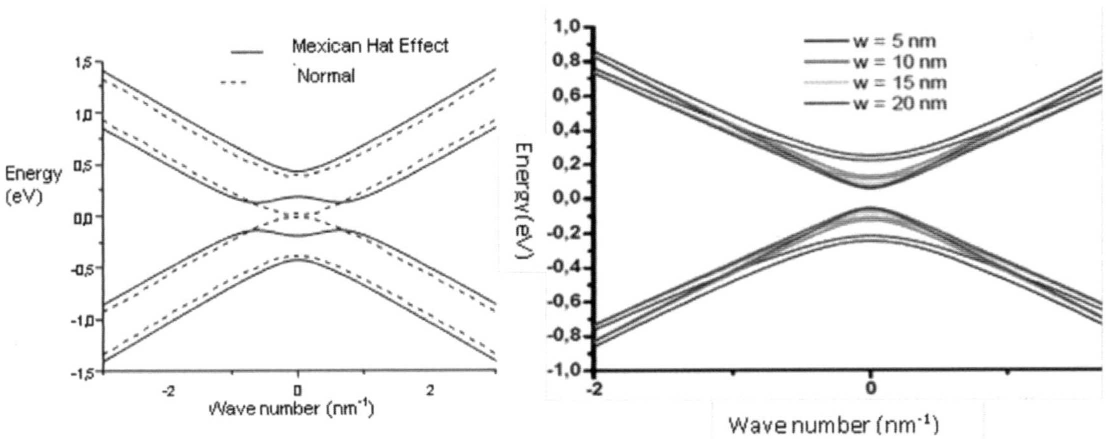

(a) (b)

Figure 3. The band gap energy of (a) bilayer graphene based on asymmetry (b) bilayer AGNR

Figure 3(b) describes the band gap of the bilayer AGNR seen at point K with four variations of the ribbon width. The picture shows that the larger the ribbon width value of the bilayer AGNR, the resulting bandgap getting smaller. In the figure, it is also seen that with the increase of AGNR ribbon width, the band gap degeneration approaches the lowest energy level. Another thing that can be discussed about the absence of Mexican Hat Effect despite high asymmetry. This condition arises because of the influence of the limited value of wave numbers in the x-direction, due to limited size width. In addition, the effective Hamiltonian role of the bilayer AGNR provides the effect of band gap changes in terms of degeneration, asymmetric structure, or in the opening of the bandgap due to changing values γ_1, γ_3, and γ_4.

4. Conclusions

By using tight binding method (TB) has been found dispersion energy relation of bilayer graphene and bilayer AGNR graphene. The difference between these two materials is seen in several ways. In the bilayer graphene, there is a Mexican Hat Effect that makes the bilayer graphene energy gap affected by the asymmetry structure while in the AGNR bilayer there is no Mexican Hat Effect and the band gap is only influenced by the bandwidth.

Acknowledgments
This research was financially supported by "PTUPT DIKTI" Research grand in the fiscal year of 2017

References

[1] Dean C R 2010 Boron nitride substrates for high-quality graphene electronics *Nat. Nanotech.* **5** pp 722–726
[2] Novoselov K S 2004 Electric field effect in atomically thin carbon films *Science.* **306** pp 666–669
[3] Neek-Amal M and Peeters F M 2010 Nano indentation of a circular sheet of bilayer graphene *Phys. Rev. B.* **81** p 235421
[4] Nair R R 2008 Fine structure constant defines visual transparency of graphene *Science.* **320** pp 1308–1311
[5] Castro Neto A H 2009 The electronic properties of graphene *Rev. Mod. Phys.* **81** pp 109–162
[6] McCann E and Fal'ko V I 2006 Landau-Level degeneracy and quantum Hall effect in a graphite bilayer *Phys. Rev. Lett.* **96** pp 086805
[7] McCann E and Koshino M 2013 The electronic properties of bilayer graphene **76** pp 056503

Synthesis of Fibers and Particles from Polyvinyl Chloride (PVC) Waste Using Electrospinning

A. Zulfi[1,2], Y. A. Rezeki[1,2], D. Edikresnha[1,2], M. M. Munir[1,2], and K. Khairurrijal[1,2]

[1] Department of Physics, Faculty of Mathematics and Natural Sciences, Institut Teknologi Bandung, Jalan Ganesa 10, Bandung 40132, Indonesia
[2] Research Center for Bioscience and Biotechnology, Institut Teknologi Bandung, Jalan Ganesa 10, Bandung 40132, Indonesia

E-mail: krijal@fi.itb.ac.id

Abstract. Synthesis of fibers from the waste of polyvinyl chloride (PVC) has been successfully done using electrospinning method. The PVC solutions were made with the solvents N, N dimethylformamide (DMF), tetrahydrofuran (THF) and dimethylacetamide (DMAc). The effects of PVC concentration on the morphology and the diameter of fibers were observed. The morphological change from particles to fibers took place along with the increasing concentration of PVC in DMF, DMAc, and THF. When DMF was used as the solvent, the average particle diameter was 2.91 μm for the concentration of 5 wt%. As the concentration was increased to 10 wt%, 15 wt%, and 20 wt%, fibers were obtained, and their mean diameters were 1.12, 1.47, and 1.63 μm, respectively. When DMAc was used, the average particle made of the precursor solutions of concentrations of 5 wt%, 10 wt%, and 15 wt% were 2.79 μm, 3.03 μm, and 3.31 μm, respectively. If the concentration was further increased to 20 wt%, then fibers were formed with the average diameter of 1.35 μm. Finally, with the use of THF solvent, the average particle diameters were 6.83 μm and 8.38 μm for the solution concentrations of 5 wt% and 10 wt%, respectively. When the concentrations were 15 wt% and 20 wt%, fibers were finally achieved with the average diameter of 1.38 μm and 2.83 μm, respectively. It was clearly found that there is a PVC critical concentration for a morphological transition from PVC particles to fibers.

Keyword. Electrospinning, fibers, and polyvinyl chloride (PVC)

1. Introduction

The development of nanotechnology has grown tremendously in recent years and has produced many applications in various fields. One of the most interesting developments in nanotechnology, due to its vast applications, is nanofiber. Nanofiber is a one-dimensional material whose form is like fine threads with a diameter of one to several hundred nanometers. Nanofiber is proven to possess many advantages such as having a large ratio of surface area over volume, high flexibility, tensile strength, and porosity, which can then be utilized in many applications [1], [2]. Some of the areas where nanofiber has been extensively researched as drug delivery medium, sensors, catalysts, filters, protective clothing, etc. [3]–[6]

•

Content from this work may be used under the terms of the Creative Commons Attribution 3.0 licence. Any further distribution of this work must maintain attribution to the author(s) and the title of the work, journal citation and DOI.
Published under licence by IOP Publishing Ltd

One of the techniques to produce nanofibers is electrospinning. A high-voltage source, a collector, a syringe pump, and a nozzle are the major components of this technique [6]. This technique utilizes electrostatic forces to form fibers, and it has the ability to produce fibers with different characteristics. The characteristics of resulting fibers can be controlled by adjusting the electrospinning parameters such as concentration of the solution, viscosity, charge density, net charge, charge neutralization, surface tension, flowrate and voltage [4], [7].

Polyvinyl chloride (PVC) is one of the most commonly used thermoplastic materials and is the second most produced thermoplastic after polyethylene. Due to its physical and chemical advantage as well as its anti-degradable properties and low production costs, PVC is widely applied as pipes, packaging materials, window frames, cables, bottles, and even in medical devices [8]–[10]. The flexibility of the PVC application then causes an accumulation of waste due to the growth of the plastic industry. As far as we know, there is still no study of electrospinning associated with PVC waste. The purpose of this study is then to provide alternative solutions in the processing of PVC waste into fibers by means electrospinning method. The solvents used for the precursor solutions were N, N dimethylformamide (DMF), tetrahydrofuran (THF) and dimethylacetamide (DMA). In this research, the relationship between the concentration of PVC in the precursor solution to the morphology of the fibers and particles as the product of electrospinning was investigated.

2. Materials and methods

2.1. Materials

PVC waste was collected from the PVC pipe waste. The main solvents used were N, N dimethylformamide (DMF), tetrahydrofuran (THF), and dimethylacetamide (DMA), which were purchased from Sigma-Aldrich.

2.2. Synthesis of fiber

To observe the effect of polymer concentration to the fibers, several concentrations of PVC solution (5, 10, 15 and 20) wt% were prepared by dissolving PVC in different solvents (DMF, DMAc, and THF).

The fiber membranes were fabricated using electrospinning apparatus (Nachriebe 600) that consists of a syringe pump and its controller, a high voltage (HV) power supply, a drum collector, and a synthesis chamber with a controlled environment, as schematically shown in Figure 1. The prepared precursor solution was filled in the plastic syringe with needle diameter of 0.8 mm. The filled syringe was mounted on the syringe pump to control the flow rate of the ejected solution. The metal needle tip was then connected to the positive terminal of the HV power supply. The grounded drum collector was coated with aluminum foil. The ejected solution from the metal needle tip, which forms a Taylor cone, was monitored using a CCD camera in the synthesis chamber.

The characterization of the fibers was done by using a digital microscope (National, DC3-163) with the magnifications of 400 times and 1000 times. The microscope images were then processed to obtain the average and the distribution of the diameter of fibers. The measurement of diameter was taken at different locations on the fibers up to 100 points. The average diameter of fibers was calculated using Equation 1.

$$CV = \frac{\sigma}{\mu} \tag{1}$$

where σ is the standard deviation and μ is the average diameter of fibers [14].

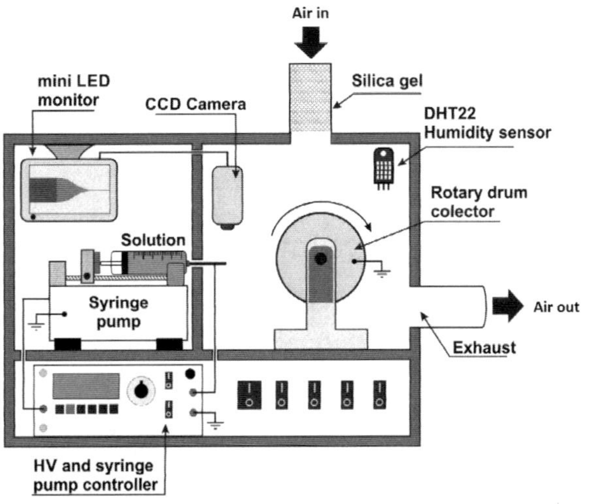

Figure 1. The schematic diagram of electrospinning (Nachriebe 600)

3. Results and discussion

It is widely known that the morphology of the fibers produced from electrospinning is affected by the polymer properties (the type and molecular weight), solution properties (polymer concentration, viscosity, conductivity and surface tension), processing parameters (voltage, distance the needle tip to collector, flow rate, and the diameter of the tip of the needle), and environmental factors (temperature and humidity). By adjusting the parameters, the microstructure of the nanofibers can be altered [6], [11]. In this research, the effects of the concentration of polymer in a DMF, DMAc, and THF solvent were studied as follows.

PVC solutions with various concentrations of (5, 10, 15, and 20) wt% were created using a DMF solvent. Figure 2 shows microscope images of the PVC fibers with DMF solvent under the digital microscope with magnifications of 400 times and 1000 times and their respected histograms showing the size distribution of fibers. During synthesis, the process parameters were kept constant. The flow rate was set at 0.5 mL·h^{-1}, the applied voltage was 12 kV, and the distance from the collector to the tip of the needle was kept at 15 cm.

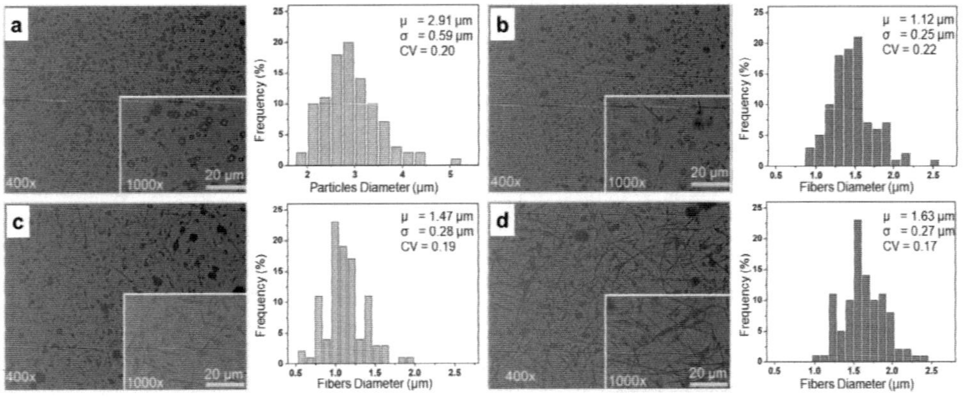

Figure 2. Microscope images of the PVC fibers produced by employing DMF solvent under the digital microscope with the magnification of 400 times and 1000 times and their corresponding histograms: (a) PVC 5 wt%, (b) PVC 10 wt%, (c) PVC 15 wt%, and (d) PVC 20 wt%

From Figure 2, it can be seen that the gradual change of microstructures from particles to fibers does exist along with the increase of the concentration of PVC waste solution. In Figure 2.a, the morphology of the particles made of the precursor solution with the concentration of 5 wt% was particles with average diameter of 2.91 μm. The fibers started to appear when the concentration of the precursor solution was 10 wt% as shown in Figure 2.b with an average diameter of fibers of 1.12 μm. Furthermore, in Figures 2.c to Figure 2.d, as the polymer concentration increased, more fibers were formed, and the diameter of the resulting fibers increased as well. The average diameter of produced fibers for the 15 wt% concentration was 1.47 μm while the average diameter made of 20 wt% concentration was 1.63 μm.

Figure 3 gives the microscope images of the PVC fibers with DMAc solvent under the digital microscope with magnifications of 400 times and 1000 times, and their respected histograms are showing the size distribution of fibers. During the synthesis, the process parameters were again kept to be the same as before.

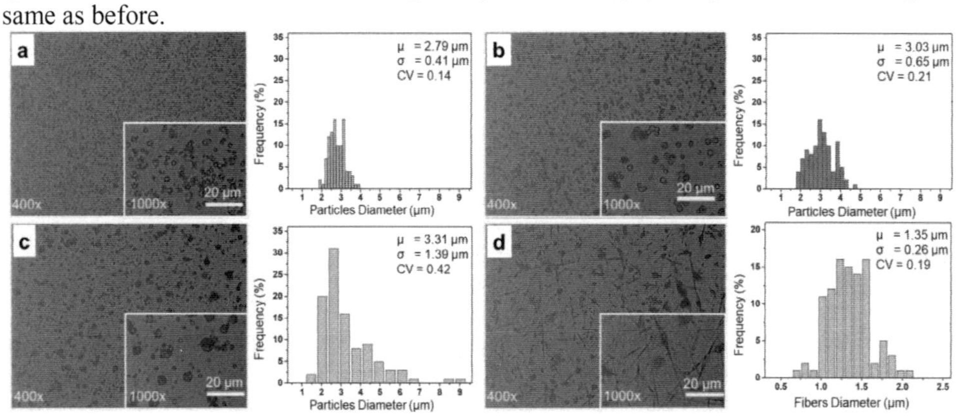

Figure 3. Microscope images of PVC fibers produced by employing DMAc solvent under the digital microscope with magnifications of 400 times and 1000 times and their corresponding histograms: (a) PVC 5 wt%, (b) PVC 10 wt%, (c) PVC 15 wt%, and (d) PVC 20 wt%

From Figure 3, there was also a gradual morphological change from particles to fibers along with increasing the concentration of PVC waste in the solution. However, in this case, the fibers started to appear at a higher concentration (20 wt%) as a contrast to DMF where the fibers were formed at lower concentration (10 wt%). The average diameters of obtained particles were (2.79, 3.03, and 3.31) μm for the concentrations of (5, 10, and 15) wt%, respectively, while the average diameter of fibers was 1.35 μm (Figure 3.d).

Figure 4 depicts microscope images of the PVC fibers obtained by using THF solvent under the digital microscope with magnifications of 400 times and 1000 times, and their respected histograms showing the size distribution of fibers. During the synthesis, the process parameters were kept constant as before.

Figure 4 shows microscope images of the PVC fibers the use of THF solvent resulting in a morphological change from particles to fibers similar to the two prior solvents (DMF and DMAc). Submicron particles were produced with average sizes of particles of 6.83 and 8.38 μm as depicted in Figures 4.a and Figure 4.b. In Figure 4.c, it can be seen that the fibers began to form with an average diameter of fibers of 1.38 μm. Finally, in Figure 4.d, the particles disappeared and were replaced by fibers with an average diameter of 2.83 μm.

Based on the obtained results, it is clearly found that the concentration of PVC in the solutions greatly affects the morphology. Particles will be formed if the concentration of the precursor solution is low for all types of solvents used since the interaction between solvent molecules with other solvent molecules is more dominant as compared to the interaction between the polymer molecules with other polymer molecules. This causes a tendency to form spherical particles aiming to reduce the surface area. Additionally, for low polymer concentrations, the surface tension is likely to have a large value. When the surface tension is much higher than the force due to the interaction of charge and also the force due to the interaction between the molecular chains of the polymer, then particles will be formed [6], [11]. Furthermore, fibers will be generated as the concentration of polymer is increased for all solvents. As the concentration is further increased, the diameter of the fibers will increase as well. If the interaction between polymer molecules is higher than the solvent molecules, the fibers will be formed when the liquid jet is attracted to the collector due to electrical force. The diameter of fibers will increase due to the high viscosity as a result of high polymer concentration. The high viscosity will restrict the deformation of a liquid jet into the fiber, and thus the stretching process will be longer. Consequently, if other parameters are kept constant, the diameter of the fibers will be larger as a result of higher polymer concentration [6], [11]–[14].

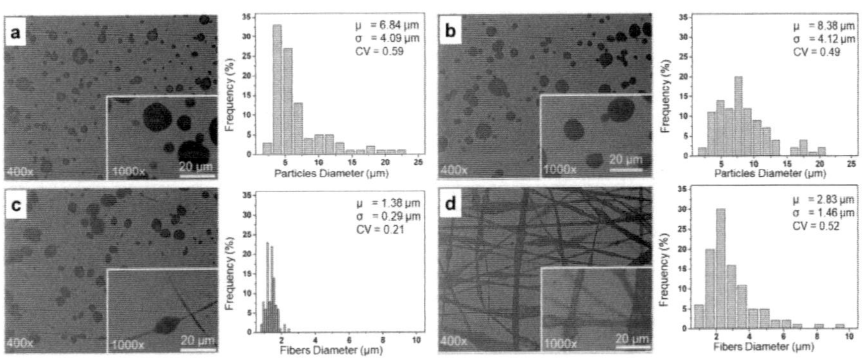

Figure 4. Microscope images of PVC fibers produced by using THF solvent under the digital microscope with the magnifications of 400 times and 1000 times and their corresponding histograms: (a) PVC 5 wt%, (b) PVC 10 wt%, (c) PVC 15 wt%, and (d) PVC 20 wt%

4. Conclusions

The morphological change of PVC particles to fibers has been successfully produced by electrospinning using PVC waste diluted in DMF, DMAc and THF solvents. From the obtained microscope images, there was a critical concentration of PVC solution in which electrospun PVC experienced a morphological change from particles to fibers. In general, PVC particles were formed for PVC concentrations lower than its critical concentration, and the average size of particles increased with the PVC concentration. For the concentrations higher than its critical PVC concentration, PVC fibers were obtained, and the average size of fibers increased with the PVC concentration.

Acknowledgments
This research was financially supported by the Ministry of Research, Technology, and Higher Education, the Republic of Indonesia under the University's Excellent Research (PUPT) Grant and the PMDSU scholarship of A. Z and Y. A. R the fiscal year 2017.

References

[1] Z.-M. Huang, Y.-Z. Zhang, M. Kotaki, and S. Ramakrishna, "A review on polymer nanofibers by electrospinning and their applications in nanocomposites," *Compos. Sci. Technol.*, vol. 63, no. 15, pp. 2223–2253, 2003.
[2] K. Yoon, B. S. Hsiao, and B. Chu, "Functional nanofibers for environmental applications," *J. Mater. Chem.*, vol. 18, no. 44, pp. 5326–5334, 2008.

[3] J. Yun, J. S. Im, Y.-S. Lee, and H.-I. Kim, "Electro-responsive transdermal drug delivery behavior of PVA/PAA/MWCNT nanofibers," *Eur. Polym. J.*, vol. 47, no. 10, pp. 1893–1902, 2011.
[4] M. M. Munir, A. B. Suryamas, F. Iskandar, and K. Okuyama, "Scaling law on particle-to-fiber formation during electrospinning," *Polymer*, vol. 50, no. 20, pp. 4935–4943, 2009.
[5] H. Widiyandari, M. M. Munir, F. Iskandar, and K. Okuyama, "Morphology-controlled synthesis of chromia–titania nanofibers via electrospinning followed by annealing," *Mater. Chem. Phys.*, vol. 116, no. 1, pp. 169–174, 2009.
[6] M. M. Munir, F. Iskandar, M. Djamal, and K. Okuyama, "Morphology controlled electrospun nanofibers for humidity sensor application," in *AIP Conference Proceedings*, 2011, vol. 1415, pp. 223–226.
[7] H. Fong, I. Chun, and D. Reneker, "Beaded nanofibers formed during electrospinning," *Polymer*, vol. 40, no. 16, pp. 4585–4592, 1999.
[8] I. Janajreh, M. Alshrah, and S. Zamzam, "Mechanical recycling of PVC plastic waste streams from cable industry: A case study," *Sustain. Cities Soc.*, vol. 18, pp. 13–20, 2015.
[9] M. Sadat-Shojai and G.-R. Bakhshandeh, "Recycling of PVC wastes," *Polym. Degrad. Stab.*, vol. 96, no. 4, pp. 404–415, 2011.
[10] V. Luciani, G. Bonifazi, P. Rem, and S. Serranti, "Upgrading of PVC rich wastes by magnetic density separation and hyperspectral imaging quality control," *Waste Manag.*, vol. 45, pp. 118–125, 2015.
[11] K. Sarkar *et al.*, "Electrospinning to forcespinning™," *Mater. Today*, vol. 13, no. 11, pp. 12–14, 2010.
[12] A. Zulfi, A. Fauzi, D. Edikresnha, M. Munir, and others, "Synthesis of High-Impact Polystyrene Fibers using Electrospinning," in *IOP Conference Series: Materials Science and Engineering*, 2017, vol. 202, p. 012010.
[13] S. Padron, A. Fuentes, D. Caruntu, and K. Lozano, "Experimental study of nanofiber production through forcespinning," *J. Appl. Phys.*, vol. 113, no. 2, p. 024318, 2013.
[14] M. R. Badrossamay *et al.*, "Engineering hybrid polymer-protein super-aligned nanofibers via rotary jet spinning," *Biomaterials*, vol. 35, no. 10, pp. 3188–3197, 2014.

Fabrication and Characterization of Monodisperse Polystyrene Latex (PSL) with Various Diameters

A Sandi[1,2], A Sawitri[3], A Rajak[1,2], A Zulfi[1,2], D Edikresnha[1,2], M M Munir[1,2] and K Khairurrijal[1,2]

[1]Department of Physics, Faculty of Mathematics and Natural Sciences, Institut Teknologi Bandung, Jalan Ganesa 10, Bandung 40132, Indonesia
[2]Bioscience and Biotechnology Research Center, Institute for Research and Community Services, Institut Teknologi Bandung, Jalan Ganesa 10, Bandung 40132, Indonesia
[3]Department of Physics, Faculty of Science and Technology, UIN Sunan Gunung Djati Bandung, Jalan A. H. Nasution 105, Bandung 40614, Indonesia

E-mail: krijal@fi.itb.ac.id

Abstract. The focus of this study was the synthesis of polystyrene latex (PSL) particles and their diameter control for the application of aerosol filtration test. Simple emulsion polymerization was used as the method to produce uniformly spherical PSL particles. The diameter of PSL particles ranging from (160 to 941) nm was obtained by controlling the reaction parameters during the polymerization process. The reaction parameters were the concentration of styrene monomer, the concentration of potassium persulphate initiator, and the solution temperature. The diameter of PSL particles was larger as the concentration of styrene monomer, and initiator concentration increased, as analyzed from their SEM images. The higher the solution temperature, the resulting particles diameter would be smaller.

Keywords. Aerosol filtration test, polymerization process, polystyrene latex (PSL), and SEM.

1. Introduction

In recent years, research about aerosol filter has become the center of attention of many researchers, especially for those working on the air filter. One efficient method for the aerosol filtration is to use a nanofiber membrane as a filter produced by electrospinning method [1, 2], and a constant current electrospinning system was employed to obtain highly uniform nanofibers [3, 4]. The efficiency of the nanofiber filter can be determined by the performance to filter particles [5, 6]. Hence, it is necessary to have appropriate test particles for the filtration investigation. On the application of air and water filtration, the commonly used test particles are polystyrene latex (PSL). PSL particles that are spherical and uniform [7–11] make them possible for applications in many areas, such as the fabrication of hollow template [8, 9] and standard particles for calibrating aerosol instrument [12, 13].

PSL particles with nanometer size can be used for air filtration while those with nanometer-micrometer size can be applied in water filtration. The need of PSL particles with various sizes was limited by the expensive production cost. Therefore, there must be an innovative way to produce PSL particles more efficiently at less expensive cost. In previous studies, many synthesis methods have

Content from this work may be used under the terms of the Creative Commons Attribution 3.0 licence. Any further distribution of this work must maintain attribution to the author(s) and the title of the work, journal citation and DOI.
Published under licence by IOP Publishing Ltd

been developed [14–17]. One commonly used method is a simple emulsion polymerization method without any addictive substance. It is a synthesis method in the liquid phase using styrene monomers being diluted in distilled water [9]. The advantage of this method is that the size of particles can be controlled by varying the parameters such as the process temperature, the concentration of styrene monomer, and the initiator concentration [8–11]. However, the polymerization process takes 24 h, which is relatively time-consuming.

This paper reports the synthesis of PSL particles that was conducted using addictive-free emulsion polymerization method for 10 h. Potassium persulphate was used as the initiator of the polymerization process in order to reach the larger size of particles instead of using 2,2-azobis dihydrochloride (AIBA) [8]. The size of PSL particles was expected to be from nanometer to micrometer, which could be achieved by adjusting the concentration of styrene monomer. Based on the data, an equation could be obtained to predict the size of particles. The predictive equation for the AIBA initiator has been obtained [9]. However, for the potassium persulphate initiator, there is no report. Therefore, we will discuss in more detail about the equation.

2. Materials and methods

2.1. Materials

The material used in this study was styrene monomer while the chemicals were sodium hydroxide (NaOH) and potassium persulphate. Styrene monomer and sodium hydroxide were purchased from PT. Brataco (Bandung, Indonesia) while potassium persulphate was obtained from Merck. The solvent was distilled water, which was obtained from Chemistry Laboratory of Institut Teknologi Bandung (ITB).

2.2. Synthesis of PSL particles

PSL particles were synthesized using the liquid phase method. This method is also called as the conventional emulsion polymerization in which the styrene monomer was dissolved in distilled water, and potassium persulphate was used as the initiator [8, 9]. The polymerization process was conducted in a batch reactor system and under the condition of nitrogen atmosphere. The batch reactor system is shown in Figure 1 consisting of a four-necked batch-glass reactor with a capacity of 300 mL, a magnetic stirrer, a mantle heater and a temperature controller, a thermometer, and nitrogen gas. The first step was to purify the styrene monomer from the inhibitor by washing it using a solution of NaOH (3 M). The NaOH solution was obtained by dissolving NaOH powder into the distilled water until homogeneous. It was then inserted into the Erlenmeyer flask and bubbled with nitrogen gas for 15 min (flow rate of 1 lpm). The styrene monomer was then mixed with the base solution in the Erlenmeyer flask and then shaken for 30 min. There were two different phases seen after shaking, the clear solution at the bottom and yellow solution on the top. The yellow solution was the pure solution of styrene monomer that will be used in the subsequent polymerization process. The distilled water was inserted into the reactor system while stirred and heated to a specified temperature under the atmospheric condition of nitrogen gas for approximately 15 min. After reaching the intended temperature, the previously purified solution of styrene monomer was then added to the reactor system and then waited for 10 min for styrene monomer to be uniformly dispersed in the solution. To start the polymerization process, potassium persulphate was added to the solution of styrene in the distilled water. The solution was maintained at a constant temperature for 10 h under nitrogen gas atmosphere. Finally, the solution was then cooled until it reached the ambient temperature.

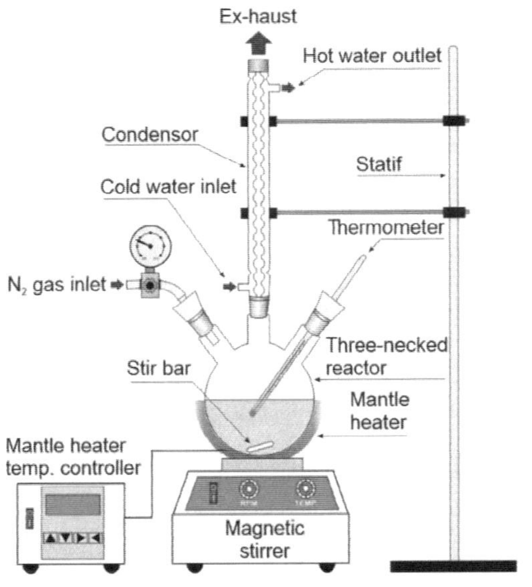

Figure 1. Scheme of the batch reactor system.

2.3. Characterization of PSL particles

The morphology and size of the obtained PSL particles were observed using a scanning electron microscope (SEM) (Hitachi, SU3500). The measurement and analysis of the average diameter and the size distribution of PSL particles were performed using SEM image analysis with the help of ImageMIF software version 3.0 developed in the Integrated Laboratory of Materials and Instrumentation, ITB. The diameter was measured at two hundred different particles in the SEM image. The size homogeneity of PSL particles could be determined by calculating the coefficient of variation (CV). Note that the coefficient of variation is the ratio of the standard deviation (σ) of diameter to the average diameter (μ) of the PSL particles. If the CV is greater than 0.3, then the particles are classified as inhomogeneous; otherwise, it will be homogeneous particles [1].

3. Results and discussion

Table 1 shows three reaction parameters, which are the concentration of styrene monomer, the concentration of potassium persulphate initiator, and the solution temperature in the reactor-batch, that were used to obtain PSL particles with several diameters. The adjustment of the parameters is very influential in controlling the size of PSL particles produced by the conventional emulsion polymerization methods [10, 11, 17]. However, this study focused on the synthesis of PSL particles by varying the styrene monomer concentration while the initiator concentration and the solution temperature were kept constant. Figure 2 shows the SEM images and the distribution of PSL particles for several different styrene monomer concentrations at the solution temperature of 80 °C. It can be seen that the morphological structure of the PSL particles was spherical and uniform. The increase in the concentration of styrene monomer had no effect on the shape of PSL particles. Similar results were also reported by previous papers [8, 9] that the shape of PSL particles was spherical and uniform, even when other parameters such as the solution temperature and the initiator concentrations were varied [8–11]. Table 1 gives the average diameters of the PSL particles along with the CVs for different styrene monomer concentrations and different temperatures. The calculated CVs implied that the obtained PSL particles were homogeneous or monodisperse.

Table 1. The sizes of PSL particles and their synthesis parameters.

Sample	Concentration of Monomer (wt.%)	Concentration of Initiator (wt.%)	Temperature (°C)	Average Diameter of Particles (nm)	Coefficient of Variation (CV)
PSL 1	4	0.04	55	281	0.05
PSL 2	9	0.04	55	295	0.03
PSL 3	11	0.04	55	451	0.08
PSL 4	12	0.04	55	941	0.14
PSL 5	0.5	0.04	80	160	0.05
PSL 6	1.2	0.04	80	216	0.13
PSL 7	2.0	0.04	80	295	0.06

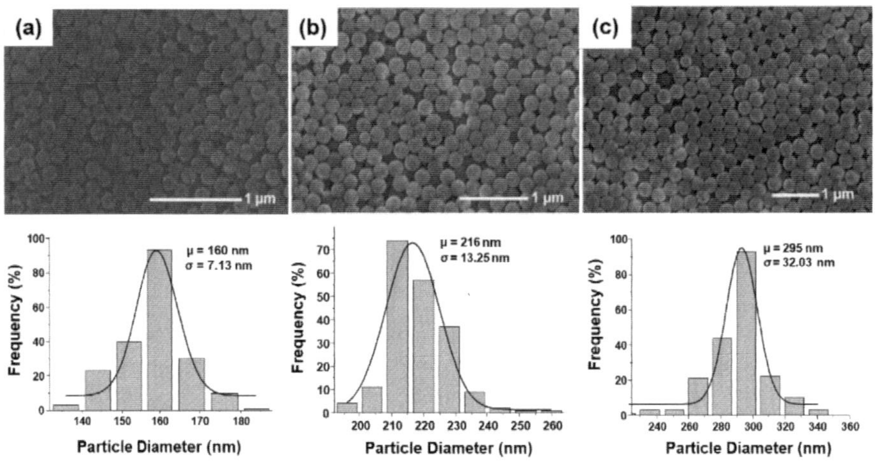

Figure 2. SEM images and distribution of PSL particles at temperature of 80 °C: (a) PSL 5, (b) PSL 6, (c) PSL 7.

Figure 3 depicts the relationship between the styrene monomer concentration and the diameter of PSL particles at different temperatures. The large PSL particles within (281 to 941) nm were obtained under the low polymerization temperature (temperature of 55 °C). Otherwise, the size of PSL particles was within (160 to 295) nm when the polymerization temperature as high as 80 °C, which are similar to the previous reports [7, 8]. However, the time and the number of styrene monomers used could be reduced by synthesizing at different temperatures.

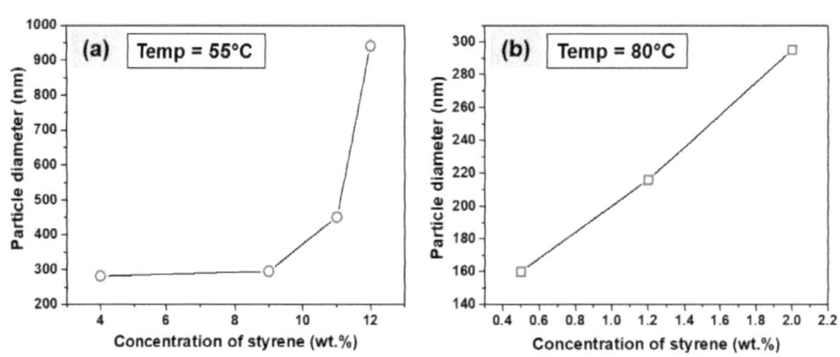

Figure 3. The particle size of PSL with varying concentration of styrene monomer at temperatures of (a) 55 °C and (b) 80 °C.

In Figure 3 (a), the high polymerization temperature (80 °C) resulted in an increased interaction and change in the reactant components, which were styrene monomer and initiator [16]. Therefore, the core or the nucleus was formed appropriately. This process requires a lot of monomers and oligomers in the early stage of polymerization since the number of free monomers and oligomers will decrease significantly. This results in the retardation in the formation of particles and production of small particles [10, 11]. For the process using the low polymerization temperature (55 °C) as depicted in Figure 3 (b), the decomposition of the initiator took a longer time that caused the formation of the oligomer to be slower and then resulted in a low number of the nuclei [18]. At this condition, the core would capture a larger amount of oligomers/monomers hence larger particles were produced [9]. It can be seen from the graph that when the temperature used was 55 °C, the size of PSL particles increased but the data obtained were an anomaly. The results agreed to a certain degree with the previous paper [9]. It was found that the use of polymerization temperature below 60 °C caused the polymerization to be imperfect and the produced trend data were less satisfying. The low polymerization temperature also led to lower interactions between components reaction (initiator, styrene monomer, distilled water) so that the process of particles formation became worst.

The size of PSL particles could be predicted by considering Equation 1 [10].

$$D_p = 71.80(S) + 15.85\ln(I) - 170.30\ln(T) + 847.87 \qquad (1)$$

where *Dp* is the particle diameter (nm), *S* is the concentration of styrene monomer (wt%), *I* am the initiator concentration (wt%), and T is the temperature of solution (°C). However, the diameters of the synthesized PSL particles were different from the predictive equation in Equation 1. The PSL particles synthesized at the polymerization temperature of 80 °C had an average diameter larger than that predicted by Equation 1. This is because the present experiment used potassium persulphate as the initiator while Equation 1 was obtained using the 2,2-azobis dihydrochloride (AIBA) initiator. This diameter difference has also been reported previously [8] where the average diameter of the PSL particles synthesized using the potassium persulphate initiator was greater than that done using the AIBA one.

4. Conclusions

Monodisperse particles of polystyrene latex (PSL) have been synthesized using a simple emulsion polymerization method (without the addictive substance) with distilled water as the solvent and potassium persulphate as the initiator. Using this method, the average diameter of PSL particles could be controlled by varying the solution temperature and the styrene monomer concentration. It has been found that the diameter of the synthesized PSL particles was in the range of (160 to 941) nm. The higher the temperature of the solution, the smaller the diameter of the PSL particles was. The greater the styrene monomer concentration, the greater the diameter of the PSL particles was. Spherical and homogeneous PSL particles can, therefore, be used as pollutants for the applications of air and water filtration systems.

Acknowledgments

This research was financially supported by Directorate of Research and Community Engagement of Ministry of Research, Technology and Higher Education, the Republic of Indonesia under the University's Excellent Research (PUPT) Grant in the fiscal year 2015–2017, and "Reset & Inovasi KK ITB" Grant in the fiscal year of 2016–2017.

References

[1] Matulevicius J, Kliucininkas L, Prasauskas T, Buivydiene D, and Martuzevicius D 2016 The comparative study of aerosol filtration by electrospun polyamide, polyvinyl acetate, polyacrylonitrile and cellulose acetate nanofiber media *Journal of Aerosol Science* **92** 27 – 37

[2] Zhang Q, Welch J, Park H, Wu C Y, Sigmund W, and Marijnissen J C M 2010 Improvement in

nanofiber filtration by multiple thin layers of nanofiber mats *Journal of Aerosol Science* **41** 230 – 236

[3] Munir M M, Iskandar F, Khairurrijal and Okuyama K 2008 A constant-current electrospinning system for production of high quality nanofibers *Review of Scientific Instruments* **79** 093904

[4] Munir M M, Iskandar F, Khairurrijal, and Okuyama K 2009 High performance electrospinning system for fabricating highly uniform polymer nanofibers *Review of Scientific Instruments* **80** 026106

[5] Gopal R, Kaur S, Feng C Y, Chan C, Ramakrisna S, Tabe S, and Matsuura T *2007* Electrospun nanofibrous polysulfone membranes as pre-filters: Particulate removal *Journal of Membrane Science* **289** 210 – 219

[6] Munir M M, Ekaputra M P, Rajak A, Rahma A, Nuryantini A Y and Khairurrijal 2015 Optimization of Solvent System and Polymer Concentration for Synthesis of Polyvinyl Alcohol (PVA) Fiber Using Rotary Forcespinning Technique *Advanced Material Research* **1123** 281 – 284

[7] Liu B, Zhang M, Wu G, and Zhang H 2016 Synthesis of large-scale, monodisperse latex particles via one-step emulsion polymerization through *in situ* charge neutralization *Colloids and Surfaces A: Physicochem Eng.* **500** 127 – 136

[8] Nandiyanto A B D, Suhendi A, Ogi T, Umemoto R, and Okuyama K 2014 Size-and charge-controllable polystyrene spheres for templates in the preparation of porous silica particles with tunable internal hole configurations *Chemical Engineering Journal* **256** 421 – 430

[9] Nandiyanto A B D, Suhendi A, Ogi T, Iwaki T, and Okuyama K 2012 Synthesis of additive-free cationic polystyrene particles with controllable size for hollow template applications *Colloids and Surfaces A: Physicochem Eng.* **396**, 96 – 105

[10] Goodwin J W, Hearn J, Ho C C, and Ottewill R H 1974 Studies on the preparation and characterisation of monodisperse polystyrene laticee *Colloid & Polymer Sci.* **252** 464 – 471

[11] Goodwin J W, Ottewill R H, and Pelton R 1979 Studies on the preparation and characterization of monodisperse polystyrene latices V.: The preparation of cationic latices *Colloid & Polymer Sci.* **257** 61 – 69

[12] Chien C, Theodore A, Wu C, Hsu Y, Birky B 2016 Upon correlating diameters measured by optical particle counters and aerodynamic particle sizers *J. Aerosol Sci.* **101** 77 – 85

[13] Raabe O G, Braaten D A, Axelbaum R L, Teague S V, Cahill T A 1988 Calibration studies of the drum impactor *J. Aerosol Sci.* **19** 183 – 195

[14] Sajjadi S 2015 Control of particle size by feed composition in the nanolatexes produced via monomer-starved semicontinuous emulsion copolymerization *J. Colloid Interface Sci.* **445** 174 – 182

[15] Zhang Q, Han Y, Wang W, Song T, and Chang J 2010 A theoretical and experimental investigation of the size distribution of polystyrene microspheres by seeded polymerization *J. Colloid Interface Sci.* **342** 62 – 67

[16] Yamamoto T, Nakayama M, Kanda Y, and Higashitani K 2006 Growth mechanism of soap-free polymerization of styrene investigated by AFM *J. Colloid Interface Sci.* **297** 112 – 121

[17] Goodwin J W, Ottewill R H, Pelton R, Vianello G, and Yateo D E 1978 Control of particle size in the formation of polymer latices *Polym. Br. J.* **10** 173 – 180

[18] Tseng C, Lu Y, Aasser M E, and Vanderhoff J 1986 Uniform polymer particles by dispersion polymerization in alcohol *J. Polym Sci. A* **24**, 2995 – 3007

IOP Conf. Series: Materials Science and Engineering **367** (2018) 012016 doi:10.1088/1757-899X/367/1/012016

Deformation of Ferrogel Based on *Carboxyl Methyl Cellulose* (CMC)/*Polyvinyl Alcohol* (PVA) Hydrogel

Sunaryono[1,2], M. N. Kholifah[1], Yudyanto[1,2], A. Taufiq[1,2], N. Mufti[1,2], R. Wulandari[3], Munasir[4], and M. Diantoro[1,2]

[1]Department of Physics, Faculty of Mathematics and Natural Sciences, Universitas Negeri Malang, Jl. Semarang No 5, Malang 65145, Indonesia
[2]Research Center of Minerals and Advanced Materials, Faculty of Mathematics and Natural Sciences, Universitas Negeri Malang, Jl. Semarang 5 Malang 65145
[3]Engineering Physics Study Program, Faculty of Industrial Technology, Institut Teknologi Bandung, Jl. Ganesha 10 Bandung 40132, Indonesia
[4]Departement of Physics, Faculty of Mathematics and Natural Sciences, Universitas Negeri Surabaya, Ketintang Surabaya 60231, Indonesia

E-mail: sunaryono.fmipa@um.ac.id

Abstract. Ferrogel is a smart material that combines the elastic properties of hydrogel and magnetic behavior from the filler of magnetic materials. In this work, ferrogel was fabricated through a freezing-thawing process with matrix materials of Carboxyl Methyl Cellulose (CMC)/Polyvinyl Alcohol (PVA) and a filler of magnetite. Magnetite nanoparticles were successfully synthesized by using a coprecipitation method at room temperature. The magnetite filler was investigated via X-ray Diffractometer (XRD) and Transmission Electron Microscopy (TEM), respectively, to analyze the structure and morphology of the magnetite. Based on the XRD analysis, the magnetite nanoparticle was matched by ICSD number of 9013529 and has an average particle size of about ~ 9.67 nm. This result is well confirmed by the TEM characterization that showed an average particle size of 9.32 nm. Furthermore, ferrogel was explored by using a Vibrating Sample Magnetometer (VSM) and an electromagnetic instrument to measure the magnetic properties and deformation of ferrogel, respectively. The VSM analysis showed that CMC/PVA hydrogel-based ferrogel has a superparamagnetic behavior due to its magnetization remanent and field coercivity that is close to zero. Interestingly, the deflection of Fe_3O_4/CMC/PVA ferrogel under the effect of the external magnetic field exhibited a good response. The deformation of ferrogel significantly increased from 4.9 cm to 7.0 cm as the filler of magnetite in the ferrogel was increasing from 14 wt% to 22 wt%. This phenomenon could be improved and explored to change the ferrogel into an artificial muscle by engineering the concentration and magnetite response in the CMC/PVA hydrogel.

Keywords. Carboxyl methyl cellulose (CMC), deformation, ferrogel, hydrogel, magnetite, and polyvinyl alcohol (PVA).

Content from this work may be used under the terms of the Creative Commons Attribution 3.0 licence. Any further distribution of this work must maintain attribution to the author(s) and the title of the work, journal citation and DOI.
Published under licence by IOP Publishing Ltd

1. Introduction

Ferrogel is a leading material of which compositions offer elastic and magnetic properties. Ferrogel based on polyvinyl alcohol (PVA) polymer is superparamagnetic and non-toxic so that it could be used for biomedical applications, such as artificial muscle and drug delivery [1]. For such applications, the nanomagnetic particles function as fillers that are added to the PVA hydrogel. The nanomagnetic particle is exciting to be developed since it has enormous potential to be used for biomedical, hyperthermia, artificial muscle, bandage, and so forth [2]. A combination of polymer with a magnetic material causes a change in ferrogel morphology when it is a magnetic field [3].

Numerous investigations on ferrogel using various polymer materials and fabrication methods have been reported. Teixeira *et al* [4] studied the effects of sodium alginate on sodium alginate-based ferrogel material combined with magnetic material as its filler. They have successfully characterized the water absorption in the gel networks. An inclusion of magnetic filler into alginate gel polymer networks can result in a higher level of water absorption and thermal stability than a pure alginate gel.

On previous research, the researchers had also successfully synthesized and characterized polyacrylic acid (PAA)/polyvinyl alcohol (PVA)-based magnetic hydrogel using Fe_3O_4 for the magnetic filler via a freezing-thawing (F-T) method [5]. The F-T process influences the distribution pattern of the magnetite particles inside the PAA/PVA hydrogel. The results of the experiment showed that the F-T process that is repeated for seven times is mostly optimum for the magnetite distribution inside the PAA/PVA hydrogel. However, research on magnetite behavior in Carboxyl Methyl Cellulose (CMC) and Polyvinyl Alcohol (PVA)-based ferrogel has rarely been reported.

Thus, fabrication of CMC/PVA matrix polymer-based ferrogel is challenging to be performed. This research focused on the deformation behavior of the nanomagnetic particles in the CMC/PVA hydrogel matrix to provide additional references regarding its potential application for artificial muscle. The selection of CMC and PVA matrix in this report cannot be separated from their superior characteristics. PVA is a polymer material with a high viscosity, solubility, and translucence. The excellence of CMC is shown by its high viscosity and resistance to low humidity. Also, CMC is stable, even more, stable than metal, though (almost as tough as cast iron), wear resistance, and it has stable chemical elements at high temperature, high strength and resistance, and corrosion resistance.

2. Materials and methods

2.1. *Synthesis of Nanomagnetic Particle and Ferrogel*

This research used iron sands taken from Wedi Ireng Beach Banyuwangi East Java for the primary material of the magnetic filler. The Fe_3O_4 nanoparticles were synthesized using a coprecipitation method at room temperature. The synthesis result of the magnetite was then prepared to be filler for ferrogel fabrication.

The fabrication of ferrogel was started by combining the hydrogel matrix materials, i.e., CMC, PVA, and water in a specific ratio. The hydrogel materials were mixed on a magnetic hot plate using a stirrer at approximately 70 °C to 90 °C. When the hydrogel matrix had been mixed up well, the magnetite filler was then distributed into the hydrogel via a stirring process until it achieved a dark-colored mixture. The last stage of the ferrogel fabrication was putting the mixture of hydrogel and magnetic filler into a mold to be further processed via a freezing-thawing method.

2.2. *Nanomagnetic Particles and Ferrogel Characterizations*

The magnetite filler obtained by the coprecipitation method was then characterized by using an X-ray Diffractometer (XRD) and Transmission Electron Microscopy (TEM). A magnetite test by XRD was done to identify the structural pattern, lattice parameter, and crystal size. TEM characterization aimed to identify the morphology of the magnetite filler while confirming the results of the XRD test.

CMC/PVA hydrogel-based ferrogel was characterized by using a Vibrating Sample Magnetometer (VSM) and magneto-elasticity test equipment. The VSM test was carried out to identify the pattern of the ferrogel magnetization representing the magnetite filler distributed in the CMC/PVA hydrogel.

The magneto-elasticity test equipment was used to identify the mechanical characteristics of the ferrogel, especially the behavioral deviations of the CMC/PVA-based ferrogel.

3. Results and discussion

3.1. XRD Pattern of Magnetite Filler

Figure 1 presents the profile of the Fe_3O_4 particles made from iron sand characterized via an XRD. The search-match results showed that all of the sample peaks had been well detected and the formed diffraction patterns were dominated by Fe_3O_4, which was in accordance with the data of Inorganic Crystal Structure Database (ICSD) model number 9013529. Further, the hkl field position of the Fe_3O_4 particle has successfully been confirmed by using the OriginPro-8 software.

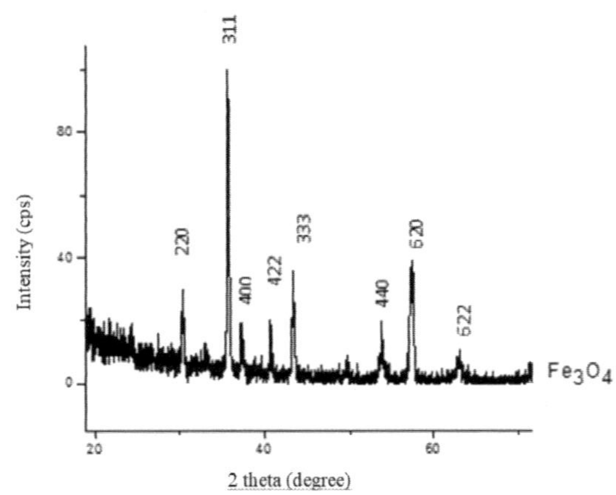

Figure 1. Position of hkl field for iron sand separation

The analysis of the Fe_3O_4 particle size based on the XRD results has been conducted successfully (Figure 2) through the Debye Scherer equation below.

$$D = \frac{K\lambda}{\beta \cos\theta} \tag{1}$$

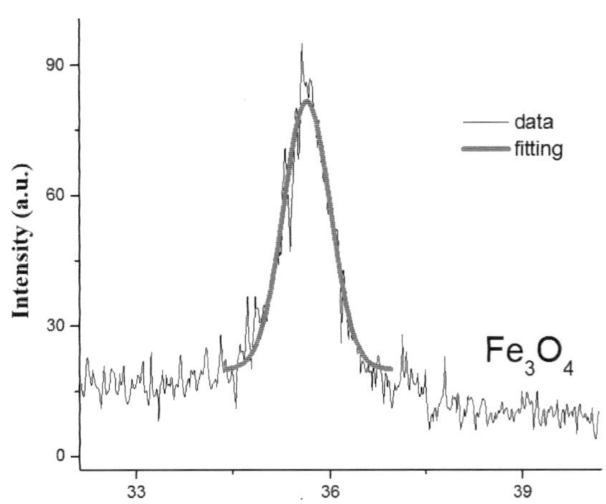

Figure 2. Fitting results of Fe_3O_4 nanoparticle diffraction peak pattern by the origin pro-8 software

Where K is constant, λ is X-ray wavelength, and β is the full width at half maximum of the peak [6]. The calculation of the average Fe_3O_4 crystal size through the Scherer equation utilized the principle of the relationship between the crystal size and the 2θ position, the half-width of the peak, and the wavelength of the XRD source. The Fe_3O_4 particle size obtained from such equation was approximately 9.67 nm. Such result is lower than that of the previous research, i.e., around 12 nm [7]. However, the magnetite particle size in this research almost corresponds to the magnetite particle size reported by Saravanan *et al* [8] who stated that magnetite particle size of around 7 nm to 9 nm.

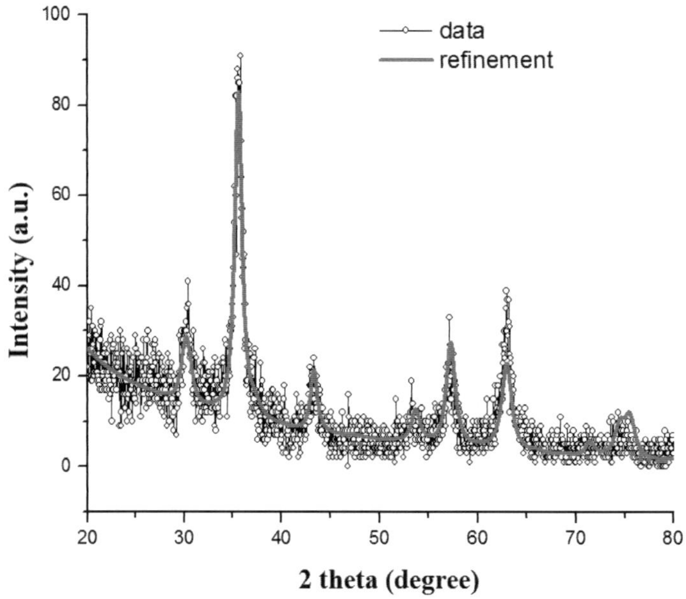

Figure 3. Fe_3O_4 nanoparticle diffraction pattern by using Rietica Software

Further, Figure 3 shows the synthesized Fe_3O_4 diffraction pattern after being analyzed by using Rietica software. Based on the results of the refinement, the Fe_3O_4 particle size was 9.46 nm with an R_{WP} value of around 14.59 and χ^2 of 0.042. Further, the average size of the particle conforms to the Debye Scherer analysis as has been explained beforehand.

3.2. TEM Characterizations

TEM characterization was done to further characterize the magnetite filler morphology and size before being combined with the CMC/PVA hydrogel. The morphology of the magnetite particles that had been successfully characterized (Figure 4) showed a moderately uniform pattern of the particle size. Such results were further analyzed using ImageJ software, and the analysis results presented that the average particle size of the Fe_3O_4 was 9.32 nm. Such results confirm the average particle size shown by the XRD results either via a refinement by Reitica software showing a 9.46 nm average value or by Scherer equation showing a 9.67 nm average value.

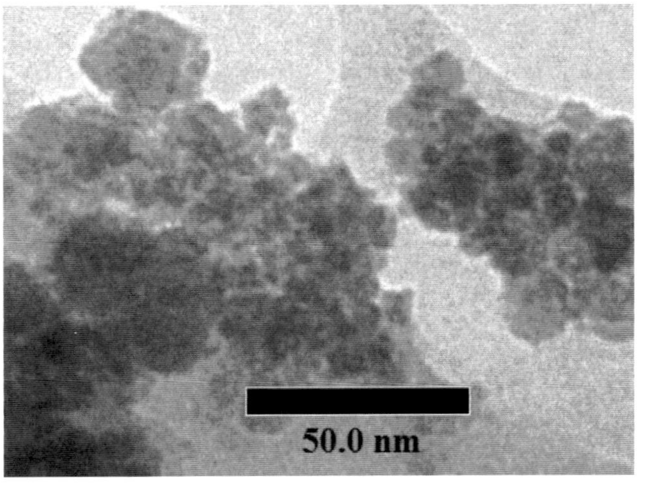

Figure 4. Fe_3O_4 morphology shown by TEM characterization

Besides, such average particle size of Fe_3O_4 is approaching the average value obtained by Shen *et al* [9], i.e., ~ 8 nm and El Ghador *et al* [10] of 10.59 nm. They claimed that their research resulted in uniform cube-shaped and monodisperse nanoparticles.

3.3. VSM Data Analysis

The magnetization characterization of the CMC/PVA hydrogel-based ferrogel via VSM had been successfully conducted. The obtained data from such VSM characterization were in the forms of magnetization remanent value (*Mr*), coercivity magnetization value (*Hc*), and saturation magnetization value (*Ms*) as displayed in Table 1.

Figure 5 shows the results of the VSM characterization refinement of the ferrogel through two treatments that employed a different number of F-T process repetitions, namely 5 and 6 times. Based on Figure 5, the ferrogel with 5 repetitions of the F-T process has the highest saturation magnetization value of 22.16 $emu \cdot g^{-1}$. Meanwhile, the ferrogel with 6 repetitions of the F-T process has a saturation magnetization value of 21.27 $emu \cdot g^{-1}$. Furthermore, the remanent and coercivity magnetization of the two ferrogel samples obtained its minimum value, and it tended to have a relatively low value that is close to zero. In accordance with the report [11], such results indicated that CMC/PVA hydrogel-based ferrogel has a superparamagnetic behavior.

The superparamagnetic behavior indicated that the magnetite filler is used for controlling ferrogel magnetization. Also, the CMC/PVA-based ferrogel has a relatively high value of saturation magnetization, namely around 21.27 $emu \cdot g^{-1}$ (with 6 times F-T) and 22.16 $emu \cdot g^{-1}$ (with 5 time F-T) when compared with the saturation magnetization of PVA-based ferrogel of 6.52 emu/g [7] and Polyacrylamide-based ferrogel of below 0.5 $emu \cdot g^{-1}$with a magnetite percentage range of around 0.75 % from its weight [12]. A high level of saturation magnetization of CMC/PVA-based ferrogel is influenced by, among others, the magnetite behavior of the CMC/PVA hydrogel. The magnetite filler tends to experience aggregation caused by its polymeric property. CMC polymer has a high viscosity level that is not affected by low humidity. Such fact requires the magnetite particles to consume high energy to be fairly distributed inside the semi-crystalline networks of the CMC/PVA hydrogel. Therefore, it prevents the particle distribution that results in aggregation causing an increase in the saturation magnetization level.

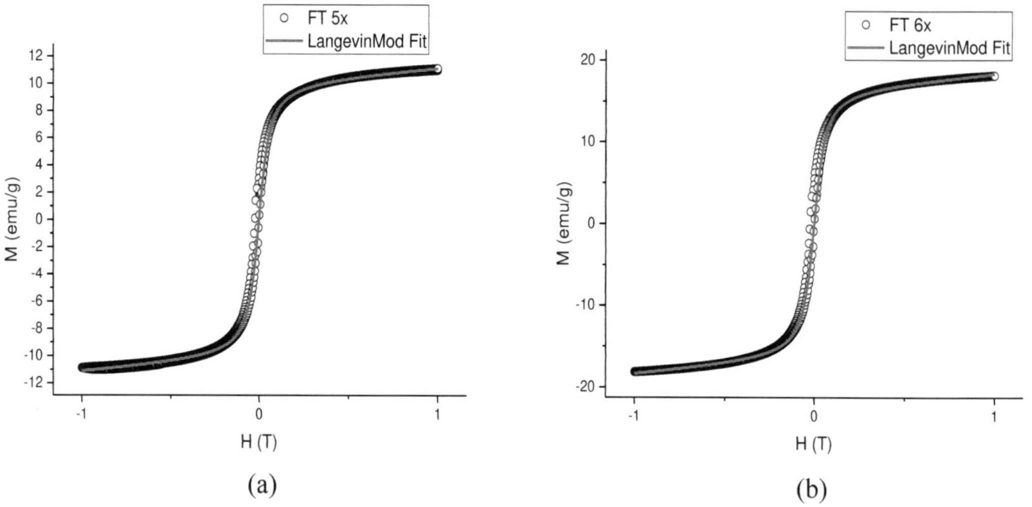

Figure 5. VSM Profile of Fe_3O_4/CMC/PVA Ferrogel based on the number of freezing-thawing cycles repetition, i.e., 5 (a) and 6 times (b). The red line is experimental data fitting by the Langevin function (L).

Table 1. Magnetization value of Fe_3O_4/CMC/PVA Ferrogel by a LangevinMod model (LM)

Sample	M_r (emu·g^{-1})	M_s (emu·g^{-1})	H_c (T)	μ (10^{-19} J/T)	D (nm)
FT 5x	0.064	22.162	0.016	8.9	6.72
FT 6x	0.007	21.273	0.081	9.2	6.89

The results of VSM characterization and refinement on the average size of the magnetite particle have also been successfully obtained. Table 1 shows the average size of the magnetite particles contained in CMC/PVA hydrogel.

$$D = \left(\frac{18 k_B T}{\pi} \frac{\chi_i}{\rho M_S} \right)^{\frac{1}{3}} \tag{2}$$

Where χi is the initial susceptibility of the sample, ρ is the mass of the magnetite filler, i.e., 5670 kg·m^{-3} (based on the Rietica analysis), and saturation magnetization (emu·g^{-1}).

The refinement showed a smaller average particle size compared with those taken from the data of XRD and TEM characterizations. It is due to two factors. First, the powder sample morphology used in the XRD and TEM characterizations while the VSM characterization employed gel sample morphology after inclusion into the CMC/PVA polymer networks. Second, the use of a dark-field mode on VSM [14] caused construction of a slightly smaller particle dimension than that caused by the bright-field mode for sample morphology observation. The results of the VSM refinement also resulted in the value of magnetic moment shown in Table 1. Based on such data, the filler magnetic moment value of Fe_3O_4 conforms to the filler magnetic moment value of Fe_3O_4 nanoparticles that was reported by Prajapat *et al* [15 – 17]. According to the theoretical review, the average particle size is already proportional to the particle magnetic moment on a single domain area. A magnetic nanoparticle size can be classified to be in a single domain area if it is below a critical size [18]. Also, according to a report by Tong *et al* [19], a critical size of a particle is around 25 nm. With such value, the magnetite filler used in this research is categorized into a single domain. Hence, the bigger the particle size, the more the magnetic moment that could be obtained. It is in excellent agreement with the data obtained from the samples with 5x FT and 6x FT. However, a correction was performed on the data of the 6x FT sample saturation magnetization, where the sample showed a reduction when its magnetic moment increased.

3.4. Data Analysis of Mechanical Properties

The potential application for artificial muscle of CMC-PVA-based ferrogel could be observed from its magnetic field threshold mechanical response under the influence of an external magnetic field. The deflection of the magnetic field threshold investigation on ferrogel is illustrated in Figure 6. The ferrogel was made static on its upper point, and it experienced a position changing as its reaction against an external magnetic field from its lower point.

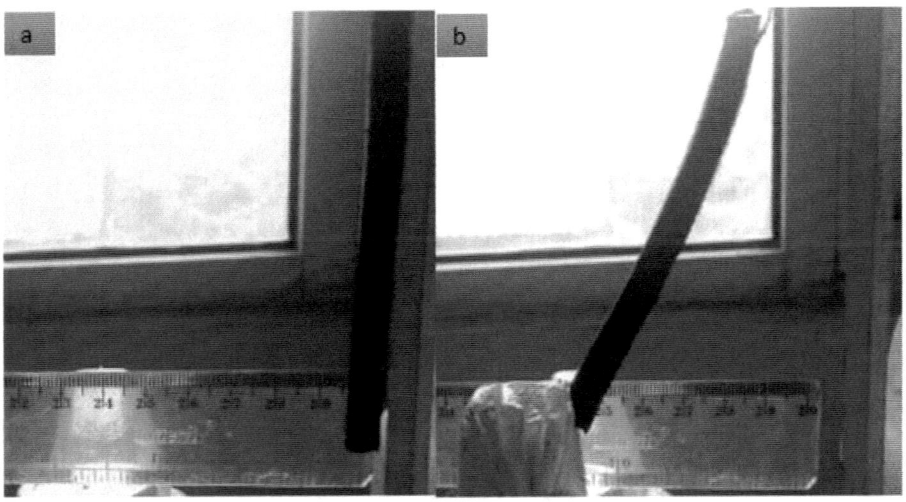

Figure 6. Deformation of Fe_3O_4/CMC/PVA ferrogel; a). Before; b). After an external magnetic field was given

The observation results of ferrogel responses to an external magnetic field are presented in Figure 7 and Table 2 by showing the threshold value of the magnetic field. Such values represent the minimum force of the magnetic field that is necessary to move the deviation to its maximum position. The ferrogel deflection will suddenly increase a moment after it goes beyond the threshold value of the magnetic field. Based on the results of the experiment, the threshold value of the magnetic field experienced a decrease along with the increasing magnetite filler concentration contained in the ferrogel. Such results conform to the results reported in previous studies conducted by Sunaryono *et al* [20] and Rmanujan *et al* [21]. Table 2 shows a 14 wt% filler concentration, with up to 26.2 mT magnetic field threshold value of the ferrogel. Such results are slightly higher than those previously reported in which the magnetic field threshold of the PVA polymer-based ferrogel was recorded as around 12.6 mT for 15 wt% magnetite filler concentration (Sunaryono *et al*) [20] and 15.2 mT for 10 wt% concentration as reported by Ramanujan *et al* [21].

A higher amount of the magnetic field threshold in a 14 wt% concentration than the previous report (Sunaryono *et al*) [20] showed that the magnetic response of the CMC/PVA-based ferrogel is much weaker than PVA-based ferrogel. It is possible under the influence of a CMC polymer that functions as a matrix in the ferrogel. A high viscosity level of CMC polymer will increase the ferrogel density, which results in a compact semi-crystalline space of the CMC/PVA polymer in ferrogel. Such condition will result in a lower magnetic response of the CMC-PVA-based ferrogel compared with the PVA-based ferrogel.

Besides the magnetic field threshold report, an analysis of the ferrogel mechanical response has also been confirmed through the ferrogel deformation test via an electromagnetic instrument. Table 2 presents the characterization results of the CMC/PVA-based ferrogel deformation with a variation of magnetite filler concentrations. The deviation pattern of ferrogel increased along with the increasing content of magnetite filler in the ferrogel. Therefore, the higher the Fe_3O_4 filler concentration, the more sensitive the ferrogel to any changes from the external magnetic field. It is in good agreement with the report stating that the concentration of Fe_3O_4 nanoparticles significantly affects the resultant changes in the composite material [22].

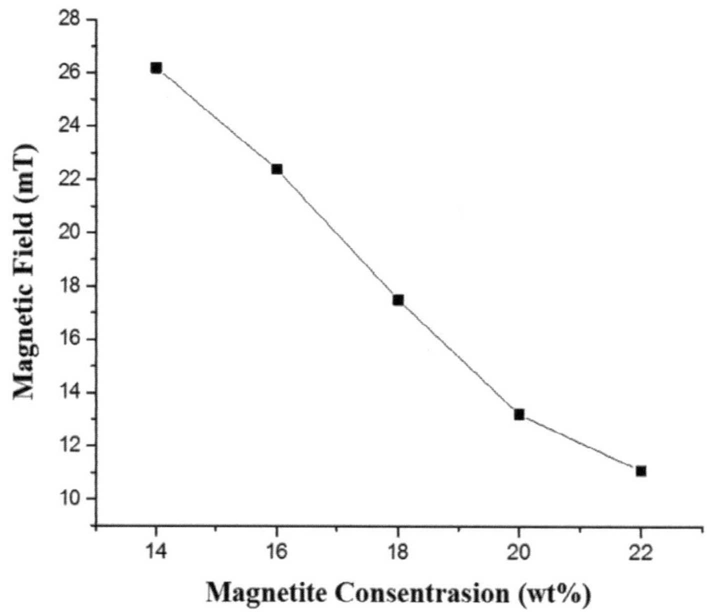

Figure 7. The magnetic field threshold value versus the Fe_3O_4 filler concentration in CMC/PVA-based ferrogel.

Ferrogel sensitivity is a crucial variable in the application of artificial muscle. In this research, the mechanical properties of the ferrogel could result in deformation from 4.9 cm to 7 cm from the average magnetite fillers of 14 wt% until 22 wt%. Such results were the maximum deformation of ferrogel as its response to the influence of an external magnetic field. Thus, the CMC/PVA-based ferrogel is a leading material with a high density and sensitivity that is potential for several technology engineering applications.

Table 2. The deflection of Fe_3O_4/CMC/FVA Ferrogel versus the magnetite filler concentration

Deflection (cm)	Magnetic Threshold (mT)	Filler Concentration (wt%)
4.9	26.2	14
5.2	22.4	16
5.5	17.5	18
6.0	13.2	20
7.0	11.1	22

4. Conclusions

The characterization of a magnetite nanoparticle that is used as filler in the CMC/PVA-based ferrogel composite has successfully been conducted, and it shows a moderately uniform pattern of the particle size with an average particle size of 9.32 nm. Such results are well confirmed by the results of the XRD analysis. The mechanical characteristic showed a declining magnetic threshold of the ferrogel from 26.2 mT to 11.1 mT along with an increasing filler concentration of 14 wt% to 22 wt%. Furthermore, the deformation of ferrogel also resulted in an optimum deviation pattern from 4.9 cm to 7 cm from the average magnetic filler of 14 wt% to 22 wt%. Such results showed the effectiveness of magnetite filler in carrying an inclusion of the magnetic property in the CMC/PVA polymer networks. Besides, the matrix characteristics of the CMC/PVA materials also affect the level of the ferrogel density that is potential for artificial muscle application.

Acknowledgments
The authors would like to thank for Rector of UM due to the funding of this Research was totally supported by *Penelitian Terapan Unggulan Perguruan Tinggi* UM 2018.

References
[1] Gonzalez J S, Hoppe C E, Muraca D, Sánchez F H, and Alvarez V A 2011 Synthesis and characterization of PVA ferrogels obtained through a one-pot freezing-thawing procedure, *Colloid. Polym. Sci.* **289** 1839–1846
[2] Cheng F, Li Y, and Chen D 2013 Energy Consumption and Morphological Development of Eucalyptus Alkaline Peroxide Mechanical Pulp by Carboxymethyl Cellulose-assisted Refining *BioResources* **8** 2173–2185
[3] Kubisz L, Skumiel A, Hornowski T, Szlaferek A, and Pankowski E 2008 The effect of temperature on the electric conductivity of poly (dimethyl siloxane) ferromagnetic gel, *J. Phys. Condens. Matter.* **20** p 204118
[4] Teixeira V F T *et al* 2014 Ion exchange kinetics of magnetic alginate ferrogel beads produced by external gelation *Carbohydr Polym.* **111** 198–205
[5] Sunaryono, Hifdziyah H, Taufiq A, Diantoro M, and Mufti N 2017 Optimization of Freezing-Thawing Process in Enhancing Magnetic Properties of Fe_3O_4/PAA/PVA Magnetic Hydrogel Composites *IOP Conf Series: Materials Science and Engineering* **202** p 012007
[6] Park J S, Kim K, Park Y G, Mo Y G, Kim H D, and Jeong J K 2009 Novel ZrInZnO Thin-film Transistor with Excellent Stability *Adv. Mater.* **21** 329–333
[7] Sunaryono *et al* 2016 Small-Angle X-Ray Scattering Study on PVA/Fe_3O_4 Magnetic Hydrogels, *Nano.* **11** pp 1650027
[8] Saravanan P, Alam S, Kandpal L D, and Mathur G N 200, Effect of substitution of Mn ion on magnetic properties of Fe_3O_4 nanocrystallites, *J. Mater. Sci. Lett.* **21** 1135–1137
[9] Shen Y F, Tang J, Nie Z H, Wang Y D, Ren Y, and Zuo L 2009 Tailoring size and structural distortion of Fe_3O_4 nanoparticles for the purification of contaminated water *Bioresour Technol.* **100** 4139–4146
[10] El Ghandoor H, Zidan H M, Khalil M M, and Ismail M I M 2012 Synthesis and some physical properties of magnetite (Fe_3O_4) nanoparticles *Int. J. Electrochem Sci.* **7** 5734–5745
[11] Shokrollahi H 2013 Structure, synthetic methods, magnetic properties and biomedical applications of ferrofluids *Mater. Sci. Eng. C.* **33** 2476–2487
[12] Blyakhman F A *et al* 2017 Polyacrylamide ferrogels with embedded maghemite nanoparticles for biomedical engineering *Results in Phys.* **7** 3624–3633
[13] Liu Z L, Liu Y J, Yao K L, Ding Z H, Tao J, and Wang X 2002 Synthesis and magnetic properties of Fe_3O_4 nanoparticles *J. Mater. Synth. Process.* **10** 83–87
[14] Yoon K Y *et al* 2016 Control of magnetite primary particle size in aqueous dispersions of nanoclusters for high magnetic susceptibilities *J. Colloid Interface Sci.* **462** 359–367
[15] Dyachenko S V, Vaseshenkova M A, Martinson K D, Cherepkova I A, and Zhernovoi A I 2016 Synthesis and properties of magnetic fluids produced on the basis of magnetite particles *Russ. J. Appl. Chem.* **89** 690–696
[16] Kumar A, Jana S, Svedlindh P, Sharma S K, Chaudhary S, and Pandya D K 2016 Anti-phase boundary-free two-dimensional epitaxial Fe_3O_4 thin films: evidence of an unquenched orbital magnetic moment at room temperature *J. Mater. Chem. C* **4** 9942–9946

[17] Prajapat C L *et al* 2016 Synthesis and magnetic study of carbon-coated iron oxide nanoparticles by laser ablation in solution *AIP Conference Proceedings* **1731** p 130011
[18] Carrey J, Mehdaoui B, and Respaud M 2011 Simple models for dynamic hysteresis loop calculations of magnetic single-domain nanoparticles: Application to magnetic hyperthermia optimization *J. Appl. Phys.* **109** p 083921
[19] Tong G, Liu Y, Cui T, Li Y, Zhao Y, and Guan J 2016 Tunable dielectric properties and excellent microwave absorbing properties of elliptical Fe_3O_4 nanorings *Appl. Phys. Lett.* **108** p 072905
[20] Sunaryono, Taufiq A, Munaji, Indarto B, Triwikantoro, Zainuri M, and Darminto 2013 Magneto-elasticity in hydrogels containing Fe_3O_4 nanoparticles and their potential applications *AIP Conference Proceedings* **1555** 53–56
[21] Ramanujan R V and Lao L L 2006 The mechanical behavior of smart magnet–hydrogel composites *Smart Mater. Struct.* **15** p 952
[22] Czaun M, Hevesi L, Takafuji M, and Ihara H 2008 A novel approach to magneto-responsive polymeric gels assisted by iron nanoparticles as nano cross-linkers *Chem. Commun.* **18** 2124–2126

Microstructure, Hardness, and Electrical Conductivity of β-TCP/Zirconia Composites Prepared from Eggshell

Hartatiek[1], Yudyanto[1], E. K. Velayati[1], Y. A. Hariyanto[1], N. Hidayat[1], A. Taufiq[1] and M. Diantoro[1]

[1]Department of Physics, Faculty of Mathematics and Natural Sciences, Universitas Negeri Malang, Jl. Semarang 5, Malang 65145, Indonesia

E-mail: hartatiek.fmipa@um.ac.id

Abstract. In this research, β-TCP composited with zirconia at different compositions were fabricated via solid-state reaction method. β -TCP was produced from eggshell. The crystal structure of the as-prepared samples was characterized using X-ray diffractometer (X-RD). Meanwhile, elemental mapping and the microstructural characteristics of the samples were characterized by scanning electron microscopy (SEM). A Vickers hardness test was performed to investigate the hardness properties. In addition, the porosity and the electrical conductivity of the sample were also analyzed. The XRD data showed that β-TCP was successfully synthesized from the eggshell. Based on Vickers hardness test results presented the sample with 40:60 wt% composition has a higher hardness that supported by SEM characterization that is homogeneous. Furthermore, the electrical conductivity with the highest value of 0.05470 ($S{\cdot}cm^{-1}$) showed by the sample with 30:70 wt% composition.

Keywords. β-TCP/zirconia, eggshell, solid-state reaction, and physical properties.

1. Introduction

To date, the study of biomaterials has been extensively done by many researchers due to their potential applications in the field of biomedicine. Tricalcium phosphate (TCP) is one of the biomaterials which commonly used in orthopedic, dental, and plastic surgeries for healing of bone defects, fracture treatment, total joint replacement, fundamentally related replacement and regeneration of bone tissue [1]. TCP with a chemical formula of $Ca_3(PO_4)_2$ is the bioceramic material with similar composition patterns with to human bone performing excellent biocompatibility, osteoconductivity and ensuring colonization of material by new bone, and bioactivity with surrounding living tissue [2, 3].

Among CaP-based bioceramics, β-TCP has excellent resorbability in a human biological environment allowing progressive replacement of the natural host tissue with gradual degradation of the implanted material to make it ideal for bone reconstruction [4, 5]. It is known that there is a trend in increasing biodegradability, osteoconductivity, and osteoinductivity when coming in contact with biological tissue [6]. β-TCP is stable at room temperature and the ideal Ca/P ratio is 1.5, it crystallizes in rhombohedral structure with a space group of $R\ 3\ c$, lattice parameters of a = 10.4121 Å; c = 37.3517 Å with Ca/P ratio similar to that of bone tissue [7]. So far, β-TCP has been synthesized by different routes, including wet-chemical, solid state synthesis, and microwave irradiation [8]. Meanwhile, calcium (Ca) that found in eggshells in the form of calcium carbonate ($CaCO_3$) with the

Content from this work may be used under the terms of the Creative Commons Attribution 3.0 licence. Any further distribution of this work must maintain attribution to the author(s) and the title of the work, journal citation and DOI.
Published under licence by IOP Publishing Ltd

composition of 90.9 % [9], presenting a prominent raw material to produce β-TCP. In addition, calcium eggshells also become a useful source of bone metabolism [10]. Therefore, utilization of eggshells to produce β-TCP is important to be conducted.

In general, β-TCP has low mechanical properties and rapid biodegradation rate [11]. The main problem, at this time, is how to control the degradation rate and improve the mechanical properties of β-TCP [12–18]. Furthermore, for tissue replacement and regeneration applications, the conformity requirement between degradation rate and mechanical properties with tissue cell growth rate is then strongly required. Therefore, it is also vital to improve the mechanical properties and control the rate of β-TCP degradation. In this work, we propose the compiling β-TCP with zirconia (ZrO_2). Theoretically, ZrO_2 can be used as a composite material due to its stable-biocompatibility and performs excellent mechanical properties [19]. Moreover, ZrO_2 is also very suitable to be used as a dental implant material due to its low thermal conductivity and aesthetic quality properties [20].

Based on the above explanation, it is essential to investigate the structural and physical properties as well as crystal structure, size and morphology, hardness, and electrical conductivity of the β-TCP/ZrO_2 composites from the eggshell.

2. Materials and methods

2.1. Synthesis and characterization of bioceramics composite β-TCP/ZrO₂ from eggshell

In this study, β-TCP was prepared from a natural eggshell and composited with zirconia for obtaining a better physical property, i.e., hardness and electrical conductivity. Regarding bone graft applications, some important of the physical properties, i.e., porosity, hardness, electrical conductivity, and microstructure ware tested. The results of the composition test of an eggshell element were conducted by X-ray fluorescence (XRF). The CaO was calcined at 1000 °C for 5 h and then evaluated by X-ray diffraction (XRD) to know the crystal structure. Furthermore, 100 mL of calcined-CaO solution (1.2 M) was reacted with 100 mL of H_3PO_4 0.8 M and stirred at 600 rpm for 5 h. The solution was deposited at room temperature for 24 h, then filtered and sintered at 1000 °C for 15 h.

Both powder β-TCP and ZrO_2 were mixed using ball milling for 2 h with β-TCP: ZrO_2 ratio 70:30; 60:40; 50:50; 40:60; 30:70 (wt%.) The series of the samples were formed in a cylindrical shape with the weight of 2 g prior sintering at 900 °C for 3 h. The sintered samples were characterized by means of XRD and scanning electron microscopy (SEM). The physical properties as well as porosity, hardness, microstructure, and electrical conductivity were also investigated. The porosity test of β-TCP / ZrO_2 composites was calculated using the following Equation 1 [21]:

$$\text{Porosity} = \left(W_{bj} - W_b\right) / W_p \times 100\% \quad (1)$$

W_{bj} :saturated burn weight, W_b : burn weight, W_p: plastic weight.

Hardness was tested with a Vickers hardness test, while the electrical conductivity determined by measuring electrical current and voltage.

3. Results and discussion

The elemental composition of the eggshell is presented in Table 1.

Table 1. Chemical composition of the calcined eggshell

Compound	S	Ca	Fe	Co	Sr	Er	Yb	Lu
Conc Unit (%)	0.040 ± 0.002	98.73 ± 0.06	0.057 ± 0.003	0.064 ± 0.001	0.20 ± 0.02	0.20 ± 0.01	0.61 ± 0.03	0.090 ± 0.001

Based on Table 1, Ca shows dominant composition up to 98.73 % comparing to other components. Therefore, it becomes a strong reason to be used as a primary source for synthesizing β-TCP. The diffraction pattern of CaO has a crystal structure match with AMCSD database with the code of 96-200-1133. It implies that the chicken eggshell can be used as raw material to produce calcium phosphate [9].

IOP Conf. Series: Materials Science and Engineering **367** (2018) 012017 doi:10.1088/1757-899X/367/1/012017

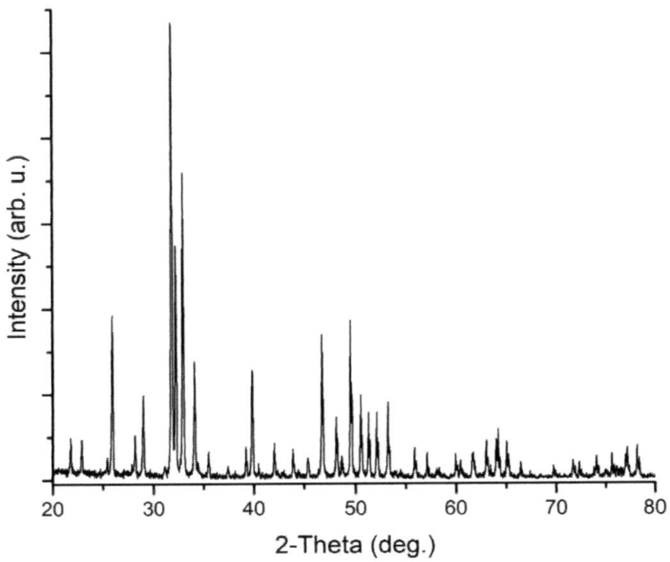

Figure 1 Diffraction pattern of β-TCP model and β-TCP after sintering at 1000 °C for 15 h

Figure 1 shows the matching sintered β-TCP diffraction pattern the β-TCP model of AMCSD database. It shows that β-TCP was successfully synthesized using the calcium source from the eggshell. This result is in a good agreement with the study reported by Aisyah [22]. Selection of sintering temperatures at 1000 °C for 15 h gives an appropriate synthesis as a reported by others [23].

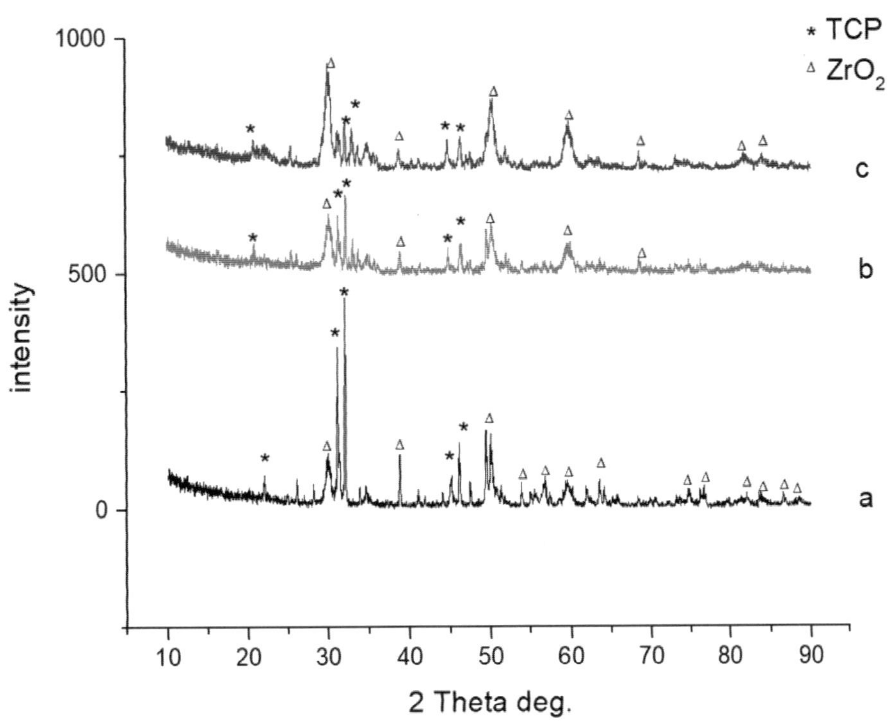

Figure 2 XRD patterns for the β-TCP : ZrO_2 compositions of (a) 1:1 (b) 0.8:1.2 and (c) 0.6:1.4

Table 2. Porosity, hardness and electrical conductivity of β-TCP / ZrO_2 composites

No.	Composition β -TCP: ZrO (wt%)	Porosity (%)	Average Hardness Vickers (kg·mm^{-2})	Electrical Conductivity (σ) (S·cm^{-1})`
1	70 : 30	14.2709	13.67	0.03112
2	60 : 40	19.1320	13.97	0.03991
3	50 : 50	15.9580	15.70	0.01999
4	40 : 60	7.9469	20.67	0.02409
5	30 : 70	6.7174	18.83	0.05470

It is shown from Table 2 that the increase of zirconia addition in the composite effect to decreases the porosity, and increases the hardness of β-TCP/ZrO_2. The increase of the hardness of the composite may originate from zirconia which has a higher hardness. The porosity of the samples exhibits values lower than 2 0% which is in the range of ceramics tolerance of 10 % to 40 % [24]. The high porosity occurs as a result of its different grain sizes; this is due to of processing at high sintering temperatures. With the addition of ZrO_2 filler may suppress the pore growth and in turn to increase the hardness. Both the porosity and the hardness properties are associated with biomedical applications such as bone and tooth implants, and orthopedic [25, 26].

The highest electrical conductivity reached up to 0.05470 S·cm^{-1} obtained from the sample with a composition of 30:70. The optimum composition of 30:70 with highest electrical conductivity is associated with the lowest porosity of 6.7174 %. We can infer that the decrease in ceramic pores causes an increase of its electrical conductivity.

Figure 3 SEM images of β-TCP:ZrO_2 composites with composition of (a) 70:30, (b) 60:40, (c) 50:50, (d) 40:60, and (e) 30:70 wt%

The SEM images show that all β-TCP/ZrO_2 composites exhibit agglomerations. The sample with the ratio of β-TCP:ZrO_2 composition of 40:60 wt% shows more homogeneous surface morphology compared to other samples. It implies that the sample with 40:60 wt% shows the highest hardness as presented in Table 2. Meanwhile, the sample with the composition of 70:30 wt% for β-TCP/ZrO_2 ratio has the lowest homogeneity resulting the lowest hardness. The surface morphology is closely related to the mechanical properties of the material. Meanwhile, zirconia would be a suitable additive material because of the stable with high biocompatibility and good mechanical properties [27].

The sample achieved the lowest electrical conductivity of 0.01999 $S{\cdot}cm^{-1}$ with the composition of 50:50. According to the SEM analyses ass in Figure 3, the sample with a ratio of 50:50 is less homogeneous than others.

4. Conclusions

We conclude that the β-TCP/ZrO_2 composites were successfully synthesized using solid-state reaction method. The addition of ZrO_2 affects to enhance the hardness and electrical conductivity of the β-TCP/ZrO_2 composites. The highest hardness of the β-TCP/ZrO_2 composites was obtained for a composition of 40:60. The optimum electrical conductivity was found in the ratio of 30:70 wt%. Furthermore, the porosity was in the range of 7 % to 19 % providing a potential application for regeneration of bone tissue.

References

[1] Kokubo T 2008 *Bioceramics and their clinical applications* (CRC Press, Boca Raton, Boston, New York)

[2] Rakovsky A, Gotman I, Rabkin E, and Gutmanas EY 2014 B–TCP–polylactide composite scaffold with high strength an enhanced permeability prepared by a modified salt leaching method *J. Mech. Behav. Biomed. Mater.* **32** 89–98

[3] Wagoner J A J and Herschler BA 2011 A review of the mechanical behavior of CaP and CaP/polymer composite for application in bone replacement and repair *Acta Biomater.* **71** 16–30

[4] Lu J *et al* 2002 The biodegradation mechanism of calcium phosphate biomaterial *J. Biomed. Mater. Res. B Appl. Biomater.* **63** 408–412

[5] Lu J X *et al* 1998 Comparative study of tissue reaction to calcium phosphate ceramic among cancellous cortical and modullar bone site in rabbits *J. Biomed. Mater. Res* **42** 357–367

[6] Ruiz–Aguilar C *et al* 2017 Synthesis and characterisation of β–TCP/bioglass/zirconia scaffolds *Advances in Applied Ceramics* **116** 452–461

[7] Massit A, Yacoubi A E, Idrissi B C E, and Yamni K 2014 Synthesis and characterization of nano–sized β–Tricalcium phosphate Effect of the aging time IOSR *J. Applied. Chemistry.* **7** 57–61

[8] Ghosh R and Sarkar R 2016 Synthesis and characterization of sintered beta–tricalcium phosphate A comparative study on the effect of preparation route *Mat. Science and Engin. C* **67** 345–352

[9] Hincke M T *et al* 2007 Purification and Immunochemistry of a soluble matrix protein of chicken eggshell *Calcif. Tissue* **56** 578–83

[10] Hirasawa T, Oni N, and Ezawa I Intestinal Ca absorption Bioinfobank *Chemistry* **116**

[11] Boslama N, Chevalir Y, Bouaziz J, and Ben Ayed F 2013 Influence of the sintering temperature on Young's modulus and the shear modulus of tricalcium–phosphate–fluorapatite composites evaluated by ultrasound techniques *Mater. Chem. Phys.* **141** 289–297

[12] Sharmin N, Hasan M S, and Parson A J 2015 Cytocompatibility mechanical an dissolution properties of high strength boron and iron oxide phosphate glass fiber reinforced bioresorbable compact *J. Mech. Biomed. Mater.* **59** 41–46

[13] Neel A, Ensanya A, and Pickup D M 2009 Bioactive functional materials a perspective on the

phosphate–based glass *J. Mater. Chem.* **6** 690–701
[14] Ahmed I, Collins C A, and Lewis M P 2004 Processing characterization and biocompatibility of iron–phosphate glass fibers for tissue engineering *Biomaterials* **25** 3223–3232
[15] Abdeghany A M, Ouis M A, And Azooz M A 2016 Spectrochim role of SrO in the bioactivity behavior of some ternary borate glasses and their glass ceramic derivatives *Acta A. Mol. Biomol. Spectrosc.* **152** 126–133
[16] Baino F, Novajra G, and Miguez–Pacheco V 2016 Bioactive glasses special application outside the skeletal system *J. Non. Cryst. Solids* **432** 15–30
[17] Rajkumar G, Aravindan S, and Rajendra V 2010 Structural analysis of zirconia–doped calcium phosphate glass *J. Non. Cryst. Solids* **356** 1432–1438
[18] Kumar A, Biswas K, and Basu B 2015 Hydroxyapatite–titanium bulk composite for bone tissue engineering applications *J. Biomed. Mater. Res. A* **103** 791–806
[19] Alizadeh A *et al* 2016 Synthesis of calcium phosphate scaffold and human endometrial adult stem cells for bone tissue engineering *Artificial cells, nanomedicine, and biotechnology* **44** 66–73
[20] Stefanic M, Krnel K, and Kosmac T 2013 Novel method for the synthesis of a β–tricalcium phosphate coating on a zirconia implant *J. European Ceramic Society* **33** 3455–3465
[21] Harsono 2006 The Effect of ceramic material base composition on physical properties of ceramics *Jurnal Natural* **4** (Universitas Brawijaya)
[22] Aisyah I 2011 *Synthesis and Characterization of β–Tricalcium Phosphate Shell–Based Chicken Eggs in Open Air with Sintering Time Variations* (Tesis Fakultas Matematika dan Ilmu Pengetahuan Alam Institut Pertanian Bogor)
[23] Barsoum M W 1997 *Fundamentals of Ceramics* (Institute of Physics Publishing, Bristol)
[24] Vallet–Regi M 2014 *Bioceramics with clinical applications* (John Wiley & Son, United Kingdom)
[25] Elghazel A, Taktak R, and Bouaziz J 2016 Combined numerical and experimental mechanical characterization of a calcium phosphate ceramic using modified Brazilian discad SCB specimen *Mater. SciEng.* A **670** 240–241
[26] Yea X *et al* 2013 Preparation and in vitro evaluation of mesoporous hydroxyapatite coated β–TCP porous scaffolds *Mater. Sci. Eng.* **33** 5001–5007
[27] Alizadeh A *et al* 2016 Synthesis of calcium phosphate–zirconia scaffold and human endometrial adult stem cells for bone tissue engineering *Artificial Nanomedicine and Biotechnology* **44** 67–73
[28] Vallet–Regi M 2014 *Bioceramics with clinical applications* (John Wiley & Sons Ltd, United Kingdom)

Influence of High Sintering Temperature Variation on Crystal Structure and Morphology of $Nd_{1.2}FeO_3$ Oxide Alloy Material by Solid-State Reaction Method

E. H. Sujiono[1], R. A. Imran[1], M. Y. Dahlan[1], A.C. M. Said[1], S. Samnur[1], and N. Ihsan[1]

[1]Laboratory of Materials Physics, Department of Physics, *Universitas Negeri Makassar,* Makassar 90224, Indonesia.

E-mail: e.h.sujiono@unm.ac.id

Abstract. $Nd_{1.2}FeO_3$ powders type perovskite structure was prepared by a solid-state reaction method. This research has been conducted with the recurring heating process at high temperature. The raw material consisted of Nd_2O_3 (99.99 %) and Fe_2O_3 (99.99 %) which characterized by XRD to confirm the phase and using SEM to identify the morphology structure of the sample. Result characterized by XRD confirms the phase of $NdFeO_3$ and Nd_2O_3 with the formation of $NdFeO_3$ having the orthorhombic structure (perovskite type). The value of FWHM and the average crystal size of $NdFeO_3$ was obtained for each sample is 0.20° and 409 nm. While SEM studies showed the surface morphology of $Nd_{1.2}FeO_3$ has homogeneous granules with grain size estimates is 0.2 μm. These results indicate that sample $Nd_{1.2}FeO_3$ was a good candidate for gas sensor materials.

Keywords. Crystal structure, morphology, sintering, $NdFeO_3$ oxide alloy, and solid state method.

1. Introduction

Many researchs have been conducted on oxide compounds to be used as a gas sensor, one of them is to use a perovskite oxide that is $NdFeO_3$ synthesized by various methods or techniques [1–5]. $NdFeO_3$ known to have type orthorhombic [1, 3–6]. The nano-perovskite oxides ABO_3 (*A*: La, Nd, Sm, and Gd; *B*: Fe, Co and Ni; and O: oxygen) have high catalytic activities and high sensitivity for gas sensor material. The $NdFeO_3$ is mainly using in gas sensing and catalysis application [2, 3, 6]. Research on $NdFeO_3$ by sol-gel citrate method obtained perovskite-type $NdFeO_3$ can be used as an H_2S gas sensor and catalytic CO gas sensor in exhaust gas environments [3, 6]. Various synthesizing techniques have been used for synthesis $NdFeO_3$ alloy oxide it such as by hydrothermal method [7], combustion [8, 9], sol-gel [10], precipitation method [11], solid-state reaction method [12] and sonication assisted precipitation [13].

The solid-state reaction is one of the oldest synthesis routes for the preparation of perovskites [12]. An advantage of this method is a cheap, simple and fast method for the synthesis perovskite. Also, the product of the reaction has high purity and good crystallinity. The properties of the perovskite materials are closely related to either the preparation or sintering conditions. In the ceramics, the sintering process is essential that effect on microstructure, grain growth, and densification [14].

Content from this work may be used under the terms of the Creative Commons Attribution 3.0 licence. Any further distribution of this work must maintain attribution to the author(s) and the title of the work, journal citation and DOI.
Published under licence by IOP Publishing Ltd

IOP Conf. Series: Materials Science and Engineering **367** (2018) 012018 doi:10.1088/1757-899X/367/1/012018

In this paper, we present perovskite oxide $Nd_{1.2}FeO_3$-based by a solid state method with varying the sintering temperature using heat treatment process. Then the crystal structure and morphology of $Nd_{1.2}FeO_3$ has been characterized by X-Ray Diffraction (XRD) and Scanning Electron Microscope (SEM).

2. Materials and methods

The powders oxide $Nd_{1.2}FeO_3$ were prepared by the solid-state reaction method. Raw materials of Nd_2O_3 (Strem Chemicals, 99.99 %) and Fe_2O_3 (Sigma-Aldrich, 99.99 %) were mixed together based on a stoichiometric calculation using a molar ratio of x = 0.2 [15] to get an oxide alloy $Nd_{1.2}FeO_3$. Then powders were grinded for 3 h and calcined using furnace at 700 °C for 6 h. Sample powders are resulting calcination then were grinded back for 5 h to get a homogeneous mixture and sintered at 950 °C for 6 h.

The synthesis process and the heating are then repeated to obtain a better sample homogeneity [16]. Then mix powders produced were grinded for 3 h and calcined at 950 °C for 6 h. Then the result of the calcined sample was grinded back for 5 h to maximize reaction and to increase the homogeneity and sintered as a variation of high temperatures 1000 °C, 1050 °C, and 1100 °C for 6 h, respectively. Finally, the samples were annealed at 450 °C for 1 h. Powders oxide $Nd_{1.2}FeO_3$ were characterized by XRD type Rigaku MiniFlex II 2θ = 10° to 70° (Cu $K\alpha$, λ = 0.154 nm) and SEM-EDS type Tescan Vega3SB to analyzed the phase composition and to confirm morphology and elemental structure.

3. Results and discussion

X-ray diffraction (XRD) pattern of samples $Nd_{1.2}FeO_3$ at high temperatures 1000 °C, 1050 °C, and 1100 °C shown in Figure 1, respectively. According to the result of synthesis $NdFeO_3$ powder obtained diffraction peaks form a single phase with a perovskite structure [4, 17]. In Figure 1 shows the formation phase of $NdFeO_3$ and Nd_2O_3. Formation of phase Nd_2O_3 produces another peak that regarded as an impurity phase and can reduce the sensitivity of the material as a gas sensor. Phases analysis using *Match!* software shows that dominant phase of $NdFeO_3$ is having the orthorhombic structure (perovskite type) with Pnma space group [18].

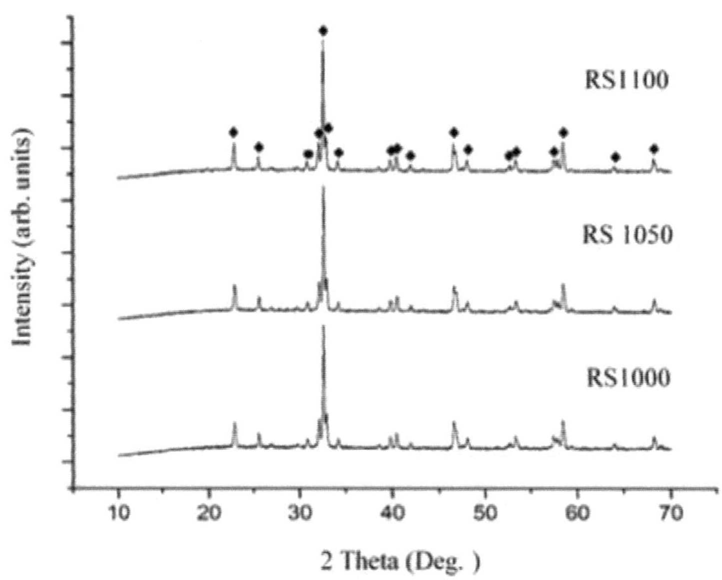

Figure 1. XRD patterns of $Nd_{1.2}FeO_3$ as variation of different termperatures (♦= $NdFeO_3$, ●= Nd_2O_3).

Table 1. Data of positions (2θ), intensities and FWHM values of $Nd_{1.2}FeO_3$ phases

Sample	Identification phase	2θ (°)	Intensity (Counts)	FWHM (°)
RS1000	$NdFeO_3$	32.54	13053.33	0.20±0.02
RS1050	$NdFeO_3$	32.56	13260.00	0.20±0.02
RS1100	$NdFeO_3$	32.52	13810.00	0.20±0.02

It can be seen from Table 1, FWHM values of $NdFeO_3$ phase same for all three samples at the dominant peak is *hkl* (121) with the peak position $2\theta = 32.5°$. Synthesis of $NdFeO_3$ has also been done by Jada Shanker *et al.* with temperature 900 °C [5], and Yabin Wang *et al.* with temperature 1000 °C [4] which get similar results that phase $NdFeO_3$ exist at $2\theta = 32.56°$ with *hkl* (121). FWHM value stated of homogeneity between atoms in a crystal which the smaller the FWHM value means the lattice more homogeneous or has good crystallinity, which means that the level of material quality is also getting better [4]. The crystal size can be calculated using *Debye-Scherrer* equation (1):

$$D = \frac{0.9\lambda}{\beta \cos\theta} \quad (1)$$

where λ is the wavelength of X-ray (0.15405 nm for Cu Kα), β is FWHM value at *hkl* (121), and θ is diffraction angle. Based on the calculation, the value of the crystal size of $Nd_{1.2}FeO_3$ which synthesized was 409.37 nm, 409.39 nm and 409.35 nm, respectively.

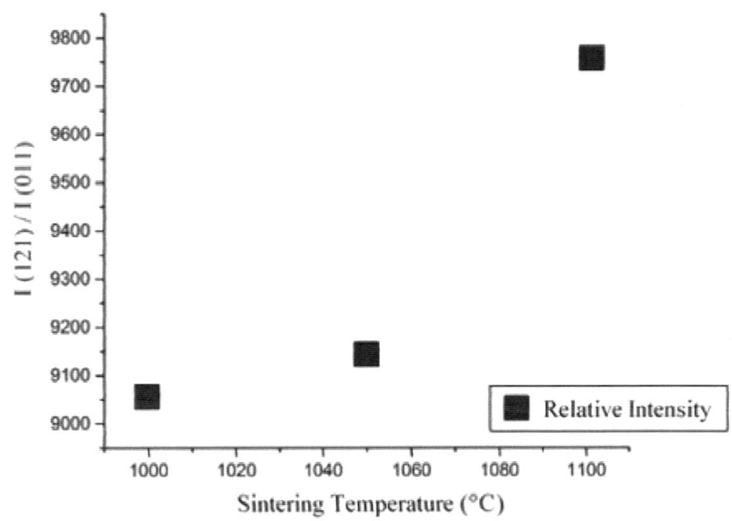

Figure 2. Relative intensity $I(121)/I(011)$ of $Nd_{1.2}FeO_3$ powders oxide material as a function of sintering temperatures.

Figure 2 shows diffraction intensity curve between the dominant of peak intensity (121) with a peak intensity of (011) for each variation of sintering temperature. In this study, we found that a significant change of peak intensity of (011) as a comparison of (121) which is indicating that the intensity of Nd_2O_3 phase decreases as the temperature of sintering increases.

IOP Conf. Series: Materials Science and Engineering **367** (2018) 012018 doi:10.1088/1757-899X/367/1/012018

Figure 3. SEM images of sample $Nd_{1.2}FeO_3$ as a variation of sintering temperature a) RS1000, b) RS1050, and c) RS1100, respectively.

Table 2. Data composition element of $Nd_{1.2}FeO_3$ samples using EDS

Compound	Norm. C [wt%]			Error (3 Sigma) [wt%]		
Norm.	RS1000	RS1050	RS1100	RS1000	RS1050	RS1100
Oxygen	16.94	16.50	16.41	5.91	5.83	6.90
Sodium	1.32	-	-	0.38	-	-
Magnesium	0.27	-	-	0.15	-	-
Aluminium	0.25	-	-	0.14	-	-
Potassium	0.04	-	-	0.09	-	-
Titanium	0.11	-	-	0.10	-	-
Iron	21.11	21.60	19.87	1.63	1.70	1.77
Copper	0.85	0.81	0.38	1.22	0.22	0.19
Neodymium	58.99	60.94	63.15	4.26	4.52	5.23
Silicon	0.12	0.15	0.18	0.11	0.11	0.13

The morphology, crystal structure and particle size of the sample were investigated by SEM. Microstructures of polycrystalline $Nd_{1.2}FeO_3$ sintered at various temperatures are shown in Figure 3. SEM results of all samples in Figure 3, it is generally assumed that the powders $Nd_{1.2}FeO_3$ oxide alloy material have a high homogeneity level with formation small uniform granules which an estimated grain size about of 0.2 μm and the color is almost evenly gray. This powder has a very high porosity, and this is an advantage for improving the gas-sensing characters, as has reported by Niu Xinshu *et al.* [2] and Ho *et al.* [3].

Table 2 shows the elemental composition of samples $Nd_{1.2}FeO_3$, were RS1000 contain Nd (58.99 wt%) and Fe (21.11 wt%), RS1050 contain Nd (60.94 wt%) and Fe (21.60 wt%), and RS1100 contain Nd (63.15 wt%) and Fe (19.87 wt%), which also contain a minor phase as shown in Table 2. That minor phase as indication due to the sample holder preparation process and no significant effect on the constituent elements of each sample.

4. Conclusions

The perovskite oxide $Nd_{1.2}FeO_3$ powders have been prepared by a solid-state method with the recurring heating process at high temperatures of 1000 °C, 1050 °C, and 1100 °C, respectively. The result of X-Ray diffraction analysis showed the phase of $NdFeO_3$ and Nd_2O_3 which phase $NdFeO_3$ is an orthorhombic structure with Pnma space group. The results indicate that $Nd_{1.2}FeO_3$ analyzed at high-temperature variation given a relatively stable quality, it can be seen the same FWHM value is 0.20° with an estimated crystalline size about of 409 nm. While SEM studies showed, the surface morphology of $Nd_{1.2}FeO_3$ has homogeneous granules and high porosity with an estimated grain size of 0.2 μm. The results indicate that $Nd_{1.2}FeO_3$ is a good candidate for gas sensor materials as has reported elsewhere.

Acknowledgments

This research was funded by Directorate Research and Community Services, Directorate General of Research and Development, Ministry of Research, Technology, and Higher Education, Republic of Indonesia, under research scheme of *Hibah Kompetensi* fiscal year 2017.

References

[1] Khorasani-Motlagh M, Noroozifar M, Yousefi M, and Jahani S 2013 Chemical synthesis and characterization of perovskite $NdFeO_3$ nanocrystals via a co-precipitation method *International Journal of Nanoscience and Nanotechnology* **9** pp.7–14.

[2] Niu X, Li H, and Liu G 2005. Preparation, characterization and photocatalytic properties of $REFeO_3$ (RE= Sm, Eu, Gd) *Journal of Molecular Catalysis A: Chemical* **232** pp.89–93

[3] Ho T G *et al* 2011 *Advanced in Natural Science: Nanoscience and Nanotechnology* **2** pp 015012–15021

[4] Wang Y S 2010 Growth rate dependence of the $NdFeO_3$ single crystal grown by float-zone technique *Journal Crystal Growth* **318** pp 927–931

[5] Shanker J, Suresh M B and Babu D S 2015 *International Journal of Scientific Engineering and Research* **3** pp 194–197

[6] Ru Z *et al* 2010 Electrical and CO-sensing properties of $NdFe_{1-x}Co_xO_3$ perovskite system *Journal of Rare Earths* **28** pp 591–595

[7] Zheng W *et al* 2010 Hydrothermal synthesis of $LaFeO_3$ under carbonate-containing medium *Material Lettters* **43** pp 19–22

[8] Manoharan S S and Patil K C 1993 Combustion route to fine particle perovskite oxides *Journal Solid State Chemical* **102** pp 267–267

[9] Chakraborty A, Dévi P S and Maiti H S 1995 Low temperature synthesis and some physical properties of barium-substituted lanthanum manganite ($La_{1-x}Ba_xMnO_3$) *Journal Materials Research* **10** pp 918–925

[10] Cui H, Zayat M, and Levy D 2006 Epoxide assisted sol–gel synthesis of perovskite-type $LaMxFe_{1-x}O_3$ (M= Ni, Co) nanoparticles *Journal of Non-Crystalline Solids* **352** pp 3035–3040

[11] Pandya H N, Kulkarni R G, and Parsania P H 1990 Study of cerium orthoferrite prepared by wet chemical method *Materials Research. Bulletin* **25** pp 1073–1077

[12] Yuan S *et al* 2011 Decisive role of oxygen vacancy in ferroelectric versus ferromagnetic Mn-doped $BaTiO_3$ thin films *Journal of Applied Physics* **109**

[13] Singh S, Yadav B C, and Dwivedi P K 2013 Fabrication of nanobeads structured perovskite type neodymium iron oxide film: Its structural, optical, electrical and LPG sensing investigations *Sensor Actuators B: Chemical* **177** pp 730-739

[14] Athayde D D *et al* 2016 Review of perovskite ceramic synthesis and membrane preparation methods *Ceramics International* **42** 6555–6571

[15] Zharvan V *et al* 2017 The Effect of Molar Ratio on Crystal Structure and Morphology of $Nd_{1+X}FeO_3$ (X=0.1, 0.2, and 0.3) Oxide Alloy Material Synthesized by Solid State Reaction Method *IOP Conf. Series: Materials Science and Engineering* **202** p 1

[16] Mir S A, Ikram M, Asokan K 2014 *Optik - International Journal for Light and Electron Optics* **125** pp 6903-6908

[17] Shanker J, Suresh M, and Babu D 2016 Synthesis, Characterization and Electrical Properties of $NdXO_3$(X=Cr, Fe) Nanoparticles *Materials Today: Proceedings* **3** pp 2091–2100

[18] Chanda S *et al* 2013 Raman spectroscopy and dielectric properties of nanoceramic NdFeO3 *Materials Research Bulletin* **48** pp 1688–1693

Fabrication of PANI/Ag/AgCl/ITO-PET Flexible Film and Its Crystallinity and Electrical Properties

M. Diantoro[1,2], F. Rohmiani[1], A. A. Mustikasari[1], and Sunaryono[1,2]

[1]Department of Physics, Faculty of Mathematics and Natural Sciences, Universitas Negeri Malang, Jl. Semarang 5 Malang 65146, Indonesia
[2]Central Laboratory of Minerals and Advanced Materials, Faculty of Mathematics and Natural Sciences, Universitas Negeri Malang, Jl. Semarang 5 Malang 65146, Indonesia

Email: markus.diantoro.fmipa@um.ac.id

Abstrak. PANI as one of the conductive polymers which have been widely using in electronics or storage devices such as a supercapacitor. PANI has recently become an option because of its potential for a broad area of application. Protonation or introduce a dopant can control the electrical properties of PANI. However, researcher facing a disadvantage since PANI also active in acidic conditions. To control the conductivity and the stability in an acidic environment, the researcher has introduced Ag/AgCl to PANI. We report the synthesis and analyses of silver nanoparticles (AgNPs), PANI, and PANI/Ag/AgCl/ITO-PET films. PANI was synthesized by chemical polymerization, while AgNPs were synthesized via a reductive chemical method using $NaBH_4$ as an $AgNO_3$ reductor. The resulting PANI was characterized using FTIR to determine the functional group, while to obtain the purity of the Ag phase was checked by using XRD. The preparation of PANi/Ag/AgCl solution was carried out by mixing method with the variation of the mass of $AgNO_3$. The precipitate was carried out by using ITO-PET substrate. PANI/Ag/AgCl/ITO-PET films were characterized by using FTIR, XRD, SEM-EDX, and capacitance meters. It has was found that crystallinity increases with the addition of Ag films to PANI/Ag/AgCl/ITO-PET. The crystallinity reached 29.85 %. It was also revealed that the dielectric constant decreased with increasing Ag in PANi/Ag/AgCl/ITO-PET films.

Keywords. Ag, AgCl, crystallinity, dielectric constant, ITO-PET, and polyaniline.

1. Introduction
Many kinds of polymers have been explored for various applications.The previous research of organic polymers worked on electronic applications such as silver doped flavonoid [1], and ZnO-cellulose acetate [2]. A specific interest of conducting polymers also being extensively studied due to their potential application in technology [3]. Among the various conductive polymers, PANi is attracting attention because of inexpensive, easy to synthesized, excellent conductivity and stability among polymers. Its electrical properties can be modified by oxidation or introducing other material such as iron and copper [4]. The use of acid in PANI provides a specific protonation state that can be used in broader applications [5]. We can use this polymer as microelectronic devices, diodes, sensors, supercapacitors, microwave absorption, and corrosion prevention [6]. Further, PANI also shows a good performance in optical devices since it has a high absorbance coefficient in visible light spectra

Content from this work may be used under the terms of the Creative Commons Attribution 3.0 licence. Any further distribution of this work must maintain attribution to the author(s) and the title of the work, journal citation and DOI.
Published under licence by IOP Publishing Ltd

as well as high load carrier [7]. The uniqueness of PANI compared to the other conductive polymers is its electrical properties which can be controlled by the transfer of doping charge and protonation [8]. However, the redox reaction of PANI is active only for acidic conditions, generally at pH < 4. It limits its application in bioelectrochemistry, which usually requires neutral pH in the environment [9]. To improve the shortcomings of PANI and to produce materials with desirable properties, various developments have been proposed. One of them through a combination of PANI with other specific materials [4, 10–11].

The previous work reported that a significant increase of electrical properties on PANI because of Ag-doped [8]. The electrical stability performance of PANI was achieved by inducing AgCl [9,12–13]. Here we report on the microstructure and dielectric properties of PANI-Ag-AgCl on ITO-PET substrate as a flexible film.

2. Materials and methods

The PANI ES was synthesized using polymerization while the AgNPs were prepared using the co-precipitation method. Both of those materials were blended as a precursor of solution for deposition on substrate flexible. The properties of a sample were characterized by FTIR, SEM-EDX, X-RD, and capacitance measurement using LCR-meter. The characterization focuses on microstructure and dielectric properties of PANI/Ag/AgCl/ITO-PET. A more detail procedure of this experiment distinctly showed below.

2.1. Synthesis of PANI

The first step by preparing two phases, namely organic and inorganic phases. The organic phase was a solution of aniline and HCl, while the inorganic phase was ammonium peroxidisulfate and distilled water. Both of the solutions were mixed for the polymerization process. This process lasts for 24 h. The result will create green sediment. The polyaniline sludge ES was cleaned using HCl to remove residual monomer, oxidant, and polymerized decomposition products and also further treatment using distilled water followed filtering and drying over a hotplate at 60 °C.

2.2. Synthesis of Ag nanoparticles

The silver nanoparticles (AgNPs) were synthesized using the chemical reduction method using Sodium Borohydride ($NaBH_4$) as the $AgNO_3$ reductor. During the reaction, we use mercaptosuccinic acid (MSA) as a stabilizer. We used distilled water and methanol as solvents. The next step, a 2.34 mL of MSA $C_4H_6O_4S$ and 800 mL of methanol were mixed and blended by using magnetic stirrer at 700 rpm. The next step, we also prepared a solution of 0.68 g $AgNO_3$, and 13.548 mL of distilled water then mixed with the previous solution until the yellowish solution appears. Further, we added a 1.52 g of $NaBH_4$ in 200 mL of distilled water until the solution was blackish brown with the help of a burette. The temperature was kept at 5 °C to 10 °C. Furthermore, the solution was precipitated and filtered with 600 mL of ethanol for 3 times. The filtrate was dried using a hotplate over a temperature of 50 °C to obtain AgNPs.

2.3. Deposition film by spin coating method

We firstly prepared a 1.5 mL of distilled water maintained at a temperature of 80 °C, then mixed 0.05 g of PVA stirred at 750 rpm for 2 h. The next step was to introduce 0.2 g of PANI with different of Ag; 0.00 g, 0.02 g, 0.04 g, 0.06 g, and 0.08 g and stirred at 750 rpm for two hours without heating. Deposition of PANI/Ag/AgCl solution on ITO-PET substrate was performed by drop casting method then rotated at 500 rpm for 30 s.

3. Results and discussion

3.1. Functional groups of PANI/Ag/AgCl

Pani ES can be transformed from PANI EB by reacting HCl as indicated by Figure 1. The carbon-carbon of the chemical configuration of PANI EB and PANI ES is distinguishable. The functional group's structure of the resulted sample is depicted in Figure 2. It demonstrated that the C-Cl aromatic out of plane vibration occur at 601 cm^{-1} as a specific characteristic of PANI ES. The absorption band of 821 cm^{-1} and 881 cm^{-1} is bending out of plane binding CH. The peak of wavenumber of 1010 cm^{-1} indicates the presence of a C-H bond, while the band at 1155 cm^{-1} showing the stretching vibration of -$NH^{+=}$. The Polaronic structure of C-N^{+*} occurs over an area of 1249 cm^{-1}, whereas for quinoid ring double bonds occur throughout the region of 1585 cm^{-1} [15]. The vibration of stretching of N-H and O-H occurs within the area of 3142 cm^{-1} and 3242 cm^{-1} [16]. The modeling approach of IR spectra indicates that PANI has been successfully synthesized.

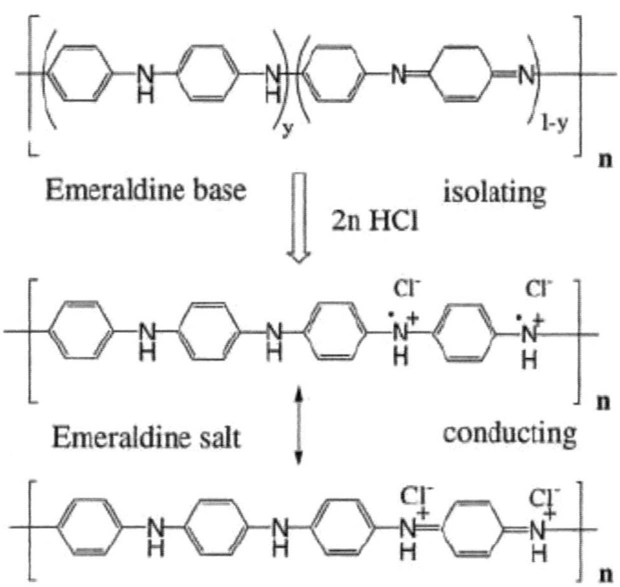

Figure 1. The chemical structure of PANI EB and PANI ES [14].

The effect of the addition of Ag on PANI is also observed from the FTIR spectra, as shown in Figure 3. A new peak shows Ag at 418 cm^{-1} indicating an AgO bond (metal-oxide) based on Figure 3.b. In addition to these bands, a slight shift according to stretching C-N bond appear from 1315 cm^{-1} to 1303 cm^{-1}. The latter indicates an interaction between PANI and Ag.

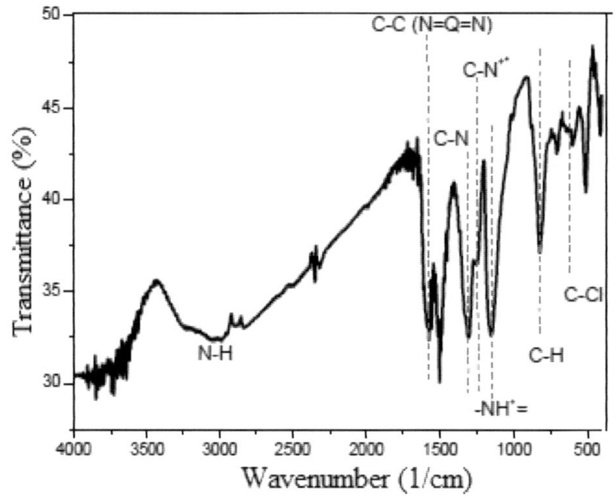

Figure 2. FTIR Spectra of PANI

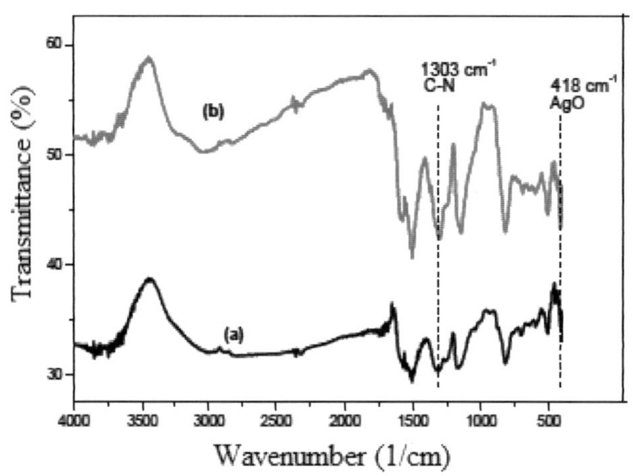

Figure 3. Characteristic bending of FTIR Spectra on (a) PANi and (b)PANI/Ag/AgCl

3.2. Crystal structure of PANI/Ag/AgCl/ITO-PET

The silver diffraction pattern of the AgNPs is shown in Figure 4. The analysis results using Cellref indicates that a single pure phase obtained, under the cubic structure. The associated peaks position at 2θ of 38.10°, 44.32°, and 64.66° are indicating Bragg's planes of (111), (200), and (220) respectively. The sample was randomly oriented. The predominant of the (111) Bragg's plane intensity does not indicate merely the grains' orientation but the crystal structure factor.

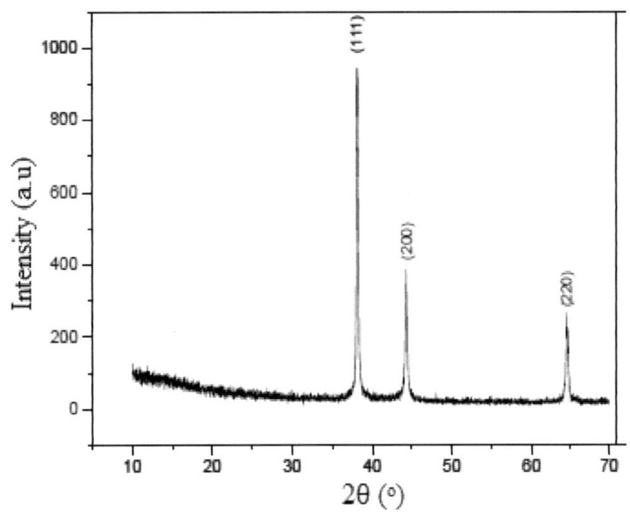

Figure 4. Silver (Ag) Diffraction Pattern.

Figure 5 shows that sample 2 to sample 5 appear Ag peaks at 2θ of 38°, 44°, 64°, 77°, and 81°. Besides the peak of Ag, there were other peaks of AgCl. AgCl peaks are observed at 2θ of 27°, 32°, 46°, 54°, 57°, 67°, 74°, 76°, and 85°. The peak at 2θ of 26° belongs to the substrate, ITO-PET. The Ag and AgCl phases belong to the cubic structure with a *Fm3m* space group. The appearance of AgCl peak occurs because of the Cl content in polyaniline. Also, AgCl may occur due to the mixing process that occurs in the liquid phase and opens for the reaction of Ag^+ and Cl^-. Based on Figure 5 it is seen that the more mass of Ag added, the higher the intensity of 2θ from Ag, and decreasing the intensity of AgCl. Lowering intensity may be originated from the more Ag^+ is reduced to AgO.

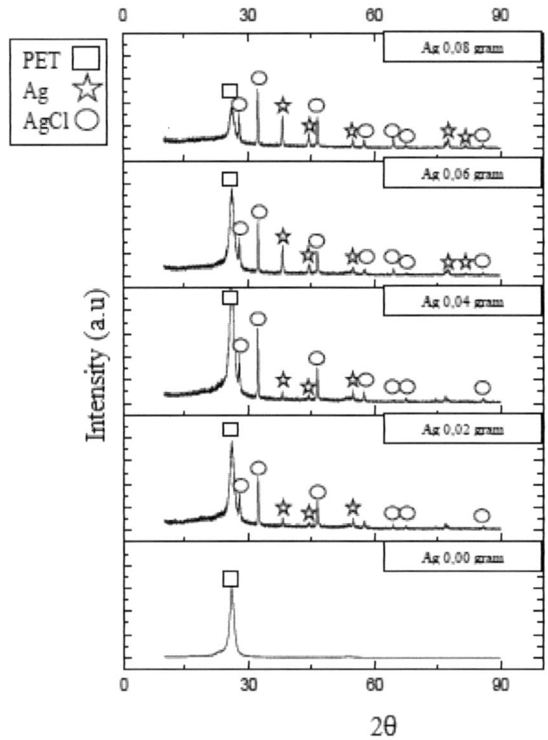

Figure 5. The diffraction patterns of PANI on ITO-PET substrate, and PANI/Ag/AgCl/ITO-PET.

Figure 6 shows the increase of crystallinity from initially amorphous, increasing the crystalline extent with increasing Ag on the composite. The maximum crystallinity degree of the film is 29.85 %. The crystal size of a material can be reviewed using the Scherrer equation through FWHM from the peaks of XRD. The grain size of Ag ranged from 28.74 nm to 41.32 nm, whereas AgCl ranged from 27.19 nm to 36.23 nm.

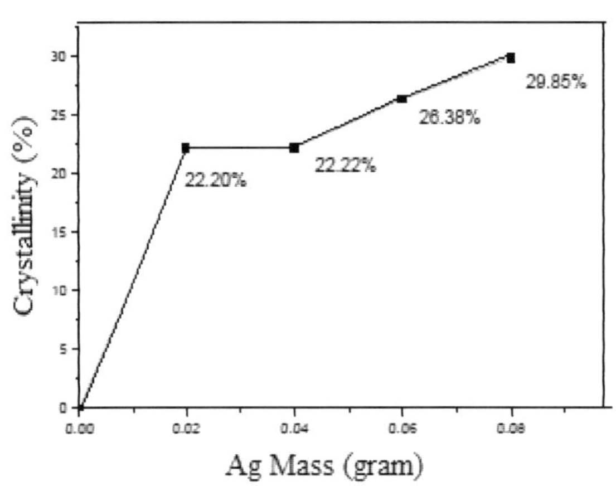

Figure 6. The Crystallinity of PANI/Ag/AgCl/ITO-PET

The increase of crystallinity of PANI/ITO-PET is depended on the doping of Ag. The crystal size of Ag impacted to higher crystallinity. This fact implied that small dopant of Ag and AgCl insert to PANI structure. Therefore, the dopant has two issues, there is a change of PANI chain and enhanced

IOP Conf. Series: Materials Science and Engineering **367** (2018) 012019 doi:10.1088/1757-899X/367/1/012019

the regularity of crystallinity of PANI [17]. The kind of PANI has been determined its crystallinity, such as PANI ES have a high crystallinity than EB. On this research, PANI ES has been formed by adding HCl. The part of Cl^- insert to the weakest chain of PANI and bound with Ag.

3.3. Morphology of PANI/Ag/AgCl/ITO-PET

The surface morphology of the sample and the elements contained in the sample has been observed using SEM-EDX. The SEM-EDAX of PANi and PANi/Ag/AgCl/ITO-PET film with the addition of 0.06 g Ag mass and mapping have been examined. From the mapping visible the distribution of elements in the composite. Figure 7 shows the morphology of PANI.

Figure 7. Morphology of PANI

The result of SEM EDX and mapping with the addition of Ag 0.06 g in PANI/Ag/AgCl/ ITO-PET film and its thickness is shown in Figure 8. The distribution of element on the surface spread well. Its shape is granular.

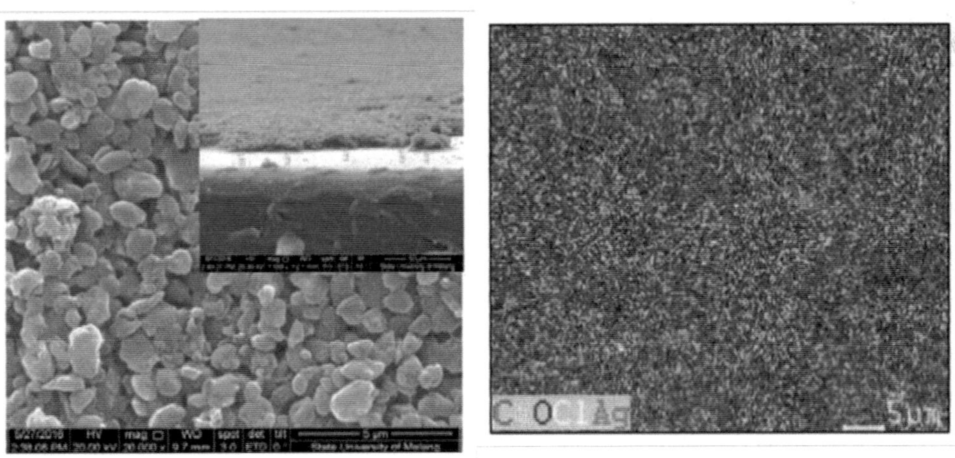

Figure 8. (a) Microstructure of PANI/Ag/AgCl/ITO-PET and (b) mapping of PANI/Ag/AgCl/ITO-PET

The microstructure of PANI/Ag/AgCl/ITO-PET is granular as indicated by Figure 8.a. The elemental analyses by EDX are shown in Table 1. The content of PANI film are C, N, O, and Cl indicated that PANI film has successfully formed.

IOP Conf. Series: Materials Science and Engineering **367** (2018) 012019 doi:10.1088/1757-899X/367/1/012019

Table 1. EDX result of PANI

Element	Wt%	At%
C	64.79	72.54
N	11.45	11.00
O	16.16	13.58
Cl	07.59	02.88

Table 2 showed that the atomic percentage of element C approximately 80.33 %. The addition of PVA to form a solution is a polymer with the dominant element C is the same as PANI. The results of the mapping also show a uniform, i.e., the C element that dominates and is evenly distributed on the sample surface, seen in Figure 8.b.

Table 2. EDX result of PANi/Ag/AgCl/ITO-PET

Element	Wt%	At%
C	70.05	80.33
N	10.54	10.37
O	07.98	06.87
Cl	03.76	01.46
Ag	07.67	00.98

3.4. Dielectric constant of PANi/Ag/AgCl/ITO-PET

The dielectric constant of the PANI/Ag /AgCl/ITO-PET is shown in Figure 9. Based on it, the dielectric constant decreases with the increasing of Ag mass in composite with maximum 17.897. The previous study showed that the addition of Ag in the composite increases the conductivity of the samples as in the PANI/Ag composite [18], although there is another phase of AgCl. The results of this measurement introduced that the decreased dielectric constant as function as the increasing of a conductive material.

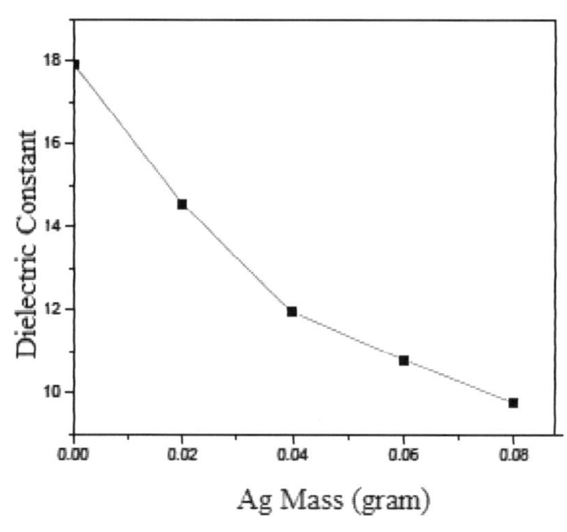

Figure 9. The Dielectric Constant of PANI/Ag/AgCl/ITO-PET and addition of Ag.

4. Conclusions

It can be concluded that the composite of PANI/Ag/AgCl/ITO-PET was successfully synthesized. We also observed that the increase of Ag inducing on the film effect to the increase of crystallinity of silver in the film but a decrease of the film dielectric constant. The decreasing of dielectric constant on PANI/Ag/AgCl/ITO-PET film indicated that the conductivity is increased. In this research, the dielectric measurement indirectly showed electrical conductivity measurement. Based on this result, it is open that PANI/Ag/AgCl/ITO-PET film is possible to be used as an electrode material for pseudo capacitor supercapacitor.

Acknowledgements
This research was partially supported by PDUPT Grand of Ministry of Research Technology and Higher Education.

References

[1] Diantoro M *et al* 2016 Electrical conductivity modification using silver nanoparticles of Jatropha Multifida L. and Pterocarpus Indicus w. extracts films *AIP Conf. Proc.* **1719** p 030034
[2] Diantoro M *et al* 2017 Microstructure and dielectric properties of cellulose acetate-ZnO/ITO composite films based on water hyacinth *J. Phys. Conf. Ser.* **853** p 012047
[3] Agarwal S *et al* 2016 Synthesis and characteristics of polyaniline/zirconium oxide conductive nanocomposite for dye adsorption application *J. Mol. Liq.* **218** pp 494–498
[4] Diantoro M *et al* 2012 The influence of iron- and copper- doped of PANi thin film on their structure and dielectric properties *AIP Conf. Proc.* **1454** pp 268–271
[5] Devarayan K *et al* 2015 Flexible transparent electrode based on PANI nanowire/nylon nanofiber reinforced cellulose acetate thin film as supercapacitor *Chem. Eng. J.* **273** pp 603–609
[6] Umare S S, Shambharkar B H, and Ningthoujam R S 2010 Synthesis and characterization of polyaniline–Fe3O4 nanocomposite: Electrical conductivity, magnetic, electrochemical studies *Synth. Met.* **160** pp 1815–1821
[7] Shumaila *et al* 2011 Synthesis and characterization of Se doped polyaniline *Curr. Appl. Phys.* **11** pp 217–222
[8] Choudhury A 2009 Polyaniline/silver nanocomposites: Dielectric properties and ethanol vapour sensitivity *Sens. Actuators B Chem.* **138** pp 318–325
[9] Xiaomiao Feng *et al* 2006 One-step synthesis of AgCl/polyaniline core-shell composites with enhanced electroactivity *Nanotechnology* **17** p 3578
[10] Al-Ahmed A *et al* 2004 Composites of polyaniline and cellulose acetate: preparation, characterization, thermo-oxidative degradation and stability in terms of DC electrical conductivity retention *Synth. Met.* **144** pp 29–49
[11] M Diantoro *et al* 2017 Effect of Fe_3O_4 on the Electro-Optic and Magneto-Electric Characteristics of (PANI/Fe_3O_4)-Ag Film *IOP Conf. Ser. Mater. Sci. Eng.* **202** p 012062
[12] Sui X, Chu Y, Xing S and Liu C 2004 Synthesis of PANI/AgCl, PANI/$BaSO_4$ and PANI/TiO_2 nanocomposites in CTAB/hexanol/water reverse micelle *Mater. Lett.* **58** 1255–1259
[13] DerMarderosian A and Beutler J A 2002 *The review of natural products: the most complete source of natural product information.* (St Louis: Facts and Comparisons)
[14] Lippert T *et al* 1999 Surface modification and structuring of electrical conducting and isolating polyaniline films *Appl. Phys. A* **69** pp S291–S293
[15] Geethalakshmi D, Muthukumarasamy N, and Balasundaraprabhu R 2015 Structural and optical study of spin-coated camphor sulfonic acid-doped polyaniline/titanium-di-oxide nanoparticles hybrid thin films *Superlattices Microstruct.* **82** pp 447–460
[16] Arora R *et al* 2014 Effect of fabrication technique on microstructure and electrical conductivity of polyaniline-TiO_2-PVA composite material *3rd Int. Conf. Mater. Process. Characterisation ICMPC 2014* **6** pp 238–243

IOP Conf. Series: Materials Science and Engineering **367** (2018) 012019 doi:10.1088/1757-899X/367/1/012019

[17] Fosong W *et al* 1988 Study on the Crystallinity of Polyaniline *Mol. Cryst. Liq. Cryst. Inc. Nonlinear Opt.* **160** pp 175–84
[18] Reda S M and Al-Ghannam S M 2012 Synthesis and electrical properties of polyaniline composite with silver nanoparticles *Adv. Mater. Phys. Chem.* **2** p 75

Swelling Behavior in Solvent Vapor Sensing based on Quartz Crystal Microbalance (QCM) Coated Polyacrylonitrile (PAN) Nanofiber

A. Rianjanu[1], S. N. Hidayat[1], T. Julian[1], E. A. Suyono[3], A. Kusumaatmaja[1,2], and K. Triyana[1,2a]

[1] Department of Physics, Universitas Gadjah Mada, Sekip Utara Yogyakarta, 55281, Indonesia
[2] Nanomaterial Research Group, Universitas Gadjah Mada, Sekip Utara Yogyakarta, 55281, Indonesia
[3] Department of Biology, Universitas Gadjah Mada, Sekip Utara Yogyakarta, 55281, Indonesia

E-mail : triyana@ugm.ac.id

Abstract. Polyacrylonitrile (PAN) nanofiber deposited on quartz crystal microbalance (QCM) substrate as solvent vapor sensing has been successfully developed. The absorption and swelling behavior has been assumed to be responsible for sensing mechanism in vapor sensing. In this study, we aim to investigate the correlation between the swelling degree (polymer-solvent affinity) and the sensor response. The PAN nanofiber has been successfully deposited on QCM substrate with relatively homogenous nanofiber diameter about (260 ± 20) nm. The tests vapor solvent was included dimethylformamide (DMF), dimethylacetamide (DMAc), ethylene glycol (EG), toluene, ethanol, and distilled water. The results indicated that the sensor response for various vapor solvent clearly influences by its polymer-solvent affinity. The highest sensor response was achieved with DMF vapor due to its highest affinity with PAN polymer. The swelling behavior of polymer can be a potential candidate for developing vapor sensors with a polymer as an active layer.

Keywords. PAN nanofiber, QCM, vapor sensing, swelling, polymer-solvent affinity.

1. Introduction

Recently, various vapor sensors have been developed by using different sensing materials, such as conducting polymers [1], inorganic thin films [2, 3], and semiconductor metal oxides [4, 5] with good results. Moreover, various vapor sensor was also have been developed based on different sensing mechanisms such as resistance sensor [6], photoelectric sensors [7], optical sensor [8], amperometric sensor [9] and acoustic wave sensor [9]. One of the kinds of acoustic wave (AW) sensor platform is quartz crystal microbalance (QCM). A QCM sensor operates based on acoustic-electric effect and mass deposition. A QCM sensor offers an accurate, convenient, and real-time detection, sensors with high sensitivity response that is widely used as vapor/gas sensors [10]. Nanostructure material especially nanofibers have received increasing attention for various applications including gas sensor application due to their large surface area, high porosity and interconnected porous structures [11,12].

Content from this work may be used under the terms of the Creative Commons Attribution 3.0 licence. Any further distribution of this work must maintain attribution to the author(s) and the title of the work, journal citation and DOI.
Published under licence by IOP Publishing Ltd

Many researchers have been used electrospun nanofiber [13, 14] and polymer modified nanofiber [10, 15] as sensing layer for various vapor/gas sensing.

The sensing mechanism of vapor sensor mainly dominated by chemical interaction between the active layer and the analytes. However, many volatile organic compounds like methanol, ethanol, and toluene are not reactive at room temperature and mild conditions. Fortunately, they may have weak physical interactions with the sensing layer, involving absorbing or swelling the polymer matrices. These interactions do not vary the oxidation levels of conducting polymers, yet can also affect the properties of the sensing materials and make these gases/vapor detectable [16]. Several studies have been done investigating the swelling behavior of various polar solvent at electrospun nanofiber membrane in order to improve their electrical or mechanical properties [17–19]. The previous study conducted by Huang [17] investigated the swelling effect of polyacrylonitrile (PAN) nanofiber after solvent vapor treatment with N,N-dimethyl formamide (DMF). The results confirmed that swelling behavior occurred when PAN nanofiber was exposure with DMF. The swelling behavior of PAN nanofiber and DMF vapor is expected due to their high polymer-solvent affinity.

Even though the swelling mechanism was believed to be one of the common mechanisms in gas sensing devices, the deep investigation of the process and its correlation was still limited. In this study, we investigate the dependence of PAN nanofiber vapor sensor with its swelling behavior in order to get a better understanding of sensing mechanism due to the swelling process of a polymer nanofiber. The affinity between polymer and solvent has been believed to be the causes of swelling solvent in polymer matrices [20]. The swelling ability of solvent and the polymer is convinced to depend on their affinity. Meanwhile, the affinity between polymer-solvent can be calculated by its relative energy density (RED) which is the ratio between solubility parameter distance (R_a) and the interaction radius ($R_o\boldsymbol{R}_0$) of the polymer. Therefore, with that knowledge, we are investigating the correlation between the relative energy density (affinity) with the PAN nanofiber sensor under influence of various solvent vapor.

2. Materials and methods

Polyacrylonitrile (PAN) polymer with a molecular weight of 150000 from Sigma–Aldrich was used for the preparation of nanofiber sensor. Non-ionic surfactant Triton x-100 was purchased from Merck, Germany. The solvents used for vapor sensing included N,N-dimethyl formamide (DMF), dimethylacetamide (DMAc), ethylene glycol (EG), toluene, and ethanol all analytical grade were purchased from Merck, Germany. AT-Cut QCM sensor with the gold electrode and 10 MHz resonant frequency were purchased from Novaetech, Italy. All above materials used as received without any further purification.

Polyacrylonitrile (PAN) solution was prepared by dissolving 0.6 g polymer PAN into 10 mL DMF followed by mechanical stirring at 1000 rpm for 4 h at ambient temperature until a homogeneous solution was achieved. After PAN fully dissolved in DMF, a small amount (1 %) of non-ionic surfactant Triton X-100 was added to the solution. The non-ionic surfactant is added to lower the surface tension of the solution. Followed by mechanical stirring at 900 rpm for 30 min to fully dissolved the surfactant. The solution then transferred into 10 mL plastic syringes for electrospinning process. When the electrospinning process was performed, a DC voltage of 10 kV was applied, and the tip to collector distance was set as 15 cm. The collector used in the electrospinning process was metal plate covered with aluminum foil with crystal QCM put in the collector. The electrospinning process was carried out for 30 s, 60 s, and 90 s to get a suitable thickness nanofiber film. Finally, the nanofiber film was placed in the desiccator for 12 h to evaporates the residual solvent before further treatment. Scanning electron microscope (SEM) JEOL-JSM-6510 with auto fine coater JEOL JEC-3000FC was used for investigating the morphologies of electrospun nanofiber PAN.

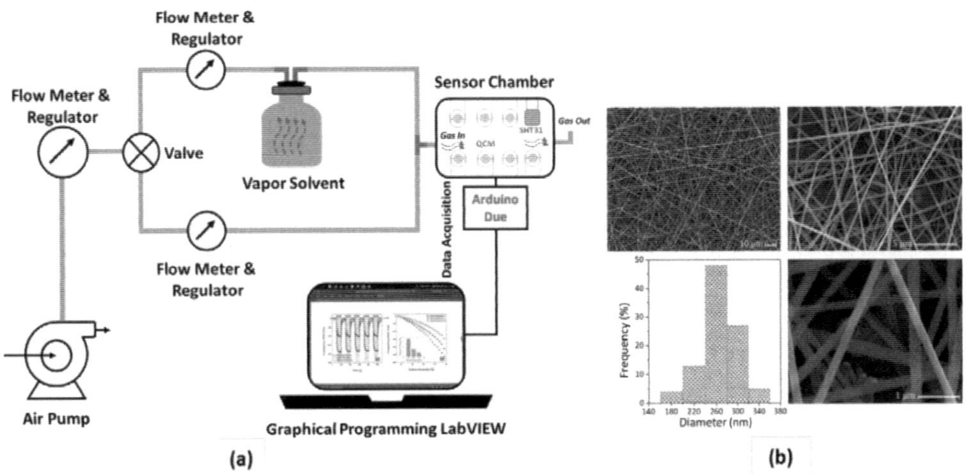

Figure 1. (a) Schematic illustration of PAN nanofiber QCM vapor sensing process, (b) SEM image of PAN nanofiber with different magnifications with nanofiber diameter distributions.

A schematic diagram of the solvent vapor sensing system is presented in Figure 1.a. The QCM sensor was installed in the testing chamber with Sensirion SHT 31 (temperature and humidity sensor) included to monitoring the humidity and the temperature of the chamber. During the experiments process, various of gas flow rate was used. The sensing of the QCM sensor was measured by a shift in the resonance frequency. The shift in the resonance frequency is directly related to mass deposition on sensing layer according to Sauerbrey equations. The resonance frequency sent to PC accompany RS232 or serial protocol with graphical programing language LabVIEW. To calculate the mass deposition of the PAN nanofiber, we used the frequency shift after coating based on Sauerbrey equations as follows [21].

$$\Delta f = -\frac{2f_0^2}{A\sqrt{\rho_q \mu_q}}\Delta m \qquad (1)$$

where Δf represents the frequency shift of crystal QCM (Hz), f_0 is the intrinsic frequency (10 MHz), Δm is the mass variation (g), A is the electrode surface area (1.130 cm^2), and ρ_q and μ_q is the shear modulus and density of quartz crystal (2.947×10^{11} g·cm^{-1}·s^{-2}) and (2.648 g·cm^{-3}), respectively.

3. Results and discussion

The PAN nanofiber vapor sensors deposited on quartz crystal microbalance (QCM) has been successfully performed. The smooth and continuous PAN nanofiber with relatively homogenous nanofiber diameter about 260 nm as shown in Figure 1.b. Nanostructured PAN nanofiber sensors offer large specific surface area, high porosity, and interconnected porous structures. As mention before that, the deposition time was varied to obtain different mass deposition. For convenience, we used the word QCM 1, QCM 2, and QCM 3 to refer the sample with deposition time 30 s, 60 s, and 90 s, respectively. The frequency shift after coating was increased for QCM 1 to QCM 3 and found about 530 Hz, 778 Hz, and 2163 Hz, respectively.

Figure 2.a shows the dynamic response of PAN nanofiber QCM sensors with various of vapor flow rate under the influence of N,N-dimethyl formamide (DMF) vapor against the background of ambient air ~ 50 % RH at room temperature. The frequency shift of the sensor shows increasing with the increase of the gas flow rate for all samples. The frequency of PAN nanofiber vapor sensor was decreased upon to increasing of vapor flow rate. Upon each injection of DMF vapor into the chamber, the sensor response showed a sharp drop and reached a steady value after several minutes. Figure 2.b shows the dynamic response (adsorption/desorption) of the sensor of PAN nanofiber sensors with 1

L/min vapor flow rate. The frequency shifts of the PAN nanofiber coated QCM sensor exposed to DMF vapor for 5 min were 12 Hz, 50 Hz, and 136 Hz for QCM 1, QCM 2, and QCM 3 respectively. The mass deposition of PAN nanofiber sensor shows in inset Figure 2.b and calculated by Sauerbrey equation in Equation 1. The mass deposition of PAN nanofiber vapor sensor was about 117 ng, 171 ng, and 476 ng for QCM 1, QCM 2, and QCM 3, respectively. The sensor response was increase with the increase of its mass deposition. Increasing mass deposition implies that the contact area between the active layer and the analysts was also increased, thus increasing the sensor response.

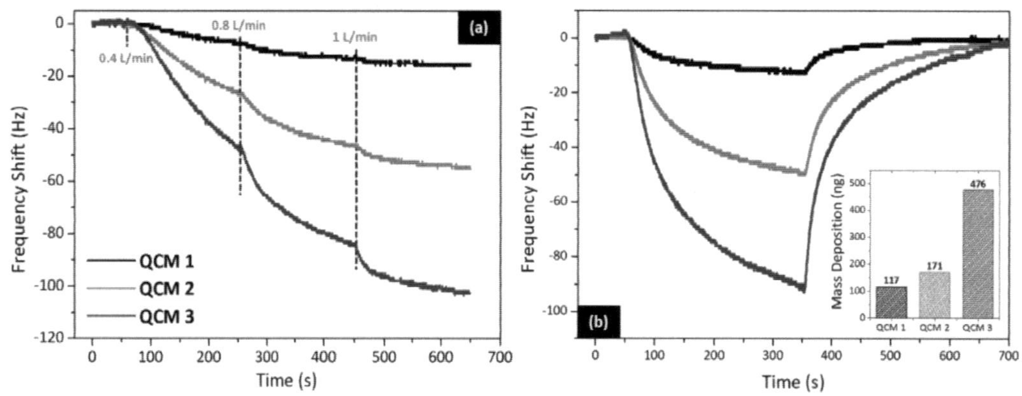

Figure 2. (a) Dynamic response of QCM sensors coated with PAN nanofiber upon exposure to increasing the vapor flow of dimethylformamide vapor, (b) One cycle of QCM response against DMF vapor (1 L/min) as a function of time for different PAN nanofiber samples (inset: mass deposition of PAN nanofiber sensors).

Figure. 3 shows the SEM image of PAN nanofiber before and after solvent evaporations. The results indicate there is a morphological change in PAN nanofiber film after evaporating with DMF vapor for one hour. The swelling behavior of PAN nanofiber under the influence of DMF vapor clearly seen at SEM images. When the PAN nanofiber film treated with DMF vapor, the film appeared to swell. This phenomenon is occurring because the vapor solvent can condense at nanofiber junction and slightly dissolved the polymer to facilitate fusion [17]. In addition to the swelling behavior, we also believe that another weak physical interaction (absorption) was also responsible for sensor response. Those two interactions occurred when PAN nanofiber under the influence of DMF vapor. The swelling behavior of PAN nanofiber under the influence of different solvent vapor is explained later.

The swelling phenomenon is commonly occurred to polymer due to absorption of solvent. The swelling degree of polymer-solvent depends on its affinity. The affinity of polymer-solvent normally express by their relative energy density (RED), which is the ratio between solubility parameter distance (R_aR_α) and the interaction radius (R_o) of the polymer. The solubility parameter distance can be calculated with equation developed by Hansen and Skaarup in literature [20].

$$(R_a)^2 = 4(\delta_{Dp} + \delta_{Ds})^2 + (\delta_{Pp} - \delta_{Ps})^2 + (\delta_{Hp} - \delta_{Hs})^2 \quad (2)$$

where, δ_D is the density parameter of dispersion interaction, δ_P is density parameter of polar cohesive energy, and δ_H is density parameter of electron exchange parameter. Meanwhile, the interaction radius (R_o) of PAN is reported in the literature about 10.9 $MPa^{0.5}$ [20].

Table 1 shows the Hansen solubility parameter and the relative energy density value of the polymer and solvent used in this study. The relative energy density 0 means there is no energy difference, RED less than 1 indicate high affinity and RED greater than 1 indicate progressively lower affinity. The polymer-solvent affinity between PAN and the solvent progressively increased from distilled water,

EG, ethanol, methanol, toluene, dimethylacetamide, and dimethylformamide. The highest and the lowest affinity was possessed by DMF and distilled water, respectively.

Table 1. Relative Energy Density (RED), Hansen Solubility Parameters of polymer-solvent used in this study

Polymer/Solvent	Hansen Solubility Parameters [a]				$RED = \frac{R_a}{R_o}$
	δ_D	δ_P	δ_H	Δ_{total}	
Polyacrylonitrile (PAN)	21.7	14.1	9.1	27.4	–
N,N–Dimethyl formamide (DMF)	17.4	13.7	11.3	24.9	0.815
Dimethyl Acetamide (DMAc)	16.8	11.5	10.2	22.7	0.936
Toluene	18.0	1.4	2.0	18.2	1.498
Ethanol	15.8	8.8	19.4	26.5 2	1.517
Ethylene glycol (EG)	17.0	11.0	26.0	32.9	1.798
Distilled water	16.6	16.0	42.3	48.2	3.191

Figure 3.c shows the correlation between the response of the PAN sensor under the influence of different solvent versus its relative energy density. The higher response was achieved by the DMF vapor with 20 Hz frequency shift, while the lowest response with 2 Hz frequency shift was possessed by distilled water vapor. The results are indicating that response of the PAN nanofiber vapor sensor directly depended on their polymer-solvent affinity. The direct correlation between the PAN nanofiber sensor and their polymer solvent affinity is suitable for our main hypothesis.

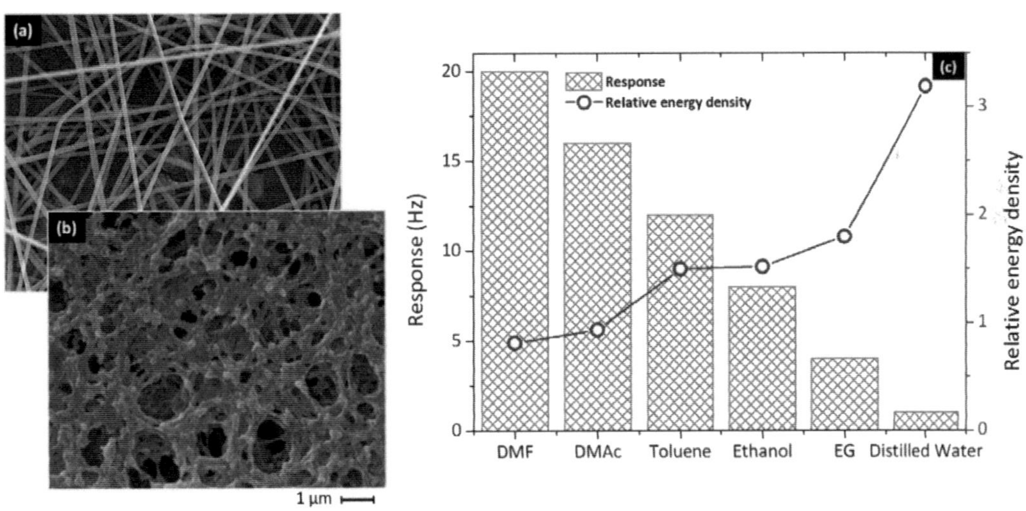

Figure 3. SEM image of PAN nanofiber sensor (a) before and (b) after vapor evaporation with DMF vapor for one hour; (c) PAN nanofiber response (QCM 1) under the influence of various solvent vapor with their relative energy density

4. Conclusions

This study has been successfully developed a vapor sensor based on PAN nanofiber deposits on QCM substrate. The PAN nanofiber structures were clearly seen by SEM images. The sensor also shows good sensitivity and medium response. The swelling behavior of PAN nanofiber under the influence of solvent vapor was evidenced by the SEM image. The swelling degree of between polymer and solvent depends on their affinity. The swelling mechanism has been confirmed to be responsible for the response of the sensor; this could be an interesting approach to developed solvent vapor sensor devices.

Acknowledgments

This work was supported by the "Penelitian Pendidikan Magister Menuju Doktor Sarjana Unggulan (PMDSU), Contract No. 1991/UN1-P.III/DIT-LIT/LT/2017" from the Ministry of Research Technology and Higher Education, the Republic of Indonesia.

References

[1] Bora A *et al* 2017 A room temperature methanol vapor sensor based on highly conducting carboxylated multi-walled carbon nanotube/polyaniline nanotube composite *Sensors Actuators B. Chem.* **253** pp 977–986
[2] Kaushik A *et al* 2015 Organic–inorganic hybrid nanocomposite-based gas sensors for environmental monitoring *Chem. Rev.* **115** pp 4571–4606
[3] Evans S D *et al* 2000 Vapour sensing using hybrid organic–inorganic nanostructured materials *J. Mater. Chem.* **10** pp 183–188
[4] Dhahri R *et al* 2017 Enhanced performance of novel calcium/aluminum co-doped zinc oxide for CO2 sensors *Sensors Actuators B. Chem.* **239** pp 36–44
[5] Hidayat S N and Triyana K 2016 *AIP Conf. Proc.* **1755** p 20001
[6] Wang J *et al* 2017 High performance hydrazine vapor sensor based on redox mechanism of twisted perylene diimide derivative with lower reduction potential *Sensors Actuators B. Chem.* **239** pp 898–905
[7] Aba L *et al* 2014 *J. Nanotechnol.* **2014** pp 1–5
[8] Evyapan M *et al* 2016 Study of the sensor response of spun metal phthalocyanine films to volatile organic vapors using surface plasmon resonance *Sensors Actuators B. Chem.* **236** pp 605–613
[9] He K *et al* 2014 Amperometric determination of hydroquinone and catechol on gold electrode modified by direct electrodeposition of poly (3, 4-ethylenedioxythiophene) *Sensors Actuators B. Chem.* **193** pp 212–219
[10] Wang N *et al* 2014 Electrospun nanofibrous chitosan membranes modified with polyethyleneimine for formaldehyde detection *Carbohydr. Polym.* **108** pp 192–199
[11] Mahendra B, Nugroho F and Yusuf Y 2018 Low-frequency electrohydrodynamic convection patterns in nematic liquid crystal aligned using parallel-oriented nanofiber *Jpn. J. Appl. Phys.* **57** p 21701
[12] Triyana K *et al* 2015 Electrospun nanofibers based on polyvinyl alcohol/chitosan and its stability in KOH solution *Mater. Sci. Forum* **827** pp 321–325
[13] Jia Y et al 2015 Phenylacetic acid-modified nanofibrous polystyrene membranes for use as highly sensitive ammonia sensors *Sensors Actuators B. Chem.* **212** pp 273–277
[14] Hidayat S N *et al* 2017 *International Seminar on Sensors, Instrumentation, Measurement and Metrology* (ISSIMM) (IEEE) pp 119–123
[15] Huang W *et al* 2013 Highly sensitive formaldehyde sensors based on polyvinylamine modified polyacrylonitrile nanofibers *RSC Adv.* **3** pp 22994–23000
[16] Bai H and Shi G 2007 Gas sensors based on conducting polymers *Sensors* **7** pp 267–307
[17] Huang L, Manickam S S, and McCutcheon J R 2013 Increasing strength of electrospun nanofiber membranes for water filtration using solvent vapor *J. Memb. Sci.* **436** pp 213–220
[18] Chotimah *et al* 2016 *Int. J. Adv. Sci. Eng. Inf. Technol.* **6** pp 675–681
[19] Nasikhudin *et al* 2017 *Mater. Sci. Forum* (MSF) **901** pp 79–84
[20] Hansen C M 2007 *Hansen Solubility Parameters A User's Handbook* (CRC Press Taylor and Francis Group, Boca Raton)
[21] Sauerbrey G 1959 *Zeitschrift für Physik* **155** pp 206–222

Detection of Vegetable Oil Variance Using Surface Plasmon Resonance (SPR) Technique

Supardianningsih[1], R. D. Panggabean[1], I. A. Romadhon[1], F. D. Laksono[1], U. Nofianti[1], and K. Abraha[1]

[1]Department of Physics, Universitas Gadjah Mada, Yogyakarta, Indonesia

E-mail : kamsul@ugm.ac.id

Abstract. The difference between coconut oil, corn oil, olive oil, and palm oil has been detected using surface plasmon resonance (SPR) technique. This is a new method in material characterization that can be used to identify vegetable oil variance. The SPR curve was measured by SPR system consisting of optical instruments, mechanical instruments, Main UNIT, and user interface (computer). He-Ne laser beam of wavelength 633 nm was used as light source, while gold (Au) thin film evaporated on half cylinder prism was used as the base so that surface plasmon polariton (SPP) waves propagate at the interface. Tween-80 and PEG-400 are used as surfactant and co-surfactant to make water-oil emulsion from each sample. The sample was prepared with the ratio of oil: surfactant: co-surfactant as 1:2:1 and then stirred on the water to make emulsions. The angle shift was measured by the change of SPR angle from prism/Au/air system to prism/Au/water-oil emulsion. The different SPR angle of each sample has been detected in the various number of spray, a method that was used for depositing the emulsion. From this work, we conclude that the saturated fatty acid component was the most significant component that changes the refractive index in the vegetable oil in water emulsion that can be used to characterize the vegetable oil variance.

Keywords. HE-Ne laser, SPR, and vegetable oil.

1. Introduction

Vegetable oils are important materials for home cooking and food industry. The composition of contained fatty acid makes the difference of vegetable oil. Some fatty acids are essential for growth and in metabolism so we must get them from the diet. Unsaturated fatty acid is good for heart-healthy because it can reduce blood pressure and tend to promote inflammation, while saturated fat increases heart disease risk and blood pressure [1]. We have to know the difference between vegetable oil variance because it will affect our healthy life.

Several works have been reported for characterizing vegetable oil such as electronic nose [2], gas chromatography [3], Fourier transform infrared [4], and digital image and pattern recognition techniques [5]. However, most of the methods are highly-sophisticated and expensive. In a developing country like Indonesia, we need a low cost and simple method for characterizing material such as surface plasmon resonance (SPR) technique. Although this method was less accurate than the other methods, SPR technique offers a simple, rapid and accurate tool for distinguished vegetable oil variance.

Content from this work may be used under the terms of the Creative Commons Attribution 3.0 licence. Any further distribution of this work must maintain attribution to the author(s) and the title of the work, journal citation and DOI.
Published under licence by IOP Publishing Ltd

SPR technique has a high sensitivity and fast response due to the refractive index changes at the sensor interface. The penetration depth or decay length explains the sensitivity at the interface of the evanescent field. Only changes of the dielectric properties in the vicinity of the interface between the dielectric and metal will affect the evanescent field. This extreme sensitivity to the surrounding dielectric media makes the SPR technique is an extraordinarily useful tool for characterizing materials in different fields [6]. In this work, we aim to distinguish vegetable oil variance using homemade SPR system.

2. Materials and methods

2.1. Material and reagents

Olive oil, corn oil, palm oil, and coconut oil were purchased from the local store Mirota Kampus. Polyethylene glycol 400/PEG 400 ($HO(C_2H_4O)_n H$, Merck) was purchased from Laboratorium Penelitian dan Pengujian Terpadu (LPPT) UGM. Polyoxyethylene 20 sorbitan monooleate which is also known as Tween 80 ($C_{64}H_{124}O_{26}$ with the molecular weight of 1,310 $g \cdot mol^{-1}$) was purchased from Gudang Kimia Jogja Chemical Shop. Au (99 %, Antam) was obtained from Laboratorium Fisika Material UGM. Double distilled water was purchased from Progo Chemical Shop.

2.2. Sample preparation

Vegetable oil in water emulsion was prepared by mixing the vegetable oil (1 mL), Tween 80 (2 mL), and PEG 400 (1 mL). Tween 80 was a hydrophilic nonionic surfactant and used as an emulsifier. PEG 400 was a cosolvent/cosurfactant which makes the stability of emulsion by wedging themselves between surfactants molecules. The ratio of oil: surfactant: cosurfactant is 1:2:1 so that a self micro-emulsifying performance was successfully obtained [7]. The next process to make an emulsion was dissolving and stirring all of them with 40 mL of double distilled water.

Pure gold (Au 99 %) with mass 20 mg was evaporated on the surface of the half cylinder prism using Vacuum Evaporator System JEE 4X EM 300023144 RA JEOL-Technic Co. Ltd, Japan to make a gold thin film called Prism/Au/Air system. This system was characterized by SPR technique to get the SPR Curve or Attenuated Total Reflection (ATR) Curve as the base curve to distinguish the vegetable oil and get the SPR Angle. After sprayed with the emulsion, the system was called Prism/Au/Vegetable oil-Water emulsion/Air system and the SPR Angle shifted to the higher number.

2.3. Sample characterization

This sample was characterized by Fourier Transform Infra-Red (FTIR) and SPR technique. FTIR transmission characterization was used to obtain the contained functional groups. All samples were made under two conditions (i) pure vegetable oil and (ii) vegetable oil in water emulsion. Both samples were analyzed with FTIR spectrophotometer 8201PC Shimadzu either in Laboratorium Kimia Organik dan Biokimia UGM (pure vegetable oil) and Laboratorium Politeknik ATK Yogyakarta (vegetable oil in water emulsion).

In the SPR characterization, the emulsion was deposited onto the surface of Au thin film using spray method. The SPR system consists of optical instruments, mechanical instruments, main UNIT, and user interface (computer). Optical instruments in this system are a He-Ne laser beam of wavelength 633 nm, a beam splitter, a half cylinder prism, and a pair of a light detector. The mechanical system was used to rotate both prism and detector. Main UNIT and user interface were used to control this measurement. SPR device set-up is shown in Figure 1.

The light from the laser was passed through a polarizer to get the p-polarized or a transverse magnetic (TM) light mode and then divided into two parts with the same wavelength and intensity by the beam splitter. The intensity of first p-polarized light was measured by the detector as the reference laser intensity. Meanwhile, the second light was reflected by the prism towards the second detector to get the reflected light intensity. The reflectivity was obtained by dividing the reflected light and the first light intensity. In this experiment, we varied the angle by rotating the prism and the second detector (angle scanning).

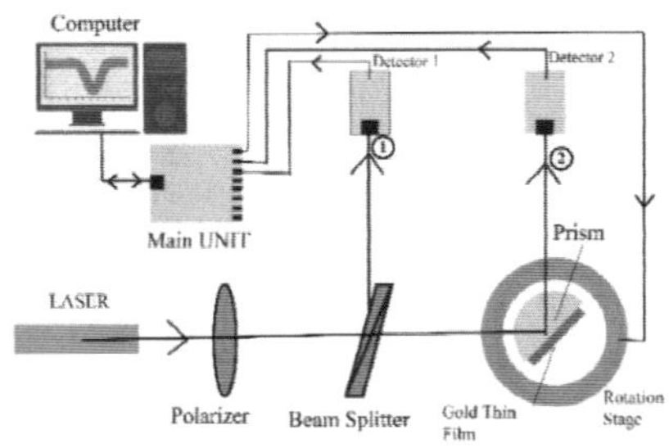

Figure 1. SPR device set-up

3. Results and discussion

3.1. FTIR Analysis

We know that every vegetable oil differs in composition, length and unsaturated degree of the fatty acid as well as their position in the chain. The double bond is considered as unsaturated fatty acid, while the single bond is considered as saturated bond. The fatty acid composition of each sample is shown in Table 1.

Table 1. Typical Fatty Acid Compositions of Some Vegetable Oil [8]

Fatty Acid Component	Corn Oil	Palm Oil	Coconut Oil	Olive Oil
Saturated	16%	48%	90%	12%
Caprylic 8:0	-	-	8%	-
Capric 10:0	-	-	7%	-
Lauric 12:0	-	-	48%	-
Myristic 14:0	-	-	16%	-
Palmitic 16:0	13%	44%	9%	10%
Stearic 18:0	3%	4%	2%	2%
Unsaturated	84%	50%	9%	86%
Oleic 18:1 n-9	31%	40%	7%	78%
Linoleic 18:2 n-6	52%	10%	2%	7%
Linolenic 18:3 n-3	1%	-	-	1%
Other	-	2%	1%	2%

The symbol for an acid includes the total number of carbons and double bonds, followed by the location of the first double bond counting from the methyl end of the chain. From that table, we know that both corn oil and olive are high in unsaturated fatty acid. The difference is that corn oil is dominant in polyunsaturated fatty acid (linoleic acid), while olive oil is dominant in monounsaturated fatty acid (oleic acid). Meanwhile, coconut oil is high in saturated fatty acid, while palm contains both saturated and unsaturated fatty acids is mostly in the same percentage.

Figure 2 (a) shows the FTIR transmission spectra of pure vegetable oil samples. The triglyceride is dominant in the spectra. The major peaks that represent triglyceride functional groups could be observed around 2924.09 cm^{-1} (C–H stretching (asymmetry)), 2854.65 cm^{-1} (C–H stretching (symmetry)), 1743.65 cm^{-1} (C=O stretching), 1458.18 cm^{-1} (C–H bending (scissoring)), 1165 cm^{-1} (C–O stretching and C–H bending), and 725.23 cm^{-1} (C–H bending (rocking)). There was a very weak peak at 1651.07 cm^{-1} (Olive Oil, Corn Oil, and Palm Oil) and 1612.49 cm^{-1} (Coconut Oil) in IR spectra, which was C=C stretching. In the other hand, the C–C bond was found at 1226.73 cm^{-1} (Coconut Oil) and 1234.44 cm^{-1} (Olive Oil, Corn Oil, and Palm Oil). These show that all of the samples have both saturated and unsaturated fatty acid components. This means FTIR show the same result of all samples (it can't detect the difference in the samples).

IOP Conf. Series: Materials Science and Engineering **367** (2018) 012021 doi:10.1088/1757-899X/367/1/012021

Figure 2.b shows the FTIR transmission spectra of vegetable oil in water emulsions. There was a dominant peak at 3313.98 cm^{-1} in the IR spectra that represent the water and PEG 400 (both have O–H bond), while the peaks of triglyceride functional groups of vegetable oil decreased. The other dominant peak was a C=C bond at 1637.30 cm^{-1}. From the FTIR spectra of pure vegetable oil, we found that the peak of the C=C bond was very weak. The increasing intensity of this C=C bond at the IR spectra of water in oil emulsion represents the Tween 80.

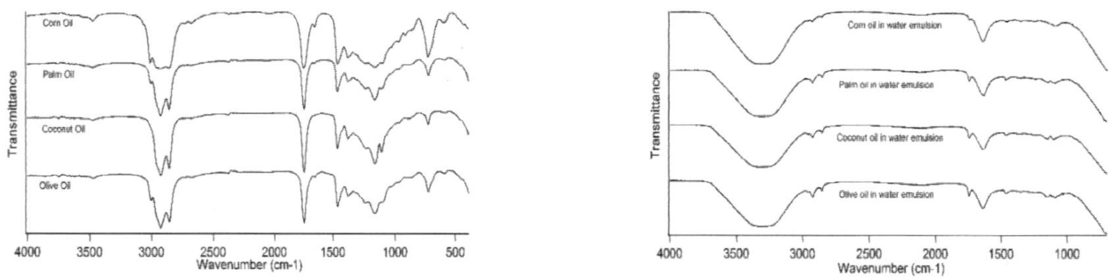

(a) (b)

Figure 2. FTIR spectra of (a) pure vegetable oils and (b) vegetable oil in water emulsions

3.2. SPR analysis

Figure 3 shows the ATR curve of vegetable oil in water emulsion. The dotted line shows the ATR curve of Prism/Au/Air system functioned as the base curve or the reference curve, while the full line shows the ATR curve of Prism/Au/Vegetable oil/Air system in multiple numbers of spray from 10x to 50x number of sprays. The shifts of ATR curve show the characteristic of each vegetable oil in water emulsion. That is caused by the increase of the refractive index at the sensor surface. The value of the refractive index increment depends on the structure of the analyte molecules [9], hence the angle shift of each sample is different. The number of shifts in SPR angle and minimum reflectance is described in Table 2 and Table 3.

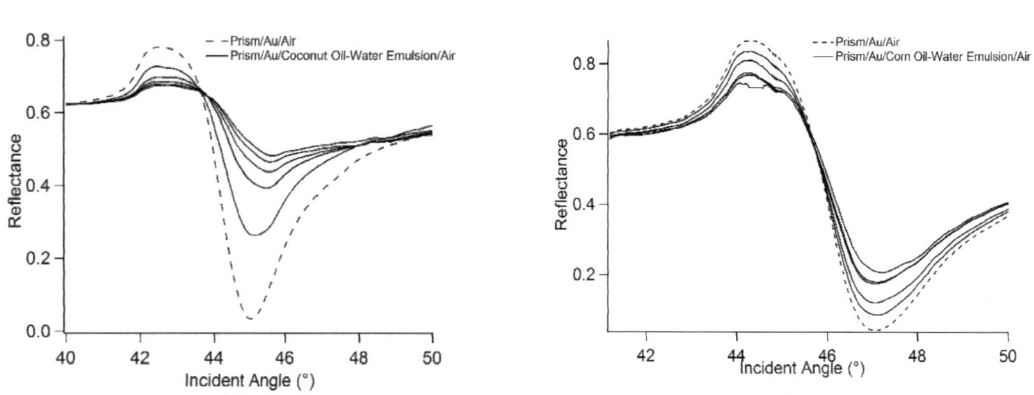

(a) (b)

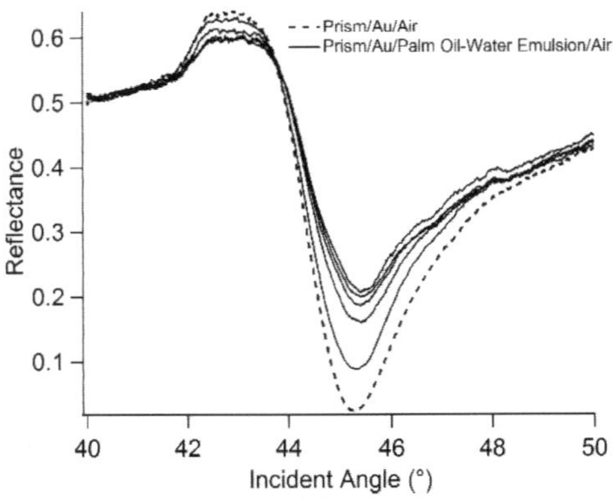

(c)

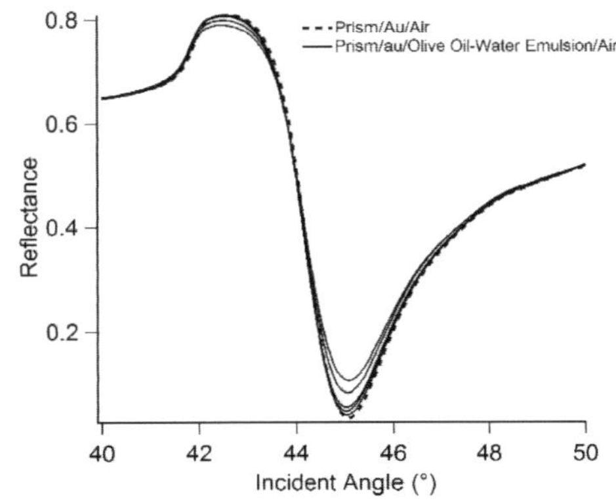

(d)

Figure 3. The ATR curve of (a) coconut oil in water emulsion, (b) corn oil in water emulsion, (c) palm oil in water emulsion, and (d) olive oil in water emulsion

Table 2. The SPR angle shift (°) of the vegetable oil variance

Number of Sprays	Coconut Oil	Corn Oil	Palm Oil	Olive Oil
10x	0	0.06	0.04	0
20x	0.11	0.06	0.16	0
30x	0.46	0.06	0.16	0
40x	0.52	0.06	0.16	0.03
50x	0.58	0.17	0.16	0.06

Table 3. The Increasing of minimum reflectance of the vegetable oil variance

Number of Sprays	Coconut Oil	Corn Oil	Palm Oil	Olive Oil
10x	0.23	0.01	0.07	0.01
20x	0.36	0.05	0.14	0.02
30x	0.41	0.13	0.17	0.02
40x	0.44	0.14	0.18	0.05
50x	0.45	0.17	0.19	0.08

From Table 2 and Table 3, we find that both of the SPR angle and the minimum reflectance are shifted to a higher number when the number of sprays is increased, but the amount of each sample was different. Coconut oil in water emulsion shows the biggest shift both in the SPR angle and the increase of minimum reflectance, while olive oil in water emulsion shows the smallest shift. According to the fatty acid composition in Table 1, coconut oil contained 90 % saturated fatty acid while olive oil contained 12 % fatty acid. Coconut oil containing the highest fatty acid component shows the biggest shift both in the SPR angle shift and the minimum reflectance. In the other hand, olive oil containing the lowest fatty acid shows the smallest shift both in the SPR angle shift and the minimum reflectance. The SPR angle of corn oil and palm oil have the similar result after sprayed to the Au thin film. Nevertheless, palm oil containing 48 % saturated fatty acid shows the bigger minimum reflectance shift than corn oil which contains 16 % saturated fatty acid.

The other result, the SPR sensor is able to quantitatively describe the interaction of oleic acid molecules on gold thin film by detecting localized small refractive index shifts [10]. The change in the amount of free fatty acids sensed by SPR correlated well with the acid value of the AOCS official method when refined vegetable oils and pressed and non-refined vegetable oils are measured. These support our result that fatty acid component, especially the saturated fatty acid, is the most significant components that change the refractive index of the vegetable oil in water emulsion. The SPR technique can distinguish coconut oil, corn oil, palm oil, and olive oil by detecting the changes of the refractive index at the sensor interface.

4. Conclusions

We have successfully identified differences between coconut oil, corn oil, olive oil, and palm oil that have been detected using SPR technique. The difference in the ATR curve of vegetable oil in water emulsion was dominantly caused by the saturated fatty acid composition of each sample. This confirms the new method for characterizing materials, especially in identifying vegetable oil variance which could be further developed in the near future.

Acknowledgments

We gratefully acknowledge support from Hibah Penelitian Dosen Departemen Fisika Universitas Gadjah Mada with funding from BPPTN/DAMAS TAHUN ANGGARAN 2017.

References

[1] H.J. Roy 2013 *Coconut Oil* (USA: Pennington Biomedical Research Center, Lousiana State University) Pub No 94

[2] Xu L 2016 A novel method for qualitative analysis of edible oil oxidation using an electronic nose *Food chemistry* **202** pp 229–235

[3] Wrona M, Pezo D, and Nerin C 2013 Rapid analytical procedure for determination of mineral oils in edible oil by GC–FID *Food chemistry* **141** pp 3993–3999

[4] Jiang X *et al* 2016 Determination of the acid values of edible oils via FTIR spectroscopy based on the OH stretching band *Food chemistry* **212** pp 585–589

[5] Milanez K D T M and Pontes M J C 2014 Classification of edible vegetable oil using digital image and pattern recognition techniques *Microchemical Journal* **113** pp 10–16

[6] Rubio A S 2015 *Modified Au-based nanomaterials studied by surface plasmon resonance spectroscopy* (Springer)

[7] Shah I 2011. *Development and characterization of oil-in-water nanoemulsions from self-microemulsifying mixtures* (The University of Toledo

[8] Gunstone F D 1996 Fatty acids—nomenclature, structure, isolation and structure determination, biosynthesis and chemical synthesis *In Fatty Acid and Lipid Chemistry* (MA: Springer) pp 1–34

[9] Homola, J., 2006. Electromagnetic theory of surface plasmons. *In Surface plasmon resonance based sensors* (Berlin: Springer) pp 3–44

[10] Cho Y J, Kim T E and Gil B 2013 Correlation between refractive index of vegetable oils measured with surface plasmon resonance and acid values determined with the AOCS official method *LWT-Food Science and Technology* **53** pp 517–521

Photoresist: Fabrication, Characterization and Its Sensitivity on the Exposures of X-Ray and Ultraviolet

Sutikno[1], Susilo[1], and H. D. Raharja[1]

[1]Department of Physics, Faculty of Mathematics and Natural Sciences, Semarang State University, D7 Building, 2nd Floor, Sekaran, Gunungpati, Semarang, Indonesia

E-mail: smadnasri@yahoo.com

Abstract. The epoxy resin-based photoresist is fabricated by mixing of resin (polymer), sodium acetate trihydrate and ethanol in mass variation using heated magnetic stirrer at 100 rpm speed and temperature of 75 °C. Sodium acetate trihydrate and ethanol each play role as photoactive compound (PAC) and solvent, respectively. Photoresist thin films were grown through spin coating method in voltage 5 V during the 60 s and heating temperature of 150 °C for 15 min. To determine photoresist sensitivity, ultraviolet and X-ray were exposed on the photoresist surfaces. The fabricated photoresist properties are densities of 1 $g{\cdot}mL^{-1}$ to 1.23 $g{\cdot}mL^{-1}$, dynamic viscosities of 7 Cp to 22 Cp and kinematic viscosities of 7 Cst to 18 Cst. The absorbances of thin films are in the wavelength range of 350 nm to 1050 nm at the maximum absorbances of 0.2 to 0.5 in the wavelength g-line, h-line, and i-line. The generated maximum current achieved (1.84×10^{-8}) A. The microstructures of epoxy-based photoresist seem homogeneous. The sensitivities of UV exposures show a photochemistry reaction on photoresist occurred, however for X-ray exposure no reaction found.

Keywords. Epoxy, photoresist, and photosensitivity.

1. Introduction

Very fast microelectronic industry development requires sustainable raw materials supply [1]. One of the raw materials in microelectronic industries is resisted material, a material which is composed of four ingredients namely resin (polymer), photoactive compound (PAC), solvent and additive [2]. The photoresist is a complex polymer and additive with a low molecular weight which plays a role as photosensitive material and it is used in manufacturing printed circuit board [3].

The photoresist is classified into two kinds namely negative and positive photoresist. For positive one, the radiation-exposed area is soluble in photoresist developer; otherwise for a negative one, the exposed area remains to stay on the photoresist surface [4]. In the manufacturing process of a semiconductor, the radiation sources which are often used in a lithography process are ultraviolet (UV) and X-ray [5].

The epoxy-based resin is selected as an alternative due they are available on the broad market either in the liquid or solid. They have good adhesion on the semiconductor surface, high sensitivity and cheap [3]. Epoxy resin is copolymer thermosetting plastic which is often used in the composite polymer matrix. The epoxy resin is produced by chemical reaction process between epichlorohydrin and bisphenol-A compound [6]. The epoxy-based resin is used to make patterns by applying UV

Content from this work may be used under the terms of the Creative Commons Attribution 3.0 licence. Any further distribution of this work must maintain attribution to the author(s) and the title of the work, journal citation and DOI.
Published under licence by IOP Publishing Ltd

lithography technique, a method to transfer patterns of the master onto the substrate. As an example of epoxy-based photoresists is negative photoresist SU-8. Epoxy resins have advantages of good mechanical properties, good chemical resistance, and good electrical isolation as well as low cost [7]. The photoresist SU-8 is used in MEMS applications. All related parameters to produce a good photoresist image need to be optimized [8]. The uses of photoresist in the microelectronics industry are very broad so that the crucial aspects need to be investigated and developed to get good properties and performances such as sensitivity, resolution and contrast parameter. They both determine significantly the fabricated device density in the integrated circuit (IC) system. The contrast level of UV-lithography technique-developed image is affected by wavelengths of radiation sources. The higher the sensitivity of the material than the lithography process better because it easily absorbs radiation energy to perform photochemical reactions in forming patterns with high contrast levels. This research focuses on the manufacturing of epoxy resin-based photoresist, characterizing and testing photoresist sensitivity to UV and X-ray exposures.

2. Materials and methods

The used equipment in this research includes digital balance, heated magnetic stirrer, thermometer, beaker glass, measuring glass, screen filter, substrate glass, spin coater, oven, viscometer Ostwald, spectrometer ocean optic Vis-NIR USB 4000, I-Vmeter elkahfi 100 [9], microscope MS-804. The printed circuit board (PCB), aluminum foil, UV light, and X-ray machine. The used materials consist of photoresist (Aldrich), developer, de-ionized water, epoxy (polymer matrix), sodium acetate trihydrate (photoactive compound, PAC), and ethanol (solvent).

The samples are prepared of five composition variations of epoxy resin, and on the other hand, sodium acetate trihydrate was fixed as shown Table 1. The ethanol mass for each solution is selected as the independent variable and made to investigate their effects on the characterized material properties.

Table 1. Composition of Photoresist

Sample	Epoxy Resin (g)	Ethanol 96% (g)	Sodium Acetate Trihydrate (g)
A	20	12	4
B	20	14	4
C	20	16	4
D	20	18	4
E	20	20	4

The photoresist was fabricated using epoxy resin as a polymer matrix, sodium acetate trihydrate as PAC and ethanol as solvent. The mixing was done using magnetic stirrer accompanied heating process up to 75 °C. The photoresist was spin-coated on the substrate at 150 °C during 15 min.

The manufactured photoresist liquids were measured to determine their densities and viscosities. The photoresist films on glass substrates were characterized to determine absorbances, generated currents and voltages, and thin film surface structures. The thin films which were tested their sensitivities and contrast spin-coated on the PCB were exposed by UV and X-ray radiation.

3. Results and discussion

Densities of photoresist solutions are plotted in Figure 1, as the ethanol mass increases the density increases as well. In this research, the ethanol amount was optimized to obey the best performance of manufactured photoresist and applicable for lithography.

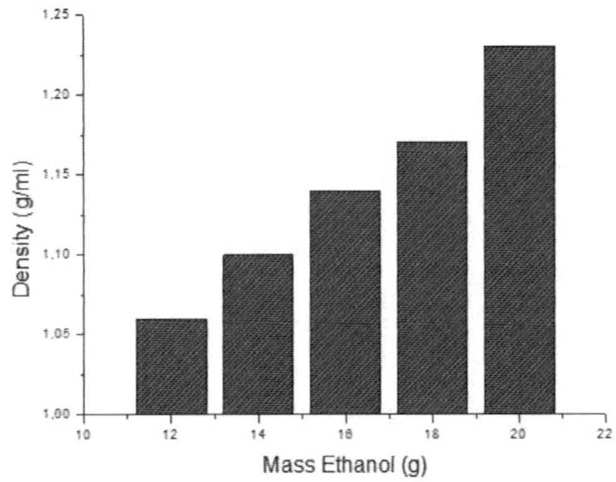

Figure 1. Graph of Densities of photoresist at different pH

Viscosities of photoresist solutions were measured using Ostwald viscometer. First, they were sucked using capillary pipe **B** (lower level) up to **A** (upper level). Then they were allowed to flow freely from line **A** down to line **B** using a stopwatch, and noted times were used to determine viscosities based on the following Equation 1.

$$\frac{\eta^{\circ}}{\eta} = \frac{\rho^{\circ} \cdot t^{\circ}}{\rho \cdot t} \quad (1)$$

η° = Standard liquid viscosity/Water viscosity (Poise)
η = Sample viscosity (Poise)
ρ° = Standard liquid density/ Water density $\left(g \cdot mL^{-1}\right)$
ρ = Sample density $\left(g \cdot mL^{-1}\right)$
t° = Average time for standard liquid movement / water movement (s)
t = Average time for sample movement (s)

Table 2. Dynamic viscosities of photoresist solutions

Sample	Dynamic Viscosity (Cp)
A	22.21
B	16.75
C	15.05
D	10.57
E	7.42

Based on Table 2, dynamic viscosity sample **A** is highest of all. Kinematic viscosity is determinable using ratio viscosity and density [10]. On Table 3, kinematic viscosity sample **A** is highest of them, and sample **E** viscosity equals to 18.05 Cst. To achieve deeper submicron device geometries, lithography plays an important role in the semiconductor manufacturing. It needs tighter resist thickness uniformity to control linewidth. The low viscosity photoresists for less than 7000 Å film thickness is available to produce the higher degree of substrate planarization and need for an improved lithographic solution. Khrisna et al. (1998) have used low viscosity photoresist (10 Cp) for lithography process [11].

IOP Conf. Series: Materials Science and Engineering **367** (2018) 012022 doi:10.1088/1757-899X/367/1/012022

Table 3. Kinematic viscosities of photoresist.

Sample	Kinematic Viscosity (Cst)
A	18.05
B	14.32
C	13.20
D	9.61
E	7

The kinematic viscosity of photoresist is affected by its ingredients, as the epoxy resin increases the viscosity increases as well. The epoxy resin increases mean the solution becomes smaller, and the otherwise conditions will occur. The photoresist strength depends on the viscosity.

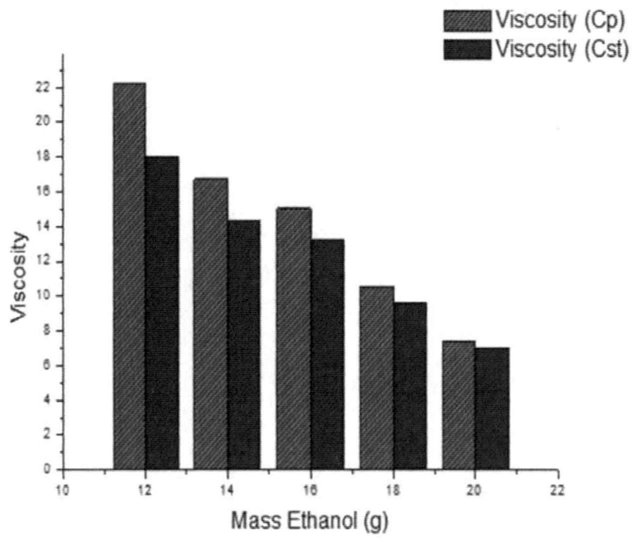

Figure 2. Dynamic and kinematic viscosities at the different mass of photoresist

In Figure 2, the higher concentration of photoresist solution the higher density. The absorbances were measured using spectrometer ocean optic Vis-NIR USB4000 and calculated based on the Equation 2.

$$OD = \log\left(\frac{1}{T}\right) = -\log(T) = A \tag{2}$$

where OD (Optical Density), A (Absorbance) and T (Transmittance). The absorbances of photoresist thin films with a different wavelength of incident light are shown in Figure 3.

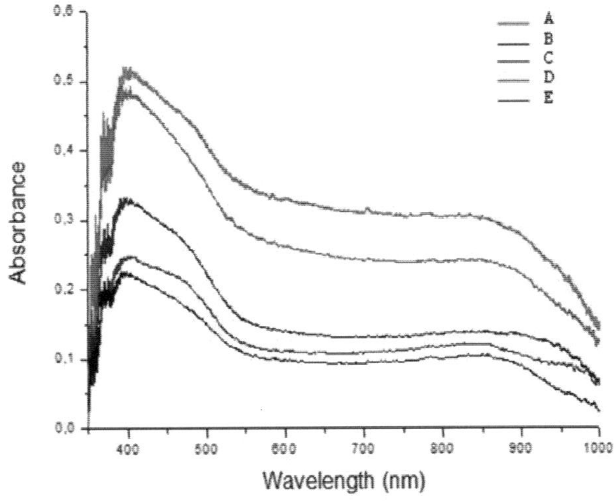

Figure 3. Graph of absorbances of photoresist thin films with a different wavelength of incident light

Based on Figure 3, absorbances of photoresist thin films are in the range of wavelengths 350 nm to 1050 nm, and the highest absorbance is in the wavelength range of 350 nm to 450 nm. Sample **A** was photoresist thin film which was heated at 75 °C and performed the highest absorbance, the maximum absorbance of light existed at 404.62 nm equals to 0.51 and sample **E** existed at 391.15 nm, it performed the lowest absorbance (0.22). The higher absorbance of photoresist thin film the higher photoresist sensitivity [12]. In the lithography, lights with wavelengths of 365 nm, 405 nm, and 436 nm were used to determine absorbances at i-line, h-line, and g-line, respectively [13] as shown Table 4.

Table 4. Absorbances for each i-line, h-line, and g-line epoxy resin-based photoresist.

Sample	i-line (365 nm)	h-line (405 nm)	g-line (465 nm)
A	0.15	0.21	0.18
B	0.14	0.24	0.21
C	0.21	0.32	0.27
D	0.32	0.47	0.41
E	0.30	0.51	0.45

The generated current of light exposed photoresist thin films are shown graphically in Figure 4. The highest one, (1.84×10^{-8}) A, is generated by sample **A**, and the lowest one, (5.71×10^{-10}) A. For the different composition of photoresist, the generated current is also different.

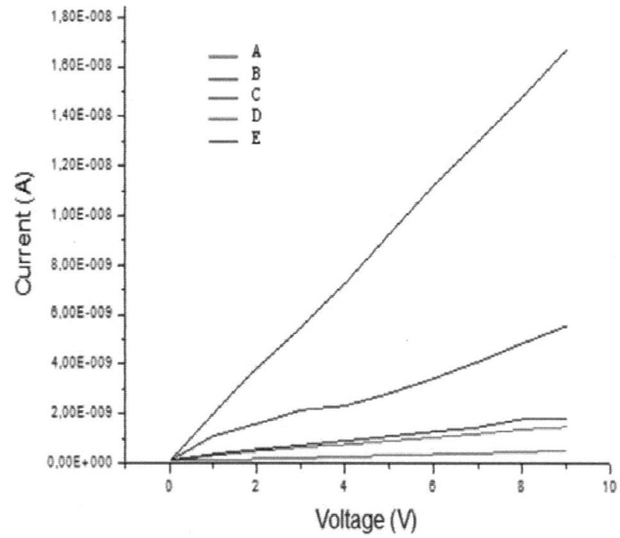

Figure 4. Graph of generated currents of photoresist thin films corresponds to source voltages

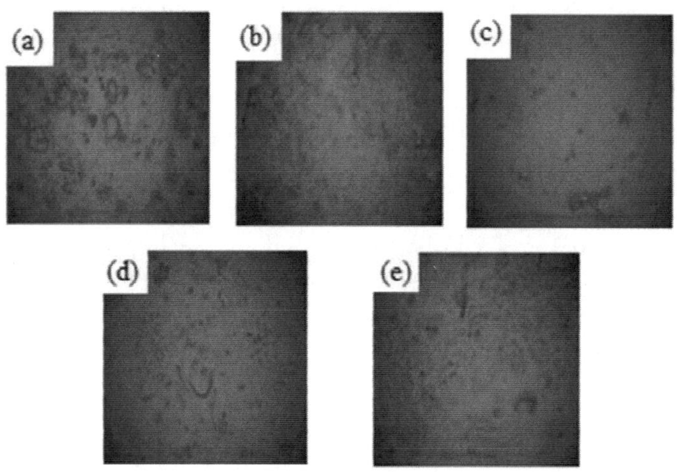

Figure 5. Structures of photoresist thin films.

Structures of photoresist thin films for all samples are shown in Figure 5. The photoresist thin films seem homogeneous. The solvent fraction increase in the solution leads to photoresist thin film coagulates on the substrate surface. Inversely, the solvent fraction decrease in the solution produces photoresist thin film more homogeneous, here sample **A** is the most homogeneous. The covered area is un-exposed, while the uncovered one is exposed, and after development, they show different results, as shown in Figure 6.

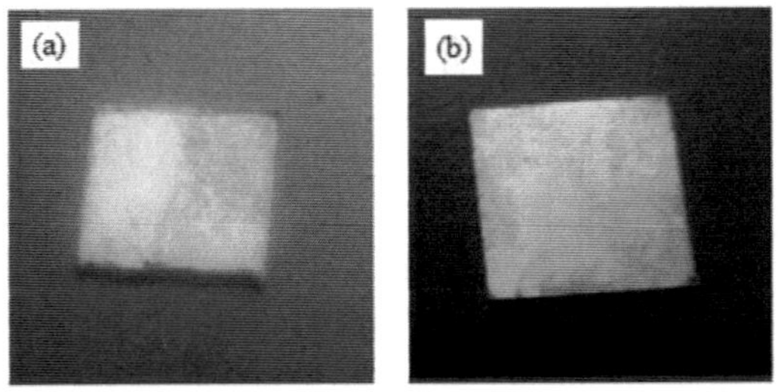

Figure 6. Photoresist spin-coated on the PCB: (a) Epoxy resin-based photoresist (Experiment result); and (b) Commercial photoresist (Sigma Aldrich).

The patterns can be developed on the epoxy resin-based photoresist and photoresist (Sigma Aldrich), although which were developed on the epoxy resin-based photoresist are less contrast. This contrast level difference is affected by solution and developer.

4. Conclusions

Epoxy resin-based photoresist (densities of 1 g·mL^{-1} to 1.23 g·mL^{-1}) is fabricated by mixing epoxy resin, ethanol, and sodium acetate trihydrate. The dynamic and kinematic viscosity of manufactured photoresist equal to 7 Cp to 22 Cp and 7 Cst to 18 Cst, respectively. The absorbances of epoxy resin-based photoresist exits in the range of 350 nm to 1050 nm with their absorbances of 0.2 to 0.5, respectively and the generated currents are as much as (5.71×10^{-10}) A to (1.84×10^{-8}) A. Their surface structures seem homogeneous and voids found of ethanol residual. The developable patterns were found on both photoresists exposed using UV. The different sensitivities depend on PAC and developer. The X-ray developed photoresist unable produce patterns.

References

[1] Sutikno *et al* 2016 Synthesis of phenolic based resist materials for photolithography *Oriental Journal of Chemistry* **1** 165 – 170
[2] Christine S *et al* 2009 Mr NIL 6000LT Epoxy based curing resist for combined thermal and uv nanoimprint lithography below 50 °C *Journal Microelectronic Engineering* **86** 722 – 725
[3] Diby A K *et al* 2007 Kinetic study of negative dry film photoresist *Express Polymer Letter* **1** 673 – 680
[4] Maneesh S *et al* 2012 Evaluation of microlithographic performance of 'deep UV' resists: synthesis, and 2D NMR studies on alternating 'high ortho' novolak resins *Jornal Chemistry Science* **124** 395 – 401
[5] Harutaka M 2015 Performance of SU-8 membrane suitable for deep x-ray grayscale lithography *Journal Micromachines* **6** 252 – 265
[6] Singla Manoj and Vikas C 2010 Mechanical properties of epoxy resin fly ash composite *Journal of Minerals & Materials Characterization & Engineering* **9** 199 – 210
[7] Xu *et al* 2014 Synthesis and characterization of a novel epoxy resin based on cyclotriphosphazene and its thermal degradation and flammability performance *Polymer Degradation and Stability* **109** 240 – 248
[8] Zhang J *et al* 2001 Polymerization optimization of SU-8 photoresist and its application in microfluidic systems and MEMS *Journal of Micromechanics and Microengineering* **11** 20 – 26
[9] Khairurrijal *et al* 2007 A simple microcontroller-based current electrometer made from LOG112 and C8051f006 for measuring current in metal–oxide–semiconductor devices *Measurement Science and Technology* **18** 3019
[10] Moradi G R *et al* 2013 Densities and kinematic viscosities in biodiesel–diesel blends at various temperatures **58** 99 – 105

[11] Khrisna *et al* 1998 Characterization of low-viscosity photoresist coating *Proceedings of the SPIE* **3333** 1323 – 1336
[12] Houlihan F M *et al* 2003 Retrospective of work at bell laboratories on the effect of fluorine substitution on the properties photoacid generator *Journal of Fluorine Chemistry* **122** 47 – 55
[13] Jipa F *et al* 2013 Femtosecond laser lithography in organic and non-organic materials *Updates in Advanced Lithography* **3**

IOP Conf. Series: Materials Science and Engineering **367** (2018) 012023 doi:10.1088/1757-899X/367/1/012023

Comparative Mechanical Improvement of Stainless Steel 304 Through Three Methods

N. Mubarok[1], H. A. Notonegoro[2] and K. A. Z. Thosin[3]

[1]PPS Materials Science, FMIPA-Universitas Indonesia, Depok 16424, Indonesia
[2]Mechanical Engineering Dept., FT-Universitas Sultan Ageng Tirtayasa, Cilegon 42435, Indonesia.
[3]Pusat Penelitian Fisika, LIPI, Serpong, Indonesia.

E-mail: naila_mubarok@yahoo.com

Abstract. Stainless Steel 304 (SS304) is one of stainless steel group widely used in industries for various purposes. In this paper, we compared the experimental process to enhance the mechanical properties of the surface SS304 through three different methods, cold rolled, annealed salt baht bronzing (ASB), and annealed salt baht boronizing-quench (ASB-Q). The phase change in SS304 due to the cold rolled process makes this method is to abandon. The increasing of the annealing time in the ASB method has a nonlinear relationship with the increases in hardness value. Comparing to the increases in hardness value of the ASB method, the hardness value of ASB-Q methods is still lower than that method.

Keywords. Annealing, ASB, and stainless Steel 304.

1. Introduction

Stainless Steel 304 (SS304) is one of stainless steel group which is widely used in industries for various purposes. Austenitic stainless steels have excellent corrosion resistance in many environments and good weldability as part of the high strength and good formability [1]. One use of SS304 is as a crude oil transport pipelines from an oil well to a processing plant in oil and gas industry. This transport method consumes less energy (compared to conveyer belt transportation). However, SS304 has many components fail in practice. Severe metallic wear frequently endures the SS304 as the effect of degradation of pipelines (erosion) by fine solids (clays), bitumen, and sands is a critical issue. In these situations, it is necessary to modify the surface properties while it is still important to maintain the nature of the SS304 itself [2]. For these purposes, various attempts have been made to engineer the surfaces of austenitic stainless steels. These efforts include cold rolled, carburizing and boronization methods. In the previous study, the deformation by the cold rolled process has transformed the austenite structure into martensitic structure [3]. Moreover, at in 20 % distortion, the value of magnetization is increased above 0.2 T with a small coercivity and caused the ability of corrosion resistant decreases by magnetic measurement [4,5]. However, Etessami *et al.* reported that the increase of annealing time of 80 % cold rolled 304 led to the formation of austenite at the edge of fine recrystallized ferrite grain boundaries [6]. Furthermore, the Enriching of the surface with boron had investigated by means of boron atoms diffuse into the metallic lattice of the surface of a workpiece, where provides wear and abrasion resistance [7].

Content from this work may be used under the terms of the Creative Commons Attribution 3.0 licence. Any further distribution of this work must maintain attribution to the author(s) and the title of the work, journal citation and DOI.
Published under licence by IOP Publishing Ltd

In this paper, we compared the experimental process to enhance the mechanical properties of the surface SS304 through three different methods. The best achievements value of among that process is reported.

2. Materials and methods

The samples specimen is taken from the 11 mm thick SS304 pipe which has compound as in Table 1. The hardening process is carried out through three different methods: cold rolled, annealed salt baht boronizing (ASB), and annealed salt baht boronizing-quench (ASB-Q). The first method, cold rolled is used to deform the sample into 20 %, 40 %, and 60 %. Second methods, the boronizing is used by soaking the sample into borax ($Na_2B_4O_7 \cdot 10H_2O$) and annealed at 750 °C dan 1000 °C for 6 h, 8 h, and 10 h. Third, the boronizing is used by annealed for 6 h of soaking sample into borax followed by quenched into the water at room temperature. The crystalline structure and the microstructure of the sample were examined using X-ray diffraction and scanning electron microscope (SEM). The enhancement of surface properties formed quantify through wear and hardness measurement.

Table 1. Chemical composition of stainless steel 304 specimen

	C	Mn	Si	P	S	Ni	Cr	Fe
Mill certificate	0.05	1.08	0.35	0.038	0.001	8.50	18.65	-
AAS	-	1.49	-	-	-	10.6	21.20	66.44

3. Results and discussion

3.1. Cold rolled methods

The XRD result of mechanical treatment via cold rolled for 20 %, 40 % and 60 % deformation shows in Figure 1 wherein 0 % treatment is used as a reference. As in the figure, the structural changes can be seen from 0 % to 60 % by identified the reveal of new peak pattern or decreases and losses of some peaks pattern. At 20 % deformation, the new peak pattern reveals below $2\theta = 30°$ which is correlated with martensitic phase present [8]. When the sample gets 40% of deformation, the presence of this peak followed by the reduction of the entire peak between $2\theta = 35°$ to 55°. Moreover, at 60 % deformation, those peak patterns below 30° are disappeared followed by the decreases of the two peak at $2\theta = 46°$ to 55° and the other peak. From these results, we see that the rise of the martensite phase occurs in 20 % to 40 % deformation. Furthermore at 60 % deformation increases the phase form has changed to another phase [8–10].

Figure 2 shows the SEM result of the sample surface of every deformation level. At 0 % deformation, it appears that the SS304 alloy consists of an austenite phase. This phase can be identified by its grain shape. The structural changes occurring in the SS304 alloy resulted in morphological changes of the grain shape. This begins with the form of wrinkles that occur on the surface of grains that have 20 % deformation. Furthermore, at 40 % deformation shows the occurrence of shift between the grains marked by the difference between the surface heights of the grains accompanied by the emergence of the needle-shaped pattern on the surface of the sample. Then at 60 % deformation, it appears that the grain shape is increasingly elongated followed by the impingement of the spacing between the grains. This form is reminiscent of the martensitic phase in steel.

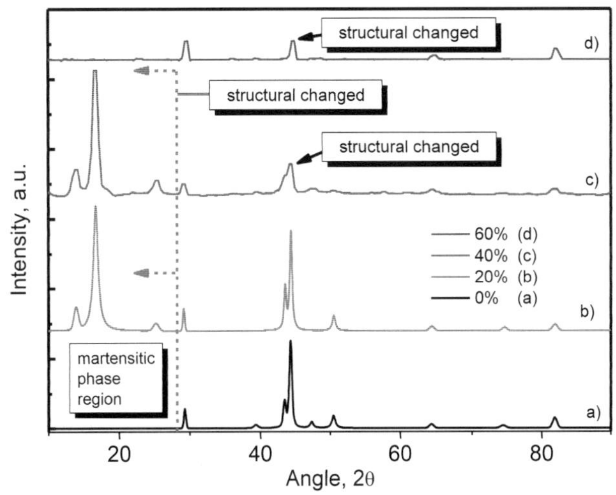

Figure 1. XRD of a) 0% deformation (black line), b) 20% deformation (pink line), c) 40% deformation (green line), d) 60% deformation (blue line). The red line arrows show the reveal of new peak appears at 20% dan 40% deformation.

Table 2. Hardness and wear test value of SS304

Deformation (%)	Hardness (HVc)	Wear Resistance
0	145	1.45
20	344	4.42
40	374	2.28
60	395	2.81

As seen in Table 2, the highest value of enhancement hardness is obtained when deformation occurs 20 % of the initial condition. Meanwhile, the highest value of enhancement friction resistance is also obtained at 20 % deformation. Furthermore, the optimum value of hardness and frictional resistance met in the 20 % of deformation. However, the emergence of the martensite phase form which indicates a phase change in SS304 due to the cold rolled process makes this method is to be abandoned.

(a) (b)

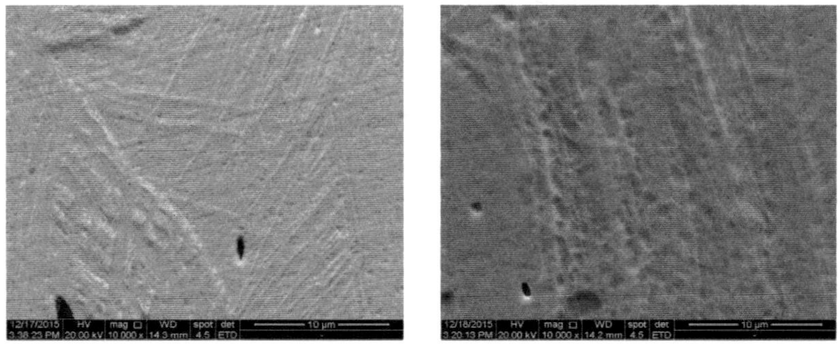

(c) (d)

Figure 2. SEM of a) 0% deformation, b) 20% deformation, c) 40% deformation, d) 60% deformation. The cold rolled methods have changed their grain shape.

3.2. Annealed salt baht boronizing (ASB) and annealed salt baht boronizing-quench (ASB-Q) methods.

Figure 3 shows morphological cross-section used scanning electron microscope from sample ASB and ASB-Q methods which represented in Figure 3a, Figure 3b and Figure 3c. The borax content had a notorious influence on the presence of boronization layer was observed in all methods. Simple scratch test for this cross-section sample is using 1200 - type sandpaper. The streaks line formed on the surface between the substrate layer and boronized layer have a correlation to the wear resistance properties of those material surface.

Table 3. The result of hardness test of ASB and ASB-Q methods

Method	Temperature (°C)	Holding time (Hours)	Hardness (HVc)
0	0	0	145
ASB	750	6	223
		8	414
		10	226
	1000	6	218
		8	227
		10	215
ASB-Q	1000	6	602

The ASB method is represented by the annealed sample at 750 °C and 1000 °C for 10 h. The result of all the ASB method shows the formation of a thin layer with a ~5 μm of thick. In Figure 3a and Figure 3b it appears that a streak line is clearly formed in both the surface area with the deepest streak occurring in the substrate area. Refer to Table 3, the hardness value formed of 750 °C annealed is higher than of 1000 °C annealed in ASB method. We can see that the increase of the annealing time has a non-linear relationship with the increases in hardness value. This condition occurs either at 750 °C or at 1000 °C of annealing temperature. From both annealed temperatures, the optimum hardness value is obtained at 750 °C for 8 h of annealing time.

The ASB-Q method is represented by the annealed sample at 1000 for 6 h which follows by a quench in water. Figure 3c shows that the streaks in the boronization area look vague compared to the streaks formed in the substrate area. Two distinct regions were identified on the boronized SS304 surface: these are (i) substrate region and (ii) boronized region which have globular matrix shape [11–13]. Refer to Table 3, the hardness value formed of this methods is 602 HVC. On the face of hardness values obtained from this method looks much higher than the previous method. Comparing to the increases in hardness value of ASB at 750 °C for 8 h anealling, this value is still lower than that

method. Therefore, the characterization of ASB-Q method at 750 °C for 8 h needs to be done to obtain more relevant benchmark results.

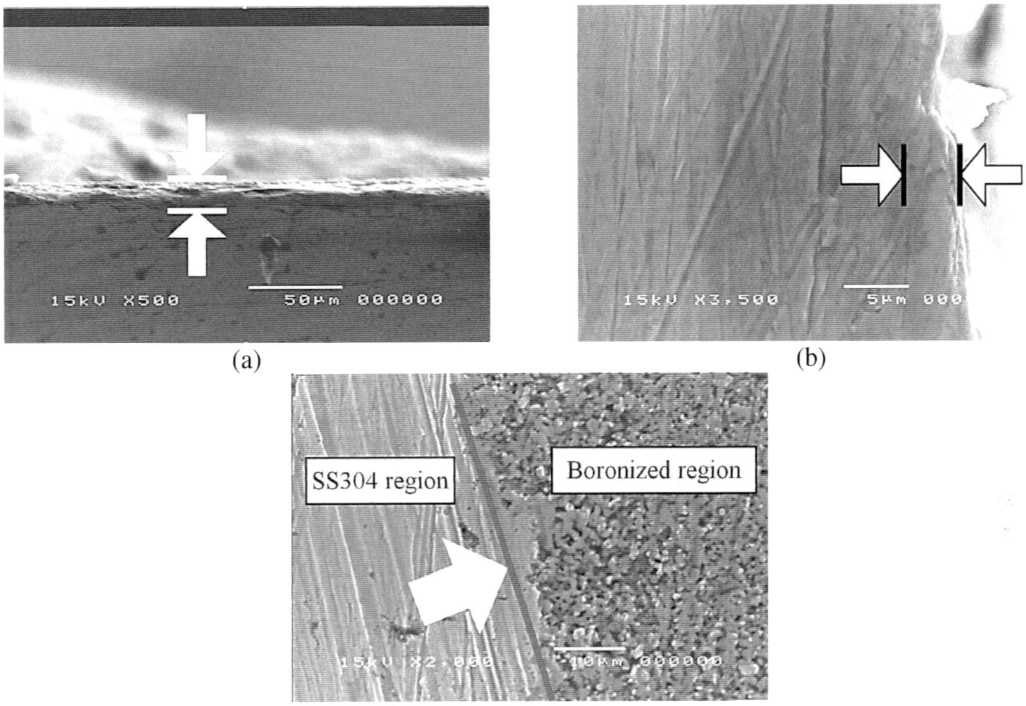

(c)

Figure 3. a) Boronization at 750 °C within 10 h with 4.5 μm layer thick, b) Boronization at 1000 °C within 10 h with 17.7 μm layer thick, c) Boronization 1000 °C within 6 h continued Quench in water, the red line shows the boundary between metal and boronized region.

4. Conclusions

Microstructure and microhardness properties of stainless steel 304 with different treatment by three methods were investigated. Methods are compared the experimental process to enhance the mechanical properties of the surface stainless steel 304. Characterization by XRD shows that cold rolled affected to decreased on structural peaks materials as long as high deformation. It can happen because structural transformation from austenite to martensitic which can be seen with increasing hardness until 390 HVc at 60 % deformation. The highest value of increment hardness is achieved by ASB-Q method 602 HVc, and the lowest value of hardness is achieved by the based material of stainless steel 304 145 HVc.

Acknowledgements

The authors wish to thank Dr. Jan Setiawan (Center for Nuclear Fuell Technology- BATAN) for helpful support.

References

[1] Marques F *et al* 2011 Influence of heat treatments on the micro-abrasion wear resistance of a superduplex stainless steel *Wear* **271** pp 1288–1294
[2] Lu B. 2013 *Erosion-Corrosion in Oil And Gas Production* **2** pp 19–60.
[3] Mubarok N *et al* 2016 The mechanical properties of austenite stainless steel 304 after structural deformation through cold work p 020022:020022
[4] Mubarok N *et al* Influence of Mechanical Deformation to the Magnetic Properties of Stainless

Steel **304** pp 20–23.
[5] Sagar SP *et al* 2005 Magnetic characterization of cold rolled and aged AISI 304 stainless steel
[6] Etesami SA, Enayati MH, and Ghatei A 2017 Austenite formation and mechanical properties of a cold rolled ferrite- martensite structure during intercritical annealing *Mater Sci Eng A* **682** pp 296–303
[7] Topuz P, Çiçek B, Akar O 2016 Section B : Metallurgy Kinetic Investigation Of Aisi 304 Stainless Steel Boronized In *Indirect Heated Fluidized Bed Furnace* **52** pp 63–68.
[8] Surkialiabad R, Hedayati A, and Alam AS 2013 Monitoring of Martensitic Transformation in Cold-Rolled 304L Austenitic Stainless Steel by Eddy Current Method *Transformation* **2** p 11
[9] Abbas G et al 2010 Cold deformation effect on the microstructures and mechanical properties of AISI 301LN and 316L stainless steels *Wear* **258** pp 605–614
[10] Lai CL *et al* 2009 Notched tensile tests of cold-rolled 304L stainless steel in 40 wt .% 80 °C MgCl 2 solution q *Corros Sci* **51** pp 380–386
[11] Meric C *et al* 2006 Investigation of the boronizing effect on the abrasive wear behavior in cast irons *Mater Des* **27** pp 751–757
[12] Luo HS and Zhao C 2013 Low temperature salt bath hardening of AISI 201 austenitic stainless Steel *Phys Procedia* **50** pp 38–42
[13] Topuz P, Çiçek B and Akar O 2016 Kinetic investigation of AISI 304 Stainless Steel boronized in indirect heated fluidized bed furnace

The Influence of Chitosan Concentration on Synthesis of Hydroxyapatite Scaffold on Crystallinity and Surface Morphology

Yudyanto[1], Hartatiek[1], R.W. Prasasti[1], Y.A. Hariyanto[1], and N. Hidayat[1]

[1]Department of Physics, Faculty of Mathematics and Natural Sciences, Universitas Negeri Malang (UM), Jl. Semarang 5, Malang 65145, Indonesia

E-mail: yudyanto.fmipa@um.ac.id

Abstract. Hydroxyapatite (HA) is a bioceramic material having a chemical formula (Ca_{10} $(P0_4)_6(OH)_2$) similar to the chemical structure of bone and hard tissue in humans. Recently, HA-scaffold has been intensively developed by many researchers due to its potential application in dealing with tissue injured. The implantation of HA-scaffold into bone defect aims to help and stimulate the growth of new bone tissue. Crystallinity and microstructure are properties that need to be considered to synthesis HA scaffold. Although, HA is biocompatible and osteoconductive but low biodegradable, to solve these problems added chitosan. Chitosan is a biopolymeric material with specific properties, in terms of biocompatible, non-toxic, osteoconductive, biodegradable, and not carcinogenc. In addition, a natural polymeric chitosan has similarities with the extracellular matrix, in terms of good biological performance and inherent cellular interactions. In this study, HA-scaffold was synthesized using $Ca(OH)_2$ as the raw material of natural calcite mineral, H_3PO_4, and a solid polyurethane polymer (sponge) as forming of pores by sol-gel method. The addition of chitosan concentration of 0 %, 2 % and 3 % in hydroxyapatite scaffold evaluated its influences on crystallinity and surface morphology. The phase, crystallinity and crystal size are evaluated by XRD. Functional groups in HA scaffold and HA compounds were evaluated with FTIR. The results showed that HA has Ca/P ratio of 1.64. The addition of chitosan concentration caused the decrease of crystallinity and increased the pore size. Meanwhile the crystal sizes were obtained: 49.99 nm (HA), 54.37 nm (HA-scaffold without chitosan), 54.98 nm (HA-scaffold with 2% chitosan), and 40.44 nm (HA-scaffold with 3 % chitosan).

Keywords. Chitosan, crystallinity, HA scaffold, and surface morphology.

1. Introduction

Hydroxyapatite (HA; $Ca_{10}(PO_4)_6(OH)_2$) is an attracting biomaterial due to its excellent biocompatibility, bioactivity, and osteoconductivity [1]. HA materials are also potential candidates for use in fluorescence labeling, cell targeting, imaging and diagnosis materials [2, 3]. The chemical composition, size, crystallinity, and morphology of the HA crystals and their aggregates play critical roles in determining their properties and potential applications [4–8]. Furthermore, HA nano-bioceramics exhibit better higher resorbability and bioactivity than in micro-scale sizes [9]. In addition, the crystal grain size can also regulate the degradation rate. Comparing with micro-size, HA

Content from this work may be used under the terms of the Creative Commons Attribution 3.0 licence. Any further distribution of this work must maintain attribution to the author(s) and the title of the work, journal citation and DOI.
Published under licence by IOP Publishing Ltd

nano-sized possesses higher dissolution rate because possess much more boundaries and the dissolution always occurs most easily on grain boundaries [10]. In the natural apatite, the low crystallinity, ion substitution, and small crystal size are important factors that are related to their good solubility [11]. The surface morphology design is considered as potent approaches to increase the bioactivity and biological response [12, 13]. In fact, the human bone mineral is composed of co-substituted essential trace elements such Na, Mg, Zn, Sr, F, Cl, Si, and CO_3^{2-} in crystal lattices that was done by Hartatiek *et al* [14, 15]. HA is measured as a potential candidate as scaffold for bone tissue engineering application, which acts as an excellent temporary substrate to allow cell in-growth, proliferation, and differentiation.

Natural biopolymers have received much more attention in the field of orthopedic and other biomedical applications due to their excellent biocompatibility and biodegradability. In recent years, chitosan and its applications in the field of tissue engineering have attracted considerable attention [16, 17]. Chitosan is a unique polysaccharide that shares a number of structural and chemical similarities with collagen and has been used as DNA and drug delivery vehicle, as skin grafting template, and as wound healing material [18, 19]. Also, chitosan is a natural polymer obtained by deacetylation of chitin, with properties are biocompatible, biodegradable, cationic and relatively non-toxic in natural [20]. Hence, chitosan is appropriate when it is composited with HA to improve its bioactivity and biological response that is related with surface morphology while crystallinity that is related with to good solubility (biodegradable). In this study, the effect of chitosan content on HA was carefully examined the effect of crystallinity and surface morphology.

2. Materials and methods

2.1. Preparation of hydroxyapatite/chitosan nanocomposite

Four synthesizing stages were conducted in order to obtain the HA-chitosan nanocomposite material as briefly listed as follows: (1) the preparation of chitosan solution, (2) synthesis of hydroxyapatite by means of co-precipitation method, (3) synthesis of hydroxyapatite-scaffold sponge, and (4) preparation of HA-chitosan composites via sol-gel route.

Chitosan powder from shrimp shells (brand Sigma-Aldrich) was dissolved in 2 % (v/v) acetic acid solution to obtain a polymer solution. Different composition of chitosan aqueous solution was prepared by dissolving the different amount chitosan powder into 100 cm^3 distilled water containing acetic acid and stirred for 3 h at 300 rpm until a good suspension was achieved and then the solution is allowed to stand for 24 h. Samples of this experiment consisted of 0 g, 0.5 g, 0.75 g of chitosan. In this experiment, HA powder was prepared by natural calcite from Druju Malang Jawa Timur. CaO powder was dissolved in 50 mL ethanol 96 % and blend with H_3PO_4 solution and stirred for 30 min at 37 °C and 300 rpm. The solution was precipitated for 24 h and then stirred at 60 °C until change be gel and calcined at 1000 °C for 3 h. HA powder was dissolved into 10 mL aquadest and stirred at 300 rpm, 37 °C for 30 min. Solid polyurethane polymer (sponge) with dimension (2 cm × 2 cm × 0.5 cm) was soaked into the gel for 5 h and then dried at 110 °C for 24 h. Finally, one gram of HA powder was dissolved into 10 mL aquadest, and then solution of chitosan was dropped and stirred with 300 rpm at 37 °C for 4 h. The solution was precipitated at room temperature for 24 h. The gel sponge was soaked into the solution and dried at 110 °C for 24 h. The sample was calcined at 1000 °C for 4 h.

3. Results and discussion

The compound groups of the resulting product was determined by an FTIR spectrometer (Prestige-21 Shimadzu). HA had OH-, HPO_4^{2-}, CO_3^{2-} and PO_4^{3-} groups. The FTIR spectroscopy has provided valuable information regarding the formation of HA powder, HA scaffold, and HA/chitosan composite. Figure 1 shows the results of FTIR spectroscopy for HA powder. Based on Figure 1, it can be inferred that all the HA groups exist in the FTIR graph indicating the formation of HA in the sample. Other clusters, such as C=C and C-H, indicate the presence of calcite compounds.

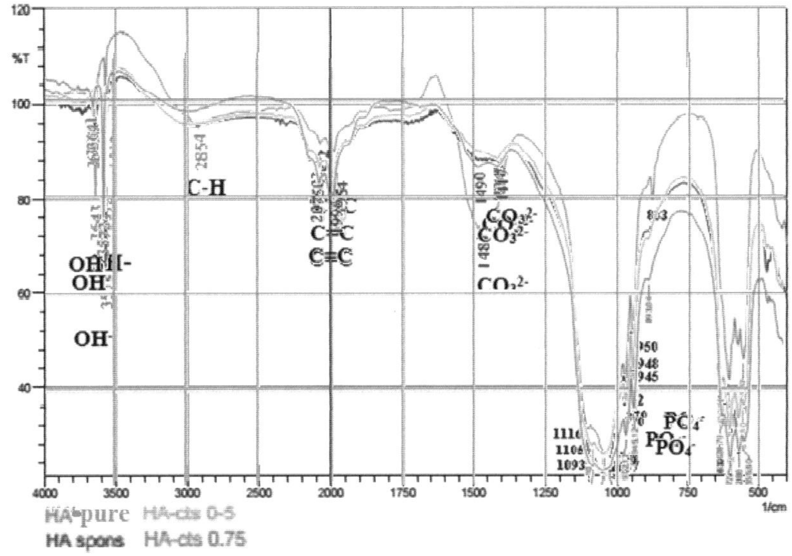

Figure1. FTIR spectra of HA powder, HA scaffold, and HA scaffold-chitosan

The phase identification and crystal structure formation were evaluated by XRD. The average crystallite size (D) was determined according to the Scherrer equation, $D = 0.9\lambda/\beta \cos \theta$, where D is the average crystallite size (Å), β is the peak broadening of the diffraction line measured at half of its maximum intensity in radian, λ is the wavelength of X-ray (nm), and θ is the Bragg's diffraction angle [21]. Figure 2 shows the results of HA, HA scaffold, and HA scaffold/chitosan.

For the HA powder (blue) the presence of 2θ peaks at about 26°, 28°, 31.07° and 32° corresponding to the diffraction planes (002), (102), (210), (221), and (300) indicates the formation of HA in the product by the co-precipitation method. There is still a $CaCO_3$ phase at 2θ in the Bragg plane of (210). The highest peaks for all the diffraction patterns, in Figure 2, were around 31° with small peaks shifts. This peak angle shift of HA indicates the addition of sponge and chitosan. This condition indicates that the synthesis of HA, HA scaffold and HA scaffold / chitosan are successfully synthesized by sol-gel method.

The crystallinity of HA was determined by the fraction of crystalline area over (crystalline crystalline area and amorphous area) × 100%. Table 1 shows the degree of crystallinity of HA scaffold and HA scaffold/chitosan.

Table 1. Cystalinity of HA scaffold and HA scaffold/chitosan

Variation	Net area	Amorphous area	Background area	Crystalline area	Crystallinity (%)
HA scaffold/Chitosan 0%	2710.21	647.14	240.18	1822.89	73.08%
HA scaffold/Chitosan 2 %	2602.36	644.89	244.12	1713.34	72.65%
HA scaffold/Chitosan 3%	2635.18	826.24	221.75	1587.19	65.77%

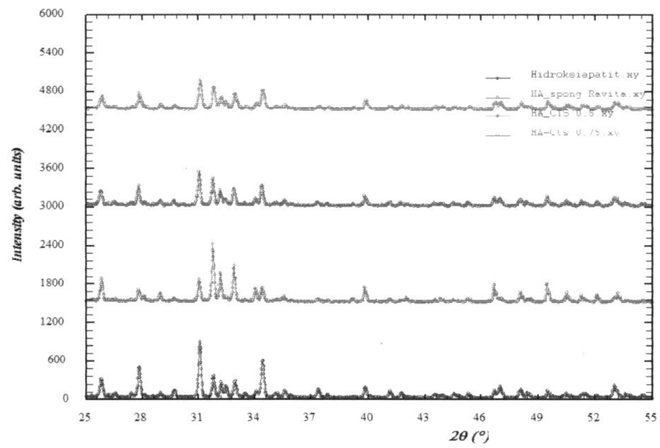

Figure 2. X-ray diffractions of HA powder, HA scaffold, and HA scaffold/chitosan (sumbu Y tidak ada angkanya. Buang angka semuanya)

Based on Table 1 it can be stated that the addition of chitosan to HA scaffold decreases the degree of crystallinity of HA-chitosan. This result corresponds to the XRD pattern in Figure 2. The shift and fall of the degree of crystallinity in the diffraction pattern indicates the composite of the HA scaffold/chitosan composite which means the presence of a bond between the HA scaffold and the polymer matrix. The 2θ peak at 31.07° (221) on the HA powder weakens after the composite was formed, also indicating the participation of HA in bonding with the polymer. This can be due to the molecular interaction between HA scaffold and chitosan very well. The change in the width of the peak on the XRD pattern relates to the size of the crystallite. Table 2 below shows the crystallite size of the HA scaffold and HA scaffold/chitosan.

Table 2. Crystallite size of Hydroxyapatite scaffold/chitosan

Properties	HA scaffold/ Chitosan 0%	HA scaffold/ Chitosan 2 %	HA scaffold/ Chitosan 3%
Crystallite size (nm)	54.37	54.98	52.32

Table 2 shows that the addition of chitosan can decrease the size of insignificant HA scaffold crystals. The size of the HA scaffold crystallite is still within the approximate size range of 50 nm, according to Kokubo [1, 22] that the crystallite size in bone ranges from 20 nm to 80 nm.

Figures 3.a to Figure 3.d indicate the SEM micrographs of surface morphology of the pure HA and composite samples with different contents of chitosan.

Based on Figure 3.a, it appears that pure HA represents homogeneously distributed morphology. The addition of chitosan causes more rough morphology and pores appear much more than pure HA. Hence composite HA scaffold/chitosan is best suited for tissue engineering. The pores provide a place for the growth of new bone tissue that can blend into the host tissue. Pores also provide a place for cell proliferation.

(a) (b) (c) (d)

Figure 3. SEM micrographs of (a) HA (b) HA scaffold/chitosan 0 % (c) HA scaffold/chitosan 2 % (d) HA scaffold/chitosan 3 %

4. Conclusions

Natural calcite from Druju Malang, Jawa Timur, has been effectively used the raw material for HA production. In this present report, chitosan was well introduced into the as prepared HA forming HA/chitosan composites. The addition of chitosan decreased the crystallinity of HA/chitosan composites. In addition, the presence of chitosan in HA caused the pores to increase so that HA/chitosan Composite is potential to be applied for regeneration of tissue.

References

[1] Cunniffe G M, O'Brien F J, Partap S, Levingstone T J, Stanton K T, and Dickson G R 2010 The synthesis and characterization of nanophase hydroxyapatite using a novel dispersant–aided precipitation method *Journal of Biomedical Materials Research Part A* **95** 1142–1149

[2] Kozlova D *et al* 2012 Cell targeting by antibody–functionalized calcium phosphate nanoparticles *J. Mater. Chem.* **22** 396–404

[3] Chen F, Huang P, Zhu Y J, Wu J, and Cui D X 2012 Multifunctional Eu^{3+}/Gd^{3+} dual–doped calcium phosphate vesicle–like nanospheres for sustained drug release and imaging *Biomaterials* **33** 6447–6455

[4] Xia L *et al* 2013 Enhanced osteogenesis through nano–structured surface design of macroporous hydroxyapatite bioceramic scaffolds via activation of ERK and p38 MAPK signaling pathways *J. Mater. Chem. B* **1** 5403–5416

[5] Lin K, Chen L, and Chang J 2012 Fabrication of dense hydroxyapatite nanobioceramics with enhanced mechanical properties via two–step sintering process *Int. J. Appl. Ceram. Technol.* **9** 479–485

[6] Wu L, Dou Y, Lin K, Zhai W, Cui W, and Chang J 2011 Hierarchically structured nanocrystalline hydroxyapatite assembled hollow fibers as a promising protein delivery system *Chem. Commun. Camb.* **47** 11674–11676

[7] Zhang N, Zhai D, Chen L, Zou Z, Lin K, and Chang J 2014 Hydrothermal synthesis and characterization of Si and Sr co–substituted hydroxyapatite nanowires using strontium containing calcium silicate as precursors *Mater. Sci Eng. C: Mater. Biol. Appl.* **37** 286–291

[8] Shen Y, Liu J, Lin K, and Zhang W 2012 Synthesis of strontium substituted hydroxyapatite whiskers used as bioactive and mechanical reinforcement material *Mater. Lett.* **70** 76–79

[9] Sadat–Shojai M, Khorasani M T, Dinpanah–Khoshdargi E, and Jamshidi A 2013 Synthesis methods for nanosized hydroxyapatite with diverse structures *Acta Biomater.* **9** 7591–7621

[10] Lin K *et al* 2011 Evaluation of host inflammatory responses of beta–tricalcium phosphate bioceramics caused by calcium pyrophosphate impurity using a subcutaneous model *J. Biomed. Mater. Res. B Appl. Biomater.* **99** 350–358

[11] Valletregi M 2004 Calcium phosphates as substitution of bone tissues *Prog. Solid State Chem.* **32** 1–31

[12] Xia L *et al* 2013 Enhanced osteogenesis through nano–structured surface design of macroporous hydroxyapatite bioceramic scaffolds via activation of ERK and p38 MAPK

signaling pathways *J. Mater. Chem. B* **1** 5403–5416

[13] Lin K *et al* 2013 Tailoring the nanostructured surfaces of hydroxyapatite bioceramics to promote protein adsorption osteoblast growth and osteogenic differentiation *ACS Appl. Mater. Interfaces* **5** 8008–8017

[14] Hartatiek H *et al* 2016 *Karakteristik hidoksiapatit doping silikon Si_{05}HA hasil reaksi keadaan padat dan potensinya sebagai material biomedis Proseding Seminar Nasional Fisika dan Pembelajrannya 2016* (Jurusan Fisika FMIPA UM) (in Indonesia)

[15] Hartatiek H *et al* 2017 Solid state reaction synthesis of Si–HA as potential biomedical material: An endavor to enhance the added value of Indonesian mineral resources Proceding 2017 *IOP Conf: Mater. Sci. Eng.* **202** p 012021

[16] Naeimi M *et al* 2016 Incorporation of chitosan nanoparticles into silk fibroin–based porous scaffolds: Chondrogenic differentiation of stem cells *Int. J. Polym. Mater.* **65** 202–209

[17] El–Sherbiny I M, Yahia S, Messiery M A, and Reicha F M 2014 Preparation and Physicochemical Characterization of New Nanocomposites Based on β–Type Chitosan and Nano–Hydroxyapatite as Potential Bone Substitute Materials *Int. J. Polym. Mater.* **63**: 213–220

[18] Zhou C and Wu Q 2011 A novel polyacrylamide nanocomposite hydrogel reinforced with natural chitosan nanofiber *Colloid Surface Bioninterf.* **84**: 155–62

[19] Johnson J W and Herschler A 2011 A review of the mechanical behavior of CaP and CaP/polymer composite for application in bone replacement an repair *Acta Biomater.* **7** 16–30

[20] Lanka D and Mittapally V K 2016 Preparation and applications of chitosan nanoparticles: a brief review Research and Reviews: *Journal of Material Science* 2321–6212

[21] Wang X, Ma J, Wang Y, and He B 2002 Bone repairs in radii and tibias of rabbits with phosphorylated chitosan reinforced calcium phosphate cement *Biomaterials* **23** p 4167–76

[22] Kokubo T 2008 *Bioceramics and their clinical applications* (CRC Press, Boca Raton, Boston, New York)

Investigation of Magnetic Properties and Mechanical Responses on Hydrogel-TMAH-Magnetite

Sunaryono[1,2], M. F. Hidayat[1], C. Insjaf[1], A. Taufiq[1,2], N. Mufti[1,2], and Munasir[3,4]

[1]Department of Physics, Faculty of Mathematics and Natural Science, Universitas Negeri Malang, Jl. Semarang 5, Malang, Indonesia 64145.
[2]Research Center of Minerals and Advanced Materials, Faculty of Mathematics and Natural Science, Universitas Negeri Malang, Jl. Semarang 5, Malang, Indonesia 64145.
[3]Department of Physics, Faculty of Mathematics and Natural Science, Universitas Negeri Surabaya, Kampus Ketintang, Jl. Ketintang, Surabaya, Indonesia 60231.
[4]Research Center for Advanced Materials, Universitas Negeri Surabaya, Kampus Ketintang, Jl. Ketintang, Surabaya, Indonesia 60231.

E-mail: sunaryono.fmipa@um.ac.id

Abstract. Hydrogel-TMAH-Magnetite (ferrogel) was fabricated by using polyvinyl alcohol (PVA) hydrogel and magnetite fluids with tetramethylammonium hydroxide (TMAH) surfactant. Iron sand as the raw material was used to synthesize magnetite nanoparticles by co-precipitation method. Magnetite nanoparticles and ferrogel were characterized using X-Ray Fluorescence (XRF) to determine the content of elements in it. To know the functional group network of magnetite nanoparticles, magnetite enclosed with TMAH and ferrogel; we investigated using Fourier Transform Infra-Red (FTIR). Meanwhile, the magnetic properties of the hydrogel-TMAH-magnetite were measured by using Vibrating Sample Magnetometer (VSM). Furthermore, the composition analysis of the ferrogels using FTIR showed that all the synthesis materials were inside the ferrogels. The saturation magnetization of the hydrogel-TMAH-magnetite with a composition of TMAH 1.2 mL (3.95 emu·g^{-1}) was higher than that of TMAH 0.8 mL (0.85 emu·g^{-1}). It exhibited that the composition of TMAH 1.2 mL was an optimum composition to produce nanoparticle magnetite-TMAH having a stable and high performance. Furthermore, the magneto-elasticity of hydrogel-TMAH-magnetite in the effect of the external magnetic field had a good response. However, the composition of the nanoparticle magnetite-TMAH in the ferrogel did not significantly affect the elongation of the gel.

Keywords. Ferrogel, hydrogel, iron sand, magneto-elasticity, and magnetite-TMAH.

1. Introduction

The use of natural material to develop the nanotechnology research has frequently been done, one of them is synthesis and characterization of magnetite nanoparticle. The researchers have successfully fabricated the magnetite nanoparticles of which raw material is local iron sand [1–3]. The magnetite nanoparticles can be synthesized through some methods such as planetary ball-milling [4, 5], co-precipitation [6, 7], hydrothermal [8], microwave [9], sol-gel, microemulsion [10, 11], polyol method

Content from this work may be used under the terms of the Creative Commons Attribution 3.0 licence. Any further distribution of this work must maintain attribution to the author(s) and the title of the work, journal citation and DOI.
Published under licence by IOP Publishing Ltd

[12, 13] or heat decomposition [14]. By the freezing-thawing route, gamma-ray irradiation, or the combination of both, the magnetite nanoparticles can be applied become a multifunction material such as ferrogel [15, 16]. From some previous reports, ferrogel can be used in health and medical fields such as medicine conduction, hyperthermia therapy, contact lens, and the others [17–19].

Furthermore, ferrogel is a unique material of which form can change with some variations of a polymer matrix and magnetic filler within it. Sahiner *et al* [20] have successfully fabricated the poly composite (acrylonitrile-co-1-vinyl imidazole) or p(AN-c-1-VI) with the magnetite nanoparticles by microemulsion method. p(AN-c-1-VI) based composite, and magnetite nanoparticles can be used as a catalyst material that can reduce the organic material. Besides, Rese´ndiz-Herna´ndez *et al* [21] also have successfully investigated the magnetization behavior of PVA polymer based ferrogel and magnetite. They reported that ferrogels with the comparison of 1:1 between filler composition and PVA polymer behaved in superparamagnetic with the value of maximum saturation magnetization was 0.6 $emu \cdot g^{-1}$. Seeing the mechanical behavior of ferrogel, the sensitivity of PVA polymer based-ferrogel move and magnetite nanoparticle move was in line with the filler concentration in the polymer network [22]. However, the investigation of PVA polymer based-ferrogel and magnetite nanoparticles enclosed with TMAH surfactant media is rarely reported.

The magnetic properties and mechanical behaviors of the PVA polymer based-ferrogel and TMAH-magnetite nanoparticles will be the focus of the study in this report. The synthesis of magnetite nanoparticles and the ferrogel fabrication in this research used co-precipitation and freezing-thawing methods, respectively. These methods were used since they were relatively fast, simple, low-cost, and easy to be conducted in the Nanomaterial Laboratory of Physics Department, State University of Malang.

2. Materials and methods

2.1. Magnetite Nanoparticles Synthesis

The sample of magnetite nanoparticles was synthesized by using co-precipitation method. It was started by the separation of natural iron sand to separate the Fe_3O_4 powder from the other particles such as Si, Ti, Mn, Zn, and so forth. The Fe_3O_4 powder as the result of separation was mixed with Hydrogen Chloride (HCl), and it was stirrer on a hotplate magnetic stirrer to fasten the separation process of Fe^{2+} ion and Fe^{3+} ion. The result of ion separation was then filtered and dropped gradually with Ammonium Hydroxide (NH_4OH) solution so that it formed a black precipitate. The precipitate as the result of titration was then washed by using distilled water so that the pH was normal. The magnetite nanoparticles as the result of synthesis were added with TMAH surfactant with the variation of 0.8 mL and 1.2 mL to prevent the aggregation of particles.

2.2. Ferrogel Synthesis

By certain comparison, PVA polymer and distilled water were mixed and stirrer on a hotplate magnetic stirrer so that the mixture became condensed. After that, the magnetite nanoparticles that had been enclosed with TMAH surfactant was put into the mixture and distributed equally over all PVA hydrogel network indicated by the change of mixture color to become black. Subsequently, the mixture was poured into a container of which length was 10 cm, and diameter was 6 mm. Finally, the Freezing-Thawing (F-T) process was conducted to get the ferrogel composite material.

2.3. Characterization

Magnetite nanoparticles and ferrogel were investigated by using X-Ray Fluorescence (XRF) PANalytical Minipal 4 to know the content of it. Meanwhile, to discover the functional group network of the magnetite nanoparticles, the magnetite enclosed with TMAH and ferrogel, a test was conducted by using Fourier Transform Infra-Red (FTIR) Shimadzu, Type: IRPrestige21. The magnetization properties of the ferrogel were tested by using Vibrating Sample Magnetometer (VSM) Oxford).

IOP Conf. Series: Materials Science and Engineering **367** (2018) 012025 doi:10.1088/1757-899X/367/1/012025

Meanwhile, the elongation characterization was investigated to get the mechanical behavior of ferrogel by using an electromagnetic instrument.

3. Results and discussion

3.1. XRF Characterization

Table 1 and Table 2 show the results of XRF test of magnetite as the result of co-precipitation and ferrogel synthesis. Based on XRF test, the purity of Fe contained in Fe_3O_4 nanoparticles, and ferrogel was 87.31 % and 66.3 %, respectively. Beside Fe, there was a polluter in the magnetite and ferrogel samples such as Ti, Ni, P, Ca and the others. These components are difficult to avoid since the main material used in this research was natural iron sand from Wedi Ireng Beach, Banyuwangi.

Table 1. XRF Characterization of Fe_3O_4

Comp (%)	Fe	Ti	Bi	Mn	Eu	V	Ca	Si	Re	P	Ni	Cr	Zn	Yb
Conc. Unit	87.31	7.63	0.95	0.90	0.82	0.50	0.49	0.48	0.35	0.26	0.10	0.09	0.09	0.0

Table 2. XRF Characterization of Ferrogel

Comp (%)	Fe	Ni	P	Ca	Ti	Si	Cu	Zn	Ba	Re	Mn	Cr	V
Conc. Unit	66.3	13.2	5.0	4.81	3.30	2.7	2.0	0.77	0.6	0.5	0.36	0.34	0.32

3.2. FTIR Characterization

The compound contained in the magnetite, the magnetite enclosed with TMAH, and ferrogel has certain energy depending on the composite compound. Every energy of the compound will be represented by a functional group of each compound. FTIR test was characterized to know and confirm the contents of the magnetite, magnetite enclosed with TMAH, and ferrogel through the amount of the energy absorbed by every functional group to vibrate.

Furthermore, this FTIR test was investigated to ensure the presence of a functional group of Fe-O that becomes a special characteristic of the Fe_3O_4 compound and ensure the other functional group that was in accordance with the additional compound in TMAH and ferrogel. Figure 1 shows the data graph as the result of FTIR test on the magnetite sample, magnetite enclosed with TMAH, and ferrogel. The functional group of Fe-O in the three samples was detected in the range of wave number for vibration of Fe-O which was (500 to 750) cm^{-1} [23–27]. Thereby, the magnetite nanoparticles were effectively spread within the ferrogel. Besides the Fe-O group, there was also the O-H group observed in the three samples in the range of (3000 to 4000) cm^{-1} [28–33]. This result indicated that magnetite powder used in this research still contained water compound. Although the magnetite nanoparticles had been heated to the temperature of 100 °C before doing FTIR test, there was still a part of water compound included in the magnetite nanoparticles.

The spectrum of TMAH surfactant enclosing the magnetite nanoparticles was confirmed well in the wave number of 1487.12 cm^{-1} [34–36] and (900 to 1000) cm^{-1} [37]. Each wave number showed the functional group of CH_3 and C-N that become the special characteristic of the wave number of TMAH surfactant. The result of FTIR test on the ferrogel sample also revealed the peak at the wave number of 2943.37 cm^{-1} that becomes one of the characteristics of PVA polymer compound, this obtained wave number was in accordance with the previous research that was the functional group of the compound was at the point of 2941 cm^{-1} [38] and in the range of (2900 to 3000) cm^{-1} [39].

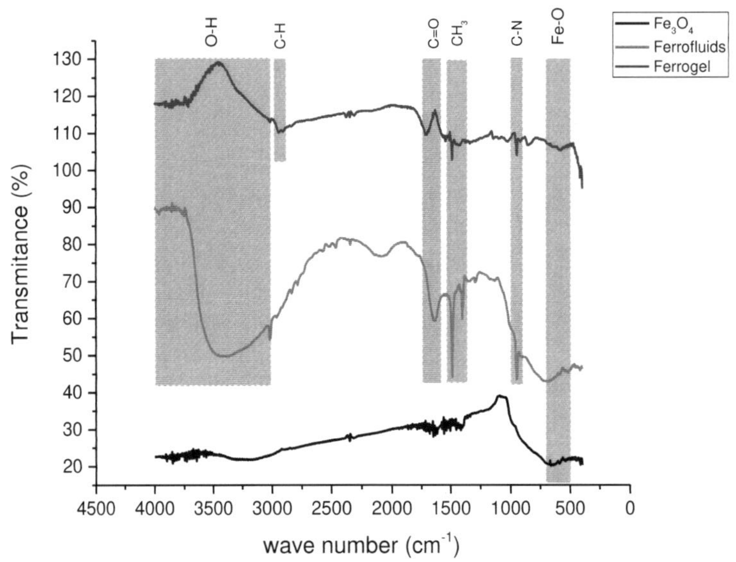

Figure 1. FTIR Characterization of the magnetite, the magnetite enclosed with TMAH, and ferrogel

3.3. VSM Characterization

VSM test was conducted to the ferrogel sample based on magnetite filler enclosed with TMAH surfactant and PVA hydrogel. This characterization aimed to know the magnetization properties of ferrogel such as saturation magnetization (M_s), the remanent magnetization (M_r), and coercivity (H_c). Besides, the magnetization data could be refined by using Langevin equation so that it resulted in magnetic moment variable (μ) and the size of magnetite nanoparticles (D).

Figure 2 displays the hysteresis curve of VSM test result in the ferrogel sample of FGT 0.8 and FGT 1.2. The sample code of FGT 0.8 and FGT 1.2 ferrogels has a meaning, each of FG means ferrogel sample, T is the TMAH surfactant enclosing magnetite nanoparticles and 0.8 and 1.2 mean the amount of TMAH concentration that was 0.8 mL and 1.2 mL enclosing the particle.

Based on the data of VSM result in Figure 2, the sample of FGT 1.2 had a higher value of saturation magnetization than the value of FGT 0.8. This case showed that the concentration of TMAH surfactant 1.2 mL had better magnetization effectiveness than the TMAH surfactant 0.8 mL. This result was also strengthened by the result of VSM data refinement showing that magnetic moment of ferrogel sample with the TMAH surfactant 1.2 mL was higher than 0.8 mL. The sample magnetization data also revealed that all ferrogel samples behaved as a superparamagnetic. This result was shown by the very small of remanent magnetization value and small coercivity field that was close to zero [40], [41]. This phenomenon was strengthened by the report of Rese´ndiz-Herna´ndez *et al* [21] stating that ferrogel with the value of maximum saturation magnetization was 0.6 emu·g^{-1} (the value that was in accordance with FGT 0.8 data) behaved as a superparamagnetic.

The result of magnetization data refinement also showed that the bigger the value of M_s the obtained particle size was getting smaller (data from Table 4). This result was confirmed well by a report stating that the increase in M_s value was caused by the decrease in the crystal size of the sample [42, 43]. If the average size of magnetite nanoparticles as the result of refinement with the XRD data or TEM data was about (11 to 15) nm [22], the value of refinement result showed the smaller value compared to XRD and TEM data. This case was possibly caused by three factors such as the presence of dead surface layer in the magnetite nanoparticles [42] and the presence of coating or enclosing the magnetite nanoparticles with another substance [44] that was present in this research in the form of TMAH surfactant and different measuring technique between XRD, TEM, and VSM.

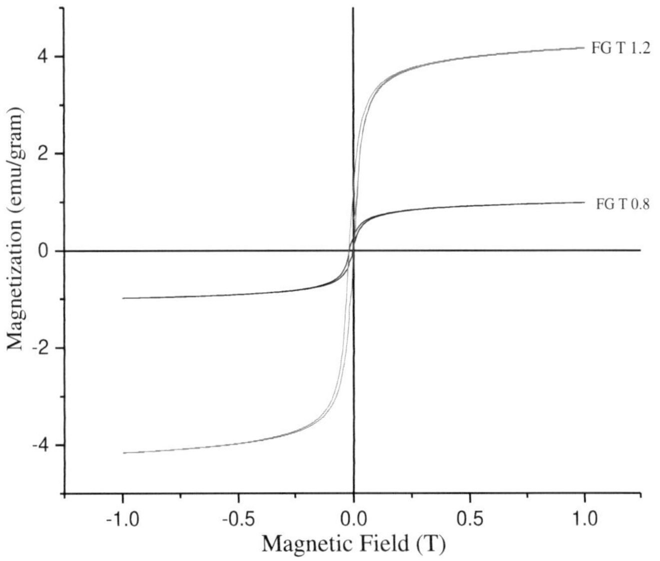

Figure 2. Hysteresis Curve of Ferrogel

Table 3. Amount of Data Fitting Result of VSM Test

Sample	M_s (emu·g^{-1})	M_r (emu·g^{-1})	H_c (T)	μ (J/T) × 10^{-19}
FG T 0.8	0.8543	0.0077	0.0104	2.217
FG T 1.2	3.9495	0.0259	0.0084	2.306

Table 4. The Particle Size of Ferrogel as the Result of VSM Analysis

Sample	Particle Size (nm)
FG T 0.8	9.38
FG T 1.2	6.99

When the sample was enclosed by surfactant, the interaction between particles would decrease so that the aggregation between particles also decreased and gave the impact on the decrease in the size of magnetite nanoparticles when it was characterized by using VSM (the sample condition was in the form of a gel). For the XRD and TEM data, the condition of the particle was in the dry condition so that TMAH surfactant could not work effectively to enclose the particles and gave impact on the big aggregation of the particle so that the size was bigger.

3.4. Magneto-Elasticity Characterization

Magneto-elasticity characterization of ferrogel aimed to know the ferrogel deformation especially the elongation pattern and ferrogel response as the result of the influence of external magnetic field. Ferrogel was arranged vertically with the top tip was made constant, and the bottom tip was let to move freely. When the bottom tip of the ferrogel was given the magnetic field, the sample response could be observed from the ferrogel elongation resulted as shown in Table 5. The ferrogel sample was varied based on the concentration of magnetite filler dispersed into the ferrogel.

Based on Table 5, the elongation level happening to the ferrogel sample was very small that was about 2 mm. This result was the same for all samples with the different concentration of filler. There was no significant difference of ferrogel elongation happening with the variation of magnetite nanoparticles filler from 10 %wt to 20 %wt. This case was not suitable for the previous research stating that the magnetic response will increase by the increase in magnetic filler within the ferrogel [22]. This phenomenon was possibly caused by some factors. For examples are the very small size of magnetic filler (in the nanometer order), the less big concentration of filler, and the too narrow area of ferrogel response. The smaller the particle size, the remanent magnetization resulted was getting smaller as well so that it caused the decrease in magnetite response. This case will be different when the filler has a micrometer size; the ferrogel response will be faster and easier in the observation process of the elongation [45, 46]. Besides, the smaller the concentration of magnetic filler and the area of magnetic response, the deformation of ferrogel move will be weaker. We know that the area of ferrogel response for the elongation only had a diameter of about 6 mm.

Table 5. Result of Elongation Test of Ferrogel with Ferrofluid Percentage

Percentage of Filler (%)	Elongation (mm)
10	0.2
12.5	0.2
15	0.2
17.5	0.2
20	0.2

4. Conclusions

Magnetite nanoparticles have been successfully synthesized from the natural sand. The magnetite nanoparticles were enclosed with TMAH surfactant to prevent and reduce the aggregation of the particles. By the variation of TMAH surfactant, the ferrogel fabrication has been successfully conducted by using the freezing-thawing method. In the result of VSM, the concentration of TMAH surfactant was 1.2 mL, and it had better magnetization effectiveness compared to 0.8 mL. This result was also strengthened by the result of VSM data refinement showing that magnetic moment of ferrogel sample with the TMAH surfactant 1.2 mL had the higher number than 0.8 mL. Referring to the data of sample magnetization, all ferrogels behaved as superparamagnetic. Meanwhile, the magneto-elasticity characterization of ferrogel showed the stable pattern; there was no significant response change between the samples of which magnetite filler was different to the influence of external magnetic field.

Acknowledgments

This Research can be conducted well due to funding support by DPRM of DIKTI. Therefore, the authors would like to thank for DPRM was partially supported by *Penelitian Strategies Nasional* (PSN).

References

[1] Taufiq A *et al* 2017 Fabrication of Magnetite Nanoparticles Dispersed in Olive Oil and Their Structural and Magnetic Investigations *IOP Conf. Ser. Mater. Sci. Eng.* **202** 012008

[2] Sunaryono A *et al* 2015 Various Magnetic Properties of Magnetite Nanoparticles Synthesized from Iron–Sands by Coprecipitation Method at Room Temperature *Mater. Sci. Forum* vol **827** 229 – 234

[3] Rahmawati R *et al* 2017 Preparation of MWCNT–Fe_3O_4 Nanocomposites from Iron Sand Using Sonochemical Route *IOP Conf. Ser. Mater. Sci. Eng.* **202** 012013

[4] Acisli O *et al* 2017 Combination of ultrasonic and Fenton processes in the presence of magnetite nanostructures prepared by high energy planetary ball mill *Ultrason Sonochem.* **34** 754 – 762

[5] De Carvalho J F *et al* 2013 Synthesis of magnetite nanoparticles by high energy ball milling *Appl. Surf. Sci.* **275** 84 – 87

[6] Peternele W S *et al* 2014 Experimental Investigation of the Coprecipitation Method: An

Approach to Obtain Magnetite and Maghemite Nanoparticles with Improved Properties *J. Nanomater.* **2014** 94:94 – 94:94
[7] L Shen *et al* 2014 Facile co–precipitation synthesis of shape–controlled magnetite nanoparticles *Ceram. Int.* **40** 1519 – 1524
[8] Z Qin *et al* 2016 Hydrothermal synthesis of superparamagnetic and red luminescent bifunctional Fe_3O_4@Mn^{2+}–doped $NaYF_4$:Yb/Er core@shell monodisperse nanoparticles and their subsequent ligand exchange in water *Appl. Surf. Sci.* **378** 174 – 180
[9] Xiao W *et al* 2012 Microwave–assisted synthesis of magnetite nanoparticles for MR blood pool contrast agents *J. Magn. Magn. Mater.* **324** 488 – 494
[10] Li Y, Jiang R, Liu T, Lv H, and Zhang X 2014 Single–microemulsion–based solvothermal synthesis of magnetite microflowers *Ceram. Int.* **40** 4791 – 4795
[11] Lu T *et al* 2013 Surfactant effects on the microstructures of Fe_3O_4 nanoparticles synthesized by microemulsion method *Colloids. Surf. Physicochem. Eng. Asp.* **436** 675 – 683
[12] Cai W and Wan J 2007 Facile synthesis of superparamagnetic magnetite nanoparticles in liquid polyols *J. Colloid. Interface. Sci.* **305** 366 – 370
[13] Liang J, Ma H, Luo W and Wang S 2013 Synthesis of magnetite submicrospheres with tunable size and superparamagnetism by a facile polyol process *Mater. Chem. Phys.* **139** 383 – 388
[14] Sun L *et al* 2014 Microemulsion synthesis and electromagnetic wave absorption properties of monodispersed Fe_3O_4/polyaniline core–shell nanocomposites *Synth. Met.* **187** 102 – 107
[15] Marinovic–Cincovic M T *et al* 2014 Physicochemical characteristics of gamma irradiation crosslinked poly(vinyl alcohol)/magnetite ferrogel composite/Fizicko–hemijska svojstva poli(vinil alcohol)/magnetit ferrogel kompozita umrezenog gama zracenjem *Hem. Ind.* **68** 743 – 754
[16] Zrínyi M, Barsi L, and A Büki 1997 Ferrogel: a new magneto–controlled elastic medium *Polym. Gels. Netw.* **5** 415 – 427
[17] Turcu R *et al* 2015 Magnetic microgels for drug targeting applications: Physical–chemical properties and cytotoxicity evaluation *10th Int. Conf. Sci. Clin. Appl. Magn. Carr.* **380** 307 – 314
[18] Priya H *et al* 2014 Smart polymers for the controlled delivery of drugs – a concise overview *Acta. Pharm. Sin. B* **4** 120 – 127
[19] Mendoza Z P *et al* 2013 Magnetic properties study of iron–oxide nanoparticles/PVA ferrogels with potential biomedical applications *J. Nanoparticle. Res.* **15**
[20] Sahiner N, Butun S, and Ilgin P 2011 Soft hydrogel particles with high functional value *Colloids. Surf. Physicochem. Eng. Asp.* **381** 74 – 84
[21] Reséndiz–Hernández P J, Rodríguez–Fernández O S, and García–Cerda L A 2008 Synthesis of poly(vinyl alcohol)–magnetite ferrogel obtained by freezing–thawing technique *VIII Lat. Am. Workshop. Magn. Magn. Mater. Their. Appl.* **320** e373 – e376
[22] Sunaryono A T *et al* 2013 Magneto–elasticity in hydrogels containing Fe_3O_4 nanoparticles and their potential applications *American Institute of Physics Conference Series* **1555** 53 – 56
[23] Clements T W *et al* 2016 Maltol–Functionalized Fe_3O_4 Nanoparticles as T2 Magnetic Resonance Imaging Contrast Agents *Chemistry Select* **1** 1602 – 1606
[24] Habila M A *et al* 2016 Synthesis and application of Fe_3O_4@SiO_2@TiO_2 for photocatalytic decomposition of organic matrix simultaneously with magnetic solid phase extraction of heavy metals prior to ICP–MS analysis *Talanta* **154** 539 – 547
[25] Phu N D *et al* 2011 Crystallization process and magnetic properties of amorphous iron oxide nanoparticles *J. Phys. Appl. Phys.* **44** 345002
[26] Prasad A S 2016 Iron oxide nanoparticles synthesized by controlled bio–precipitation using leaf extract of Garlic Vine (Mansoa alliacea) *Mater. Sci. Semicond. Process.* **53** 79 – 83
[27] Salamun N, How X N, Triwahyono S, Jalil A A, and Karim A H 2014 Synthesis and characterization of Fe_3O_4 nanoparticles by electrodeposition and reduction methods *Malaysian Journal of Fundamental and Applied Sciences* **7**

[28] Ahmadi R *et al* 2013 Kinetics of magnetite nanoparticles formation in a one step low temperature hydrothermal process *Ceram. Int.* **39** 4999 – 5005

[29] Karimi M H et al 2017 Investigation of the effect of magnetic field on mass transfer parameters of CO_2 absorption using Fe_3O_4–water nanofluid *AIChE Journal* ***63.6*** 2176 – 2186

[30] Marinescu R, Laptoiu D, Antoniac I, and Petcu C 2014 Development of Modified Viscoelastic Solution with Magnetic Nanoparticles – Potential Method for Targeted Treatment of Chondral Injuries *Key. Eng. Mater.* **583** 145 – 149

[31] Matei G, Airinei A, and Creanga D 2006 Submicron Structure in Biocompatible Ferrofluids *Acta. Phys. Pol. A.* **109** 405 – 409

[32] Nabiyouni G, Ghanbari D, Karimzadeh S, and Samani B 2014 Ghalehtaki Sono–chemical Synthesis Fe_3O_4–$Mg(OH)_2$ Nanocomposite and Its Photo–catalyst Investigation in Methyl Orange Degradation *J. Nanostructures.* **4** 467 – 474

[33] Tju H, Taufik A, and Saleh 2017 Photocatalytic sonocatalytic and photosonocatalytic of Fe_3O_4/CuO/ZnO nanocomposites with addition of 2 different types of carbon *AIP. Conf. Proc.* **1788** 030131

[34] Andrade Â L et al 2012 Effect of Tetramethylammonium Hydroxide on Nucleation Surface Modification and Growth of Magnetic Nanoparticles *J. Nanomater.* **2012** 15:15 – 15:15

[35] Dutta S *et al* 2013 Molecular characterization of fossil palynomorphs by transmission micro–FTIR spectroscopy: Implications for hydrocarbon source evaluation *Int. J. Coal. Geol.* **115** 13 – 23

[36] Feng X and Lou X 2015 The effect of surfactants–bound magnetite (Fe_3O_4) on the photocatalytic properties of the heterogeneous magnetic zinc oxides nanoparticles *Sep. Purif. Technol.* **147** 266 – 275

[37] Andrade A L *et al* 2013 Preparation of composite with silica–coated nanoparticles of iron oxide spinels for applications based on magnetically induced hyperthermia *Hyperfine Interact* **218** 71 – 82

[38] Nafee S S, Hamdalla T A, and Shaheen S A 2017 FTIR and optical properties for irradiated PVA–$GdCl_3$ and its possible use in dosimetry *Phase Transitions* **90** 439 – 448

[39] Mahapure P D, Gosavi S W, and Aiyer R C 2015 Studies on PVP PVA and their nAg composites based humidity sensors *Physics and Technology of Sensors (ISPTS), 2015 2nd International Symposium on. IEEE* pp 181 – 186

[40] Liu T Y, Hu S H, Liu T Y, Liu D M, and Chen S Y 2006 Magnetic–sensitive behavior of intelligent ferrogels for controlled release of drug *LANGMUIR* **22** 5974 – 5978

[41] Singh K K, Senapati K K, and Sarma K C Synthesis of superparamagnetic Fe_3O_4 Nanoparticles coated with green tea polyphenols and their use for removal of dye pollutant from aqueous solution *J. Environ. Chem. Eng.* **5** 2214 – 2221

[42] Amirabadizadeh A, Salighe Z, Sarhaddi R, and Lotfollahi Z Synthesis of ferrofluids based on cobalt ferrite nanoparticles: Influence of reaction time on structural morphological and magnetic properties *J. Magn. Magn. Mater.* **434** 78 – 85

[43] Kucheryavy P *et al* 2013 Superparamagnetic Iron Oxide Nanoparticles with Variable Size and an Iron Oxidation State as Prospective Imaging Agents *Langmuir* **29** 710 – 716

[44] Sood A *et al* 2015 Ascorbic acid–mediated synthesis and characterisation of iron oxide/gold core–shell nanoparticles *J. Exp. Nanosci.* 5 370 – 382

[45] Cezar C A *et al* Biphasic Ferrogels for Triggered Drug and Cell Delivery *Adv. Healthc. Mater.* **3** 1869 – 1876

[46] Zubarev A 2013 Magnetodeformation of ferrogels and ferroelastomers Effect of microstructure of the particles' spatial disposition *Phys. Stat. Mech. Its. Appl.* **392** 4824 – 4836

The Effect of an Ultrasound Radiation on the Synthesis of 4A Zeolite from Fly Ash

H. Susanto[1], N. A. C. Imani[1], N. R. Aslamiyah[1], T. Istirokhatun[2] and M. H. Robbani[1]

[1]Department of Chemical Engineering, Faculty of Engineering, Diponegoro University, Indonesia Jl Prof. Sudarto-Tembalang, Semarang, Indonesia
[2]Department of Environmental Engineering, Faculty of Engineering, Diponegoro University, Indonesia Jl Prof. Sudarto-Tembalang, Semarang, Indonesia

E-mail: heru.susanto@che.undip.ac.id

Abstract. The use of coal as a fuel source generates a lot of solid waste fly ash. Thus, efforts to reduce or utilize the amount of fly ash are urgently needed. This paper presents zeolite synthesis from fly ash. The fly ash was activated by using NaOH solution prior to fusing process with a weight ratio of 1:2 and mixing with distilled water at a weight ratio of 1:5. Thereafter, the addition of alumina with a concentration of 0.71 %, 1.42 %, 2.12 %, and 2.8 % w/v was performed. The effects of heating and ultrasound radiation on the characteristic of zeolite products were investigated. The results showed that the addition of alumina 2.8 % w/v resulted in the Si/Al ratio of zeolite 4A is ~1. SEM images demonstrated that the presence of ultrasound wave resulted in crystals structure morphology as also supported by XRD characterization. The average crystal size for the ultrasonic treatment was 42.46 nm.

Keywords. Fly ash, zeolite, and ultrasonic.

1. Introduction

In the last decades, coal has become the second largest energy source in the world. However, coal has also contributed around 45 % of all energy-related carbon emissions and air quality degradation [1]. The large quantities of coal consumption generate a high amount of coal fly ash, which becomes the major solid waste of coal combustion. The utilization of coal fly ash has widely developed in various applications such as construction industry, ceramic industry, catalysis, and environmental uses [2].

On the other hand, many researchers have developed several techniques to synthesize zeolite from coal fly ash. Coal fly ash may be a suitable substrate for the zeolite synthesis because of its composition, where a major amount of Al and Si is found [1]. The conventional hydrothermal process was the oldest technique to synthesize zeolite from coal fly ash [3]. Although this method could provide the highest conversion, the resulted supernatant as a waste of this method was very harmful to the environment.

Park *et al.* [4] used the molten salt technique to reduce the by-product and water consumption. However, the molten salt method produced irregular product morphology with lower yield and conversion. Several studies have performed to increase the yield and the properties of zeolite from molten stages [5–7]. Nevertheless, the modified molten stages or multi-step treatment methods require

Content from this work may be used under the terms of the Creative Commons Attribution 3.0 licence. Any further distribution of this work must maintain attribution to the author(s) and the title of the work, journal citation and DOI.
Published under licence by IOP Publishing Ltd

a higher cost with longer preparation time and also complex procedures. Shigemoto and co-workers [8] introduced alkali fusion method, which could produce better property with higher yield and lower process time. However, the loss of weight during fusion process could affect the final product properties.

Querol *et al.* [9] revealed that microwave-assisted hydrothermal process could reduce activation time in alkali fusion method. Recent studies also reported that microwave heating could reduce the process cost and produce smaller zeolite crystals [10,11]. However, this method still provides lower yield and needs a significant study to apply the process in industrial scale. Musyoka *et al.* [12] reported that an ultrasonic-assisted hydrothermal technique could improve zeolite crystallization and decrease the time process. Other studies documented the use of sonication could produce more stable zeolite and reduce the required energy compared to conventional hydrothermal process [13,14]. However, the ultrasonic-assisted hydrothermal process also needed a high cost and further study to up-scale this method.

Previous studies suggested that the ultrasonic-assisted hydrothermal process showed a proper product despite its high cost. Therefore, this research tries to reduce the cost by using a simple ultrasonic apparatus with a low-frequency wave to synthesize 4A zeolite. The effects of alumina addition, heating and ultrasound wave on the characteristic of 4A zeolite product were investigated.

2. Materials and methods

2.1. Materials
F class fly ash (47.26 % Si and 8.13 % Al) was obtained from Paiton coal-fired power plant located in Probolinggo, Indonesia. NaOH and Al_2O_3 were purchased from Merck (Germany). Aquadest was provided by Membrane Research Center (Mer-C), Diponegoro University (Indonesia).

2.2. Zeolite synthesis
Two different methods of synthesis were used, i.e. (i) a conventional hydrothermal process, and (ii) an ultrasonic assisted hydrothermal process. The conventional process was started with fusing fly ash and NaOH (1:2 weight ratio) then the mixture was heated at 550 °C for 1.5 h. The resulting solid was crushed, and the fine powder was mixed with aquadest (1:5 weight ratio) and stirred for 2 h. The liquid phase from the resulting slurry was separated using decantation and filtration. The separated liquid phase was mixed with alumina solution (0.64 g solid alumina and 4.67 g NaOH in 34.5 mL aquadest) to get precursors solution. Finally, the precursor solution was heated and stirred at 80 °C for 3 h. The ultrasonic assisted hydrothermal process was performed after precursor solution was obtained. The ultrasonic wave (40 kHz, using local production ultrasonic transducer) was applied in the precursor solution with and without heating (80 °C) for 3 h. Detail composition and operating condition of each sample are presented in Table 1. The resulting solid from synthesis process was separated using decantation and filtration then dried at 60 °C for 48 h.

Table 1. Detail composition and operating condition of synthesis

Sample	Condition			
	Ultrasonic (kHz)	**Time (h)**	**Temperature (°C)**	**Alumina Addition (%w/v)**
1	40	3	80	0.71
2	40	3	80	1.42
3	40	3	80	2.12
4	40	3	80	2.8
5	-	3	80	0.71
6	40	3	25	0.71

IOP Conf. Series: Materials Science and Engineering **367** (2018) 012026 doi:10.1088/1757-899X/367/1/012026

2.3. Characterization

Si/Al ratio of produced 4A zeolite was characterized using atomic absorption spectrophotometry (AAS; Shimadzu AA-7000) with 640 nm (Si) and 475 nm (Al) wavelength. Morphological investigation of the zeolite was performed by scanning electron microscopy (SEM; JEOL JSM-6510LA). Crystalline structure analysis was carried out by using X-ray diffraction (XRD; Shimadzu XRD-7000).

3. Results and discussion

3.1. Effect of alumina addition on the 4A zeolite Si/Al ratio

Firstly, the effect of different alumina addition on the 4A zeolite characteristic was investigated. Figure 1 demonstrates the Si/Al ratio of the synthesized 4A zeolite as the function of alumina concentration.

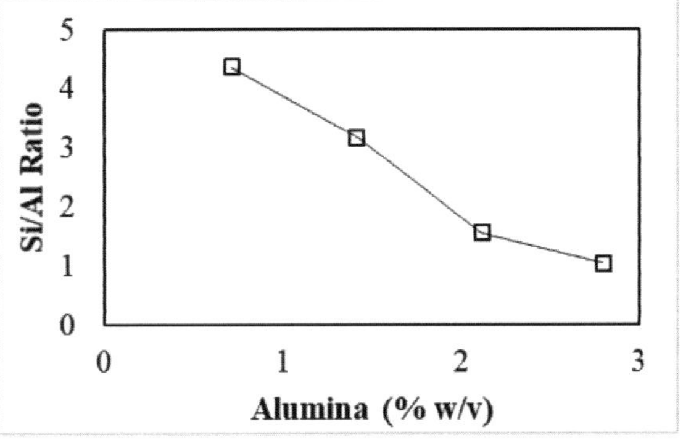

Figure 1. The Si/Al ratio of synthesized 4A zeolite in various alumina concentrations

The Si/Al ratio decreased with the increase in alumina concentration. The results also showed that the addition of alumina at 2.8 % w/v resulted in the Si/Al ratio of zeolite ~1. This phenomenon indicates that the sufficient amount of alumina added would lead to initiate better zeolite crystallization. The addition of alumina was required to initiate the precursor solution which in super saturation state to drive the crystallization process [13]. Another study also revealed that addition of alumina in hydroxide or sodium state could increase the amount of alumina in precursor solution and regulate the required Si/Al ratio [15]. Furthermore, this result suggests that the optimum alumina addition for 4A zeolite synthesis using ultrasonic-assisted hydrothermal process was 2.8 % w/v.

3.2. Effect of heating and ultrasonic wave on the 4A zeolite Si/Al ratio

The investigation was performed to compare the ultrasonic-assisted process, conventional heating process, and the combination of both process in zeolite crystallization process. The 40 kHz of ultrasonic wave was applied in the ultrasonic-assisted process while the 80 °C of temperature was applied in the conventional process. Figure 2 shows the Si/Al ratio of produced 4A zeolite with different synthesis processes.

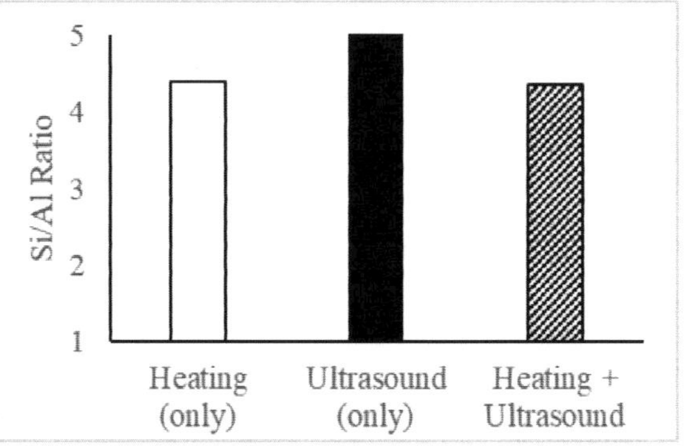

Figure 2. The Si/Al ratio of synthesized 4A zeolite with various preparation methods

As seen in Figure 2, the ultrasonic-assisted without heating process showed the highest Si/Al ratio while the conventional process (heating only) displayed similar Si/Al ratio with the ultrasonic-assisted with heating. These results indicated that the heating process has a vital role in zeolite crystallization. Belviso *et al.* [16] reported that the function of ultrasonic to bind the Si and Al on solid particle would not work optimally at room temperature. Another study also revealed that the minimum temperature for zeolite crystallization process was 50 °C [17], while the electro-mechanic wave from ultrasonic could not provide sufficient heat to fulfill the required temperature. Figure 2 also shows that the combination of both processes obtained similar Si/Al ratio with heating process only. Musyoka and co-workers [15] documented that ultrasound wave could initiate the cavitation phenomenon which leads a nucleus seed forming then the seed would grow to be a crystal core. Other research also elaborated that the ultrasonic treatment could accelerate Al and Si dissolution from amorphous alumina silicate state, improve the polycondensation process, and enhance the transition of semi-crystalline to crystalline phases [13]. Hence, the combination of ultrasonic-assisted and conventional hydrothermal process gave a better characteristic in 4A zeolite synthesis due to its synergistic effect on zeolite crystallization process.

3.3. SEM Results

The morphology of synthesized 4A zeolite was visualized using SEM. As seen in Figure 3a, the morphology of coal fly ash was glassy microsphere with hundred micrometers in size. This result was similar to a previous study [14]. Then, when coal fly ash was transformed to precursor solution, the morphology changed to be irregular crystal shape (Figure 3b). The shape of a solid phase of precursor solution suggests that the crystal growth has started at this stage.

Figure 3c displays the square crystal structure of the synthesized 4A zeolite, which was prepared using ultrasonic-assisted hydrothermal process. The 4A zeolite shape was completely formed and had the same tendency with previous studies [13,15]. These results demonstrate that the impact of the ultrasound wave on crystal structure growth was observed. The formed square crystals had a bit longer shape compared with other studies because of shorter process duration in this study. A longer duration of synthesis process would produce a shorter crystal due to a more Si-Al dissolution. Figure 4d is the image of the 4A zeolite with the addition of 2.8 % w/v alumina, and Si/Al ratio was ~1. The shape of the resulting 4A zeolite showed bolder form than the produced 0.71 % w/v alumina addition 4A zeolite (Figure 3c). Overall, these images confirmed that the optimum alumina addition in ultrasonic-assisted hydrothermal process was 2.8 % w/v.

Figure 3. The SEM images of (a) coal fly ash; (b) solid phase of precursor solution; (c) synthesized 4A zeolite - sample 1; and (d) synthesized 4A zeolite - sample 4.

3.4. XRD Results

XRD analysis was employed to characterize the mineralogical and crystalline structure of synthesized 4A zeolite. Figure 4 presents the XRD analysis of 4A zeolite with the addition of 2.8 % w/v alumina, and Si/Al ratio was ~1 (sample 4) prepared using ultrasonic-assisted hydrothermal process. XRD data demonstrated that crystals of sodium alumina silicate ($NaAlSiO_4$) were obtained. The peaks were identified at 2Θ value: 26.3, 27.8, 30.3, 33.1, 34.4, 34.7, 35.5, 38.2, 40.1, 41.7, 41.9, 44.3, 46.5, and 48.4. The presence of identified crystals indicates that the 4A zeolite has really been synthesized [5, 14]. The calculation using Debye Scherrer equation [18] showed that the size of synthesized 4A zeolite which produced using ultrasonic-assisted hydrothermal process was 42.46 nm.

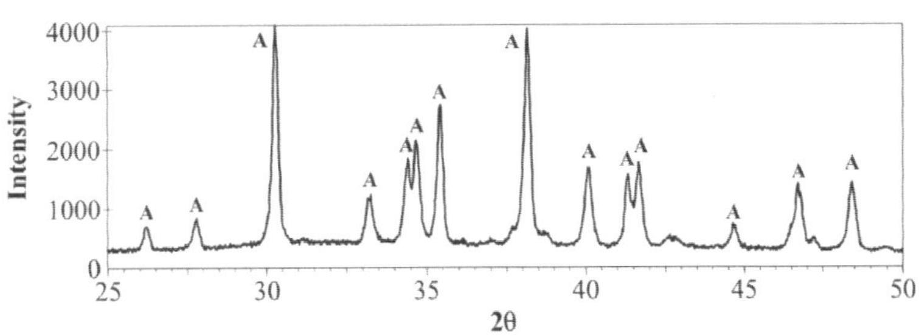

Figure 4. XRD result of synthesized 4A zeolite using ultrasonic-assisted hydrothermal process.

4. Conclusions

This study suggests that the optimum alumina addition on 4A zeolite synthesis using ultrasonic-assisted hydrothermal process was 2.8 % w/v. The alumina had an important role to initiate the zeolite crystallization process. The combination of ultrasonic-assisted and conventional hydrothermal process resulted in a better characteristic in 4A zeolite synthesis. SEM images demonstrated that the presence of ultrasound wave resulted in crystals structure morphology. Types of crystals obtained were indicated by XRD characterization, which were crystals of sodium alumina silicate at level 4. The average crystal size for the ultrasound treatment was 42.46 nm.

Acknowledgements

The authors thank Universitas Diponegoro for the financial support. The authors also thank smart city project by USaid Shera for the opportunity given to increase the quality of this paper.

References

[1] Belviso C 2017 State-of-the-art applications of fly ash from coal and biomass: A focus on zeolite synthesis processes and issues *Prog. Energy Combust. Sci.* **65** 109–35
[2] Hemalatha T and Ramaswamy A 2017 A review on fly ash characteristics–Towards promoting high volume utilization in developing sustainable concrete *J. Clean. Prod.* **147** 546–559
[3] Fukasawa T, Horigome A, Tsu T, Karisma A D, Maeda N, Huang A N and Fukui K 2017 Utilization of incineration fly ash from biomass power plants for zeolite synthesis from coal fly ash by hydrothermal treatment *Fuel Process. Technol.* **167** 92–8
[4] Park M, Choi C L, Lim W T, Kim M C, Choi J and Heo N H 2000 Molten-salt Method for the Synthesis of Zeolitic Materials I. Zeolite Formation in Alkaline Molten-Salt System *Elsavier* 81–9
[5] Huiping S, Huaigang C, Zepeng Z and Fangqin C 2014 Adsorption properties of zeolites synthesized from coal fly ash for Cu (II) *J. Environ. Biol.* **35** 983–8
[6] Jha B and Singh D N 2014 A three step process for purification of fly ash zeolites by hydrothermal treatment *Appl. Clay Sci.* **90** 122–9
[7] Du Plessis P W, Ojumu T V., Fatoba O O, Akinyeye R O and Petrik L F 2014 Distributional fate of elements during the synthesis of zeolites from South African coal fly ash *Materials* **7** 3305–18
[8] Shigemoto N, Hayashi H and Miyaura K 1993 Selective formation of Na-X zeolite from coal fly ash by fusion with sodium hydroxide prior to hydrothermal reaction *J. Mater. Sci.* **28** 4781–6
[9] Querol X, Moreno N, Umaa J C, Juan R, Hernndez S, Fernandez-Pereira C, Ayora C, Janssen M, Garca-Martnez J, Linares-Solano A and Cazorla-Amoros D 2002 Application of zeolitic material synthesised from fly ash to the decontamination of waste water and flue gas *J. Chem. Technol. Biotechnol.* **77** 292–8
[10] Aldahri T, Behin J, Kazemian H and Rohani S 2017 Effect of microwave irradiation on crystal growth of zeolitized coal fly ash with different solid/liquid ratios *Adv. Powder Technol.* **28** 2865–74
[11] Fukasawa T, Karisma A D, Shibata D, Huang A N and Fukui K 2017 Synthesis of zeolite from coal fly ash by microwave hydrothermal treatment with pulverization process *Adv. Powder Technol.* **28** 798–804
[12] Musyoka N M, Petrik L F and Hums E 2011 Ultrasonic assisted synthesis of zeolite A from coal fly ash using mine waters (acid mine drainage and circumneutral mine water) as a substitute for ultra pure water *Proc. Int. Miner. Water Assoc. Aachen Ger.* 423–8
[13] Bukhari S S, Rohani S and Kazemian H 2016 Effect of ultrasound energy on the zeolitization of chemical extracts from fused coal fly ash *Ultrason. Sonochem.* **28** 47–53
[14] Ojumu T V., Du Plessis P W and Petrik L F 2016 Synthesis of zeolite A from coal fly ash using ultrasonic treatment - A replacement for fusion step *Ultrason. Sonochem.* **31** 342–9

[15] Musyoka N M, Petrik L F, Hums E, Baser H and Schwieger W 2012 In situ ultrasonic monitoring of zeolite A crystallization from coal fly ash *Catal. Today* **190** 38–46
[16] Belviso C, Cavalcante F and Fiore S 2013 Ultrasonic waves induce rapid zeolite synthesis in a seawater solution *Ultrason. Sonochem.* **20** 32–6
[17] Ojha K, Pradhan N C and Samanta A N 2004 Zeolite from fly ash: synthesis and characterization *Bull. Mater. Sci.* **27** 555–64
[18] Monshi A, Foroughi M R and Monshi M R 2012 Modified Scherrer equation to estimate more accurately nano-crystallite size using XRD *World J. Nano Sci. Eng.* **2** 154–60

IOP Conf. Series: Materials Science and Engineering **367** (2018) 012027 doi:10.1088/1757-899X/367/1/012027

Local structure examination of mineral-derived Fe_2O_3 powder by Fe K-edge EXAFS and XANES

H. Husain[1,2], B. Hariyanto[1], M. Sulthonul [1], P. Thamatkeng[3], S. Pratapa[1*]

[1]Department of Physics, Faculty of Natural Sciences, Institut Teknologi Sepuluh Nopember (ITS) Jl. Arief Rahman Hakim, Surabaya 60111 Indonesia
[2]Department of Physics, Faculty of Mathematics and Natural Sciences, Universitas Negeri Makassar (UNM), Jl. A. Pangeran Pettarani, Makassar 90222 Indonesia
[3]Synchrotron Light Research Institute (Public Organization), Nakhon Ratchasima, Thailand

E-mail: suminar_pratapa@physcs.its.ac.id

Abstract. Fe K-edge XANES (X-ray Absorption Near Edge Structure) and EXAFS (Extended X-ray Absorption Fine Structure) spectra have been used to investigate the local structures of hematite (Fe_2O_3) powders which were synthesized from local iron stone. The formation of hematite with various calcination temperature (500, 650, 800) °C was observed from XRD data of each powder, and Fe K-edge EXAFS and XANES were performed to sample with calcination temperature 800 °C. The XANES spectra confirmed the oxidation state of the powders. The energy position of the first pre-edge peak at (7113.5 ± 0.5) eV confirm Fe^{3+}. The absorption edges of Fe_2O_3 powders was 7123.41 eV. The first EXAFS data analysis showed that Fe_2O_3 powder exhibited nearest-neighbor distances R_{Fe-O} = 1.56449 Å and R_{Fe-Fe} = 3.01449 Å. The XRD lattice parameter values for the sample is in a fair agreement with the nearest-neighbor distances according to the EXAFS data.

Keywords. EXAFS, XANES, oxidation state, local structure, iron stone mineral, and Fe_2O_3.

1. Introduction

Applications of materials are determined by their characteristics. The characteristics of the materials depend on the internal structure. The internal structure consists of atoms and associates with neighboring atoms in crystals, molecules, and microstructures [1]. There are many tools for material analysis especially for structure studies such as x-ray diffraction (XRD) and x-ray absorption spectroscopy (XAS). XRD is usually used for crystal (phase) study, while XAS is utilized in the so-called "local structure" assessment. XAS can be divided into XANES (x-ray absorption near edge structure) and EXAFS (extended x-ray absorption fine structure). XANES covers the area of the spectrum which dominated by strong photoelectron scattering within a specific absorbing atom, while EXAFS comprises an area of weak photoelectron scattering of about 40 eV above the absorption edge value which can be correlated with neighboring atomic interactions. XANES is used for qualitative analysis, particularly the oxidation state of the absorbing atoms which makes it offers results that x-ray diffraction cannot yield [2]. Meanwhile, EXAFS uses quantitative analysis to obtain local information

Content from this work may be used under the terms of the Creative Commons Attribution 3.0 licence. Any further distribution of this work must maintain attribution to the author(s) and the title of the work, journal citation and DOI.
Published under licence by IOP Publishing Ltd

structure, one being the distance between the nearest atoms. The structural data from EXAFS are believed more accurate than those from XRD [3].

XAS has been used for characterization Fe-based materials, such as $BiFeO_3$, $Bi_2Fe_4O_9$ and $La_{1-x}Sr_xFeO3$ [4–5], with Fe_2O_3 (hematite) as one of the raw material. Usually, Fe_2O_3 was obtained from commercial materials. Meanwhile, the natural iron stone mineral has been used as a source of Fe compounds for the synthesis of iron oxide [6] but synthesized Fe_2O_3 from the mineral need a further investigation because the purity can not be granted only by using XRD measurements. This is because Fe atom tends to absorb Cu radiation [7]. Therefore, XAS was expected to confirm the formation of Fe_2O_3 from the mineral through the oxidation state and nearest-neighbor distances using XANES and EXAFS data analysis, respectively.

In this paper, we reported the structure characteristic of Fe_2O_3 which was synthesized from a local iron stone mineral. The characterization includes XRD for phase identification and lattice parameter determination and XANES and EXAFS to analyze the electronic and local structure. In addition, we compared the XANES spectra of our hematite sample and hematite model to investigate the quality of our sample.

2. Materials and methods

The Fe_2O_3 was synthesized from natural iron stone by using co-precipitation [6]. The synthesis followed several steps. Firstly, an iron stone (6 g) was dissolved in 80 mL HCl by stirring at 70 °C for 1 h. Then, the solution was filtered two times using a filter paper. Next, titration was performed by adding NH_4OH in the solutions until pH = 5 was achieved. Then, the solution was washed several times. The precipitate was dried at 110°C using a hot plate for 7 hours. The powders were calcined at 500 °C, 650 °C, and 800 °C to achieve a pure Fe_2O_3.

The XRD patterns were collected using X-Ray diffractometer XPert MPD ($\lambda_{CuK\alpha}$ = 1.54060 Å) with step size 0.0170°, and generator set on 40 kV and 30 mA. The patterns were analyzed for phase identification and then subjected to the Rietica software for lattice parameter determination. Fe K-edge XANES and EXAFS spectra were recorded with X-Ray Absorption Spectroscopy (XAS) instrument at beam line 8 SLRI (Synchrotron Light Research Institute), Thailand. Fe K-edge spectra were recorded with 0.2 eV step size in fluorescence mode with silicon drift detector and fixed-exit double crystals X-ray monochromator. XANES and EXAFS spectra were analyzed by using Athena and Artemis software [8]. XANES was used for investigation of oxidation state. Meanwhile, EXAFS was used to determine the local structure parameters, i.e., distance between nearest atoms and average of structure irregularity (σ^2), were determined by fitting of radial structure function between EXAFS data from the sample and the model from Fe K-edge spectrum.

3. Results and discussion

The XRD patterns of the synthesized powders can be seen in Figure 1. The patterns reveal that pure phase of Fe_2O_3 has been formed (PDF No. 96-210-1169) with the strongest peak at $2\theta = 33^o$ (Figure 1.a, Figure 1.b, Figure 1.c). It is clear, however, that the quality of the data is rather poor as shown by the low "signal-to-noise" ratio since Fe atom tends to absorb Cu radiation [7]. This may give impact on the quality of the extracted data from the patterns after pattern-fitting analysis, e.g., using the Rietveld method.

Rietveld phase analyses using Rietica software revealed the structure data as shown in Table 1. The fitting between model and observed data (Figure 1) were acceptable as shown by GoF values which were less than 2 %. The quality of the data, however, is fair as shown by the R_{wp} values which are larger than 15 %. As a result of this quality, the lattice parameters of hematite in the samples are also fair. Table 1 shows the lattice parameter and cell volume of the synthesized hematite. The lattice parameters and cell volumes of hematite in all samples were relatively similar, but the sample with relatively large standard deviations. In general, the standard deviation of less than 1/50000 acceptable [9], but in our case, the values are only in the range of 1/10000 and 1/20000. Therefore, the XRD data quality gave a significant impact on the extracted lattice parameter values. Since these values are important for explaining the physical characteristics of the bulk samples, there is a requirement to measure them more accurately. The XAS experiment was then expected to provide structural data with better accuracy and also oxidation number of the absorbing atom.

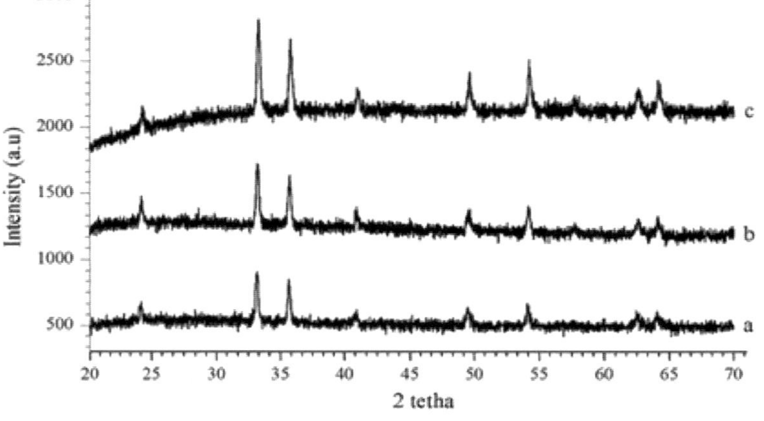

Figure 1. XRD patterns (CuKα radiation) of Fe_2O_3 which were synthesized at (a) 500 °C, (b) 650 °C, and (c) 800 °C.

Table 1. Lattice parameters of the synthesized hematite from a local iron stone.

Sample	GoF	R_{wp}	Lattice Parameter (Å)		Volume
			a=b	c	
Fe_2O_3 (500 °C)	1.52	5.42	4.968079	13.569098	290.040 ± 0.166
Fe_2O_3 (650 °C)	1.66	5.31	4.967348	13.563683	289.839 ± 0.149
Fe_2O_3 (800 °C)	1.44	4.40	4.971446	13.572015	290.491 ± 0.095

Figure 2 shows the XANES plot of the 800 °C Fe_2O_3 powder sample and the Fe_2O_3 standard sample. The pre-edge peak (noted by 1 in Figure 2) for both spectra is ± 7113.5 eV which is consistent with Fe^{3+} [10–13]. The data for the synthesized sample qualitatively support the conclusion about the present of Fe^{3+} with no reduction to Fe^{2+} during the process. Furthermore, the absorption edges (E_o) for both powders was 7123.41 eV (noted by line 2 in Figure 2). E_o can be interpreted as the required energy to raise the photoelectrons just at the outmost energy band. These two results show that the synthesized powder has the same oxidation number with the standard, confirming that it exhibits Fe_2O_3. We also noted that there was a slightly different value of the second peak (noted by @ and *) between the samples, indicating different absorption energy when the photoelectrons interact with the neighboring atom. Meanwhile, the EXAFS region started at 40 eV above the absorption edge (noted by 3 in Figure 2).

Furthermore, the structure information from the XAS measurement about the nearest-neighbor distances was obtained from EXAFS data analysis by using Artemis software. Figure 3 presents the fitting of radial structure function between the model and the EXAFS data for 800 °C Fe_2O_3 sample. Fitting was made to the first shell distances of Fe-O and Fe-Fe. The fitting was acceptable as evidenced by the low values of the reliability factor σ^2 refers to the average of structure irregularity caused by thermal vibration [13]. The analysis showed that Fe_2O_3 powder exhibited nearest-neighbor distances $R_{Fe\text{-}O}$ = (1.56449 ± 0.01000) Å and $R_{Fe\text{-}Fe}$ = (3.01449 ± 0.01000) Å. On the other hand, the Rietica-derived lattice parameters and atomic positions gave nearest-neighbor atom between Fe-O and Fe-Fe have been calculating, 1.9195 Å and 2.8582 Å, respectively. We notice here that the EXAFS and XRD nearest-neighbor distances are quite different, therefore, for the assurance of more accurate measurements we need further measurement, for example using Standard Reference Materials (SRM) which have known lattice parameters. Moreover, the values of σ^2 are 0.00936 and 0.00503 for Fe-O and Fe-Fe, respectively, explaining that thermal vibration does not significantly raise structure irregularity.

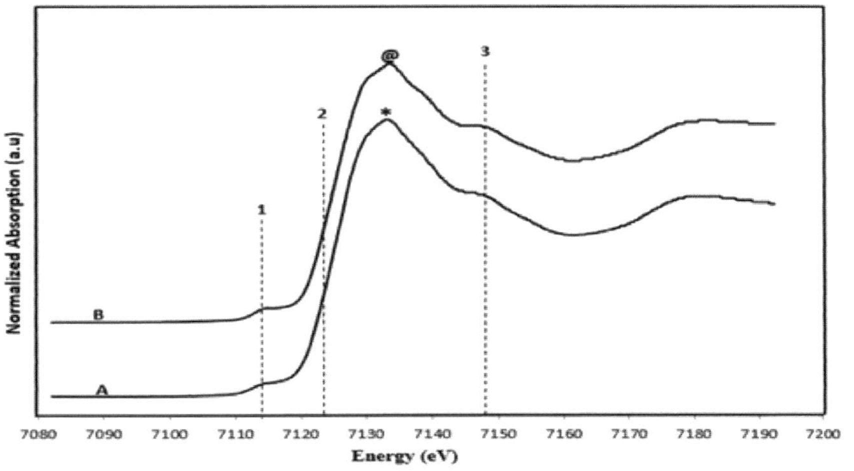

Figure 2. Fe K-edge absorption of (A) Fe_2O_3 standard sample, (B) Fe_2O_3 synthesized powder.
Peak Position: @= 7134.03 eV; * = 7133.20 eV

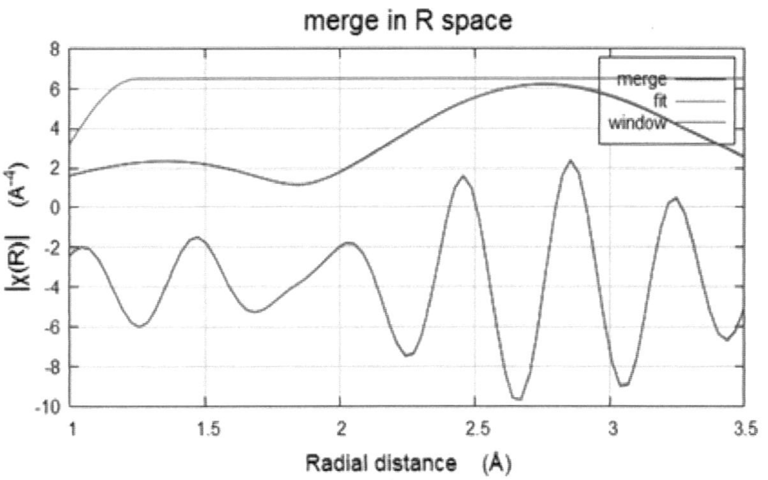

Figure 3. Fitting EXAFS Fe_2O_3 sample and model from Fe K-edge spectrum

4. Conclusions

Hematite has been successfully synthesized with various calcination temperature by co-precipitation method from local iron stone. XRD analysis showed the pure hematite for all sample variation, with slightly different lattice parameter values. The XANES and EXAFS experiment was performed for the Fe_2O_3 sample with calcination temperature 800 °C. The pre-edge peak of the spectra confirmed the presence of the only Fe^{3+}, in conjunction with hematite and supporting the phase identification results from XRD. The nearest-neighbor distances from The EXAFS data analysis were R_{Fe-O} = 1.56449 Å and R_{Fe-Fe} = 3.01449 Å. The accuracy of the results, however, requires further investigation.

Acknowledgments
The author (H.H.) would like to thank Indonesian Endowment Fund for Education (LPDP) scholarship to support finance of the author. The authors would like to thank SLRI, Thailand for providing XAS beamtime (Project ID 2618).

References
[1] Gleiter H, 2000 Nanostructured materials: basic concepts and microstructure *Acta. Mater.* **48**, 1–29
[2] Hiroki O *et al* 2012 Local structure of magnetite and maghemite and chemical shift in Fe K-edge XANES *J. Mineralogical and Petrological Sciences* **107** 127-132
[3] Hennig C *et al* 2003 *J. Sieler. Z. Krostallogr* **218** 37–45
[4] Park, T J *et al* 2008 Electronic structure and chemistry of iron-based metal oxide nanostructured materials: a NEXAFS investigation of $BiFeO_3$, $Bi_2Fe_4O_9$, α-Fe_2O_3, γ-Fe_2O_3, and Fe/Fe_3O_4 *J. Phys. Chem. C* **112** 10359–10369
[5] Abbate M, de Groot F M F, Fuggle J C, and Fujimori A 1992 *Physics Rev. B* **46** 45111
[6] Sunaryono *et al* 2015 *Mater Sci. Forum* **827** 229–234
[7] Zevin L S 2012 *Quantitative x-ray diffractometry* (Springer-Verlag, New York)
[8] Muchlis K *et al* 2017 Synchrotron WAXS and XANES studies of silica SiO_2 powder synthesized from Indonesian natural sands *J. Phys. Conf. Ser.* **795** 012007
[9] Herbstein F H 2000 How precise are measurements of unit-cell dimensions from single crystals? *Acta. Crystallogr. B* **56** 547–557
[10] Bingham PA *et al* 2014 Selective behaviour of dilute Fe^{3+} ions in silicate glasses: an Fe K-edge EXAFS and XANES study *J. Non-Cryst. Solids* **387** 47–56
[11] Wilke M *et al* 2006 Structural environment of iron in hydrous aluminosilicate glass and melt-evidence from X-ray absorption spectroscopy *Chem. Geol.* **229** 144–161
[12] Magnien V *et al* 2008 Kinetics and mechanisms of iron redox reactions in silicate melts: The effects of temperature and alkali cations *Cosmochim Acta.* **72** 2157–2168
[13] Fornasini P and Grisenti R 2015 Fornasini, P., & Grisenti, R. (2015). On EXAFS Debye-Waller factor and recent advances *J. Synchrotron Radiat* **22** 1242–1257

The Influence of Photoinitiator on the Characteristics of Electrospun Poly (vinyl alcohol)/Chitosan Based Nanofibers

H. Susanto[1], M. W. Faz[1], M. P. H. Rani[1], and M. H. Robbani[1]

[1]Department of Chemical Engineering, Faculty of Engineering, Diponegoro University, Indonesia
Jl. Prof. Sudarto-Tembalang, Semarang, Indonesia

E-mail: heru.susanto@che.undip.ac.id

Abstract. Poly (vinyl alcohol)/chitosan (PVA/CS) based nanofibers have non-toxic, compatible and antimicrobial properties that can be used in various applications. This research focused on the impact of benzophenone (BP) as a photoinitiator on the characteristics of electrospun PVA/CS based nanofibers. The electrospun PVA/CS nanofibers were crosslinked by the combination of immersion in glutaraldehyde (GA) solution and UV light irradiation with/without the addition of BP. The influence of BP was studied by scanning electron microscopy (SEM), fourier transform infrared (FTIR) spectroscopy, and pure water permeability analysis. The results showed that the average diameter and swelling degree of photocrosslinked nanofibers were decreased, while the pure water flux increased due to BP addition. FT-IR data confirmed that the BP was successfully blended with the presence of ketones group peak on 1655 cm^{-1}. The change in morphology was observed by SEM investigations.

Keywords. Benzophenone, chitosan, electrospinning, nanofibers, and poly(vinyl alcohol).

1. Introduction

Electrospinning has been rapidly developed in recent years because it can generate consistently polymer fibers ranging from tens to microns in diameter, depending on the polymer and processing conditions. The nanofibers prepared by electrospinning method have many advantages such as very large surface area-to-volume ratio, high porosity with very small pore size and malleability to produce different morphologies [1,2]. Due to these advantages, electrospun nanofibers have widely been investigated in the last ten years for biomedical, sensors, filter media, electrochemical energy storage and other industrial applications [2–6].

An electrospinning system consists of three major components, (i) a high voltage power supply, (ii) a spinneret, and (iii) a grounded collecting plate [2]. In an electrospinning process, a high voltage is applied to the tip of the polymer solution or melted polymer. As the intensity of the electric potential is increased, the droplet formed at the tip deforms into a conical shape. Once polymer surface tension at the top of the cone is reduced to a critical level, an electrically charged jet of fine fluid is ejected. The solvent within the ejected solution begins to evaporate, or the polymer melt begins to solidify. As a result, nanofibers are produced and eventually deposited on a grounded collector [7].

Content from this work may be used under the terms of the Creative Commons Attribution 3.0 licence. Any further distribution of this work must maintain attribution to the author(s) and the title of the work, journal citation and DOI.
Published under licence by IOP Publishing Ltd

Chitosan (CS), a (1-4)-linked 2-amino-2-deoxy-ù-D-glucopyranose, is one of the most abundant natural polysaccharides. It has excellent biocompatibility, biodegradable property, nontoxicity and ease of solubility in organic acids [1]. However, electrospinning of pure chitosan has some constraints, such as difficult to form continuous fiber jets during electrospinning. This is due to the repulsive forces between the ionic groups and high surface tension of the polymer solution as a result from the high viscosity of chitosan even in moderate concentration [7]. To overcome this constraint, chitosan is commonly blended with another synthetic polymer prior to electrospinning, e.g., poly(vinyl alcohol) (PVA). PVA is a non-toxic, water-soluble, biocompatible, and biodegradable synthetic polymer, which is widely used in the biomedical field. PVA has a high fiber forming and hydrophilic properties [1,8].

In this study, PVA/CS blend nanofiber membranes were prepared by electrospinning method. In order to improve stability, PVA/CS nanofibers were further treated by the post-treatments. The electrospun PVA/CS nanofibers were crosslinked by the combination of immersion in glutaraldehyde (GA) solution and UV light irradiation with/without the addition of BP.

2. Materials and methods

2.1. Materials
Polyvinyl alcohol (Mw =130.000 g/mol, 99 % hydrolyzed) and glutaraldehyde solution (1% in water) were purchased from Sigma Aldrich. Chitosan (Mw = (1,680,000 to 1,750,000) g/mol, 80.4% deacetylated) was purchased from local Indonesian company. Benzophenone was purchased from Fluka Analitycal, while acetic acid was purchased from Merck Chemical.

2.2. Electrospinning and Post-treatment
PVA solution with 6 wt%, 7 wt%, and 8 wt% concentrations were prepared using distilled water at 80 ^{o}C and stirred for 2 h. A 2 wt% chitosan solution using 2 % acetic acid aqueous solution was also prepared. These two solutions were subsequently mixed with PVA/chitosan ratio was 3:1.

Electrospinning of the polymer solutions was carried out using Nachriebe 600 Electrospinning by a set-up consisting of a syringe (3 mL) and needle (3-cm long and flat tip), a ground electrode and a high voltage supply (18 kV). Polymer solutions were electrospun at a feeding rate of 0.7 mL/h and a tip-to-collector distance of 12 cm. A grounded stationary rectangular metal collector covered by a piece of aluminum foil was used for the nanofibre deposition. In the post-treatment experiments, electrospun PVA/CS nanofibres were immersed in 3 % glutaraldehyde solution for 2 h at room temperature. Then the treated samples were dried at room temperature for 24 h. The dried samples were exposed to UV light at 256 nm wavelength with or without addition of benzophenone.

2.3. Characterization
The viscosity of the polymer solutions was measured using Brookfield DV-E viscometer with spindle-type 5 at 50 rpm. The morphology of the electrospun PVA nanofibers was characterized using scanning electron microscopy (SEM, JEOL-JSM 6510LA). Fiber diameters were determined using Image J (IJ 1.46r) software. They were measured by drawing lines perpendicular to the fiber axis on the SEM images. The pixel values of the lies were converted to diameter values. Fourier transform infrared spectroscopy (FT-IR) analysis was carried out using IRPrestige-21 Shimadzu spectrometer. The permeability of nanofiber membrane was measured using amicon cell with dead-end filtration mode at a pressure of 0.3 kg/cm^2.

3. Results and discussion

3.1. Effect of PVA/CS Concentration on Viscosity of Electrospinning Solution
Blending of chitosan and PVA electrospinning solution directly affected the resultant properties of the solutions. Figure 1 shows the dependence of the viscosity on its concentration. It was found that its

IOP Conf. Series: Materials Science and Engineering **367** (2018) 012028 doi:10.1088/1757-899X/367/1/012028

viscosity increased linearly with increasing concentration. This result is in accordance with what has been conducted by Paipitak *et al.* [8]. This result is essential for the optimization of nanofiber formation of this composite material.

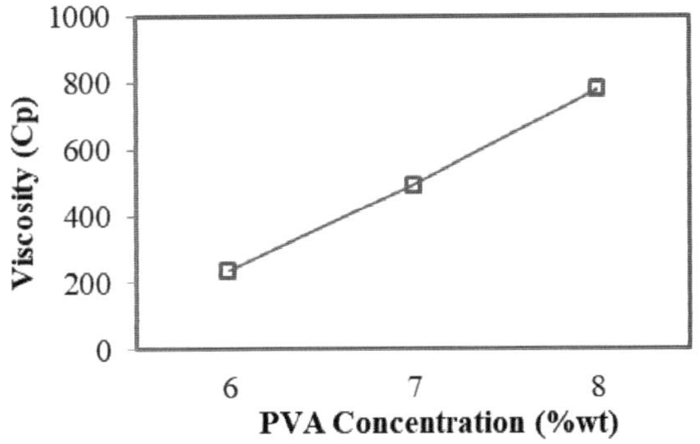

Figure 1. The viscosity of PVA/CS electrospinning solution as a function of PVA concentration.

3.2. Effect of PVA/CS Concentration and Photoinitiator on Morphology of Electrospun Nanofibers

SEM micrographs and fiber diameter distribution histograms of the nanofibres result are presented in Figure 2. We found that increasing the viscosity increased the average diameter of the fiber. Similar results were reported by several authors [9,10]. The higher onentration used, the thiker average diameter was resulted, i.e., 222 nm, 315 nm and 461nm for PVA 6 %, 7 %, and 8 %, respectively.

Electrospun nanofiber with PVA concentration of 7 % had a narrower size distribution compared to 6 % and 8 % concentrations. It means PVA 7 % nanofiber had more homogenous fiber diameters. However, the addition of concentration led to increasing bead sizes in electrospun nanofiber.

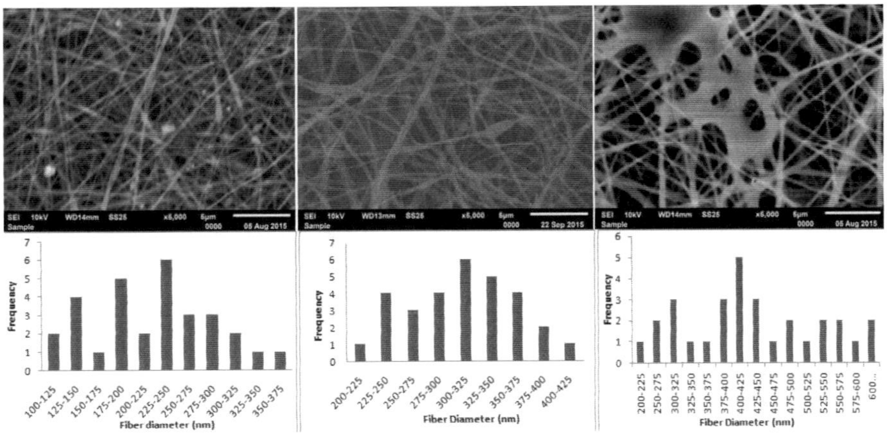

Figure 2. SEM images of the samples after electrospinning and fiber diameter distribution histograms (left to right: 6 %, 7 %, and 8 % concentration)

Post-treatment of the PVA/chitosan nanofibers with glutaraldehyde and UV irradiation showed morphological changes in the microstructure of the membrane (Figure 3). Combination of physical and chemical crosslink could affect the diameter of the fibers. This impact was shown by both post-treatments where mean fiber diameter decrease to 150 nm to 200 nm. In the other hand, addition of benzophenone as photoinitiator decreased the average diameter of the fibers. This phenomenon may be caused by benzophenone which initiates surface polymerization by giving a radical surface [11]. Major photocrosslink processes occurred on membrane surface causing shrinkage of fibers. The radical surface from benzophenone allowed the photocrosslink process take places and reduced the average diameter of the fibers [12].

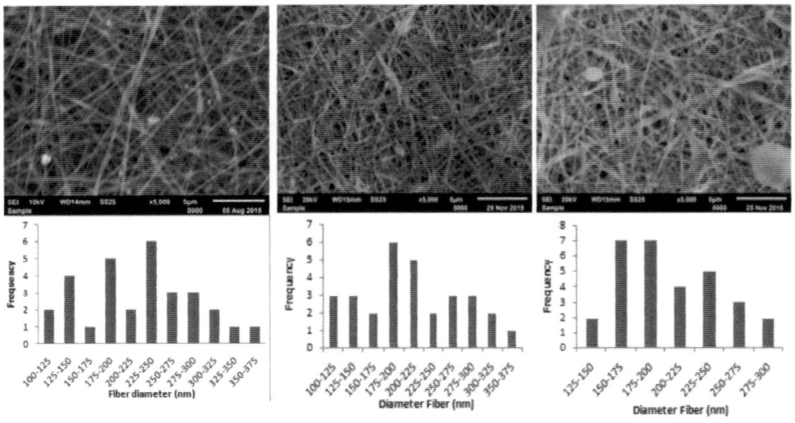

Figure 3. SEM images of the samples after post-treatment and fiber diameter distribution histograms (left to right, nonposttreatment; without photoinitiator; with photoinitiator)

3.3. Effect of Photoinitiator on Permeability of Electrospun Nanofibers

It is seen from Figure 4 that the post-treatment caused membrane became insoluble in water. Hence permeability analysis could be investigated. The edge of nonpost-treatment membrane dissolved in the water, so the permeability analysis could not be examined. The results showed that the photoinitiator addition increased the permeability of the nanofibers. UV irradiation caused the polymer chain broke and formed free radical which led to cross-link reaction. The photoinitiator formed more free radical and the crosslinking process which led to the improvement of pore size of nanofibers and flux increase, respectively [13]. SEM results also confirmed that the electrospun nanofibers with the addition of photoinitiator have a larger pore size which leads to increase the water flux.

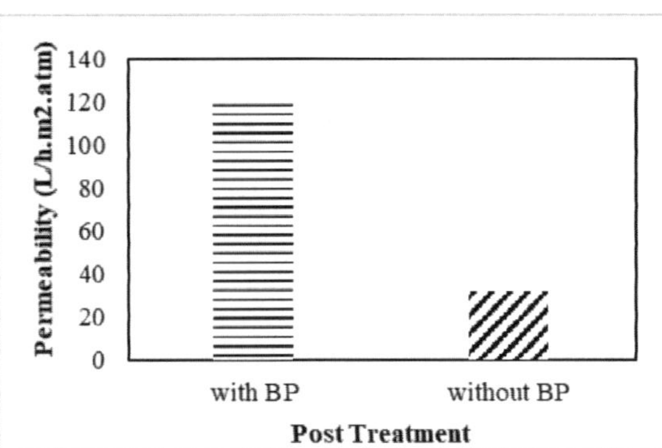

Figure 4. Permeability of water in nanofibers

3.4. Effect of Photoinitiator to Chemical Characteristic of Electrospun Nanofibers

The FTIR spectra exhibit overlapping broad bands of OH group and NH group at (3200 to 3600) cm^{-1}, CH at ~2923 cm^{-1}, CH_3 at ~1375 cm^{-1}, CH_2 bend at ~1465, cm^{-1}, and CN group at (1025 to 1200) cm^{-1} which is characteristic of CS and PVA blend nanofiber (Figure 5). The peak of glutaraldehyde as post-treatment solution appears in 1643 cm^{-1}. This band probably imine group that usually appears in ~1661 cm^{-1}. This imine resulted from the reaction between the amine group in chitosan and the aldehyde group on the glutaraldehyde [14]. The IR spectra also show the increase in corresponding groups at

~1993 cm^{-1} that indicate ether/acetal (-COC) group. The appearance of the ether/acetal peak indicates that the acetylation reaction occurred between aldehyde and hydroxyl in the PVA [4]. The acetal groups peak as the sign of crosslinking process between PVA, chitosan, and glutaraldehyde appeared in 1140 cm^{-1}. The addition of benzophenone as photoinitiator caused the appearance of 1655 cm^{-1} peak which related to the presence of ketones group of benzophenone [10].

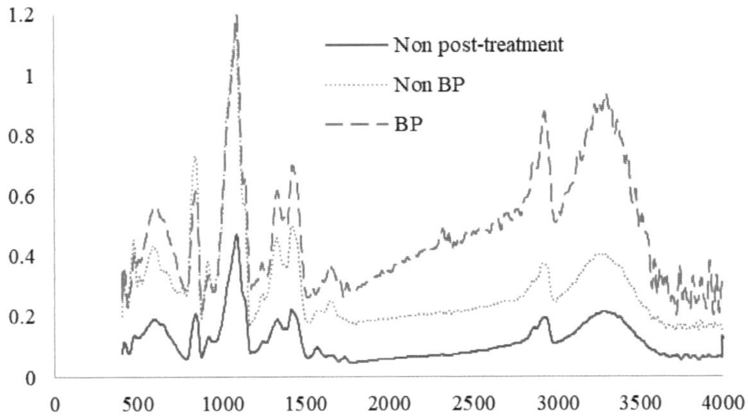

Figure 5. FTIR Spectra results of Electrospun nanofibers

4. Conclusions

The electrospun PVA/CS nanofibers have been prepared. The increase in polymer viscosity increased the diameter of the fiber and formed larger beads. The addition of benzophenone as photoinitiator decreased the average diameter of the fibers. The photoinitiator addition increased the permeability of the nanofibers. FTIR analysis exhibited the existence of relevant functional groups of PVA, chitosan, glutaraldehyde, and benzophenone in the composites. The crosslinked structure was observed for the electrospun PVA/Chitosan nanofibers treated with benzophenone as a photoinitiator.

Acknowledgments

The authors acknowledge the support of the Ministry of Research, Technology and Higher Education, the Republic of Indonesia for funding this project. The authors also thank smart city project by USaid Shera for the opportunity given to increase the quality of this paper.

References

[1] Aliabadi M, Irani M, Ismaeili J, Piri H and Parnian M J 2013 Electrospun nanofiber membrane of PEO/Chitosan for the adsorption of nickel, cadmium, lead and copper ions from aqueous solution *Chem. Eng. J.* **220** 237–243

[2] Çay A, Miraftab M and Kumbasar E P A 2014 Characterization and swelling performance of physically stabilized electrospun poly (vinyl alcohol)/chitosan nanofibres *Eur. Polym. J.* **61** 253–262

[3] Ding Y, Wang Y, Su L, Bellagamba M, Zhang H and Lei Y 2010 Electrospun Co_3O_4 nanofibers for sensitive and selective glucose detection *Biosens. Bioelectron.* **26** 542–548

[4] Farid O, Mansour F, Habib M, Robinson J and Tarleton S 2016 Investigating the sorption influence of poly (vinyl alcohol)(PVA) at different crosslinking content *J. Environ. Chem. Eng.* **4** 293–298

[5] Jia Y-T, Gong J, Gu X-H, Kim H-Y, Dong J and Shen X-Y 2007 Fabrication and

characterization of poly (vinyl alcohol)/chitosan blend nanofibers produced by electrospinning method *Carbohydr. Polym.* **67** 403–409
[6] Li X, Zhang Y, Li H, Chen H, Ding Y and Yang W 2014 Effect of oriented fiber membrane fabricated via needleless melt electrospinning on water filtration efficiency *Desalination* **344** 266–273
[7] Liang D, Hsiao B S and Chu B 2007 Functional electrospun nanofibrous scaffolds for biomedical applications *Adv. Drug Deliv. Rev.* **59** 1392–1412
[8] Paipitak K, Pornpra T, Mongkontalang P, Techitdheer W and Pecharapa W 2011 Characterization of PVA-chitosan nanofibers prepared by electrospinning *Procedia Eng.* **8** 101–105
[9] Ki C S, Baek D H, Gang K D, Lee K H, Um I C and Park Y H 2005 Characterization of gelatin nanofiber prepared from gelatin–formic acid solution *Polymer* **46** 5094–5102
[10] Jung I H, Yu J, Jeong E, Kim J, Kwon S, Kong H, Lee K, Woo H Y and Shim H-K 2010 Synthesis and Photovoltaic Properties of Cyclopentadithiophene-Based Low-Bandgap Copolymers That Contain Electron-Withdrawing Thiazole Derivatives *Chem.- Eur. J.* **16** 3743–3752
[11] Deng J, Xu Z, Xiang C, Liu J, Zhou L, Li T, Yang Z and Ding C 2017 Comparative evaluation of maceration and ultrasonic-assisted extraction of phenolic compounds from fresh olives *Ultrason. Sonochem.* **37** 328–334
[12] ZhangB, Kang F, Tarascon J M and Kim J M2015 *J.K. RRecent. Prog. Mater. Sci* **76** 319–80
[13] Nur A P, Sari D K and Susanto H *Jurnal Teknologi Kimia dan Idustri* **2** 189–97
[14] Susanto H, Samsudin A, Rokhati N and Widiasa I 2013 Immobilization of glucose oxidase on chitosan-based porous composite membranes and their potential use in biosensors *Enzyme Microb. Technol.* **52** 386–392

The Fabrication of Bulk Heterojunction P3HT: PCBM Organic Photovoltaics

D. Darwis[1], E. Sesa[1], D. Farhamza[1], and Iqbal[1]

[1]Centre for Organic Electronics Tadulako University,
Jl. Soekarno Hatta KM.9, Tondo, Mantikulore, Kota Palu, Sulawesi Tengah 94148 Indonesia

E-mail: Darmawati.Darwis@uon.edu.au

Abstract. Bulk heterojunction Organic photovoltaic (OPV) devices are gaining a lot of interest due to their potential for ease of processing and lower manufacturing cost sustainable energy generation. In consequence, the number of studies into the properties and characteristics of organic solar cell devices has been increased to improving their power conversion. A further advancement over past decade has shown that improved efficiency could be obtained by mixed of poly(3 - hexylthiophene) (P3HT) and [1] – phenyl - C61-butyric acid methyl ester (PCBM) as an active layer. A series of optimizations of this P3HT: PCBM blends, such as the mixture ratio variation, the annealing treatments, and solvent treatment, have been emerged to improve the efficiency of the OPV. As a result, significant improvements were achieved. Here, we report the fabrication heterojunction devices of 2.9 % efficiency. This result has been achieved using the configuration of a typical heterojunction solar cell modules consists of layered glass/ITO/PEDOT: PSS/active layer/cathode interlayer

Keywords. Bulk heterojunction, efficiency, OPV, P3HT, and PCBM.

1. Introduction

Renewable energy sources are recognized as a technology that plays a significant role in lowering the global warming effect by reducing CO_2 emissions to the atmosphere. The trend of industries reveals that renewable energy investment has been increasing which reflect the urgency of study into these type of energy source. The capacity of the renewable power generating showing a largest annual enhancement ever in 2016, with an estimated capacity added of 161 GW [1, 2]. Compare to 2015; the total renewable power capacity increased up to nearly 2.017 GW at years end [1]. Solar cells indicated for the first time contributed to greatest additional energy capacity than any other generating technology (34 %)[2].

Today, crystalline silicon solar cells dominate the solar power market. Global production of this type photovoltaics and modules rose significantly. This technology has been impacting the world which encourages the expansion of energy source from fuel to green energy. Recently, silicon solar cells have reached a conversion efficiency of 25.6 % [3]. However, the shortcoming facing researcher of this type photovoltaics is not only their efficiency conversion but also the expensive of fabrication cost, both regarding energy and money.

Content from this work may be used under the terms of the Creative Commons Attribution 3.0 licence. Any further distribution of this work must maintain attribution to the author(s) and the title of the work, journal citation and DOI.
Published under licence by IOP Publishing Ltd

IOP Conf. Series: Materials Science and Engineering **367** (2018) 012029 doi:10.1088/1757-899X/367/1/012029

One solution that possibly overcomes these drawback is organic photovoltaics (OPV) [4, 5]. In most straightforward term, an organic photovoltaic is a device that deals with organic materials for absorption of light and transport of charge in producing electricity directly from the sun by the effect of photovoltaics. Here we report the fabrication of OPV using blended of poly(3- hexylthiophene) (P3HT) and [6,6][6]-phenyl-c61-butyric acid methyl ester (PCBM) as an active layer. Investigation of this P3HT: PCBM mixed, using UV-VIS spectroscopy is also presented.

2. Materials and methods

2.1. Device Fabrication

The structure of the OPV based P3HT: PCBM that was used in this research is shown below in Figure 1a. The anode consisted of glass slides pre-patterned with indium tin oxide (ITO) were obtained from Kintec Company, Hong Kong. The ITO glasses were prepared using the following cleaning procedure: (1) ITO substrate was immersed in acetone for 5 min and then sonicated for 20 min. (2) Each substrate was then dried under a flow of Nitrogen. (3) Furthermore, the substrates were then stored in isopropyl alcohol (IPA) for sonication for further 15 min. (4) The glasses were then again dried under a Nitrogen flow. (5) Finally, the pre-cleaned substrates were put in the UV – ozone for 10 min ready for use. A 70 nm layer of poly(3,4-ethylene dioxythiophene)/ poly(styrene sulfonate) (PEDOT: PSS) were deposited on the top of the ITO pattern and were annealed at 140 °C for 30 min. This layer seals the anode layer from oxygen and prevents the immigration of electron to the cathode. Also, the PEDOT: PSS film acts as a layer for hole transport [7]. A weight ratio of 1:1 was used for P3HT (sanitized in the COE of Newcastle University) and PCBM (purchased from Lumtec) blends. The solution of P3HT: PCBM blends (20 mg/mL in $CHCl_3$) was stir sonicated for 25 min and then immediately spin coated at 2000 rpm for 1 min resulting in a 70 nm thick of an active layer. The P3HT: PCBM blended films were dried at 140 °C for 4 min and then introduced to a nitrogen glove box, where they are placed into a substrate holder for cathode evaporation, as shown in Figure 1b. This holder can accommodate 54 substrates.

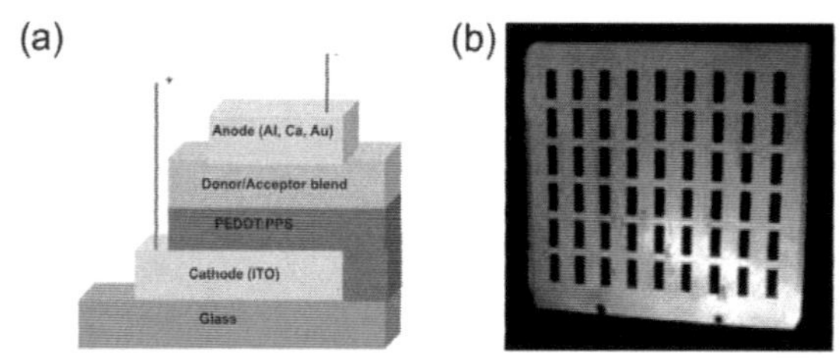

Figure 1. (a). The structure of bulk heterojunction OPV, (b). The sample holder

Aluminium/Al electrodes were then evaporated onto the central region of the slides at a deposition rate of 2 A/s and a pressure lower than (2×10^{-6}) Torr. The final thickness of this Al electrode was 110 nm.

2.2. Device Characterization

During this project, A Varian Cary 6000i UV-vis spectrophotometer with a scanning wavelength of 200 nm to 900 nm was used to obtain the absorption of the P3HT: PCBM blend films. The relevant films were spin cast on quartz substrates. A critical procedure was performed before any measurement which is a baseline reference scan with no sample present to remove artifacts introduced by spectrophotometer.

AM 1.5 solar simulator product in a nitrogen atmosphere is used in obtaining a spectrum of light approximating natural sunlight to test OPV devices. For this testing, the instrument is used to get the electrical characteristics of the cells which lead to the quality of the photovoltaic device. The solar simulator consists of three main components: associated power supply and a light source; filters and

IOP Conf. Series: Materials Science and Engineering **367** (2018) 012029 doi:10.1088/1757-899X/367/1/012029

optics to modify the output beam to meet the classification requirements; and the necessary controls to operate the simulator and adjust irradiance. The filter was used for changing the Xenon (Xe) lamp spectrum to be more like sunlight. The target irradiance is 1000 V/m^2 (1 sun). Finally, the current, voltage and irradiance signals are recorded by the system simultaneously. The power efficiency and fill factor are determined directly by Labview software. To obtain I-V data, we used the following step. The cell was loaded into a holder and then connected to a Keithley 2400 Sourcemeter to systematically sweep the potential between the two electrodes of the devices. During this process, the current produced by the device under test was recorded, and I-V curve was then generated.

3. Results and discussion

3.1. Absorbance Characteristic

Figure 2 showing the UV-vis absorbance spectrum of the bulk P3HT: PCBM film as a spin-coated film and indicated that the UV-vis spectra of the film presenting the expected typical spectra with a superposition of both component which PCBM-based bands and absorption P3HT-based band [8, 9]. It establishes that the absorption P3HT-based peak at 520 nm provides information on conjugation degrees of the polymer chains while the presence of the relative size of the P3HT vibronic shoulder at 610 nm indicated the degree interchain order [9–11].

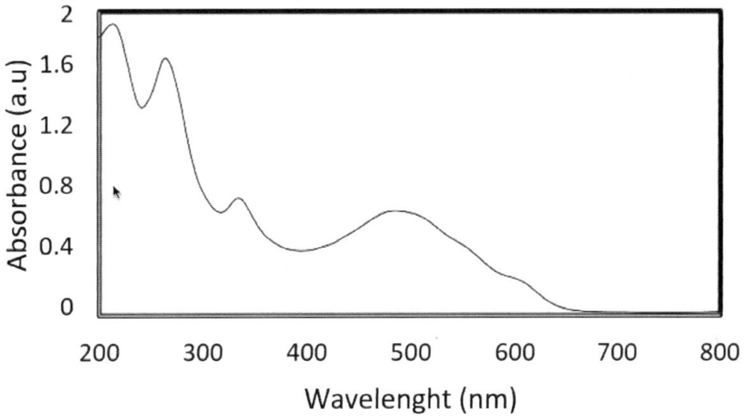

Figure 2. The UV-vis absorbance spectrum of the bulk P3HT: PCBM film

3.2. Bulk Heterojunction OPV Characteristic

Bulk heterojunction P3HT: PCBM were fabricated with a layer structure of glass /ITO/ PEDOT: PSS /P3HT: PCBM/ Ca/Al. Current versus voltage (I-V) characterization as shown in Figure 3 has been performed under AM 1.5 simulated with the key parameters is listed in Table 1.

The device performance is showing a power conversion efficiency (PCE) of 2.9%, short-circuit current density (Jsc) of –7.6 mA·cm^{-2}, open circuit voltage (Voc) of 0.6 V, and fill factor (FF) of 0.7. The characteristics of this cell are comparable to those obtained for devices fabricated with unannealed P3HT: PCBM organic solar cells [11, 12]. The low V_{OC} can be associated with the interface between the P3HT and PCBM molecules [5, 13] and the loss charge carriers at the electrodes. Also, The Fill Factor that appeared is lower than the ideal value could be assigned to the voltage dropping due to the high of the series resistant and the current leakage through the shunt resistant of the device [5, 14].

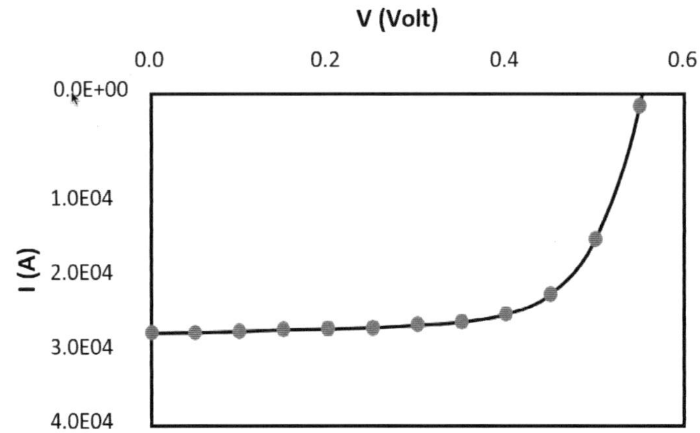

Figure 3. IV curve of the bulk heterojunction P3HT: PCBM

Table 1. The key parameters of the OPV

Parameter	Value
Efficiency (%)	2.9
V_{OC}(V)	0.6
I_{SC}(A)	-2.9×10^{-4}
J_{SC}(mAcm^{-2})	−7.6
Fill Factor	0.7

4. Conclusions

In conclusion, we have produced a bulk heterojunction P3HT: PCBM organic photovoltaics devices with the efficiency of 2.9 %. Absorbance characteristic of polymer P3HT/PCBM film for use in thin-film electronics is also presented, which shows the expected typical spectra with a superposition of both components which PCBM-based bands at 260 nm and 330 nm and absorption P3HT-based band at 520 nm. These results could be an essential milestone in evaluating the performance of bulk heterojunction P3HT: PCBM organic solar cells for further improvement.

Acknowledgments

This study was conducted by the support of a Directorate General of Higher Education Indonesia (DIKTI) project (International Research Collaboration and Scientific Publication) and the Centre for Organic Electronics Newcastle University.

References

[1] Anonim 2016 *Global Trends in Renewable Energy Investment 2016* (http//fs-unep-centreorg/publications)

[2] REN21 2017 *Renewables 2017 global status report* (http//wwwren21net/gsr-2017)

[3] Green M A *et al* 2017 Solar cell efficiency tables (version 50) *Progress in Photovoltaics Research and Applications* **25** p 668-676

[4] Gutiérrez-González I *et al* 2014 Structural and electronic properties of the P3HT–PCBM dimer A theoretical Study *Chemical Physics Letters* **612** p 234-239

[5] Kadem B Y, Al-Hashimi M K, and Hassan A K 2014 The Effect of Solution Processing on the Power Conversion Efficiency of P3HT-based Organic Solar Cells *Energy Procedia* **50** p 237-245

[6] Darwis D *et al* 2014 Surfactant-free nanoparticulate organic photovoltaics *Solar Energy Materials and Solar Cells* **121** p 99-107

IOP Conf. Series: Materials Science and Engineering **367** (2018) 012029 doi:10.1088/1757-899X/367/1/012029

[7] Darwis D *et al* 2011 Surfactant-Free P3HT / PCBM Nanoparticles for Organic Photovoltaics (OPV) *AIP Conference Proceedings* **1415** p 120-123
[8] Hu Z *et al* 2010 Correlation between spectroscopic and morphological properties of composite P3HT/PCBM nanoparticles studied by single particle spectroscopy *Journal of Luminescence* **130** p 771-780
[9] Brown P J *et al* 2003 Effect of interchain interactions on the absorption and emission of poly(3-hexylthiophene) *Phys. Rev. B* **67** p 064203
[10] Ulum S *et al* 2013 Determining the structural motif of P3HT PCBM nanoparticulate organic photovoltaic devices *Solar Energy Materials and Solar Cells* **110** p 43-48
[11] Xue B *et al* 2010 Vertical Stratification and Interfacial Structure in P3HT PCBM Organic Solar Cells *The Journal of Physical Chemistry C* **114** p 15797-15805
[12] Cowan S R A R, Anshuman and Heeger A J 2010 Recombination in polymer-fullerene bulk heterojunction solar cells *Phys. Rev. B* **82** p 245207
[13] Rauh D *et al* 2011 Relation of open circuit voltage to charge carrier density in organic bulk heterojunction solar cells *Applied Physics Letters* **98** p 133301
[14] Al-Ibrahim M *et al* 2005 Effects of solvent and annealing on the improved performance of solar cells based on poly(3-hexylthiophene) Fullerene *Applied Physics Letters* **86** p 201120

Oxygen Plasma Effect on QCM Sensor Coated Polystyrene Film

N. F. Khusnah[1], S. P. Sakti[1], and D. J. D. H. Santjojo[1]

[1]Physics Department, Universitas Brawijaya, Jl. Veteran Malang, 65145, Indonesia
Collaborative Research Center for Advanced Systems and Material Technology, Universitas Brawijaya, Jl. Veteran Malang, 65145, Indonesia

E-mail: sakti@ub.ac.id

Abstract. Hydrophobicity property of polystyrene (PS) thin film is one of the essential factors to be considered in the development of quartz crystal microbalance (QCM) biosensor using polystyrene as matrix layer. Many methods were developed to improve the immobilization rate of the biomolecule on the sensor surface without affecting the QCM essential works. Surface modification of the sensor surface aims to modify the physical and or chemical property of the surface. A straightforward method, the fast, environmentally-friendly, and low-cost solution to modify the sensor surface coated with polystyrene film is using oxygen plasma. In this experiment, the polystyrene film was spin-coated on both surface of QCM electrodes and then heated at 100 °C. The specimen is then placed for 5 min long in a chamber filled with oxygen plasma generated by 2 MHz RF-DC high-density plasma system. The relationship between DC-bias used and the changes in morphology properties of the coated film was characterized by Topography Measurement System (TMS) and Contact Angle Measurement. The electrical characteristic of QCM was also characterized using Impedance Analyzer. It was revealed that the contact angle of oxygen plasma treated film is changed and depicted the hydrophobic character. Also, there is an increasing resonance frequency of the sensor after oxygen plasma treatment indicates an etching mechanism occurs during plasma treatment.

Keywords. Oxygen plasma effect, polystyrene film, QCM sensor

1. Introduction

A quartz crystal microbalance (QCM) is a piezoelectric mass-sensing device which has been used widely for physical sensors, chemical sensors, and biosensors. The typical applications of the QCM sensor are based on the gravimetric methods as described by the Sauerbrey equation. The frequency shift of the sensor is proportional to the amount of mass deposited on the sensor surface. The main advantageous of QCM that support its rapid development in various disciplines, i.e., low-cost, modifiable substrate, label-free, real-time measurement, and quick response.

In the chemical and biosensor application, the QCM sensor surface is covered with a layer which is selective to the target molecule being detected. The selective layer can be coated directly on the sensor surface, i.e., on top of the sensor electrode or a matrix layer. The behavior of the selective layer and also any matrix layer must satisfy the requirements to works in gravimetric modes.

The polymer film can be used as a matrix layer or as a sensitive layer. Its function depends on the polymer surface property as well as the nature of the target molecule. Polystyrene can be used as a

Content from this work may be used under the terms of the Creative Commons Attribution 3.0 licence. Any further distribution of this work must maintain attribution to the author(s) and the title of the work, journal citation and DOI.
Published under licence by IOP Publishing Ltd

matrix layer in the biosensor application. On top of the polystyrene layer, a particular biomolecule was immobilized. Furthermore, the polystyrene was interested to be studied as coating layer because it acted as the protective film for the sensor electrode. However, in some conditions, the surface behavior of the polystyrene need to be modified to a specific condition which satisfies to the application. To be used as a coating film of QCM sensor in the application for chemical sensor or biosensor, the surface property such as hydrophilic/hydrophobicity, roughness, and chemical composition of the polymer is one of the crucial factors to be considered. For example, to increase the sensitivity of QCM sensor to hydrocarbons pollutant, hydrophobic polymer coatings are often chosen [1].

One important aspect to be considered during polymer film surface modification is the rigidity of the film. The shear moduli of the film must be high enough to keep the coating layer in a glassy condition to avoid damping to the sensor. The excellent bulk physical and chemical properties, inexpensive, and easy to process are a few factors that made polystyrene successfully applied in many QCM sensor developments. Surface modification is needed to alter the polystyrene chemical and physical properties without affecting the bulk properties.

Among many methods, low-pressure plasma processing is widely used to modify polymer material due to their ability to alter the outermost substrate without affecting its bulk properties [2, 3]. This treatment has been employed to achieve some purposes such as to produce specific functional group, increase surface energy, increase hydrophobicity/hydrophilic, modify surface morphology, and much more. Oxygen plasma treatment, for example, can increase the surface energy of polymers which can alter the hydrophobicity of the surface. For the biosensor application, plasma treatment has been used to improve the binding efficiency of the biomolecule on the sensor surface [4].

A variety of oxygen functional groups such as C-O, C=O, O-C=O, and C-O-O will be produced at the polymer surface when oxygen plasma or oxygen-containing plasma is employed for polymer modification. Functionalization [4] and etching [5] process may occur during the oxygen plasma treatment of polymers. The process depends on plasma parameter. Etching of the polymer is occurred through the reaction of atomic oxygen with the carbon atoms at the surface of a film, giving the volatile product such CO. Surface functionalization such as a functional group of oxygen is formed through the reaction between the active species from plasma and the surface atoms.

Polystyrene has been used as a coating material in QCM sensor development for biosensor application. This film used with or without any surface modification. Polystyrene acts as a matrix layer during immobilization of some biomolecule. Biomolecule immobilization on top of polystyrene surface works through physical adsorption. Physical adsorption consists of hydrophobic interactions or hydrophobic/ionic interactions between the biomolecule and the polystyrene surface. However, the degree of the hydrophobicity needs to be optimized for the immobilization. Therefore it is essential to explore a method to modify the hydrophobicity of the polystyrene surface

In this experiment, polystyrene thin film is adjusted by oxygen plasma to alter the hydrophobicity and surface roughness of polystyrene as a coating layer of the QCM sensor. For the biosensors application, the change of the surface hydrophobicity and roughness alters the biomolecule immobilization. Therefore understanding the DC bias effect of oxygen plasma generated by 2 MHz RF-DC high-density plasma system to hydrophobicity and surface roughness of the polystyrene coated QCM is an attractive method as it can be done with better controllability.

2. Materials and methods

The polystyrene film was prepared by solving a polystyrene grain with a molecular weight of 192 kDa in Toluene solution at 3 w/w%. The solution then was ultra-sonicated for about 10 min to accelerate the dissolution process. A 50 μL of the polystyrene solution was coated onto 10 MHz AT-cut silver electrode QCM sensor using spin coating method at 3000 rpm. The coating was applied to both sides of the sensor. The polystyrene-coated sensor was heated at 100 °C for an hour in an oven at the atmospheric condition to vaporize the solvent.

The experiment was carried out by exposing one side of polystyrene coated QCM sensor in 2 MHz RF-DC high-density plasma chamber generator as shown in Figure 1 for 5 min. This system is a capacitively coupled discharge which upper electrode (anode rod) was connected to RF generator power supply, and the lower plate electrode (cathode plate) was connected to DC generator power supply. The oxygen plasma is generated with RF voltage of 60 Volt and current 0.5 A while the DC bias voltage was varied from –100 V to –500 V. The oxygen inlet flows into the chamber was controlled at 50 sccm at 60 Pa. The negative DC bias at the lower electrode where the sample was put is aimed to accelerate the generated ion by the upper electrode to bombard the polystyrene film surface.

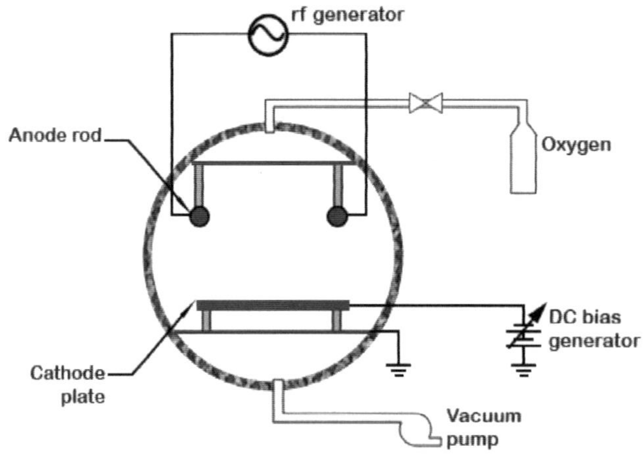

Figure 1. 2 MHz RF-DC High-Density Plasma System Schematic.

The water surface contact angle, surface roughness, and resonance frequency of the sensor before and after plasma treatment were measured. The measurement of the water surface contact angle was carried out by using self-developed goniometer based contact angle measurement system [6]. Surface roughness was measured using non-destructive surface profilometer TMS-1200 TopMap μLab from Polytec. The plasma spectrum also measured without and during sample treatment using fiber optic spectrometer Aurora 4000 from CNI Laser with a wavelength scanning range from 200 nm to 1000 nm with an optical resolution of 0.75 nm.

3. Results and discussion

Oxygen plasma is mostly used in polymer modification by functionalizing it surface with polar oxygen group and slow etching. The typical species formed in oxygen plasma glow discharge generated by 2 MHz RF-DC plasma system is shown in Figure 2. Two atomic lines are detected, i.e., 775 nm and 842 nm which corresponding to OI [7,8]. During plasma treatment of polystyrene, there are no peaks changes identified from plasma spectrum except decrease of intensity.

The plasma spectrum which remains before and during the plasma treatment indicated that there were no new species existed during the polystyrene plasma treatment. These indicated there was no atom nor molecule released from the polystyrene surface to the chamber.

The effect of the plasma treatment to the polystyrene surface property was analyzed by measuring the contact angle of the surface before and after plasma treatment. Figure 3 shows the water contact angle measurement of polystyrene film before and after oxygen plasma treatment. This graph shows after plasma exposure the sample hydrophobicity increased slightly compared to native polystyrene film (from around 85° to 95°). The trend of the graph shows the surface became more hydrophobic by an increase of negative DC Bias starting from –200 V to –500 V. However, different case happened for low DC bias, i.e., –100 V. The contact angle measurement showed the polystyrene coating became hydrophilic. It is probably due to low ion flux and energy bombardment to the polystyrene surface. Because of it, the oxygen ions/atoms and the outermost atoms are chemically reacted or diffuse. While the changes of a more hydrophobic surface are probably due to functionalization or etching mechanism that works dominantly so the surface profile of the polystyrene coating changes during the ion bombardment.

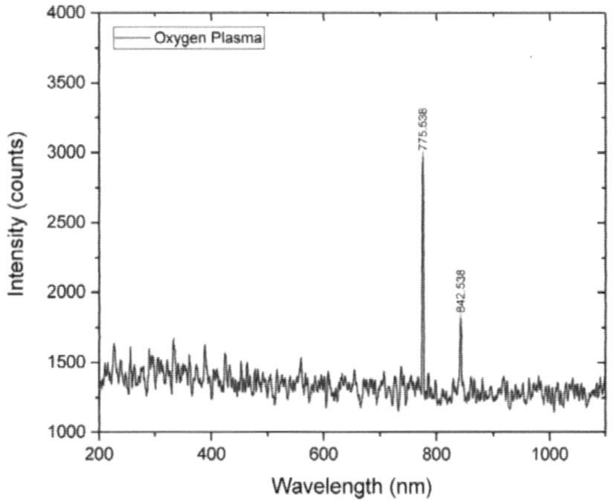

Figure 2. Typical Plasma Oxygen Spectrum

Further analysis was performed by measuring the surface roughness of the coating layer before and after plasma treatment. Figure 4 shows the surface roughness of the polystyrene coating before and after plasma treatment. This graph showed that the increase of surface roughness only happens from DC bias by –200 V to –500 V but slightly decrease for –100 V. This results correlated with the surface hydrophobicity as shown in Figure 3. The increase of surface roughness is as if confirm that the formation of surface structure occurred during the process. These possibly occur through functionalization and etching mechanism. The etching mechanism occurred as the ion has adequate energy caused by the DC bias voltage which accelerates the ion. Higher DC bias voltage results in the higher kinetic energy of the ion in the plasma chamber.

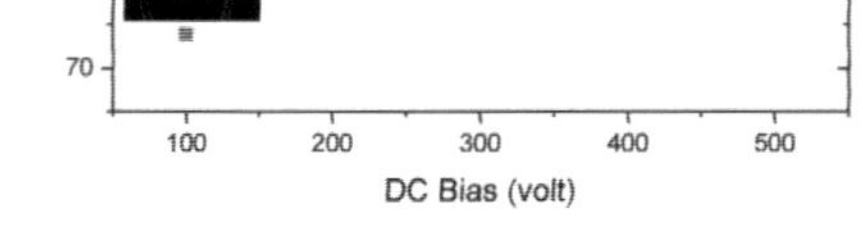

Figure 3. Water contact angle surface of the polystyrene coating before and after plasma treatment.

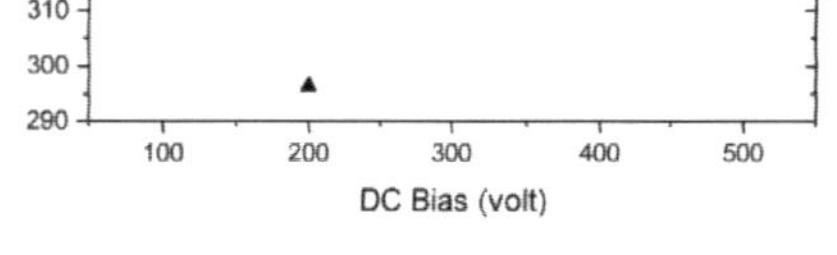

Figure 4. Polystyrene surface roughness before and after plasma treatment.

The impedance measurements were performed before and after plasma treatment to investigate the effects of the plasma treatment on the QCM sensor coated with polystyrene. Figure 5 and Figure 6 shows the minimum impedance and series resonant frequency trends of the sensor before and after plasma treatment. Almost all the sample, except with parameter –300 V, shows an increase of frequency which could be associated with deposited mass reduction. As described by Sauerbrey equation for the gravimetric condition, the resonance frequency of the sensor decreases when the deposited mass on the sensor is increased and vice versa [9]. The condition is also valid as long as the deposited mass, in the case of the experiment was polystyrene coating, behaves as a rigid material [10].

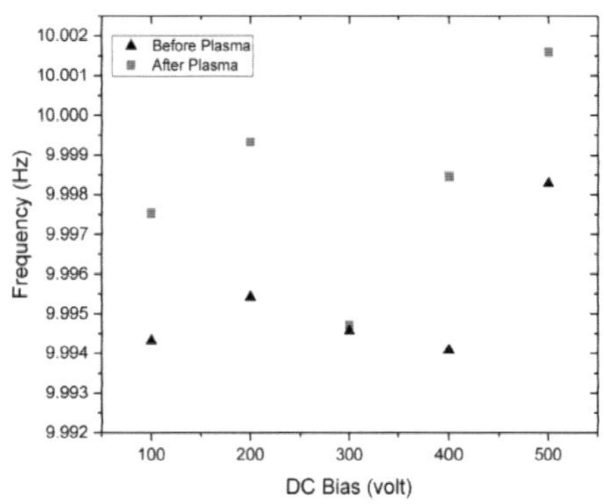

Figure 5. Series resonant frequencies of the QCM coated polystyrene before and after plasma treatment.

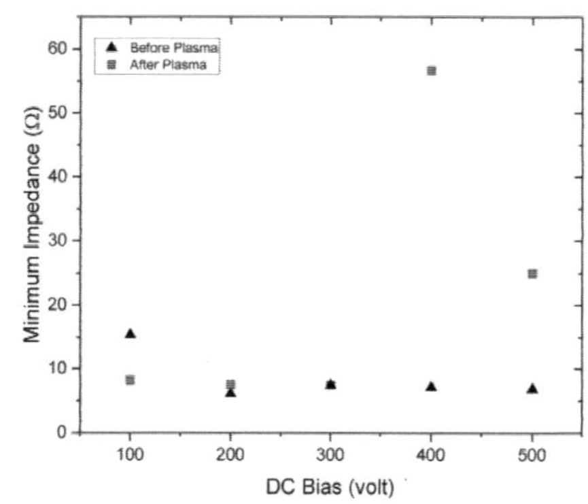

Figure 6. QCM sensor minimum impedance before and after plasma treatment.

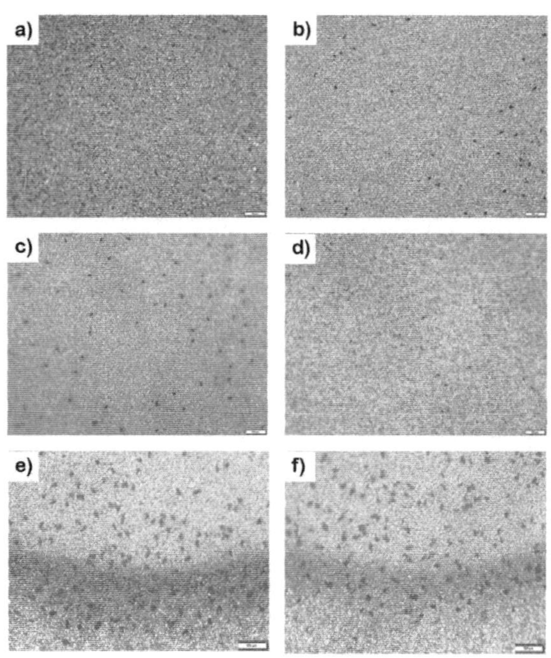

Figure 7. Optical Photograph of QCM coated polystyrene electrode area a) before O_2 plasma treatment and b), c), d), e), f) after O_2 plasma treatment with -100 V, -200 V, -300 V, -400 V, -500 V of DC bias respectively.

The mechanical bulk property of polystyrene layer after ion bombardment in the plasma chamber remained. It was supported by the data as depicted in Figure 6 which showed that the minimum impedance of the sensor after plasma treatment remained low. The low impedance of the sensor indicated that the coating layer remains rigid. It means that ion bombardment of the oxygen plasma did not change the shear modulus of the coating layer.

Qualitatively the changes to polystyrene coating surface could be observed by using an optical microscope. Some black dots are formed on the surface of QCM especially on the electrode area as shown in Figure 7. The increase of the DC Bias proportionally increases the number of black dots itself. This black dots as if indicates the ion bombardment strongly occurs when higher DC Bias is used. The number of black dots also seems to affect the changes to surface hydrophobicity.

4. Conclusions

This experiment revealed that processing Polystyrene coating surface in RF-DC oxygen plasma by varying the DC Bias leads to the changes of its hydrophobicity. The water contact angle dependence on DC bias is highly related to surface chemical composition and the surface morphology. Thus, functionalization and etching mechanisms are suggested taking place during the treatment. Increases of the sensor resonance frequency indicated etching mechanism after plasma treatment which is related to decreasing of deposited mass on the sensor surface. Higher energy of oxygen ion bombardment to the polystyrene surface leads caused by DC bias increased the surface roughness.

Acknowledgements

Part of this works was funded by the Ministry of Research, Technology and Higher Education of the Republic of Indonesia under the PUPT research grant scheme.

References

[1] Huang X and Elias A L 2014 Surface modified polymer thin films with enhanced sensitivity to a naphthenic acid model compound: A study by quartz crystal microbalance *Sensors Actuators B Chem.* **198** 7–14

[2] Häidopoulos M, Horgnies M, Mirabella F, and Pireaux J J 2008 Angle-resolved XPS study of plasma-deposited polystyrene films after oxygen plasma treatment *Plasma Process Polym.* **5** 67–75

[3] Dupont-Gillain C C, Adriaensen Y, Derclaye S, and Rouxhet P G 2000 Plasma-oxidized polystyrene: Wetting properties and surface reconstruction *Langmuir.* **16** 8194–200

[4] Guo B, Li S, Song L, Yang M, Zhou W, Tyagi D, and Zhu J 2015 Plasma-treated polystyrene film that enhances binding efficiency for sensitive and label-free protein biosensing *Appl. Surf. Sci.* **345** 379–86

[5] Ting Y H, Liu C C, Park S M, Jiang H, Nealey P F, and Wendt A E 2010 Surface roughening of polystyrene and poly(methyl methacrylate) in Ar/O_2 plasma etching *Polymers (Basel)* **2** 649 63

[6] Sakti S P, Aji R Y, and Amaliya L 2017 Low-Cost Contact Angle Measurement System for QCM *Sensor TELKOMNIKA* **15** 560–9

[7] Kramida A, Ralchenko Y, Reader J, and NIST ASD Team *2016 NIST Atomic Spectra Database (version 54)* (Gaithersburg MD)

[8] Krstulović N, Labazan I, Milošević S, Cvelbar U, Vesel A, and Mozetič M 2006 Optical emission spectroscopy characterization of oxygen plasma during treatment of a PET foil *J. Phys. D Appl. Phys.* **37** p 3799

[9] Lucklum R, Behling C, and Hauptmann P 2000 Gravimetric and non-gravimetric chemical quartz crystal resonators *Sensors Actuators B Chem.* **65** 277–83

[10] Vogt B D, Lin E K, Wu W I, and White C C 2004 Effect of film thickness on the validity of the sauerbrey equation for hydrated polyelectrolyte films *J. Phys. Chem. B* **108** 12685–12690

Transport Properties and Magnetoresistance of $La_{0.8}Ca_{0.13}Ag_{0.07}MnO_3$ Perovskite Manganite Synthesized by Sol-Gel Method

B. Kurniawan[1], F. Ruli[1], A. Imaduddin[2], and R. Kamila[1]

[1]Department of Physics, FMIPA Universitas Indonesia, Kampus UI Depok, Depok 16424, Indonesia
[2]Metallurgical Research Center, Indonesian Institute of Sciences, Puspiptek-Serpong 15314, Indonesia

E-mail: bkuru@fisika.ui.ac.id

Abstract. In this paper, we investigate the transport properties and magnetoresistance effect of $La_{0.8}Ca_{0.13}Ag_{0.07}MnO_3$ perovskite manganite synthesized by sol-gel method. The XRD pattern of the sample shows a rhombohedral perovskite structure with space group $R\bar{3}c$. The EDX analysis confirms that the sample contains all expected chemical elements without any additional impurity. The temperature dependence of electrical resistivity was measured using a cryogenic magnetometer. The results show a metal-insulator transition temperature (T_{M-I}) at 280 K. The resistivity of the sample increases with an increase of temperature below T_{M-I}. Theoretical analyses of the temperature dependence of resistivity suggest that the resistivity due to electron-electron scattering is predominant below T_{I-M}. The resistivity of the sample decreases when applied magnetic field 1 T at a temperature range of 10 K to 300 K. The magnetoresistance of $La_{0.8}Ca_{0.13}Ag_{0.07}MnO_3$ emanates from spin-polarized tunneling process at the grain boundary.

Keyword. Magnetometer, manganite, perovskite, sol-gel method, and transport properties.

1. Introduction

Perovskite manganite materials with general formula ABO_3 (A = La, B = Mn) are an antiferromagnetic and insulator material [1]. Mangan and oxygen ions in perovskite lanthanum manganite form MnO_6 octahedral geometry [2]. Partial substitution in A site with divalent or monovalent ions leads to distorting MnO_6 geometry and oxidize Mn^{3+} to Mn^{4+} which influence the magnetic and transport properties [1, 3–6]. Zener double exchange interaction qualitatively explained correlation among structural, magnetic and transport properties, transfer electrons from Mn^{3+} to Mn^{4+} through O^{2-} [7]. Hence, mixed valence Mn^{4+}/Mn^{3+} ions play an essential role in the electric and magnetic properties of lanthanum manganite [8].

Skini *et al* reported that the perovskite manganite $La_{0.8}Ca_{0.2}MnO_3$ exhibits high magnetoresistance (MR) with a metal-insulator transition temperature (T_{M-I}) below room temperature [9]. Guo *et al* also reported that the perovskite manganite $La_{0.8}Ca_{0.2}MnO_3$ possess high magnetocaloric effect (MCE) with Curie temperature (T_C) below room temperature [10]. Magnetoresistance effect in perovskite

Content from this work may be used under the terms of the Creative Commons Attribution 3.0 licence. Any further distribution of this work must maintain attribution to the author(s) and the title of the work, journal citation and DOI.
Published under licence by IOP Publishing Ltd

manganite emanates to two contributions. First, the intrinsic MR originates from suppression of spin fluctuation when the magnetic field is applied [11]. The intrinsic MR reaches a maximum around the paramagnetic-ferromagnetic transition temperature [12]. Second, extrinsic MR emanates from spin-polarized tunneling process at grain boundary which is dominant at low temperature [11, 12]. For wide application at room temperature, metal-insulator transition temperature and Curie temperature need to increase near room temperature. Pi *et al* reported that substitutions La site with Ag in perovskite lanthanum manganite increase the $T_{M\text{-}I}$ and T_C [13]. In this present work, the authors partially substitute Ca site with Ag to form perovskite manganite $La_{0.8}Ca_{0.13}Ag_{0.07}MnO_3$ and investigate the structural and transport properties of that perovskite material.

2. Materials and methods

The material $La_{0.8}Ca_{0.13}Ag_{0.07}MnO_3$ was synthesized by sol-gel method using La_2O_3, $Ca(NO_3)_2.4H_2O$, $AgNO_3$, $Mn(NO_3)_2.4H_2O$ as precursor materials. La_2O_3 was dissolved in nitric acid solutions. Then, lanthanum nitrate solution was mixed with other metal nitrate solution. Citric acid was added to the solution with a molar ratio of 1.2 between citric acid and metal nitrate. The pH value of the solution was adjusted to 7 by adding ammonia solution while stirring the solution at a temperature of 80 °C until the gel was formed. The gel was heated at 120 °C for dehydration. It was calcined at 600 °C for 5 h to remove produced organic materials. The materials obtained from this process were ground by using an agate mortar. The powders were pressed into pellets and then heated at 900 °C for 24 h.

The crystal structure of the sample was determined by X-ray diffractometer (XRD) using monochromatic CuKα radiation. The composition of elements was analyzed using energy dispersive X-ray spectroscopy (EDX). The measurement of resistivity versus temperature was determined by cryogenic magnetometer.

3. Results and discussion

The structural properties of the sample were determined by XRD at room temperature. The XRD pattern of the sample is shown in Figure 1.a. The refinement result using Rietveld method indicates that the sample has a rhombohedral structure with space group of $R\bar{3}c$. The quality of refinement was determined by the value of goodness of fit (χ^2) [14]–[16]. Lattice parameters, unit cell volume, R_p, R_{wp} and the value of χ^2 are summarized in table 1. The EDX analysis result shown in Figure 1.b exhibits La, Ca, Ag, Mn and O peaks corresponding to elements of the sample. The single phase of XRD result and EDX result indicate the successful substitution of Ca site with Ag.

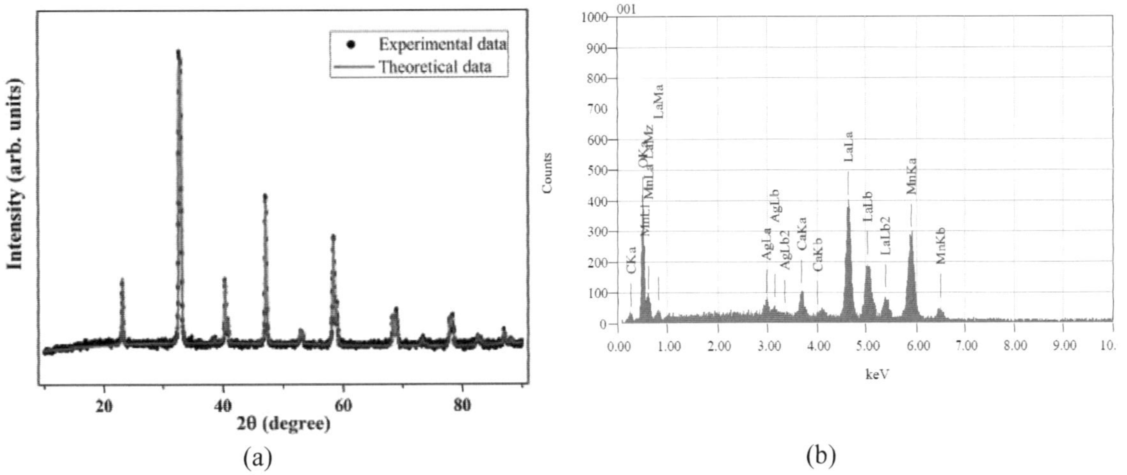

Figure 1. (a) XRD pattern of $La_{0.8}Ca_{0.13}Ag_{0.07}MnO_3$. (b) EDX result of $La_{0.8}Ca_{0.13}Ag_{0.07}MnO_3$.

Table 1. Lattice parameters, unit cell volume, R_p, R_{wp} and goodness of fit (χ^2)

Parameters	$La_{0.8}Ca_{0.13}Ag_{0.07}MnO_3$
a (Å)	5.497
b (Å)	4.497
c (Å)	13.314
Volume (Å^3)	348.385
R_p	8.279
R_{wp}	10.743
Goodness of fit (χ^2)	1.154

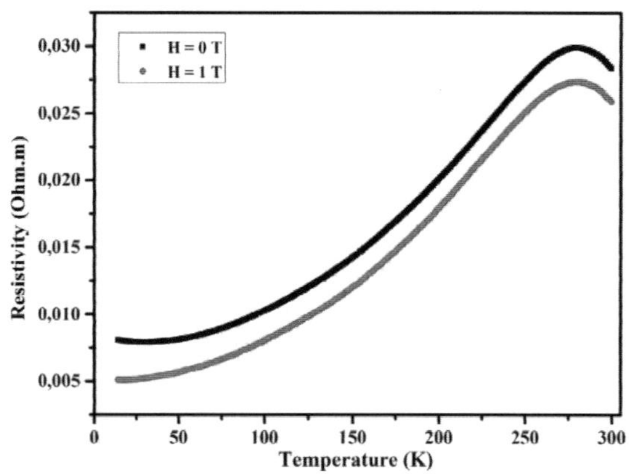

Figure 2. The temperature dependence of $La_{0.8}Ca_{0.13}Ag_{0.07}MnO_3$ for measurement without magnetic field and with 1 T magnetic field.

The measurement result of resistivity versus temperature of $La_{0.8}Ca_{0.13}Ag_{0.07}MnO_3$ at zero applied a magnetic field and applied magnetic field 1 T is shown in Figure 2. Below 280 K, the resistivity of $La_{0.8}Ca_{0.13}Ag_{0.07}MnO_3$ increases with increment temperature indicating metallic behavior. The transport mechanism at the metallic region was analyzed by fitting the experimental data of $La_{0.8}Ca_{0.13}Ag_{0.07}MnO_3$ using equation [17]–[19]:

$$\rho(T) = \rho_o + \rho_2 T^2 + \rho_{4.5} T^{4.5} \quad (1)$$

Where ρ_o, ρ_2 and $\rho_{4.5}$ are resistivities due to grain/domain boundary scattering, electron-electron scattering, and combinations of electron-electron, electron-magnon and electron-phonon scattering processes, respectively [17]–[19]. The quality of fitting data was s evaluated by the square of linear correlation coefficients R^2 [17]. Parameters of fitting data and the value of R^2 are summarized in Table 2. Figure 3 shows the fitting result of experimental data using equation 1. The ρ_2 higher than that of $\rho_{4.5}$ indicates the resistivity of $La_{0.8}Ca_{0.13}Ag_{0.07}MnO_3$ dominantly originated from electron-electron scattering. The highest value of ρ_o suggests that grain/domain boundary plays a vital role in transport mechanism of $La_{0.8}Ca_{0.13}Ag_{0.07}MnO_3$.

Table 2. Fitting parameters of electrical resistivity data of polycrystalline $La_{0.8}Ca_{0.17}Ag_{0.03}MnO_3$

	ρ_o (ohm.m)	ρ_2 (10^{-3} ohm·m·K^{-2})	$\rho_{4.5}$ (10^{-6} ohm.m·$K^{-9/2}$)	Adj. R-Square
H = 0	0.047	0.223	2.116	99.95%
H = 1	0.019	0.054	1.252	99.99 %

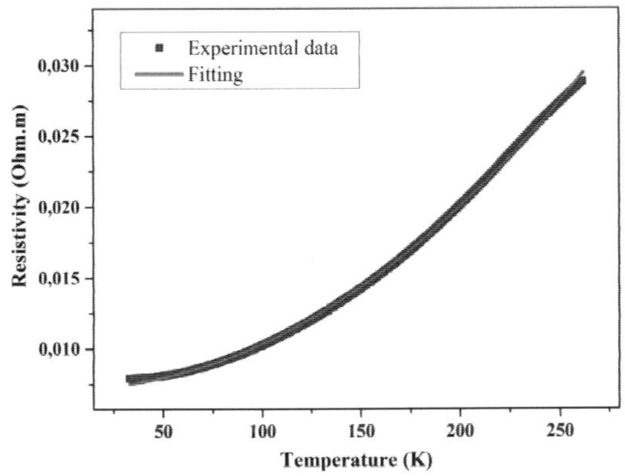

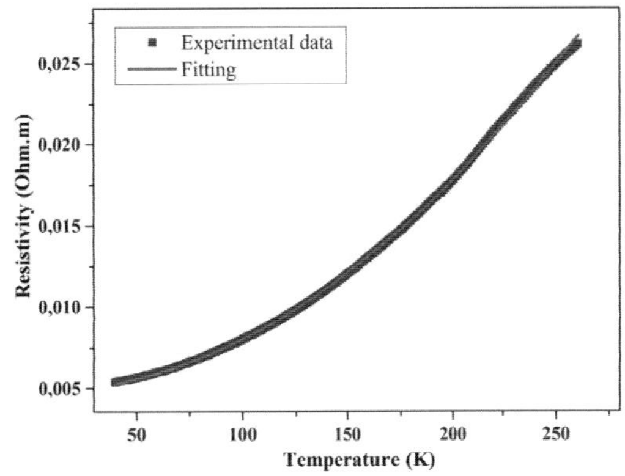

Figure 3. Theoretical fitting of resistivity data in the metallic region at (a) zero magnetic field, and (b) applied magnetic field 1 T.

Figure 3 shows that the resistivity of $La_{0.8}Ca_{0.13}Ag_{0.07}MnO_3$ decreases when applied magnetic field 1 T at range temperature of 10 K to 300 K. The value of magnetoresistance of $La_{0.8}Ca_{0.13}Ag_{0.07}MnO_3$ is calculated using Equation (2)

$$MR(\%) = \frac{\rho_o - \rho_H}{\rho_o} \times 100\% \qquad (2)$$

Where ρ_0 and ρ_H are the resistivities without and with 1 T magnetic fields, respectively. Figure 4 shows the magnetoresistance versus temperature of $La_{0.8}Ca_{0.13}Ag_{0.07}MnO_3$. The value of magnetoresistance of $La_{0.8}Ca_{0.13}Ag_{0.07}MnO_3$ decreases with increment temperature indicating that the decrement of resistivity of $La_{0.8}Ca_{0.13}Ag_{0.07}MnO_3$ emanates from spin-polarized tunneling process at the grain boundary. The magnetic spins at grain boundary are randomly oriented [20]. The spins align to the applied magnetic field direction resulting in a decrease of electron scattering by the grain boundary and increase electron tunneling [9, 11].

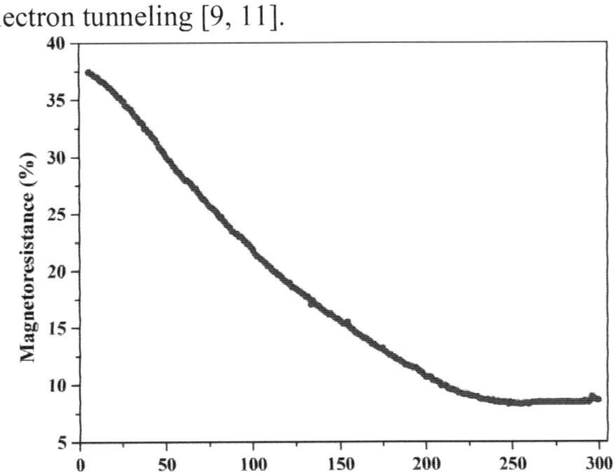

Figure 4. The magnetoresistance versus temperature of $La_{0.8}Ca_{0.13}Ag_{0.07}MnO_3$.

4. Conclusions

The transport properties and magnetoresistance of $La_{0.8}Ca_{0.13}Ag_{0.07}MnO_3$ synthesized by sol-gel method were investigated. The XRD analyses exhibit a single phase with rhombohedral structure. The EDX results confirm that the sample contains all expected chemical elements. The temperature dependence of $La_{0.8}Ca_{0.13}Ag_{0.07}MnO_3$ at the metallic region which is originated from electron-electron scattering. The magnetoresistance of $La_{0.8}Ca_{0.13}Ag_{0.07}MnO_3$ emanates from spin-polarized tunneling process at the grain boundary.

Acknowledgements

The authors gratefully acknowledge the support from Department of Physics, Universitas Indonesia for the research facilities. This work was granted by the Universitas Indonesia under research grant contract no. 633/UN2.R3.1/HKP.05.00/2017 and Indonesia Endowment Fund for Education (LPDP).

References

[1] C. N. R. Rao and R. Mahesh, "Giant magnetoresistance in manganese oxides," *Curr. Opin. Solid State Mater. Sci.*, vol. 2, no. 1, pp. 32–39, 1997.

[2] J. A. Alonso, M. J. Martínez-Lope, M. T. Casais, and M. T. Fernández-Díaz, "Evolution of the Jahn-Teller distortion of MnO_6 octahedra in $RMnO_3$ perovskites (R = Pr, Nd, Dy, Tb, Ho, Er, Y): A neutron diffraction study," *Inorg. Chem.*, vol. 39, no. 15, pp. 917–923, 2000.

[3] S. Ravi and M. Kar, "Study of magneto-resistivity in $La_{1-x}Ag_xMnO_3$ compounds.pdf," *Phys. B*, p. 169, 2003.

[4] S. L. Ye *et al.*, "Large room-temperature magnetoresistance and phase separation in $La_{1-x}Na_xMnO_3$ with $0.1\leq x\leq 0.3$," *J. Appl. Phys.*, vol. 90, no. 6, pp. 2943–2948, 2001.

[5] S. Agrestini, N. L. Saini, and A. Bianconi, "Determination of local distortions in the charge ordered and CMR la 1-xCaxMnO3 manganites," *Journal of Magnetism and Magnetic Materials*, vol. 272–276, no. I. pp. 454–455, 2004.

[6] M. Pissas, "Phase diagram of $La_{1-x}Ca_xMnO_3$ for $x\geq 0.5$," *Journal of Magnetism and Magnetic Materials*, vol. 272–276, no. III. pp. 1821–1822, 2004.

[7] C. Zener, "Interaction between the d-shell in the Transition Metals. II. Ferromagnetic Compounds of Manganese with Perovskite Structure," *Phys. Rev.*, vol. 82, pp. 440–444, 1951.

[8] G. H. Jonker and J. H. Van Santen, "Ferromagnetic compounds of manganese with perovskite structure," *Physica*, vol. 16, no. 3, pp. 337–349, 1950.

[9] R. Skini, M. Khlifi, M. Wali, E. Dhahri, E. K. Hlil, and P. Lachkar, "Electrical transport and giant magnetoresistance in La0.8−x□xCa0.2MnO3 (x=0, 0.1 and 0.2) oxides," *Journal of Magnetism and Magnetic Materials*, vol. 363. pp. 217–223, 2014.

[10] Z. B. Guo, Y. W. Du, J. S. Zhu, H. Huang, W. P. Ding, and D. Feng, "Large Magnetic Entropy Change in Perovskite-Type Manganese Oxides," *Phys. Rev. Lett.*, vol. 78, no. 6, pp. 1142–1145, 1997.

[11] H. Y. Hwang, S. Cheong, N. P. Ong, and B. Batlogg, "Spin-Polarized Intergrain Tunneling in La2/3Sr1/3MnO3," *Phys. Rev. Lett.*, vol. 77, no. 10, pp. 2041–2044, 1996.

[12] S. Das and T. K. Dey, "Electrical conductivity and low field magnetoresistance in polycrystalline La1-xKxMnO3 pellets prepared by pyrophoric method," *Solid State Communications*, vol. 134, no. 12. pp. 837–842, 2005.

[13] L. Pi, M. Hervieu, A. Maignan, C. Martin, and B. Raveau, "Structural and magnetic phase diagram and room temperature CMR effect of La1-xAgxMnO3," *Solid State Communications*, vol. 126, no. 4. pp. 229–234, 2003.

[14] M. Koubaa, W. Cheikhrouhou-Koubaa, and A. Cheikhrouhou, "Effect of Ag doping on structural, magnetic and magnetocaloric properties of La(0.65)Ca(0.35-x)Ag(x)MnO(3) manganites," *J. Alloy. Compd.*, vol. 473, no. 1–2. pp. 5–10, 2009.

[15] S. P. Altintas, A. Amira, N. Mahamdioua, A. Varilci, and C. Terzioglu, "Effect of Eu doping on structural and magneto-electrical properties of," *J. Alloys Compd.*, vol. 509, no. 13, pp.

IOP Conf. Series: Materials Science and Engineering **367** (2018) 012031 doi:10.1088/1757-899X/367/1/012031

4510–4515, 2011.
[16] D. Fatnassi, J. L. Rehspringer, E. K. Hlil, D. Niznansky, M. Ellouze, and F. Elhalouani, “Structural and magnetic properties of nanosized La0.8Ca0.2Mn1-xFexO3 particles (0 ≤ x ≤ 0.2) prepared by sol–gel method,” *J. Supercond. Nov. Magn.*, vol. 28, no. 8, pp. 2401–2408, 2015.
[17] S. O. Manjunatha, A. Rao, and G. S. Okram, “Investigation on structural, magneto-transport, magnetic and thermal properties of La0.8Ca0.2−xBaxMnO3 (0⩽x⩽0.2) manganites,” *Journal of Alloys and Compounds*, vol. 640. pp. 154–161, 2015.
[18] S. . Manjunatha, A. Rao, V. P. S. Awana, and G. S. Okram, “Investigation on magnetic, electrical and thermoelectric power of Bi-substituted La0.8Ca0.2MnO3 manganites,” *J. Magn. Magn. Mater.*, vol. 394, pp. 130–137, 2015.
[19] S. Hcini, S. Zemni, A. Triki, H. Rahmouni, and M. Boudard, “Size mismatch, grain boundary and bandwidth effects on structural, magnetic and electrical properties of Pr0.67Ba0.33MnO3 and Pr0.67Sr0.33MnO3 perovskites,” *J. Alloys Compd.*, vol. 509, no. 5, pp. 1394–1400, 2011.
[20] S. P. Isaac, N. D. Mathur, J. E. Evetts, and M. G. Blamire, “Magnetoresistance of artificial La0.7Sr0.3MnO3 grain boundaries as a function of misorientation angle,” *Appl. Phys. Lett.*, vol. 72, no. 16, pp. 2038–2040, 1998.

The Compatibility of NiO, CeO_2 and NiO-CeO_2 as a Coating on $La_{0.6}Sr_{0.4}Co_{0.2}Fe_{0.8}O_{3-\delta}$, $La_{0.7}Sr_{0.3}Co_{0.2}Fe_{0.8}O_{3-\delta}$ and $La_{0.7}Sr_{0.3}Mn_{0.3}O_{3-\delta}$ Ceramic Membranes and Their Mechanical Properties

R. M. Iqbal[1], S. D. Nurherdiana[1], M. S. Sahasrikirana[1], L. Harmelia[1], W. P. Utomo[1], E. P. Setyaningsih[1], and H. Fansuri[1]

[1]Department of Chemistry, Institut Teknologi Sepuluh Nopember, Kampus ITS Sukolilo, Surabaya 60111, Indonesia

E-mail: h.fansuri@chem.its.ac.id

Abstract. Metal oxides and perovskites have got tremendous attention as a good catalyst for partial oxidation of methane (POM). In the present work, the materials were cast into a double layer planar membrane using the dry pressing method in order to enhance its' catalytic activity. Unfortunately, the majority of metal oxides and perovskites have different thermal expansion coefficient (TEC). Therefore, this work was aimed to study the compatibility between catalyst metal oxides (NiO, CeO_2) and perovskite oxides $La_{0.6}Sr_{0.4}Co_{0.2}Fe_{0.8}O_{3-\delta}$ (LSCF 6428), $La_{0.7}Sr_{0.3}Co_{0.2}Fe_{0.8}O_{3-\delta}$ (LSCF 7328) and $La_{0.7}Sr_{0.3}MnO_{3-\delta}$ (LSM 73) when they were made as double layer membranes. The result shows that the density and volume shrinkage of LSCF 7328 was the highest, followed by LSM 73 and LSCF 6428. It was also found that the trends of hardness and density in the perovskites and metal oxides follows their thermal expansion trends. According to thermomechanical analysis (TMA) results, TEC of LSCF 6428, LSCF 7328, LSM 73, NiO-CeO_2, NiO and CeO_2 were 22.16; 9.31; 10.14; 12.21; 20.86 and 21,11 ppm/°C, respectively. From the TEC data, Large differences in TEC cause the delamination of NiO, CeO and mixtures of NiO-CeO_2 from perovskite surface except those who were prepared using amylum as a binder in the preparation of double layer membranes.

Keywords. Ceramic membrane, mechanical properties, perovskite oxide, and partial oxidation of methane.

1. Introduction

POM reaction is a promising way to convert methane to syngas (a mixture of H_2 and CO). By the Fischer-Tropsch reaction, the syngas can be converted to liquid fuels and other hydrocarbons [1]. However, a successful POM reaction needs a controlled supply of oxygen. If there is too much oxygen in the reaction, the results will be CO_2 and H_2O rather than syngas. The use of an oxygen transport membrane is a promising way to supply a controlled amount of oxygen to the partial oxidation of methane reaction in order to improve the selectivity of the reaction.

The use of mix ionic and electronic conducting (MIEC) membrane to control oxygen supply has been reported by many researchers. Scott *et al* [2] reported that $La_{0.6}Sr_{0.4}Co_{0.2}Fe_{0.8}O_{3-\delta}$ (LSCF 6428)

Content from this work may be used under the terms of the Creative Commons Attribution 3.0 licence. Any further distribution of this work must maintain attribution to the author(s) and the title of the work, journal citation and DOI.
Published under licence by IOP Publishing Ltd

could be used as an oxygen transport membrane with excellent stability on POM at high temperature (500 to 900) °C. Schlehuber *et al* [3] added that LSCF 6428 membrane is active for POM reaction during 3000 h operation and it shows good catalytic activity. Another lanthanum-based oxygen transport membrane, i.e. $La_{0.7}Sr_{0.3}MnO_{3-\delta}$ (LSM 73), has also been reported to have good ion stability at oxidant environment [4]. However, lanthanum-based perovskite oxide oxygen transport membranes generally have low oxygen permeation flux.

One method that can be used to increase the oxygen permeation of lanthanum-based perovskite membranes is by combining it with a noble metal catalyst. Yacou *et al* [5] reported that LSCF membrane covered by Pd catalyst have an improved oxygen flux. The addition of Ir and Rh on Al_2O_3 membrane surface also showed good catalytic activity on POM reaction [6]. The drawback of using noble metals as a catalyst is their high price.

Several transition metals have been reported to have comparable activity to the noble metals. Nickel has been widely known as a cheaper alternative to noble metal as an oxidative catalyst. It has also reported being active and highly stable in methane oxidation reaction [7]. Another promising material is CeO_2 which also has good activity in a POM reaction due to its high specific area [8]. In addition, both catalysts have an ability to reduce carbon deposition during POM reaction which led to increasing methane conversion and syngas selectivity on POM reaction.

One concern in using nickel and cerium catalyst to improve oxygen permeation of lanthanum-based perovskite membranes is that they react with the perovskite to produce another phase at a high temperature. The POM reaction is normally carried out at high temperature (between 500 °C and 900 °C). Both catalysts might also incompatible with the membrane when they are applied as a thin layer catalyst on the surface of the membrane. Therefore, the compatibility between the catalyst with lanthanum-based perovskite membrane is an interesting subject to be studied. This paper reports our study on the compatibility study of both catalysts as a thin layer coating to the surface of LSCF dan LSM membranes.

2. Materials and methods

2.1. Materials

The materials that were used in this research follows those reported by Nurherdiana *et al* [9] ie: using p.a. grade of La_2O_3 (Merck, 99.5 %), $SrCO_3$ (Merck, 99.9 %), Co_3O_4 (Merck, 99.5 %), Fe_2O_3 (Merck, 97 %) and $MnCO_3$ (Merck, 99.9 %) as precursors and methanol. Powders of p.a grade NiO (Aldrich, 99 %) and CeO_2 (Merck) without further treatment were used as the catalyst.

2.2. Method

A stoichiometric amount of La_2O_3, $SrCO_3$, Co_3O_4, Fe_2O_3 and $MnCO_3$ powders were ball milled for 2 h at 200 rpm using methanol as a dispersant. After milling, the methanol was separated by decantation and then the slurry was dried at 100 °C for 2 h and then calcined at 1000 °C for 2 h with a heating rate of 3 $°C \cdot min^{-1}$. The resulted perovskite was then characterized by X-ray diffraction (XRD) method using Cu-Kα radiation (λ = 1.5406 Å) which was generated by applying 30 mA of current and 40 kV of Voltage. Powders of NiO (Aldrich, 99 %) and CeO_2 (Merck) were used as the catalyst without further treatment.

The perovskites were made into single and double layer ceramic membranes by the dry pressing method. For single layer membranes, 0.93 g of perovskite powder was cast into a disc-shape green membrane (diameter = 13 mm) by applying 8-ton pressure for 15 min. The double layer membrane was made with the same method as the single layer with two variations of catalyst layer application procedure. The first was made by consecutive pressing. In this method, 0.93 g of perovskite powder was cast by applying 4 t of pressure. After that, a powder catalyst was added on the top of the casted disk and then pressed again with 8 t of pressure. The other was made exactly the same as the single layer disk-shaped membrane. However, after the casting of the membrane, a suspension of catalyst powder in an amylum solution was coated on the surface of the membrane.

All green membranes were initially calcined at 800 °C. After 3 h at 800 °C, the temperature of the furnace was increased to 1250 °C and held at the temperature for 2 h. In both heating procedure, the temperature of the furnace was increased at a rate of 3 °C·min^{-1}. The morphology and membrane composition were characterized by SEM-EDX. The hardness of the membranes was measured by Vicker's microhardness method while thermal expansion coefficient was measured by a thermomechanical analyzer (TMA).

3. Results and discussion

$La_{0.6}Sr_{0.4}Co_{0.2}Fe_{0.8}O_{3-\delta}$ (LSCF 6428), $La_{0.7}Sr_{0.3}Co_{0.2}Fe_{0.8}O_{3-\delta}$ (LSCF 7328) and $La_{0.7}Sr_{0.3}MnO_{3-\delta}$ (LSM 73) were successfully synthesize by solid state method. The color of oxides and carbonates mixtures were changed from brown and light brown to black and dark blue after calcination for LSCF and LSM, respectively. The color change is a clear indication that the precursors reacted together during calcination process to form a new phase.

XRD analysis result of the LSCF and LSM product is shown in Figure 1. The diffractograms of LSCF 6428, LSCF 7328, and LSM 73 are similar to JCPDS No 089-1268 (LSCF 7337). The results indicate that LSCF 6428, LSCF 7328, and LSM 73 have the same crystal structure to LSCF 7337 which is a perovskite oxide. The sharpness of all peaks and flat background indicated that the perovskites are highly crystalline. The diffractograms also show no other peaks except perovskite which indicate that there are no impurities in the resulted perovskites. Therefore, it can be concluded that the solid state method was successful when it was used to produce the perovskites.

Powders of NiO and CeO_2 were also characterized by XRD and the result shows a specific peak of NiO appearing at 2θ = (37.23°, 43.28°, 62.83°, 75.38°, and 79.36°) as well as CeO_2 with its characteristic 2θ at (28.54°, 33.08°, 47.49°, 56.35°, 59.09°, 69.42°, 76.70°, and 79.07°). As in the perovskite case, diffractograms of NiO and CeO_2 show no impurities.

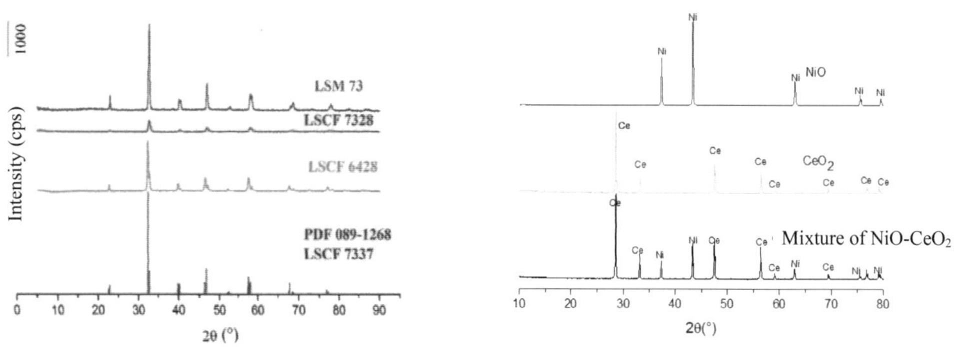

(a) (b)

Figure 1. Diffraction pattern of: (a) perovskites and (b) the catalyst

The resulted single layer membrane is shown in Figure 2. Due to shrinkage, the shape of the membranes is slightly distorted from their original shape before sintering. All single layer membrane is dense which is confirmed by water drop test method that has been reported elsewhere [10]. A drop of water was put on the surface of the membrane and if the membrane is dense then the water drop stays unchanged for a long time. In this research, the drop was stayed undispersed. When the membrane is porous, the water drop will disperse quickly.

Figure 3 shows the TMA results of the single layer membranes. The TEC (Thermal Expansion Coefficient) of LSCF 6428, LSCF 7328, LSM 73, NiO-CeO_2, NiO and CeO_2 are (22.16, 9.31, 10.14, 12.21, 20.86 and 21.11) ppm·°C^{-1}, respectively. The TEC of LSCF 6428 is the highest of all perovskites. However, its TEC is at the same order of magnitude as the catalysts (NiO and CeO_2) while LSCF 7328 and LSM 73 are one order lower. Based the TEC, the catalyst seems to be compatible only with LSCF 6428. However, when NiO and CeO_2 are mixed together, the mixture will be compatible with LSCF 7328 and LSM.

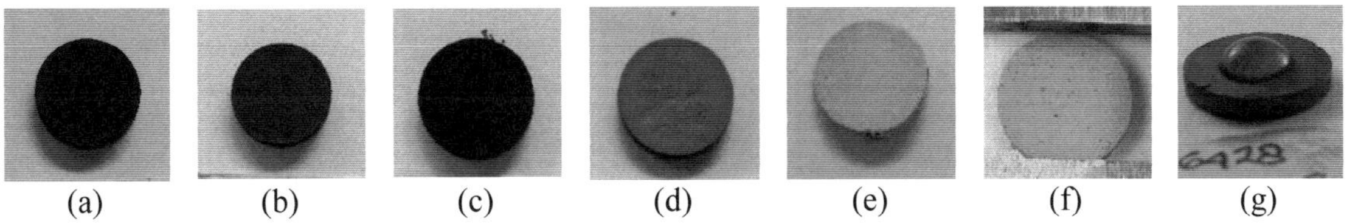

Figure 2. Single layer membrane of (a) LSCF 6428, (b) LSCF 7328, (c) LSM 73, (d) NiO, (e) CeO_2, (f) NiO-CeO_2 and (g) example of drop water test result

Huang *et al* [11] reported that there are two things which affect the expansion of perovskite material. The first is repulsion force among cations or anions in the crystal structure of perovskites. When the temperature increase, the bonding in the structure is weakened and the repulsive force becomes more dominant which make the cations and anions are pushed away from their original position. The second is the reduction of cations, especially the transition metal, to maintain the charge balance in the crystal structure due to the loss of oxygen at high temperature. The size of reduced transition metal cations are larger than the size of the oxidized one, thus it led to the increase in the size of the crystals' unit cell. The second cause of thermal expansion is called chemical expansion [12]. For that reason, there is a possibility that LSCF 6428 is easier to be reduced by heating at high temperature than LSCF 7328 and LSM. NiO and CeO_2 are also easily reduced but when they are mixed together, they become more difficult to be reduced.

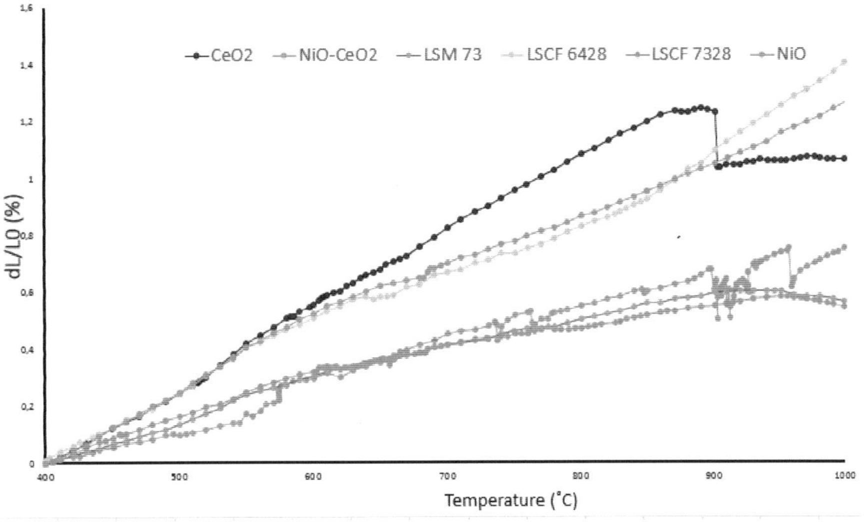

Figure 3. Thermal expansion of single layer membranes

Vicker's microhardness test results of membranes are shown in Figure 4. It is apparent that LSCF 7328 is the hardest among all perovskite's membranes. When it is compared to the trend in TEC, the hardness of membranes are more likely related to their microstructure as shown by SEM micrographs of their cross-section in Figure 5. The density of LSCF 7328 is appeared to be the highest one and therefore it is the hardest membrane. On the other hand, although LSCF 6428 is appeared to be more porous than the LSM, its' hardness is higher. The similar results are also supported by Mobius et al.

[13] who reported that the densification of LSM membrane was 92 % while LSCF 7328 and LSCF 6428 membranes were 95 % and 94 %, respectively although they were sintered at slightly different temperature.

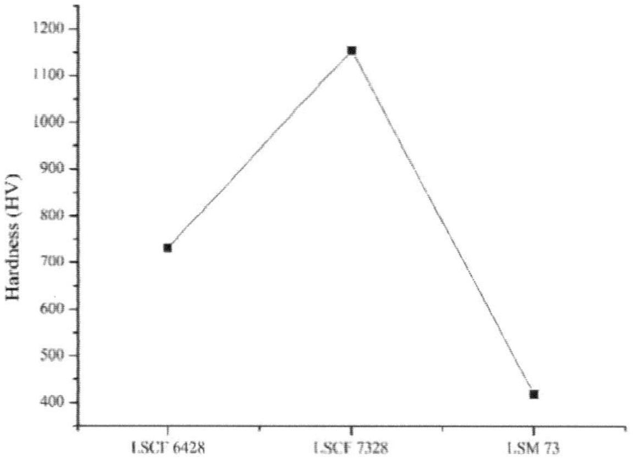

Figure 4. Hardness of single layer membranes

Meanwhile, the SEM images of the cross-section of NiO, CeO_2 and NiO-CeO_2 membranes show that the membranes are not as dense as the perovskite membranes. The porous catalyst membranes might be caused by lower sintering temperature that was applied to the membranes in this research. Some references reported that in order to make dense NiO, CeO_2 and NiO-CeO_2 membranes, the sintering temperature must be between 1400 °C and 1600 °C [14, 15].

Figure 5 shows the cross-section of double layer membranes. The figure shows that the catalyst (NiO, CeO_2 and NiO-CeO_2) were successfully coated on the surface of perovskite membrane. In the figure, the dense layer represents the perovskite part while the catalyst is represented by porous phase or particles on the surface of the perovskite membranes and it is shown more clearly by the element mapping using SEM-EDX as in Figure 6.

Figure 5. Cross section of single layer membrane (a) LSCF 6428, (b) LSCF 7328, (c) LSM 73, (d) NiO, (e) CeO_2 and (f) NiO-CeO_2

Nickel (Ni) might be diffused into perovskite membranes due to its' similarity to cobalt during the sintering process at high temperature. Although the diffusion of Ni and Ce into LSCF membranes was relatively very small, the diffusion reached all area of membranes as shown in Figure 6.d.

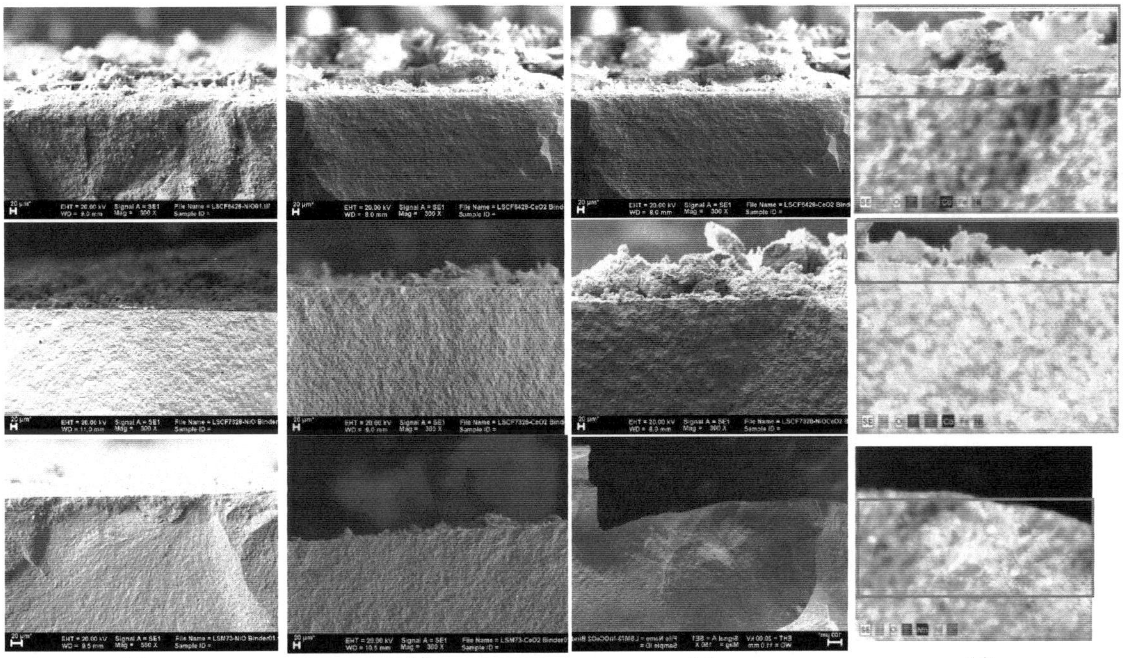

(a) (b) (c) (d)

Figure 6. Cross section of double layer membranes (top to bottom : LSCF 6428, LSCF 7328, LSM 73) coated by: (a) NiO, (b) CeO_2 (c) NiO-CeO_2; (d) elemental mapping by SEM-EDX

The visual appearances of double layer membranes after sintering process are shown in Figure 7. The figure shows that NiO, CeO_2, NiO-CeO_2 was successfully coated on the perovskite surfaces. However, the color of the catalyst was turned into black which means that the catalyst might be reacted with perovskites. In addition to the reaction, the coating also made the surface of the membrane become rougher compared to the membrane before sintering. The roughness might be caused by the decomposition of amylum as a binder that was used to apply the catalyst layer on the top of the perovskite membranes.

(a) (b) (c)

Figure 7. Double layer after sintering process: (a) NiO/Perovskite (b) CeO_2/Perovskite (c) NiO-CeO_2/Perovskite

4. Conclusions

Thermal expansion coefficient (TEC) of perovskites and catalyst material is an indicator of the compatibility of the catalyst with La-based perovskite membranes, especially when the coating process was applied to form double layer membranes. TEC values of LSCF 6428, LSCF 7328, LSM 73, NiO, CeO_2, NiO-CeO_2 are (22.16, 9.31, 10.14, 12.21, 20.86, and 21.11) ppm·°C^{-1}, respectively. The different of TEC might be caused by the reduction of the oxidation state of the transition metal cations and the porosity of the membranes. The different order of magnitude in TEC of the membranes and the catalyst may cause delamination or cracking of the coats. Hence, the coated catalyst on the surface of perovskite membrane using amylum as a binder has fabricated successfully to form a double layer planar membrane.

Acknowledgments
This work was financially supported by Institut Teknologi Sepuluh Nopember under Lab Based Education (LBE) research grant in 2017 and the Ministry of Research, Technology and Higher Education in the form of PMDSU scholarship.

References
[1] Meng B, Zhang H, Zhao Z, Wang X, Jin Y, and Liu S 2015 A novel $LaGa_{0.65}Mg_{0.15}Ni_{0.20}O_{3-\delta}$ perovskite catalyst with high performance for the partial oxidation of methane to syngas *Catal. Today* **259** 388–392
[2] Scott S P, Mantzavinos D, Hartley A, Sahibzada M, and Metcalfe I S 2002 Reactivity of LSCF perovskites *Solid State Ionics* **152–153** 777–781
[3] Schlehuber D, Wessel E, Singheiser L, and Markus T 2010 Long-term operation of a $La_{0.58}Sr_{0.4}Co_{0.2}Fe_{0.8}O_{3-\delta}$-membrane for oxygen separation *J. Membrane Sci.* **351** 16–20
[4] Zhang Q, Nakagawa T, and Saito F 2000 Mechanochemical synthesis of $La_{0,7}Sr_{0,3}MnO_3$ by grinding constituent oxides *J. Alloy Compd.* **308** 121–125
[5] Yacou C, Sunarso J, Lin C X C, Smart S, Liu S, and da Costa J C D 2011 Palladium surface modified $La_{0.6}Sr_{0.4}Co_{0.2}Fe_{0.8}O_{3-\delta}$ hollow fibres for oxygen separation *J. Membrane Sci.* **380** 223–231
[6] Mark M F and Maier W F 1996 CO_2-Reforming of Methane on Supported Rh and Ir Catalysts *J. Catal.* **164** 122–130
[7] Colomer M T and Kilner J A 2015 Effect of sintering time on structural, microstructural and chemical composition of Ni-doped lanthanum gallate perovskites *J. Solid State Chem.* **228** 167–173
[8] Bulfin B, Lowe A J, Keogh K A, Murphy B E, Lubben O, Krasnikov S A, and Shvets I V 2013 Analytical Model of CeO_2 Oxidation and Reduction *J. Phys. Chem C* **117** 24129–24137
[9] Nurherdiana S D, Sholichah N, Iqbal R M, Sahasrikirana M S, Utomo W P, Akhlus S, Nurlina and Fansuri H 2017 Preparation of $La_{0.7}Sr_{0.3}Co_{0.2}Fe_{0.8}O_{3-\delta}$ (LSCF 7328) by Combination of Mechanochemical and Solid State Reaction *Key Eng. Mat.* **744** 399–403
[10] Fansuri H *et al* 2017 Preparation of dense $Ba_xSr_{1-x}Co_{0.8}Fe_{0.2}O_3$ membranes: Effect of Ba2+ substituents and sintering method to the density, hardness and thermal expansion coefficie *Adv. Mat. Lett* .**8** 799–806
[11] Huang B X *et al* 2010 Mechanical aspects of ferro-elastic behavior and phase composition of $La_{0.58}Sr_{0.4}Co_{0.2}Fe_{0.8}O_{3-\delta}$ *J. Membrane Sci.* **349** 183–184
[12] Perry N H, Kim J J, Bishop S R, and Tuller H L 2015 Strongly coupled thermal and chemical expansion in the perovskite oxide system Sr(Ti,Fe)$O_{3-\alpha}$ *J. Mater. Chem. A* **3** 3602–3611
[13] Möbius A, Henriques D, and Markus T 2009 Sintering behaviour of $La_{1-x}Sr_xCo_{0.2}Fe_{0.8}O_{3-\delta}$ ($0.3\leq x\leq 0.8$) mixed conducting materials *J Eur Ceram Soc* **29** 2831-2839
[14] Batista R and Muccillo E 2011 Densification and grain growth of 8YSZ containing NiO *Ceram. Int* **37** 1047-1053
[15] Lipińska–Chwałek M, Schulze–Küppers F, and Malzbender J 2015 Mechanical properties of pure and doped cerium oxide *J. Eur. Ceram. Soc.* **35** 1539–1547

IOP Conf. Series: Materials Science and Engineering **367** (2018) 012033 doi:10.1088/1757-899X/367/1/012033

The Role of pH in Controlling Size and Distribution of Silver Nanoparticles using Biosynthesis from *Diospyros discolor Willd.* (Ebenaceae)

M. Nurfadhilah[1], I. Nolia[1], W. Handayani[1], and C. Imawan[2]

[1]Department of Biology, FMIPA Universitas Indonesia, Kampus UI Depok 16424, Indonesia
[2]Department of Physics, FMIPA Universitas Indonesia, Kampus UI Depok 16424, Indonesia

E-mail: windri.h@sci.ui.ac.id

Abstract. The silver nanoparticles generated by biosynthesis have a quite diverse result, both in size and shape. Structures of silver nanoparticles can be controlled by modifying the parameters of the biosynthesis such as the ratio between the precursors and reducing agents, as well as pH of the solution. In this study, the pH of *Diospyros discolor* (Bisbul) leaves aqueous extract was varied to 4, 7, 9, and 11. The extract then was added to 1 mM $AgNO_3$ precursor (1:2; v/v ratio). The result of the silver nanoparticles characterized using spectrophotometer UV-Vis to find if there was any absorbance peak formed between 400 nm to 500 nm. TEM characterization was used to determine the size and shape of silver nanoparticles, and PSA was used to see their size distribution and stability. The higher pH tends to produce smaller silver nanoparticles rapidly. The synthesis parameters that were varied in this research have affected the size, size distribution patterns, and stability of silver nanoparticles.

Keywords. Biosynthesis, *diospyros discolor,* pH, and silver nanoparticles.

1. Introduction

Silver nanoparticles (Ag NPs) have a unique characteristic and have the ability to efficiently absorb and scatter light because of the surface plasmon resonance (SPR) [1]. Ag NPs are also known to have applications in various fields such as sensors [2], biomedical and health [3]. Based on their unique characteristic, scientists develop some different methods to synthesize silver nanoparticles, which are chemical, physical and biological methods [4, 5]. Biological methods for nanoparticles synthesis can be done by using single cell microorganism such as bacteria or fungi through intracellular processes [6]. However, the intracellular biosynthesis requires a long reaction time. In contrast, the synthesis of silver nanoparticles using plant extract as an extracellular extruded agent requires shorter reaction time than the intracellular biosynthesis.

Diospyros discolor Willd. (bisbul) is known for containing triterpenes [7], alkaloids, flavonoids, and tannins [8]. The specific compound from those group can be used as reducing agents [9]. Silver nanoparticles characteristics such as size, shape, dispersion, and stability can be influenced by temperature, the ratio of precursor concentrations with reducing agents, and pH during the biosynthesis [10]. The degree of acidity (pH) can affect the size of the nanoparticles. The higher the

Content from this work may be used under the terms of the Creative Commons Attribution 3.0 licence. Any further distribution of this work must maintain attribution to the author(s) and the title of the work, journal citation and DOI.
Published under licence by IOP Publishing Ltd

pH value, the nanoparticles formed tend to be smaller and can even reach the size of 5 nm [11]. Therefore, it is expected that the adjustment of pH in this study during the biosynthesis using *D. discolor* leaves aqueous extract can affect the size, shape, and stability of Ag NPs.

2. Materials and methods

2.1. Plant collection and extract preparation

Leaves aqueous extract of *D. discolor* was prepared using raw materials collected from Universitas Indonesia garden collection. The raw materials were firstly washed to remove the dirt. After that, the samples were dried in the oven at a temperature of 40 °C until it had constant dry weight, then it was ground into powder. Two grams of leaves powder were boiled for (10 to 15) min with 100 mL of bidistilled water (2 % w/v). Then, the water extract filtered with Whatman filter paper No 1. As the experiment variable, pH of the extract was adjusted to 4, 7, 9, 11, using 1 M NaOH and 0.1 M HCl. One sample was prepared for control without pH adjustment. LAQUA Horiba PH1100 was used to measure the pH solution. A solution of 1 mM $AgNO_3$ (Merck) was developed from $AgNO_3$ powder.

2.2. Biosynthesis and characterization

To perform the biosynthesis, 30 mL of leaves aqueous extract with various pH, respectively were mixed with 60 mL AgNO3 solution (1:2; v/v ratio) at room temperature. The silver nanoparticles formed were characterized using spectrophotometer (GENESYS 10SUV-VIS) to confirm the absorbance peak from the Ag NPs in between 200 nm and 800 nm. Beside the spectroscopy result, the change of the solution's color between 24 h was also observed.

The Ag NPs resulted from biosynthesis was evaluated by observing their size and shape morphology. To observe the size and shape we used transmission electron microscope (TEM). Meanwhile, the Particle Size Analyzer (PSA) was used for particle size and zeta potential determination. This experiment was using TEM FEI Tecnai G2 Supertwin and PSA (Malvern Zetasizer) from Balai Besar Pascapanen, Departement Pertanian, Bogor and TEM JEOL JEM 1400 from Department of Chemistry, FMIPA UGM, Yogyakarta.

3. Results and discussion

3.1. Absorption spectrum analyses

Biosynthesis of Ag NPs with pH variation was performed by mixing $AgNO_3$ and aqueous extract of the *D. discolor* leaves. The color of Ag NPs colloid was examined after 24 h of reaction. The results showed that each of Ag NPs colloid changed its color during the reaction after 24 h (Figure 1). Biosynthesis using leaves extract at pH 4 resulted in a yellowish solution. Meanwhile, the Ag NPs without any pH adjustment the color tended to be brownish and got darker in pH 7, pH 9 and pH 11. The colloid color becomes darker as the pH value increases. Color variations in colloidal silver nanoparticles are associated with surface plasmon resonance and optical characteristic of silver nanoparticles [12]. The interaction between silver nanoparticles with light will result in a combined oscillation of electrons on the surface of nanoparticles when excited by light at certain wavelengths [13]. The reflectance light makes colloidal silver nanoparticles look yellow to brown [14]. Therefore, the changes of colloid's color indicate the formation of silver nanoparticles.

Based on UV-Vis spectrophotometer analyses, Ag NPs-control samples using aqueous extract leaves showed a specific peak around the 400 nm wavelength. The peak indicates the presence of silver nanoparticles [15]. During the 24-hour incubation period, there is a reaction between *D. discolor* leaves aqueous extract and $AgNO_3$. The reaction occurs when silver ions (Ag^+) derived from $AgNO_3$ reduces to neutral silver (Ag^0) [9]. The metallic silver (Ag^0) subsequently undergoes a nucleation phase until silver nanoparticles formed which are showed by the existence of spectrum absorbance between 350 nm to 500 nm (Figure 2).

Figure 1. Ag NPs colloid color from various pH value of the aqueous extract of *D. discolor* leaves using biosynthesis with silver precursor after 24 h reaction time.

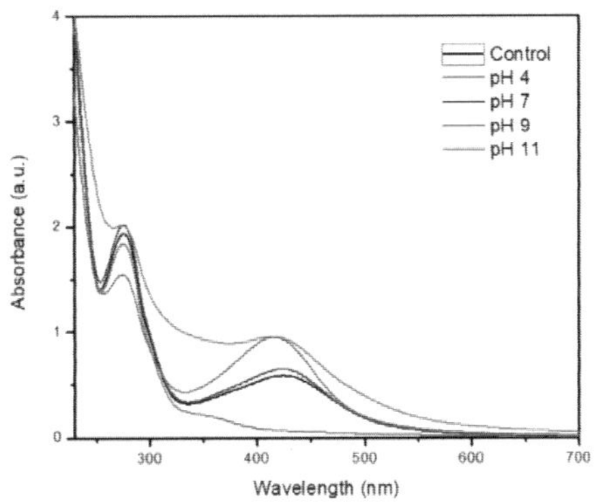

Figure 2. Spectrum UV-Vis result from silver nanoparticles synthesized using *Diospyros discolor* leaves extract pH after 24 h.

The UV-Vis spectrum also shows a difference in absorbance values that qualitatively indicates the number of silver nanoparticles [16]. Ag NPs at pH 4 showed no significant peak at 24 h reaction; it was assumed that the Ag NPs not yet formed during that time. The Ag NPs samples with pH 7, pH 9, and pH 11 show the higher significant peak. The absorbance values tended to be larger along with the higher pH values. Biosynthesis results from control sample tended to be higher than at pH 4, the pH value of aqueous extract at sample control was 6.5. The peak on the Ag NPs sample pH control and pH 7 have the same pattern. According to the absorbance spectra, the formation of Ag NPs tends to be faster in the alkaline condition.

The λ_{max} from spectrum absorbance of Ag NPs also showed varied value, where the sample control showed λ_{max} at 425 nm, pH 7 at 424 nm, pH 9 at 416 nm and pH 11 at 414 nm. This phenomenon suggests that those Ag NPs tend to have a smaller size at lower λ_{max}. The absorbance value in the UV-Vis spectrum qualitatively shows the number of silver nanoparticles formed, as well as the specific wavelength range, indicates the size of the resulting Ag NPs [17]. The size of nanoparticles tends to be smaller when the peak shifted to smaller wavelength range. The degree of acidity (pH) given an effect on binding of silver ions and biomolecules that have an impact on the silver ions reduction process and nucleation growth into silver nanoparticles [17].

3.2. Silver nanoparticles morphology, distribution, and stability

The distribution and shape of the Ag NPs have been confirmed using TEM results. The results showed that Ag NPs formed in spherical shapes and varied in sizes (Figure 3). For the size of Ag NPs based on images of the TEM results sized changed from 5 nm to 80 nm. Ag NPs-control tends to have more significant dimensions, ranging from 20 nm to 70 nm. The biosynthesis results using leaves extract at pH (4 to 7) showed that the Ag NPs exhibits a size ranging 50 nm (Table 1). Meanwhile, in pH 9 and pH 11, the dimensions of Ag NPs reach up to (15 to 30) nm. The result showed that pH could control the size of Ag NPs. In Ag NPs-pH 11, Ag NPs formed in small sizes (23 ± 8) nm. The TEM picture showed that after the Ag NPs created, some of Ag NPs seem to aggregate. The aggregation may be originated from the attraction between tiny particles. Furthermore, it could also cause due to the rapid reducing of Ag^+ where the capping agent can no longer cap the growing nanoparticles. Therefore, it can be expected that in the alkaline condition the nanoparticles have smaller size and poor balance between nucleation and growth processes could be the cause of this result [18]. Meanwhile, the result of Particles Size Analyzer (PSA) showed Polydispersity Index (PDI) and zeta potential (Table 1). The PDI value of Ag NPs tends to be higher along with increasing of pH. Ag NPs-control and Ag NPs-pH 4 have PDI values 0.364 and 0.412 indicating a moderate disperse level. Meanwhile, the PDI value of Ag NPs at pH (7 to 11), subsequently 0.480, 0.556, and 0.739 are within highly polydisperse range.

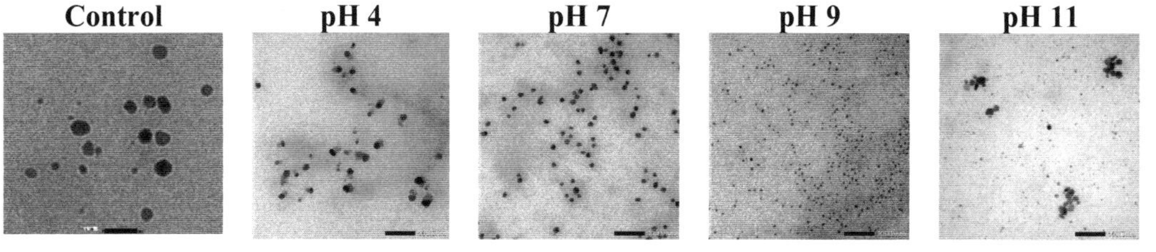

Figure 3. TEM result of silver nanoparticles morphology and distribution from biosynthesis using an aqueous extract of *Diospyros discolor* leaves (scale bar showed 100 nm scale).

Table 1. Polydispersity index (PDI), sizes and zeta potential values from Ag NPs from biosynthesis using an aqueous extract of *Diospyros discolor* leaves.

Sample	PDI	Sizes (nm)	Zeta Potential (mV)
Control	0.364	45 ± 24	–14.3
pH 4	0.412	54 ± 20	–22.7
pH 7	0.480	35 ± 12	–25.1
pH 9	0.556	21 ± 5	–28.1
pH 11	0.739	23 ± 8	–30.0

The stability level of Ag NPs during biosynthesis was indicated by its zeta potential (Table 1). It is shown that the zeta potential is higher as pH increase. Ag NPs-control has the smallest zeta potential of 14.3 mV, which indicates that the nanoparticles are less stable. The samples synthesized at pH 7 to pH 11 are in the range of moderate stable of (20 to 30) mV. The addition of NaOH to the *D. discolor* aqueous extract was found to influence the reduction rate of silver ions. The balance between the nucleation process and the growth of nanoparticles affects the shape of the nanoparticles [18]. Also, under very alkaline conditions, phenol compounds contained in *D. discolor* leaves extracts might be deprotonated, which cause the functional group on the mixture becoming unstable and unable to prevent the aggregation of silver nanoparticles formed [19].

4. Conclusions

Silver nanoparticles resulted from biosynthesis using an aqueous extract of *D. discolor* leaves after treatment with various pH modification in solution showed different characteristic in size, distribution, and stability. In this experiment, alkaline condition tended to control the size and distribution of the Ag NPs. The size of Ag NPs at pH below 7, at pH 7, and at pH 9 and pH 11 are in the range of 80 nm, under 50 nm, under 35 nm respectively. Some aggregation might happen due to lack of capping agent that can prevent attraction between the Ag NPs in colloid solution. In this experiment, pH shows potential as a dependent parameter in biosynthesis to get the desired Ag NPs in size and stability.

References

[1] Budhiraja N, Sharma A, Dahiya S, Parmar R, and Vidyadharan V 2013 Synthesis and optical characteristics of silver. nanoparticles on different substrates *ILCPA* **19** 80–88

[2] Cepriá G, Pardo J, Lopez A, Peña E, and Castillo J R 2016 Selectivity of silver nanoparticle sensors: Discrimination between silver nanoparticles and Ag^+ *Sensors and Actuators B* **230** 25–30
[3] Naidu K S B, Govender P, and Adam J K 2015 *JPAM* **9** 1–9
[4] Khatoon U T, Rao K V, Rao J V R, and Aparna Y 2011 Synthesis and characterization of silver nanoparticles by chemical reduction method *ICONSET* 97–99
[5] Zhang X, Liu Z, Shen W, and Gurunathan S 2016 Silver Nanoparticles: Synthesis, Characterization, Properties, Applications, and Therapeutic Approaches *Int. J. Mol. Sci.* **17** 1534–1568
[6] Hasan S 2015 *Res. J. Recent. Sci.* **4** 1–3
[7] Ragasa C *et al* 2009 *Nat. Prod. Res.* **13** 1252–1258
[8] Howlader S I *et al* 2012 Antioxidant and antidiarrhoeal Potentiality of Diosypros blancoi *Int. J. Pharmacol.* **8** 403–409
[9] León E R *et al* 2013 *Nanoscale Res. Lett.* **8** 318–327
[10] Veerasamy R *et al* 2011 Biosynthesis of silver nanoparticles using mangosteen leaf extract and evaluation of their antimicrobial activities *J. Saudi Chem. Soc.* **15** 113–120
[11] Agnihotri S, Mukherji S, and Mukherji S 2014 A systematic and detailed study **for** size-specific antibacterial efficacy of silver nanoparticles (AgNPs) synthesized using a co-reduction approach *RSC* **4** 3974–3983
[12] Ahmed S, Ahmad M, Swami B L, and Ikram S 2016 A review on plants extract mediated synthesis of silver nanoparticles for antimicrobial applications: A green expertise *J. Adv. Res.* **7** 17–28
[13] Mavani K and Mihir Shah 2013 *IJERT* **2** 1–5
[14] Ramteke C, Chakrabarti T, Sarangi B K, and Pandey R 2012 Characterization and evaluation of antibacterial activities of chemically synthesized iron oxide *J. Chem.* 1–7
[15] Jyoti K, Baunthiyal M, and Singh A 2015 A. Characterization of silver nanoparticles synthesized using Urtica dioica Linn. leaves and their synergistic effects with antibiotics *J. Radiat. Res. Appl. Sci.* **9** 217–227
[16] Saion E, Gharibshahi E, and Naghavi K 2013 Size-Controlled and Optical Properties of Monodispersed Silver Nanoparticles Synthesized by the Radiolytic Reduction Method *Int. J. Mol. Sci.* **14** 7880–7896
[17] Borase H P *et al* 2014 Plant extract: **a**promising biomatrix for ecofriendly, controlled synthesis of silver nanoparticles *Appl. Biochem. Biotechnol.* **173** 1–29
[18] Alqadi M K *et al* 2014 *Mater. Sci-Poland.* **32**1 107–111
[19] Darroudi M *et al* 2010 Effect of Accelerator in Green Synthesis of Silver Nanoparticles *Int. J. Mol. Sci.* **11** 3898–3905

The Influence of SnO_2 Nanoparticles on Electrical Conductivity, and Transmittance of PANI-SnO_2 Films

M. Diantoro[1,2], Kholid[1], A. A. Mustikasari[1], and Yudiyanto[1,2]

[1]Department of Physics, Faculty of Mathematics and Natural Sciences, Universitas Negeri Malang, Jl. Semarang 5 Malang 65146, Indonesia
[2]Central Laboratory of Minerals and Advanced Materials, Faculty of Mathematics and Natural Sciences, Universitas Negeri Malang, Jl. Semarang 5 Malang 65146, Indonesia

E-mail: markus.diantoro.fmipa@um.ac.id

Abstract. The study of the potency of conductive polymer and its combination with a various dopant of metals, oxides, and other radicals is significantly improved. We report the results of the study to determine the effect of SnO_2 addition on the microstructure, absorbance, and electric conductivity of PANI-SnO_2 films. In this work, we also report the effect of variations of light intensity on its electrical conductivity of the same films. The results showed that the addition of mass SnO_2 affecting the structure of the PANI-SnO_2 bonding function group in which indicated by the appearance of Sn-O-Sn peak at the wave number of 601.79 cm^{-1}. The addition of SnO_2 affects the crystallinity ratio of PANI-SnO_2 films shown by the crystallinity of PANI-SnO_2 films by 53.16 %. The addition of SnO_2 reduces the porosity of morphological shape of the PANI-SnO_2 film surface. It is surprising that addition of mass SnO_2significantly increases the electric conductivity of PANI-SnO_2 films by (4.24×10^2) $S \cdot cm^{-1}$. The addition of SnO_2 increases the transmission by 35.48 %. Light irradiance on PANI-SnO_2 film decreases the conductivity of the PANI-SnO_2 film.

Keywords. Electrical conductivity, nanoparticle, PANI, SnO_2, transmittance.

1. Introduction

Optoelectronic materials of organic materials have been widely developed as substitutes of inorganic-based optoelectronic materials [1]. The use of inorganic materials poses a problem because it cannot be deciphered naturally by soil microbes [2]. Polyaniline (PANI) is one of the conductive organic polymers which can be controlled by electronic and optical properties as optoelectronic materials [3]. So far PANI has been applied as an optoelectronic material resulting in low conductivity and transmittance [4, 7]. Therefore, it is necessary to increase the conductivity and transmittance of PANI. Moreover,the increasing electrical properties of PANI can be done by inducing another material to its such as PANI-Fe_3O_4 [8]. One of them with the addition of a class of oxides is Tin Oxide (SnO_2) [9, 11]. Based on the other works, the synthesized PANI-SNO_2showed allow conductivity, wave number absorbance, and band gap respectively of (6.4×10^{-3}) $S \cdot cm^{-1}$, 300 nm, and 4.1 eV [12]. Therefore, further research is needed to improve the conductivity and transmittance of PANI-SnO_2 film. So far, there is no study of the synthesis of SnO_2 incorporated into one-phase of PANI reported

Content from this work may be used under the terms of the Creative Commons Attribution 3.0 licence. Any further distribution of this work must maintain attribution to the author(s) and the title of the work, journal citation and DOI.
Published under licence by IOP Publishing Ltd

IOP Conf. Series: Materials Science and Engineering **367** (2018) 012034 doi:10.1088/1757-899X/367/1/012034

comprehensively. A composite material which was synthesized from different materials intrinsically consist of two materials which will show a combined property of the constituent [13]. So that PANI-SnO_2 system proposed in this research uses an in-situ polymerization method for the development of an optoelectronic material which shows a high conductivity and high transmittance. The fabrication of PANI-SnO_2 in the form of the film will increase the crystallinity [14]. The growth of this crystallinity will enhance the desired properties of the material.

2. Materials and methods

PANI-SnO_2 films have been prepared for the following procedure. First, we developed a solution made of 0.927 g of aniline hydrochloride, 80 mL of distilled water and 14 g $SnCl_4.5H_2O$. The second solution consists of 2.28 g $(NH_4)_2S_2O_8$ which were dissolved in 20 mL deionized water and then stirred it for 30 min. Subsequently, both aniline and $SnCl_4$ solutions were mixed with $(NH_4)_2S_2O_8$ into one container, stirred using a magnetic stirrer for one hour then precipitated for six hours. This precipitation for aniline polymerization process becomes PANI ES 1. The result of polymerization of the precipitate was then filtered and rinsed using distilled water for three times and deionized water for three times then followed by an acid-base chemical process to form PANI EB. The preparation of PANI EB is done by adding 15 % ammonia in PANI-SnO_2 solution, after which the solution was kept to stand for 24 h. After that, the blue sediment filtered and rinsed using distilled water as much as three times and deionized water as much as three times. Then the resulting PANI-SnO_2 precipitate was dried at 180 °C with an XD-1700 electric furnace for 1 h. The final result of the process was PANI-SnO_2 powder (PANI ES II) purplish blue.

Next, the film PANI-SnO_2 using a spin coating. First, is 0.15 % of the yield of PANI-SnO_2 powder added 20 mL of distilled water and 0.05 % PVA of PANI-SnO_2 powder mixed in one container [2]. Before mixing, PVA is first dissolved in distilled water and stirred for 6 h at room temperature. Addition of PVA will make PANI-SnO_2 in the form of ready-to-use solution to obtain the PANI-SnO_2 film. Preparation of the film used glass substrate size 2 cm × 2 cm was washed using acetone for 15 min, then dried at 80 °C heated temperature using the hot plate. The next step is the growth of thin film, PANI-SnO_2 solution dripped on the glass substrate (2 to 3) drops, and then spinner is rotated at 1500 rpm for 1 min. After the growth process is complete, the coating is dried with an automatic furnace at 100 °C for 1 h. We further analyzed its crystallinity, conductivity, and transmittance of the films.

3. Results and discussion

The addition of SnO_2 to PANI-EB produces the FT-IR PANI-SnO_2 spectrum shown in Figure 1. PANI-EB synthesis into PANI ES (II) is a doping process performed by addition of SnO_2 doping to PANI. The addition of doping makes Sn^{4+} atoms entering the bond. This bond is indicated by FT-IR peak appear at 601.79 cm^{-1}, 505.35 cm^{-1}, and 410.84 cm^{-1}. At the band of 601.79 cm^{-1} suggests the occurrence of Sn-O-Sn vibration [15], 505.35 cm^{-1} is the Sn-O-Sn bond of the asymmetry strain [12]. While at the peak of 410.84 cm^{-1} occurs the bond of Sn-OH [13].

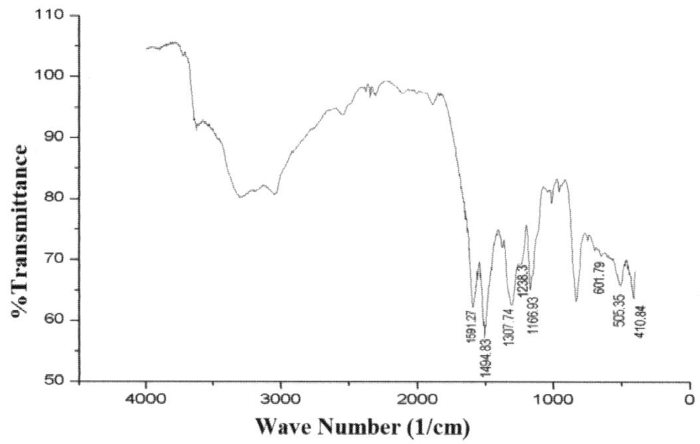

Figure 1. FTIR spectra of PANI-SnO_2Film

Based on the result of explanation of FT-IR data above can be concluded match PANI-SnO_2 based on reference data from the previous research article. There is an insignificant shift in the order of 0.1 from the database. So, it can be concluded PANI-SnO_2 successfully synthesized in this study through in-situ polymerization method. The Sn bond shown in the aniline bond is shown in Figure 2.

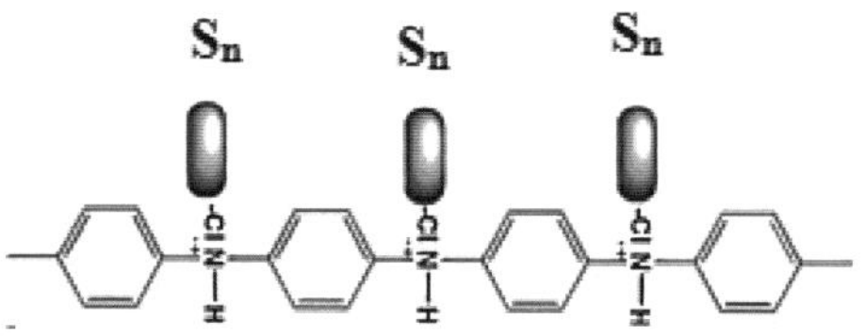

Figure 2. Possible Sn bond on the aniline chain in the PANI chain [16].

3.1. The diffraction of Polyaniline-SnO_2 analyses

The diffraction pattern of the PANI-SnO_2 film with the mass variation of SnO_2 doped PANI is shown in Figure 3. The phase analysis of PANI-EB formed at peak = 19.68° and 22.5° with the size of lattice parameter a = 7.7 Å, b = 5.8 Å, c = 4.8 Å with orthorhombic crystal structure [17, 18] using cell ref software. The model data is used as a refinement of PANI-SnO_2 data where SnO_2 itself has the peak intensity of 2θ at 26.49°, 33.77°, 54,78°, 57,95°, and 61,86° [19]. Based on the refinement results obtained the peak of refinement results on PANI-SnO_2 at 19.05°, 23.06°, 26.65° with error R^2 of 0.04821. There is no other peak of SnO_2 appear. It reflects the formation of a single phase. The grain size was obtained from FWHM data using Scherrer formula. The result of the calculation was performed under Gaussian profile-fitting analysis using software Origin 8. The average size was about 14.25 nm. The crystallinity of the material was calculated using the formula of Equation 1.

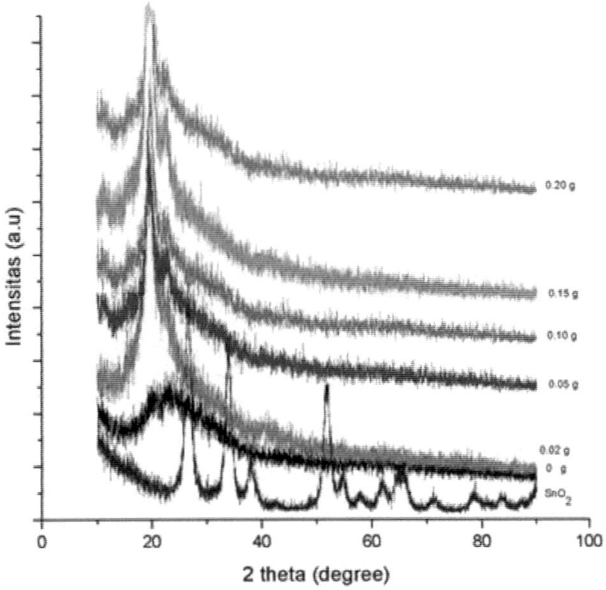

Figure 3. Pattern of XRD Result a) SnO_2, b) PANI-EB (aniline HCL), c) PANI-SnO_2 doping 0.02 g, d) PANI-SnO_2 doping 0.05 g, e) PANI-SnO_2 doping 0.10 g, f) PANI -SnO_2 doping 0.15 g, g) PANI-SnO_2 doping 0.2 g.

$$X_c = \frac{A_{kristal}}{A_{kristal} + A_{amorf}} \times 100\% \quad (1)$$

We found that the crystallinity of PANI-EB film is 22.79 %, and increase up to 53.169 % for doped PANI-SnO_2 film. The structural changes this research demonstrates, are closely related to the improvement of the resulting physical properties. This result, revealing the fundamental nature of Polyaniline where the features of the crystal structure can be controlled through doping.

3.2. Micro structure analyses

PANI-SnO_2 film image based on the result of SEM characterization and EDAX element shown in Figure 4. Based on the above SEM image can be presented a form of PANI morphology of fibers having pivot structures interconnected with other fiber networks [20]. The fiber network interconnected due to the result of polymerization, which results in additional new aniline molecules interacting with the new subsequently aniline groups. The protonation caused by the acid compound makes the structure óf the fiber hollow [21]. Based on Figure 4.b the addition of SnO_2 to the PANI film results in the closing of PANI polymerization cavities. The inclusion of SnO_2 in the PANI cavity reduces the porosity of microstructure of PANI surface morphology. The microstructure shows that Sn is entering on the surface morphological structure of the PANI-SnO_2 film. EDAX supports the presence of a Sn element in the film on PANI-SnO_2 film shown in Figure 4.c. Based on the cross-section of SEM images, it is obtained that PANI-SnO_2 film thickness is in the order of 16.32 μm.

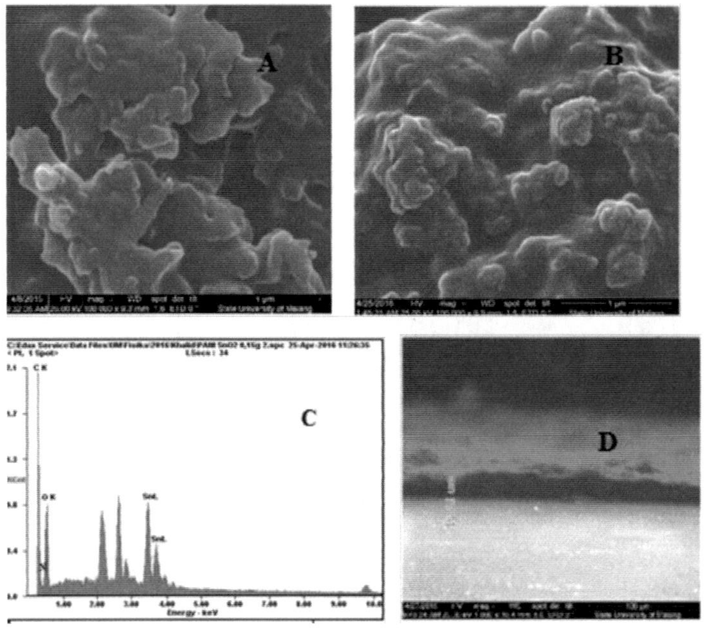

Figure 4. Morphological structure of a) PANI film, b) PANI-SnO_2, c) EDAX PANI-SnO_2 film, and d) PANI-SNO_2 film's thickness.

3.3. *Electric conductivity*

The results of the PANI-SnO_2 thin film conductivity of four-point probe measurements shown in Table 1.

Table 1. Electrical Conductivity of PANI-SnO_2 Film

SnO_2 mass (g)	conductivity ($S{\cdot}cm^{-2}$)
0	0.21
0.02	2.6×10^2
0.05	3.1×10^2
0.10	3.37×10^2
0.15	4.12×10^2
0.20	4.24×10^2

Based on Table 1, it is shown that the electrical conductivity of undoped PANI films was 0.21 $S{\cdot}cm^{-1}$ as other previous reports. The induced of SnO_2 in the PANI-SNO_2 film increase to (2.6×10^2) $S{\cdot}cm^{-1}$. The highest conductivity was reached by the sample with 0.2 g dopant with a conductivity of (4.24×10^2) $S{\cdot}cm^{-1}$. The conductivity of PANI-SnO_2 films is higher compared to PANI due to the increased of its crystallinity of the system supported by previous XRD analyses.

3.4. *Transmittance and UV-Vis measurement*

The result of characterization of UV-vis PANI-SnO_2 with various levels of addition of SnO_2 is shown in Figure 5. The addition of SnO_2 resulted in the decrease in absorbance of PANI-SNO_2 film. The highest absorbance result is inversely proportional to the transmittance, the higher the absorbance turns, the lower film's transmittance. Table 2 shows the calculation of transmittance based on UV-Vis data in Figure 5.

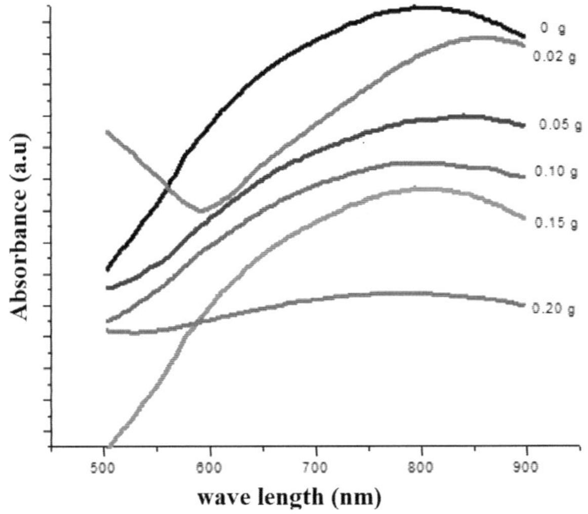

Figure 5. UV-VIS spectra of PANI-EB, and various SnO_2 doped PANI

The increased of PANI-SnO_2 film transmittance is caused by several factors, including the number of doping atoms that enter the film grid, the grain size, the thickness and the roughness of the film. Those factors are presumably present due to the addition of oxide level dopant that makes the transparency of material also increase. Based on the results of Table 2, we found that the highest transmittance is obtained by 0.2 g PANI-SnO_2 film with the transmittance of 35.48 %.

Table 2. Results of Transmittance Calculations from Various Levels of PANI-xSnO_2.

Mass (g)	Wave number (nm)	Absorbance max.	%Transmittance
0	769	0.67	21.3796
0.02	758	0.61	24.5471
0.05	763	0.61	24.5471
0.10	756	0.52	30.1995
0.15	742	0.49	32.3594
0.20	745	0.45	35.4813

3.5. Optoelectronic properties of PANI-SnO_2

The effect of light intensity on the PANI-SnO_2film for all ranges of dopant level is drawn in Figure 6. The measurement condition was performed similarly to the result as presented in Table 1 with additional various light intensity impinged on to the samples. Figure 8 tells us that the higher the dopant give rise to the increase of electrical conductivity as in Table 1. Further measurement by exposing light makes the conductivity significantly reduced and strictly independent of intensity at the higher intensity. This phenomenon may be due to the increase of thermal energy in the film PANI-SnO_2. The decrease of electrical conductivity looks to follow the electronic mechanism of conductivity due to the rise in energy level of electrons in the oxide semiconductor.

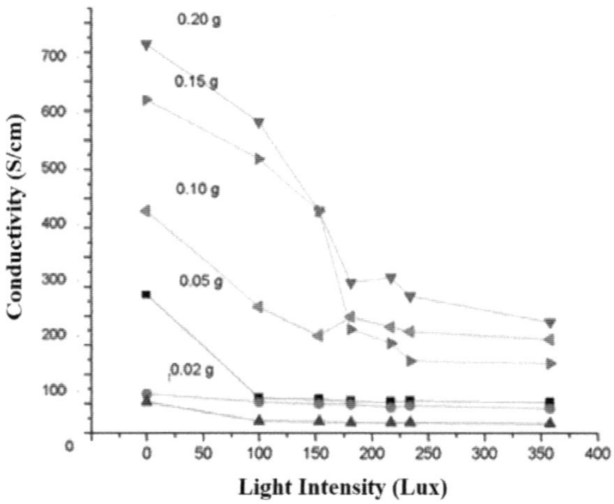

Figure 6. Effect of Light Intensity on PANI-SnO_2 film.

4. Conclusions

Based on the analysis, we can draw the following conclusions. The addition of mass SnO_2 affects the rise of SnO_2 intensity peaks, as well as the crystallinity of PANI-SnO_2. The addition of SnO_2 also changes the morphological structure of the PANI-SnO_2 film, i.e., decreases its porosity. A significant increase of transmittance and electrical conductivity as the increase of SnO_2. On the other hand, we obtain that the electrical conductivity of PANI-SnO_2 films decreases linearly with increasing intensity of light.

Acknowledgements

This research was partially supported by PDUPT Grand of Ministry of Research Technology and Higher Education.

References

[1] Franco Jr F C and Padama A A B 2016 DFT and TD-DFT study on the structural and optoelectronic characteristics of chemically modified donor-acceptor conjugated oligomers for organic polymer solar cells *Polymer* **97** pp 55–62

[2] Sarmah S and Kumar A 2013 Electrical and optical studies in polyaniline nanofibre–SnO_2 nanocomposites *Bulletin of Materials Science* **36** pp 31–36

[3] Politakos N *et al* 2013 Optical, structural and electrical properties of polyaniline systems doped with C60 and small gap C60 fullerenes *Materials Chemistry and Physics* **142** pp 387–394

[4] Mostafaei A and Nasirpouri F 2014 Epoxy/polyaniline–ZnO nanorods hybrid nanocomposite coatings: Synthesis, characterization and corrosion protection performance of conducting paints *Progress in Organic coatings* **77** pp 146–159

[5] Khairy M and Gouda M E 2015 Electrical and optical properties of nickel ferrite/polyaniline nanocomposite *Journal of advanced research* **6** pp 555–562

[6] Kumari K *et al* 2011 DC conductivity and spectroscopic studies of polyaniline doped with binary dopant $ZrOCl_2$/AgI *Bulletin of Materials Science* **34** pp 1237–1243

[7] Anwer T, Ansari M O and Mohammad F 2013 Dodecylbenzenesulfonic acid micelles assisted in situ preparation and enhanced thermoelectric performance of semiconducting polyaniline–zirconium oxide nanocomposites *Journal of Industrial and Engineering Chemistry* **19** pp 1653–1658

[8] Chen W *et al* 2009 High-efficiency dye-sensitized solar cells based on the composite photoanodes of SnO_2 nanoparticles/ZnO nanotetrapods *The Journal of Physical Chemistry A* **114** pp 3127–3138

[9] Dou X *et al* 2011 Hydrothermal synthesis of high electron mobility Zn-doped SnO_2 nanoflowers as photoanode material for efficient dye-sensitized solar cells *Chemistry of Materials* **23** pp 3938–3945
[10] Gubbala S *et al* 2008 Band-Edge Engineered Hybrid Structures for Dye-Sensitized Solar Cells Based on SnO2 Nanowires *Advanced Functional Materials* **18** pp 2411–2418
[11] Alam M *et al* 2013 Optical and electrical conducting properties of Polyaniline/Tin oxide nanocomposite *Arabian Journal of Chemistry* **6** pp 341–345
[12] Li X 2012 Synthesis of polyaniline/tin oxide hybrid and its improved electrochemical capacitance performance *Electrochimica Acta* **85** pp 9–15
[13] Newham Robert E 2005 *Propeties Of Materials* (Pennsylvania State University: Oxford University Press)
[14] Setianingsih E 2014 Karakteristik Struktur, Sifat Optik dan Sifat Listrik Film Tipis Polianilin Doping H_2So_4 yang Ditumbuhkan dengan Metode Spin Coating (Doctoral dissertation, Universitas Negeri Semarang)
[15] Manivel P *et al* 2013 Optical and electrochemical studies of polyaniline/SnO_2 fibrous nanocomposites *Materials Research Bulletin* **48** pp 640–645
[16] Sniechowski M 2005 *Structure and Dynamics of Conducting Poly (aniline) based Compounds* (Doctoral dissertation, Université Joseph-Fourier-Grenoble I)
[17] Sarno D M, Manohar S K and MacDiarmid A G 2005 Controlled interconversion of semiconducting and metallic forms of polyaniline nanofibers *Synthetic Metals* **148** pp 237–243
[18] Rajpure K Y *et al* 2000 Effect of Sb doping on properties of conductive spray deposited SnO_2 thin films *Materials Chemistry and Physics* **64** pp 184–188
[19] Tai H L, Jiang Y D and Xie G Z 2010 Preparation of Surfactants Directed PANI/In_2O_3 Nanocomposite Thin Films and Its NH3-Sensing Properties *Journal of Electronic Science and Technology* **8** pp 154–159
[20] Goodhew P J, Humphreys J and Beanland R 2000 *Electron Microscopy and Analysis* (CRC Press)
[21] Khuspe G D *et al* 2013 Facile method of synthesis of polyaniline-SnO_2 hybrid nanocomposites: Microstructural, optical and electrical transport properties *Synthetic Metals* **178** pp 1–9

Effect of Polyethylene Glycol (PEG-4000) on Dielectric Properties of $Mn_{0.5}Zn_{0.5}Fe_2O_4$ Nanoparticles

L. Armitasari[1], A. Yusmar[1], and E. Suharyadi[1]

[1]Department of Physics, Gadjah Mada University, Yogyakarta, INDONESIA

E-mail : esuharyadi@ugm.ac.id

Abstract. Dielectric properties of $Mn_{0.5}Zn_{0.5}Fe_2O_4$ nanoparticles (NPs) encapsulated by polyethylene glycol (PEG) has been studied as a function of frequency in the range 5 kHz to 120 kHz at room temperature. PEG-encapsulated $Mn_{0.5}Zn_{0.5}Fe_2O_4$ NPs were synthesized by co-precipitation method with various PEG concentration. X-ray diffraction (XRD) pattern confirmed that $Mn_{0.5}Zn_{0.5}Fe_2O_4$ NPs have cubic spinel structure. The crystallite size of the sample was 12.5 nm and then decreased with the increase of PEG concentration. It is due to the agglomeration reduced after PEG encapsulation. The real dielectric constant (ε') of the sample was 320.2 and decreased with the increase of PEG concentration, it is expected because of a decrease of crystallite size. The imaginary dielectric constant (ε') and loss tangent (tan δ) of the samples have a value 360.2 and 1.1 respectively and decreased with the increase of PEG concentration, which is due to the decrease in drift mobility of electric charge carries. The AC conductivity (σ) of the sample was (1.1×10^{-4}) kΩ^{-1} and decreased with the increase of PEG concentration, this may be due to the change in cation distribution of the sample. The capacitance of the sample was (4.5×10^{-11}) Farad and decreased with the increase of PEG concentration, it is also caused by the decrease of crystallite size. Furthermore, the real dielectric constant (ε') and imaginary dielectric constant (ε') of the samples have a value 521.1 and 393.8 respectively and decreased with increasing applied frequency; it is due to the fact that above certain frequencies the electronic exchange between the ferrous and ferric ions does not follow the applied field. The AC conductivity of the sample was (2.1×10^{-4}) kΩ^{-1} and increased with increasing applied frequency, contrary the capacitance (C) of the sample was (1.3×10^{-10}) Farad and decreased with increasing applied frequency. This can be attributed to the conductive grains become more active at the higher frequency, hence increasing the hopping frequency of electron Fe^{3+} and Fe^{2+} ions. The refractive index (n) of the sample was 24.2 and increases with the increase of dielectric constant at low frequency.

Keywords. Dielectric properties, nanoparticles, encapsulation, and PEG-400.

1. Introduction

Manganese-zinc (Mn-Zn) ferrite is one an example of magnetic nanoparticles that can be applied to transformers, inductors, loudspeakers and other electromagnetic devices. The well-known properties of Mn-Zn ferrite among these are nanoparticles having low curie temperatures, small residual magnetization and low magnetic anisotropy [1]. Mn-Zn ferrite is included in soft ferrite material. Its properties depend on microstructures such as grain size and shape, and it is also applicable in the many fields [2]. During the synthesis process, the sample was not perfectly dispersed [3, 4]. Therefore,

Content from this work may be used under the terms of the Creative Commons Attribution 3.0 licence. Any further distribution of this work must maintain attribution to the author(s) and the title of the work, journal citation and DOI.
Published under licence by IOP Publishing Ltd

encapsulation of nanoparticles may be performed. Polyethylene glycol (PEG) frequently used as encapsulation agent because of non-toxic properties, and it can increase the dispersibility of nanoparticles. PEG is advantageous to prevent agglomeration and to obtain small particles size [5, 6].

An external electric field with various frequency affects the dielectric properties due to the ability of polarization of materials, e.g., interfacial polarization occurs at the frequency range (0-10^4) Hz and dipole polarization at the frequency range (10^4 to 10^9) Hz [7]. Previously, dielectric properties of $MnFe_2O_4$ nanoparticles has been investigated at the frequency range of (5 to 120) kHz [8]. The change of frequency also influenced the value of conductivity and capacitance in the dielectric material. The results can be explained on the basis of Fe^{3+} and Fe^{2+} ions in the octahedral subspace of spinel ferrite. By adding PEG as encapsulation agent in nanoparticles (NPs) may influence the dielectric properties of NPs. The focus of this research is to investigate the effect of frequency and PEG concentration on dielectric properties of PEG-encapsulated $Mn_{0.5}Zn_{0.5}Fe_2O_4$ NPs. Effect of frequency on AC conductivity, capacitance and refractive index of $Mn_{0.5}Zn_{0.5}Fe_2O_4$ ferrite/PEG with various PEG concentration is also investigated.

2. Materials and methods

$Mn_{0.5}Zn_{0.5}Fe_2O_4$ nanoparticles have been fabricated by coprecipitation method. $ZnSO_4{\cdot}H_2O$ + $MnCl_2{\cdot}H_2O$ (Merck, Germany) were dissolved in distilled water. The mixture was stirred until the solution homogeneously, with stirring duration of 2 min and stirring speed of 500 rpm. $FeCl_3{\cdot}6H_2O$ (Merck, Germany) was dissolved in distilled water (20 mL), stirring duration for 2 min at a speed of 500 rpm at room temperature. The solution was added HCl of 3.37 mL. The mixture was gently dropped into an aqueous NaOH as co-precipitant with stirring speed of 1000 rpm at 120 °C during 60 min. The powders were dried in a furnace with a temperature of 90 °C for 4 h. X-ray diffractometer (XRD) was used to characterize the crystal structure. Transmission electron microscopy (TEM) was used to identify the nanoparticles morphology and diffraction patterns. Fourier transforms infrared (FTIR) was conducted to determine the functional group bonding samples. $Mn_{0.5}Zn_{0.5}Fe_2O_4$NPs was encapsulated by PEG-4000 with various concentration. The sample in powder was compacted into the pellets of 1.24 cm in diameter at the compressive force of 50000 N. Dielectric properties were measured using computerized impedance spectroscopy in the frequency range of 5 kHz to 120 kHz.

3. Results and discussion

Figure 1 shows the diffraction patterns of the sample $Mn_{0.5}Zn_{0.5}Fe_2O_4$ and $Mn_{0.5}Zn_{0.5}Fe_2O_4$ with PEG-4000 encapsulation of 50%. The peaks at (2θ) 29.9°, 35.3°, 42.9°, 56.8°, and 62.4° represents (220), (311), (400), (511), and (440) planes, respectively. The result was confirmed with the JCPDS $Mn_{0.5}Zn_{0.5}Fe_2O_4$ (*powder diffraction number*: 10-0467), corresponding to the mix spinel structure [5]. There were characteristic peaks of another phase at $2\theta = 31.7°$ in both samples identified by α-Fe_2O_3, which is due to the oxidation process. The crystallite size of the $Mn_{0.5}Zn_{0.5}Fe_2O_4$ was 12.5 nm and then decreased with the increase of PEG concentration.

Figure 1 shows the diffraction patterns of the sample $Mn_{0.5}Zn_{0.5}Fe_2O_4$ and $Mn_{0.5}Zn_{0.5}Fe_2O_4$ with PEG-4000 encapsulation of 50%. The peaks at (2θ),29.9°, 35.3°, 42.9°, 56.8°, and 62.4° represents (220), (311), (400), (511), and (440) planes, respectively. The result was confirmed with the JCPDS $Mn_{0.5}Zn_{0.5}Fe_2O_4$ (*powder diffraction number*: 10-0467), corresponding to the mix spinel structure [5]. There were characteristic peaks of another phase at $2\theta = 31.7°$ in both samples identified by α-Fe_2O_3, which is due to the oxidation process. The crystallite size of the $Mn_{0.5}Zn_{0.5}Fe_2O_4$ was 12.5 nm and then decreased with the increase of PEG concentration.

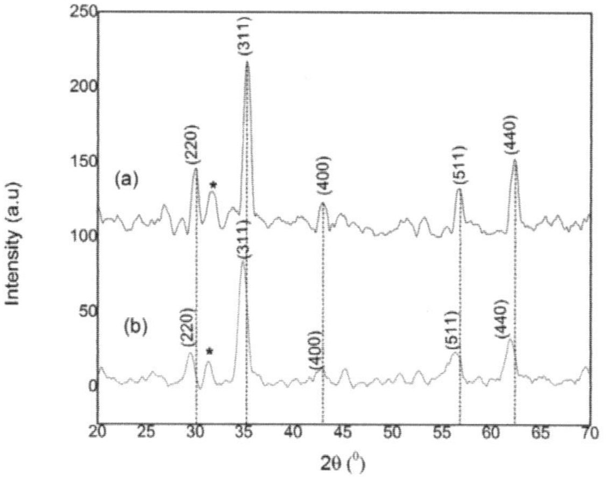

Figure 1. XRD patterns of the samples; (a) $Mn_{0.5}Zn_{0.5}Fe_2O_4$ nanoparticles and (b) PEG/$Mn_{0.5}Zn_{0.5}Fe_2O_4$

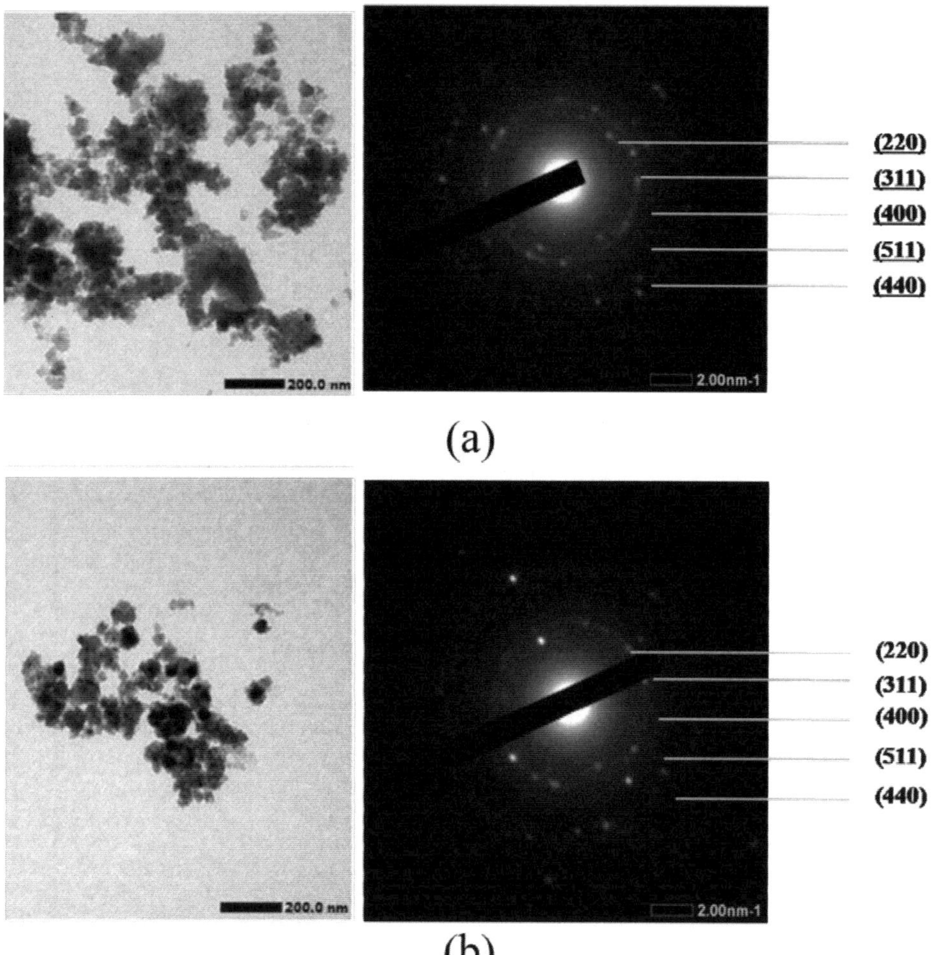

Figure 2. TEM images of (a) $Mn_{0.5}Zn_{0.5}Fe_2O_4$ and (b) PEG-encapsulated $Mn_{0.5}Zn_{0.5}Fe_2O_4$

IOP Conf. Series: Materials Science and Engineering **367** (2018) 012035 doi:10.1088/1757-899X/367/1/012035

Figure 2 shows the TEM image and diffraction patterns of $Mn_{0.5}Zn_{0.5}Fe_2O_4$ nanoparticles and PEG-encapsulated $Mn_{0.5}Zn_{0.5}Fe_2O_4$ nanoparticles. Both samples are agglomerated. Agglomeration due to the higher magnetic and interaction electrostatic forces of nanoparticles with small size and large surface.

Figure 3 shows the decrease of dielectric constant with the increase of frequency. It can be explained on the basis of interfacial polarization as predicted by Maxwell-Wagner. The conducting grains are found to be more effective at the higher frequency, and poor conducting are more effective at low frequency [6]. On the other hand, the dielectric constant decreases as the increase of PEG concentration. It is due to the decrease of grain size of PEG-encapsulated $Mn_{0.5}Zn_{0.5}Fe_2O_4$ nanoparticles.

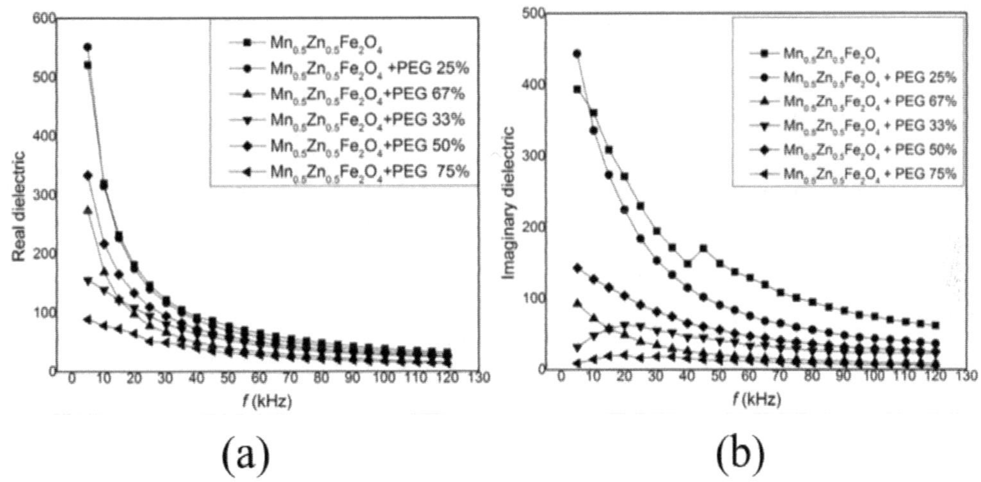

Figure 3. Frequency dependence of real (a) and imaginary (b) dielectric value for PEG-encapsulated $Mn_{0.5}Zn_{0.5}Fe_2O_4$ with various concentration of PEG

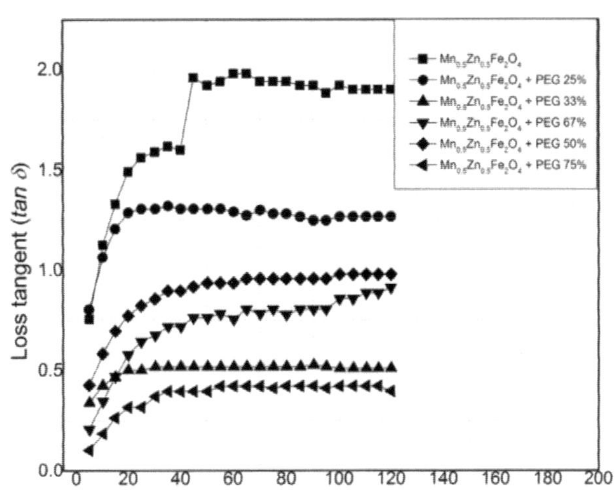

Figure 4. Frequency dependence of loss tangent for PEG-encapsulated $Mn_{0.5}Zn_{0.5}Fe_2O_4$with various concentration of PEG

Figure 4 shows the enhancement of loss tangent by the increase of frequencies 5 kHz to 30 kHz. It is due to the interfacial polarization. However, the loss tangent reaches a constant value when the frequency increased at 30 kHz to 120 kHz. This show that dipole moments cannot follow the external electric field at high frequency. Furthermore, the lost tangent decreases with the increase of PEG concentration. It is due to the decrease of grain size and charge mobility.

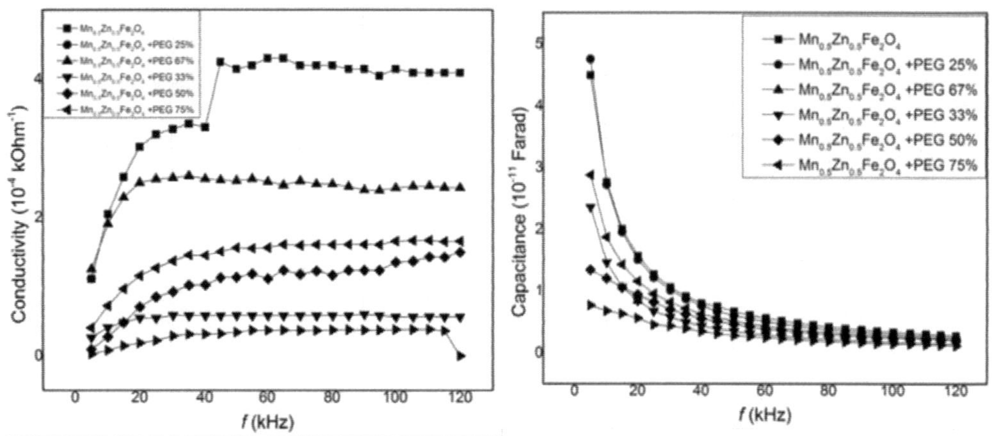

(a) (b)

Figure 5. Frequency dependence of conductivity (a) and capacitance (b) for PEG-encapsulated $Mn_{0.5}Zn_{0.5}Fe_2O_4$ with various concentration of PEG

The conductivity increases with the increase of frequency, as shown in Figure 5.a. The conduction mechanism of the ferrite can be explained on the charge carries between the Fe^{2+} and Fe^{3+} ions on the octahedral sites [9]. The capacitance decrease and tends to be constant at high frequency, as shown in Figure 5.b. It is explained that the exchange of Fe^{3+} and Fe^{2+} ions cannot follow the alternating current field [6]. The conductivity and capacitance decrease with the increase of PEG concentration. Figure 6 shows that the refractive index decreases significantly from 5 kHz to 30 kHz and tends to be constant above 30 kHz. These results confirm that the linear correlation between the refractive index and the dielectric constant.

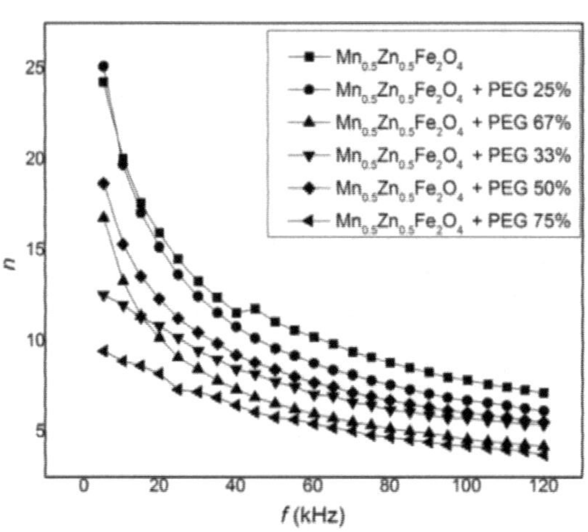

Figure 6. Frequency dependence of refractive index for PEG-encapsulated $Mn_{0.5}Zn_{0.5}Fe_2O_4$

4. Conclusions

In the present study, $Mn_{0.5}Zn_{0.5}Fe_2O_4$ nanoparticles (NPs) encapsulated by polyethylene glycol was synthesized by coprecipitation method. The dielectric constants loss, tangent, conductivity, capacitance and the refractive index of the samples decreases with the increase of PEG concentration. The dielectric constants decreased with the increase of frequency. This is due to the interfacial polarization at the low frequency. The loss tangent increased with the increase of frequency.

Acknowledgments

Research Grant of "Penelitian Berbasis Kompetensi," Ministry of Reseach, Technology and Higher Education of the Republic Indonesia (*Kementerian RistekDikti*).

References

[1] Mathew, D. S., & Juang, R. S. (2007). An overview of the structure and magnetism of spinel ferrite nanoparticles and their synthesis in microemulsions. *Chemical Engineering Journal*, *129*(1-3), 51-65.

[2] Cao, X., Liu, G., Wang, Y., Li, J., & Hong, R. (2010). Preparation of octahedral shaped Mn0. 8Zn0. 2Fe2O4 ferrites via co-precipitation. *Journal of Alloys and Compounds*, *497*(1-2), L9-L12.

[3] Indrayana, I. P. T., Siregar, N., Suharyadi, E., Kato, T., & Iwata, S. (2016, November). The calcination temperature dependence of microstructural, vibrational spectra and magnetic properties of nanocrystalline $Mn_{0.5}Zn_{0.5}Fe_2O_4$. In *Journal of Physics: Conference Series* (Vol. 776, No. 1, p. 012021). IOP Publishing.

[4] Siregar, N., Indrayana, I. P. T., Suharyadi, E., Kato, T., & Iwata, S. (2017, May). Effect of Synthesis Temperature and NaOH Concentration on Microstructural and Magnetic Properties of Mn0. 5Zn0. 5Fe2O4 Nanoparticles. In *IOP Conference Series: Materials Science and Engineering* (Vol. 202, No. 1, p. 012048). IOP Publishing.

[5] Kareem, S. H., Ati, A. A., Shamsuddin, M., & Lee, S. L. (2015). Nanostructural, morphological and magnetic studies of PEG/$Mn_{(1-x)}Zn_{(x)}Fe_2O_4$ nanoparticles synthesized by co-precipitation. *Ceramics International*, *41*(9), 11702-11709.

[6] Köseoğlu, Y. (2013). Structural, magnetic, electrical and dielectric properties of $Mn_xNi_{1-x}Fe_2O_4$ spinel nanoferrites prepared by PEG assisted hydrothermal method. *Ceramics International*, *39*(4), 4221-4230.

[7] Sudirham S and Utari N 2012 *Study on Material Properties* (Bandung: InstitutTeknologi Bandung) pp 164–166

[8] Kurnia, Heriansyah and E. Suharyadi 2017 Study on The Influence of Crystal Structure and Grain Size on Dielectric Properties of Manganese Ferrite ($MnFe_2O_4$) Nanoparticles. *IOP Conf. Series: Materials Science and Engineering* **202** pp 012046.1–012046.7

[9] Berchmans, L. J., Selvan, R. K., Kumar, P. S., & Augustin, C. O. (2004). Structural and electrical properties of $Ni_{1-x}Mg_xFe_2O_4$ synthesized by citrate gel process. *Journal of Magnetism and Magnetic Materials*, *279*(1), 103-110.

Electrosprayed Polyvinylpyrrolidone (PVP) Submicron Particles Loaded by Green Tea Extracts

Kamaruddin[1,2], I. Sriyanti[1,2], D. Edikresnha[1,2], M. M. Munir[1,2] and K. Khairurrijal[1,2]

[1]Department of Physics, Faculty of Mathematics and Natural Sciences,
[2]Bioscience and Biotechnology Research Center, Institute for Research and Community Services, Institut Teknologi Bandung, Jalan Ganesa 10, Bandung 40132, Indonesia

E-mail: mmmunir.itb@gmail.com

Abstract. Electrospraying technique has been successfully used to synthesize composite submicron particles of polyvinylpyrrolidone (PVP) and green tea extract (GTE). The precursor solutions were PVP in ethanol (15 wt%) and GTE in ethanol (10 wt%), which were then mixed at varying ratio. The mixed solution then underwent electrospraying process at an applied voltage of 15 kV, a distance of collector to the nozzle at 15 cm, and a flow rate of 3 μL/min. The composite submicron particles of PVP-GTE showed smooth and fine spherical morphology without fibers or beaded fibers. To a certain degree, the increase of GTE content in the PVP-GTE mixed solution decreased the average diameter of PVP-GTE composite particles. Moreover, the analysis of the FTIR spectra confirmed the existing molecular interaction between PVP and GTE in the composite submicron particles as shown by the shift of PVP wavenumber towards GTE, which has typically smaller wavenumber.

Keyword. Electrospraying, FTIR, green tea extract (GTE), and polyvinylpyrrolidone (PVP).

1. Introduction

Green tea has a typical taste and flavor [1] that has many benefits such as high antioxidant activity [2, 3], ability to inhibit the proliferation of cancer cells [4], and many others. However, there are some problems existing in the application of green tea that must be solved as well. The phenolic compound of green tea, i.e., catechin, has low bioavailability in the body, low solubility in water, and large molecular weight [5]. These problems have to be unraveled partly by reducing the size down to nanoscale. The solid polymeric nanoparticles, at a size range of (10 – 1000) nm, have been studied thoroughly to be used as carriers of various molecules of drugs in many types of research related to drug delivery system [6]. By encapsulating the green tea extract (GTE) into the nanoparticles, it is then expected the problem regarding the low availability can be solved. Previously, there were already efforts made to reduce the insolubility of the compound by mixing it with a polymer, as the carrier [7] or reduced the size down to nanoscale as previously mentioned. From the literature, there are reports on the reduction of the size of green tea leaves by grinding [8] and green tea extract composites nanostructures [9].

Content from this work may be used under the terms of the Creative Commons Attribution 3.0 licence. Any further distribution of this work must maintain attribution to the author(s) and the title of the work, journal citation and DOI.
Published under licence by IOP Publishing Ltd

In more than two decades, electrohydrodynamic atomization (EHDA) has been used widely to produce fibers, specifically named as electrospinning, as well as particles (electrospray) [10, 11] and has been applied for drug delivery systems. For creating micro-particles to nano-particles, electrospraying is preferred that the conventional techniques (such as nanoprecipitation, emulsion-diffusion, double-emulsification, and layer by layer) due to its simplicity, low-price, low use of a solvent, while the particles can be produced in only a single step [11]. Electrospraying technique can be potentially used to encapsulate bioactive molecules such as drugs, DNA, and many functional active ingredients in nanostructures [12] particularly in the shape of spherical particles. However, the production of drug delivery media for natural plant extract, used for many therapeutic purposes, is still very rare.

In this research, the polymer used as the carrier was polyvinylpyrrolidone (PVP) that has been standardized by the Food and Drug Administration (FDA) and widely applicable for various plant extracts such as *Garcinia mangostana* extracts [13, 14] and *Anredera cordifolia* leaves extract [15], in which all were in nanofibers form. Previously, GTE has been successfully encapsulated in nanocomposites as a combination of agglomerated beads (less spherical particles) and nanofibers [9]. Thus, this paper reports on the synthesis of composite solid particles of GTE-loaded PVP produced by electrospraying and presents the formation of purely composite spherical particles that have not been reported yet. The morphology and the size were analyzed by scanning electron microscopy (SEM) images. The existing functional groups and the intermolecular interactions of the composite particles were detected using a Fourier Transform Infra-Red (FTIR) spectrometer.

2. Materials and methods

2.1. Materials

Polyvinylpyrrolidone (PVP) with a molecular weight of 40000 g·mol^{-1} was purchased from Sigma-Aldrich. The leaves of green tea were obtained from a local market in Bandung, Indonesia. The technical grade ethanol with purity 96 % and deionized water were obtained from Bratachem, Indonesia.

2.2. Preparation of green tea extract (GTE) and precursor solution

The leaves of green tea were extracted using maceration method. The dried and grinded leaves were immersed in ethanol solution for two days with frequent stirring to dissolve the leaves. The liquid taken from the extracted leaves was then filtered and evaporated to achieve powdered extract of green tea. The extraction process was conducted at the Laboratory of Natural Product, Bioscience, and Biotechnology Research Centre, ITB. PVP was dissolved into an ethanol solvent diluted in deionized water (at ratio 7:3 w/w) with the concentration of PVP solution of 15 wt%. On the other hand, the GTE has dissolved in ethanol the concentration of 10 wt%. The PVP solution was then mixed with the GTE solution at several ratios by asymmetric composition. Each PVP and GTE solution was stirred for about one hour, and subsequently, the mixed PVP-GTE solutions were stirred using a magnetic stirrer for 2 h at the room temperature. The homogeneous solution of the PVP-GTE solution was then inserted into a syringe to undergo electrospraying process to produce PVP-GTE composite particles.

2.3. Solution properties

The characteristics of the precursor solution, including the viscosity, conductivity, and surface tension, were then measured. All solutions, which are the PVP, GTE, and PVP-GTE solutions, were measured using Fenske-Ostwald viscometer (Fisher), du Noüy tensiometer (Fisher), and conductometer (Mettler-Toledo), for their viscosities, surface tensions, and conductivities, respectively.

2.4. Electrospraying process, morphology, and size

The PVP solution and PVP-GTE solutions were placed in a 1-mL syringe (Terumo), and a high voltage (15 kV) was applied so that the electrospraying process took place. A plate collector was used as the ground electrode with an aluminum foil coated on the surface. The collector was put at 15 cm from the needle whose diameter was 27 gauge. The humidity was kept at (60 to 70) % while the temperature was at the room temperature. The flow rate of the electrospraying process was three μL/min which was pushed by the syringe pump connected to a PC and controlled by software. During electrospraying process, a high-speed digital camera (Panasonic, M2050) was operated to detect the formation of spraying modes. The morphology of the particles was observed using a scanning electron microscope (SEM) (Hitachi, SU3500). The SEM was operated at an excitation voltage of 5 kV with a magnification of 4000 times. The geometric mean diameter of the particles was measured with Image-MIF v3.0 software taken at 300 measurement points, and each graph was fitted by Origin 9.0 software to find the particle size distribution.

2.5. Fourier transform infra-red spectroscopy

Fourier Transform Infra-Red (FTIR) spectroscopy was used to identify the existing functional groups and the intermolecular interactions between the PVP and the GTE as PVP-GTE composite particle. The FTIR spectrometer (Bruker, Alpha 1-176-396) was performed with the wavenumber in the range of (500 to 4000) cm^{-1}.

3. Results and Discussion

3.1. Solution mixture and characteristics

The PVP and GTE solutions were mixed by asymmetric combination ratios. The precursor solution colors of PVP and GTE was white transparent and dark brown, respectively. The color of PVP-GTE solution varied according to the amount of addition of GTE from light to dark brown following the color of GTE. The homogeneous solutions of PVP-GTE were at ratios of 9:1, 8:2, 7:3, 6:4, while other compositions were not chosen since there was residue on the bottom part indicating that the solutions were not homogeneous. The selected solutions were then measured for their solution properties and then underwent electrospraying process for producing composite particles of PVP-GTE. Table 1 shows the measurement results on the viscosity, conductivity and surface tension of the solutions subjected to the electrospray.

Table 1. The viscosity, conductivity, and surface tension of various solutions at room temperature

PVP-GTE Solution Ratio (w/w)	Concentrations (wt%)	Viscosity (centiPoise)	Conductivity (μS/cm)	Surface Tension (dyne/cm)
10:0	15.0	12.189	0.63	35.3
9:1	14.5	12.835	0.84	35.2
8:2	14.0	11.265	1.12	34.2
7:3	13.5	9.092	1.90	33.4
6:4	13.0	7.368	2.36	32.1
0:10	10.0	1.112	4.56	29.8

The selection of polymer concentration PVP 15 wt% (Mw 40 $kg{\cdot}M^{-1}$) was an approximation by referring to the previous report [10] that correlates to particle formation. Meanwhile, the GTE concentration of 10 wt% was selected by the solubility of the GTE solution. As similar principle processing between electrospinning and electrospraying, the different morphology resulting from the electrospraying/electrospinning was found to depend strongly on the solution parameters [10].

3.2. Spraying modes, morphology, and size

The spraying mode, SEM images, and the diameter distribution of PVP particles and the PVP-GTE composite particles are shown in Figure 1. During the electrospraying process, the spraying mode of PVP (inset Figure 1.a) solution was classified as silver bullet mode that was very promising for the production of the particles and was already used in previous reports [16, 17]. The spray modes

IOP Conf. Series: Materials Science and Engineering **367** (2018) 012036 doi:10.1088/1757-899X/367/1/012036

transformed as the GTE ratio was changed in the PVP-GTE solutions. When the ratio of GTE increased, the spray mode was unstable, and the multi-jet mode was constructed (inset Figure 1.b to Figure 1.e) since the addition of GTE affected to the characteristics of the solution. This finding proves that the solution characteristic is one of the critical parameters for determining the spraying mode in electrospray process. According to the solution characteristics, the viscosity and the surface tension of the solution decreased along with the addition of GTE. On the contrary, the conductivity of the solution was increased following the amount of GTE ratio.

The SEM images and the histogram of diameter distributions are presented with a magnification of 4000 times and log-normal fit, respectively. The SEM image of PVP electrospray (Figure 1.a) shows the agglomeration or coalesced particles near the droplet source, and therefore the diameter of the particles is difficult to measure while the distribution of diameter is difficult to calculate. Supposedly, it appears due to the low molecular weight [16] and hygroscopic characteristic of the PVP [18]. The images of particles made from the precursor solution with a ratio of PVP-GTE of 9:1 (Figure 1.b) shows less coalesced particles compared to that in Figure 1.a. The existing GTE was then able to reduce the coalescent of the particles although there were still coalesced since the hygroscopic polymer PVP was still largely dominant over GTE. Otherwise, the composite particles made from precursor solutions of PVP-GTE at ratio of 8:2, 7:3, 6:4 (Figure 1.c, Figure 1.d, and Figure 1.e, respectively) shows composite particles with smooth spherical shape and perfectly free of fibers or beaded-fibers with wide range size from micro to the nanometer.

The measurement of the diameter from the SEM images of the electrosprayed PVP-GTE composite particles made from precursor solution at ratios of 9:1, 8:2, 7:3 and 6:4 was distributed within the range of (312 to 1932) nm, (298 to 2127) nm, (298 to 2234) nm, and (195 to 2190) nm, respectively. The geometric mean diameters of PVP-GTE at a ratio of 9:1, 8:2, 7:3 and 6:4 were 895 nm, 858 nm, 775 nm, and 684 nm, respectively. These results show that the average diameter of the particle reduced as the GTE content increased. The reason is that the concentration of PVP decreased as the amount of GTE increased, therefore, reducing the diameter [14, 19]. The geometric standard deviation of PVP-GTE composite particles with a ratio of 9:1, 8:2, 7:3 were 1.37, while for 6:4 was 1.41, respectively. From the value of geometric standard deviation, the particles can thus be categorized as non-uniform (it is categorized as uniform if the standard deviation geometry was ≤ 1.22; otherwise it is non-uniform [20]). This result was related to the spraying modes of PVP-GTE solutions that were in multi-jet modes, which were strongly influenced the production of the inconsistent size of particles [18].

3.3. Theoretical models

We also compare the empirical results of the particles diameter to the theoretically predicted size. The scaling law models of the electrosprayed particle diameter to the solution flow rate and conductivity is given as [16, 17]:

$$D_{drop} = G(\kappa)\left(\kappa\varepsilon_0 \frac{Q}{K}\right)^{1/3} \tag{1}$$

$$D_{comp} = G(\kappa)\left(\varphi\kappa\varepsilon_0 \frac{Q}{K}\right)^{1/3} \tag{2}$$

where D_{drop} is the diameter of the droplet and D_{comp} is the diameter of the electrosprayed composite particles, κ is the dielectric constant, Q is the flow rate, ϕ is the volume fraction of the combined polymer and bioactive substances, and ε_0 is permeability. K is the electrical conductivity, and $G(\kappa)$ is the dimensionless function given by:

$$G(\kappa) = -10.9\kappa^{-6/5} + 4.08\kappa^{-1/3} \tag{3}$$

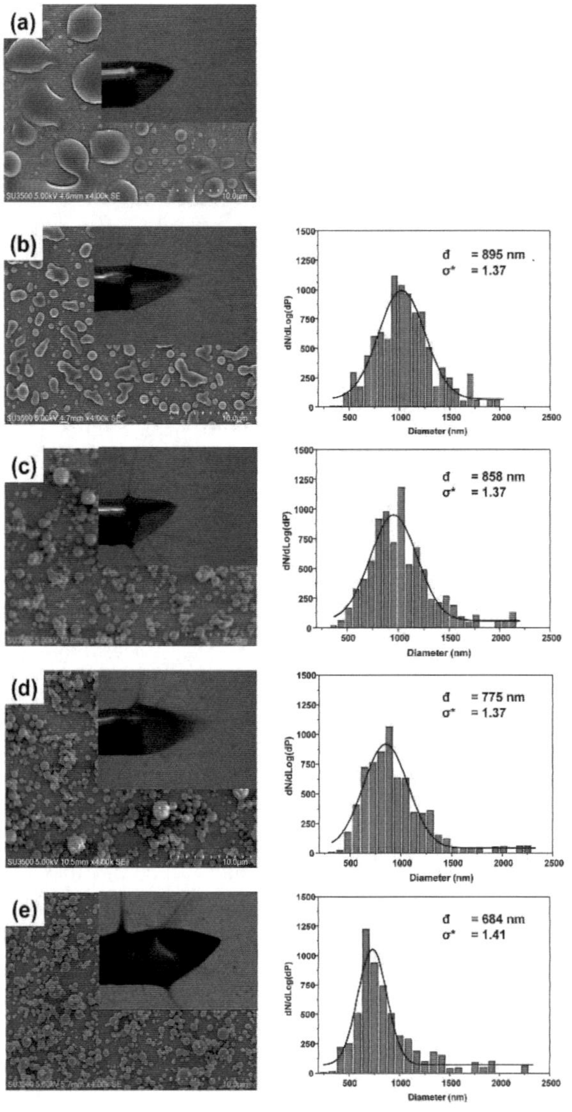

Figure 1. The spraying modes (inset), SEM images, and the histogram of the distribution of diameters of the electrospray (a) PVP and PVP-GTE with different ratios (b) 9:1, (c) 8:2, (d) 7:3, and (e) 6:4.

As seen in Figure 2, the dashed lines show the approximate size of the droplets as predicted from Equation 1 and Equation 2. The square, the circles, and the triangles are the data obtained from polymer-pigment-nanoparticles composites [17], pure polymer [16], as well as the present experiment by plotting the log of diameter against the log of Q/K, respectively.

Based on Figure 2, it was found that the experimental data, approximately, match to the theoretical model. For this research, the experimental data fitted to the theoretical model for Q/K approximately from 10^4 to 10^5, which is about ten times of magnitude. Therefore, the average diameter of the electrosprayed particles can be estimated by the prediction equations, and the diameter can thus be controlled by adjusting the flow rate and the properties of the precursor solution.

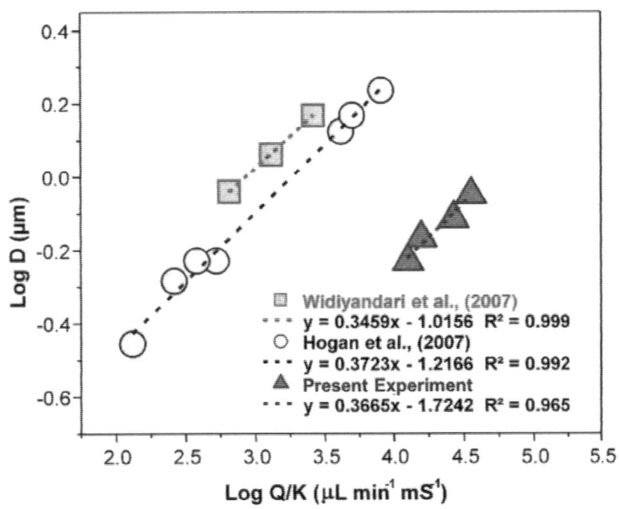

Figure 2. The log of droplet diameter from both the empirical data and theoretical calculation as a function of the solutions flowrate to electrical conductivity *(Q/K)*

3.4. Fourier transform infra-red (FTIR) spectra

The FTIR spectroscopy was performed to investigate the compatibility between GTE as the active ingredient and PVP as the polymeric matrix. The FTIR characterization of PVP, GTE, and PVP-GTE composite particle for identifying the functional groups and their peaks are shown in Figure 3. The specific peaks of PVP at 3406 cm^{-1} and 2954 cm^{-1} indicates the O-H stretching vibration and C-H asymmetric stretching vibration (21), respectively. The peak at 1652 cm^{-1} is carbonyl groups (C=O stretch vibration) [21], the peak at 1290 is CH wagging [9, 18], and the peak at 1074 is the C-N vibration [21]. The peak at 570 cm^{-1} indicates N-C=O bending [9, 18]. The specific peaks of GTE at 3193 cm^{-1} indicates the stretching vibrations of O-H groups in water, alcohol, and phenols as well as N–H stretching in amines [22]. These indicate the presence of polyphenol in GTE. The GTE peaks at 2927 cm^{-1} indicate the C-H stretch in alkanes [22]. The specific band at 1692 cm^{-1} to 1602 cm^{-1} indicates the C=C stretch in the aromatic ring and C=O stretch in polyphenols [9, 22]. The Peak at 1342 cm^{-1} is C–N stretch of amide-I in protein [22] and the peak of GTE at 1036 cm^{-1} indicates C-O stretching in amino acid [9, 22]. The functional groups of both PVP and GTE as shown by the FTIR spectra in this experiment are already similar to the previous reports. The composite particles of PVP-GTE, for all ratios, also show the peak characteristics as contributed by PVP and GTE contribution. The existing intermolecular interactions of these composite particles are confirmed by the shifting of the band wavenumber between PVP and GTE.

The PVP-GTE composite particles at ratios of 9:1, 8:2, 7:3, and 6:4 have the O-H stretching peaks at 3402 cm^{-1}, 3396 cm^{-1}, 3376 cm^{-1}, and 3372 cm^{-1}, respectively. These peaks are lower than that of PVP at 3406 cm^{-1} and higher than that of GTE at 3193 cm^{-1}. It implies that the O-H stretching peak of PVP-GTE composite particle shifted to a lower wavenumber along with the increase of GTE content. Therefore, these shifted peaks indicate that the PVP-GTE particle composites with the ratios of 9:1, 8:2, 7:3, and 6:4 do have molecular interactions between carbonyl groups of PVP and hydrogen bonds of GTE [18]. The shifted peaks of PVP-GTE composite particles with the ratios of 9:1, 8:2, 7:3, and 6:4 were 1074 cm^{-1}, 1050 cm^{-1}, 1047 cm^{-1}, and 1045 cm^{-1}, respectively, which are lower than that of PVP (1072 cm^{-1}) and higher than that of GTE peak (1036 cm^{-1}). These peaks indicate that the particles do have molecular interaction between the C-N vibration of PVP and the C-O stretching in an amino acid of GTE in PVP-GTE composite particles. These results are similar to our previous work that construct the nanostructures of PVP-GTE [9]. Overall, FTIR spectra in this experiment also confirmed that the GTE was successfully loaded into PVP as the composite submicron particles.

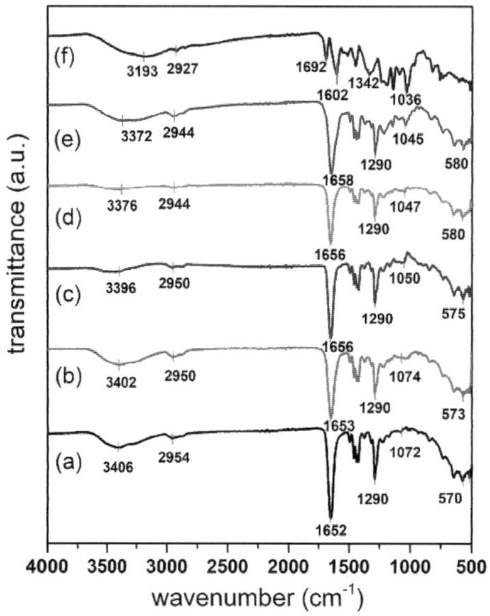

Figure 3. FTIR spectra of (a) PVP, composite particle of PVP-GTE with the ratios of (b) 9:1, (c) 8:2, (d) 7:3, (e) 6:4, and (f) GTE.

4. Conclusions

The electrospray polyvinylpyrrolidone (PVP) submicron particles loaded by green tea extract (GTE) as the bioactive substance has successfully been done. The spray mode of PVP was classified as a single jet silver bullet mode. It was then changed to unstable multi-jet by the introduction of GTE with different contents. As a result, it also affected the size and distribution of PVP-GTE composite particles. The SEM images showed that PVP-GTE composite particles have a purely smooth and spherical shape. The average diameter of PVP-GTE composite particles decreased along with the increase of the GTE content. Moreover, the average diameter of the particles was also approximately similar to the theoretical model developed from the previous reports. The FTIR spectra of PVP-GTE composite particles have confirmed the existing intermolecular interaction between PVP and GTE.

Acknowledgments

This research was financially supported by Directorate of Research and Community Engagement of Ministry of Research, Technology and Higher Education, the Republic of Indonesia under the University's Excellent Research (PUPT) Grant in the Fiscal year 2016 – 2017, Institut Teknologi Bandung, the ITB Research Grant in the fiscal year 2016. The first author (KK) acknowledges the Endowment Fund of Education (LPDP) of Republic of Indonesia for his scholarship.

References

[1] Susanti E, Ciptati, Ratnawati R, Aulanni'am, and Rudijanto 2015 A Molecular Docking of Catechins with Lxrα and Lxrβ as Potential Inhibitor Atherogenesis *Int. J. PharmTech. Res.* **8** 340 – 6

[2] Lee L-S *et al* 2013 Optimization of Ultrasonic Extraction of Phenolic Antioxidants from Green Tea Using Response Surface Methodology *Molecules* **18** 13530 – 45

[3] Kerio L C, Wachira F N, Wanyoko J K, and Rotich M K 2013 Total polyphenols catechin profiles and antioxidant activity of tea products from purple leaf colored tea cultivars *Food Chem.* **136** 1405 – 13

[4] Manea A, Vasile B S, and Meghea A 2014 Antioxidant and antimicrobial activities of green tea extract loaded into nanostructured lipid carriers *Comptes Rendus Chim.* **17** 331 – 41

[5] Naumovski N, Blades B L, and Roach P D 2015 Food Inhibits the Oral Bioavailability of the Major Green Tea Antioxidant Epigallocatechin Gallate in Humans *Antioxidants* **4** 373 – 93
[6] Jong W H and De Borm P J A Drug delivery and nanoparticles Applications and hazards *Int. J. Nanomedicine.* **3** 133 – 49
[7] Goonoo N, Bhaw-luximon A, and Jhurry D 2014 Drug Loading and Release from Electrospun Biodegradable Nanofiber *J. Biomed. Nanotechnol.* **10** 2173 – 99
[8] Zaiter A, Becker L, Karam M, and Dicko A 2016 Effect of particle size on antioxidant activity and catechin content of green tea powders *J. Food Sci. Technol.* **53** 2025 – 2032
[9] Kamaruddin, Edikresnha D, Sriyanti I, Munir M M, and Khairurrijal 2017 Synthesis of Polyvinylpyrrolidone (PVP)-Green Tea Extract Composite Nanostructures using Electrohydrodynamic Spraying Technique *IOP. Conf. Series. Materials Science and Engineering* **2017** pp 1 – 7
[10] Munir M M, Bagus A, Iskandar F, and Okuyama K 2009 Scaling law on particle-to-fiber formation during electrospinning *Polymer* **50** 4935 – 43
[11] Tapia-Hernández J A *et al* 2015 Micro- and Nanoparticles by Electrospray Advances and Applications in Foods-Review *J. Agric. Food. Chem.* **63** 4699 – 4707
[12] Zamani M, Prabhakaran M P, and Ramakrishna S 2013 Review Advances in drug delivery via electrospun and electrosprayed nanomaterials *Int. J. Nanomedicine.* **8** 2997 – 3017
[13] Sriyanti I, Edikresnha D, Munir M M, Rachmawati H, and Khairurrijal 2016 Electrospun Polyvinylpyrrolidone (PVP) Nanofiber Mats Loaded by Garcinia mangostana L Extracts *Mater. Sci. Forum.* **880** 11 – 4
[14] Sriyanti I, Edikresnha D, Rahma A, Munir M M, Rachmawati H, and Khairurrijal K 2017 Correlation between Structures and Antioxidant Activities of Polyvinylpyrrolidone/Garcinia mangostana L Extract Composite Nanofiber Mats Prepared Using Electrospinning *J. Nanomater.* **2017** p 10
[15] Nuryantini A Y, Edikresnha D, Munir M M, and Khairurrijal K 2015 Electrospun Polyvinylpyrrolidone as a Carrier for Leaves Extracts of Anredera Cordifolia (Ten) Steenis *Mater. Sci. Forum.* **827** 91 – 4
[16] Hogan C J *et al* 2007 Controlled size polymer particle production via electrohydrodynamic atomization *Colloids Surface A. Physicochem Eng. Asp.* **311** 67 – 76
[17] Widiyandari H, Hogan C J, Yun K M, Iskandar F, Biswas P, and Okuyama K 2007 Production of Narrow-Size-Distribution Polymer-Pigment-Nanoparticle Composites via Electrohydrodynamic Atomization *Macromol. Mater. Eng.* **292** 495 – 502
[18] Rahma A, Munir M M, Khairurrijal, Prasetyo A, Suendo V, and Rachmawati H 2016 Intermolecular Interactions and the Release Pattern of Electrospun Curcumin-Polyvinyl(pyrrolidone) Fiber *Biol. Pharm. Bull.* **39** 163 – 73
[19] Vongsetskul T, Chantarodsakun T, Wongsomboon P, Rangkupan R, and Tangboriboonrat P 2015 Effect of Solvent and Processing Parameters on Electrospun Polyvinylpyrrolidone Ultra-fine Fibers *Chiang Mai J. Sci.* **42** 436 – 42
[20] Fromen C A *et al* 2013 Synthesis and Characterization of Monodisperse Uniformly Shaped Respirable Aerosols *Am. Inst. Chem. Eng. J.* **0** 1–11
[21] Koczkur K M, Mourdikoudis S, Polavarapu L, and Skrabalaka S E 2015 Polyvinylpyrrolidone (PVP) in nanoparticle synthesis *R. Soc. Chem.* **44** 17883 – 905
[22] Senthilkumar S R and Sivakumar T 2014 Green tea (Camellia sinensis) mediated synthesis of zinc oxide (ZnO) nanoparticles and studies on their antimicrobial activities *Int. J. Pharm. Pharm. Sci.* **6** 461 – 5

IOP Conf. Series: Materials Science and Engineering **367** (2018) 012037 doi:10.1088/1757-899X/367/1/012037

Influence of Annealing Time Variation on Crystal Structure and Morphology of Oxide Material $Nd_{1.2}FeO_3$ by Solid-State Reaction Method

E. H. Sujiono[1, *)], A.C. M. Said[1], M. Y. Dahlan[1], R. A. Imran[1], and S. Samnur[1]

[1]Laboratory of Materials Physics, Department of Physics, *Universitas Negeri Makassar*, Makassar 90224, Indonesia.

E-mail: e.h.sujiono@unm.ac.id

Abstract. $NdFeO_3$ is one of the oxide material can be detected various gases, like S/O_2, CO, H_2S, etc. In this research, $Nd_{1.2}FeO_3$ as oxide material have been synthesized by solid-state reaction with a variation of annealing time. Characterized by XRD shows that the samples have form crystal perovskite structure with dominant phase and peak intensity correspond to *hkl* (121). FWHM value for the dominant peak was 0.22°. The crystallite of the samples was determined using Debye Scherer formula were 393.08, 393.10, and 393.10 nm, respectively. While the SEM characterized showed the morphology of the samples was homogenous with grain size estimates of 0.2μm. These results indicate the variation of annealing time 1, 2, and 3 hours did not significantly affect the crystallinity and morphology of $Nd_{1.2}FeO_3$ oxide material.

Keywords. Annealing time, crystallinity, morphology, $Nd_{1.2}FeO_3$ oxide material, and solid-state reaction.

1. Introduction

Currently in the modern chemical industry mixed metal oxide including perovskite oxide structure with general structure ABO_3 where A was a rare-earth element and B was 3D transition metal had been used remain prominent [1]. Iron was metal in the first transition series that has an orthorhombic structure with perovskite-type that was a subject of extensive investigated [2–4]. A number of research shown their utility in a wide range of applications such as a catalytic gas sensor, magnetic material, pigment material, fuel cells, gas sensor, etc. [2–6]. As a gas sensor, $NdFeO_3$ was very efficient for H_2S, CO and LPG detection [6–8].

Material oxide could be synthesized in several ways, such as ceramic method, chemical precipitation process, float zone technique, sol-gel and solid-state reaction [3,8–12]. The solid-state reaction was one of the most conventional methods, which was an easy and inexpensive method to a synthesized oxide material. The method was mixed different metal oxide alloys at high temperatures [13]. In this research, $Nd_{1.2}FeO_3$ oxide material was synthesized by solid-state reaction. Characteristic of the crystalline structure of oxide material was studied using x-ray diffraction (XRD) and the morphology using Scanning Electron Microscopy (SEM). The full width at half maximum (FWHM) is one effective method to confirm the

Content from this work may be used under the terms of the Creative Commons Attribution 3.0 licence. Any further distribution of this work must maintain attribution to the author(s) and the title of the work, journal citation and DOI.
Published under licence by IOP Publishing Ltd

crystal quality and the morphology of the material. The crystal quality has category higher due to the smaller of FWHM value and the crystal size in dimension to nanometer [14].

In this paper, presented influence of annealing time (t) as a variation of 1 h, 2 h, 3 h on crystal structure and morphology of $Nd_{1.2}FeO_3$ synthesized by solid-state reaction method. Variation of annealing time has an effective treatment for removing structural grain boundaries of ionic defects [15]. It is an indication that annealing process would increase the crystal quality of the material. The composition of $Nd_{1.2}FeO_3$ used has been reported elsewhere [16]. Therefore, this paper would be a focus on appearing results of variation annealing time on crystal quality and morphology.

2. Materials and methods

The $Nd_{1.2}FeO_3$ have been synthesized by solid stated reaction method with repeated heating treatment. The synthesized process was started by the mixed stoichiometric powder of Nd_2O_3 99.99 % (*Strem Chemicals*) and Fe_2O_3 99.99 % (*Sigma-Aldrich*) and grinded them for 3 h [16]. The samples were calcinated at 700 °C for 6 h. After that sample were grinded for 5 h and sintered at 950 °C for 6 h. This process was called the first heat treatment. Then, the sample was grinded for 3 h and calcined at 950 °C for 6 h. The last, the sample was grinded for 5 h and continued using *in situ* annealed at 450 °C for 1 h, 2 h, and 3 h, respectively. This process was called the second heat treatment.

The synthesized powder samples of $Nd_{1.2}FeO_3$ were characterized by using x-ray diffraction (Rigaku Miniflex II CuKα, λ = 0.154 mm) and Scanning Electron Microscopy (Tescam Vega-SB) for analyzed crystal structure and the morphology of the samples.

3. Results and discussion

In this research, the samples have characterized by X-ray diffraction (XRD). The intensity was observed over 2θ range 10° to 70°. The XRD pattern of the $Nd_{1.2}FeO_3$ oxide material is shown in Figure 1. The pattern has formed a crystal with the orthorhombic phase of perovskite-type (JCPDS File no 25 – 1149) with the highest peak at 2θ = 32.5° and correspond to *hkl* (121) plane. This result has similar as reported by Satyendra Singh et al., with annealing temperature at 450 °C for 2 h [17].

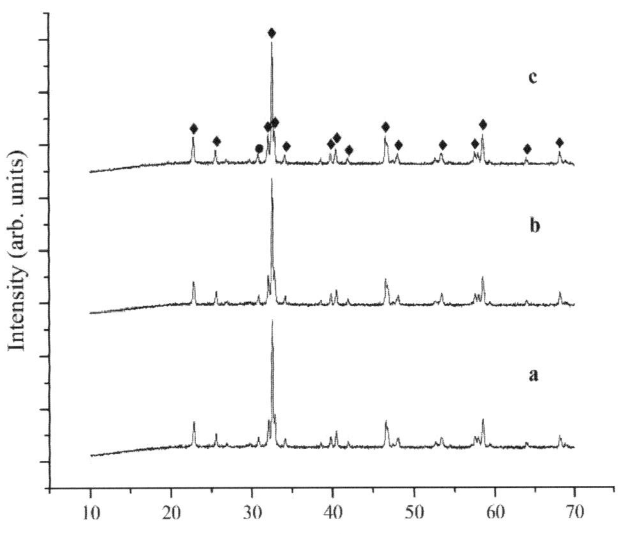

Figure 1. XRD pattern of $Nd_{1.2}FeO_3$ as variation of annealing time a = 1 h, b = 2 h, and c = 3 h, respectively (♦= $NdFeO_3$, ●= Nd_2O_3).

In this research, we also found that the highest peak intensity is corresponding to *hkl* (121). The peak was a very sensitive peak as a gas sensor, reported by Niu Xinshuet et al. The *hkl* (121) has high sensitivity, excellent selectivity, and quick response and recovery behavior to H_2S [6]. The (121) peak also suitable for detecting the CO gas as reported by Truong Giang Ho [2]. Based on this result and references indicate that $Nd_{1.2}FeO_3$ was a good material as a gas sensor.

The Rietveld analysis using Rietica software had refinement results of the lattice parameters for each sample. Lattice parameters for samples with annealed time varied for 1 h, 2 h, and 3 h has value is a = 5.58 Å, b = 7.76 Å, c = 5.45 Å, this similar as reported [10]. The average *Goodness of Fit* (GoF) for samples was 0.91%, profile (Rp) 4.63 %, Weighted Profile (Rwp) 6.09 %, and Expected (Rexp) 6.39 %, this data indicates a good agreement within the experiment result and database calculation as reported elsewhere [18].

Phase analysis using Match software obtaining two phases that contain each $Nd_{1.2}FeO_3$ oxide materials as a variation of annealing time. There is was observed in Figure 1, peak with attribute circle correspond to Nd_2O_3 and diamond to $Nd_{1.2}FeO_3$. The samples suspected were not completely reacted and did not form a single phase of $NdFeO_3$ oxide material. Increasing intensity of peak $Nd_{1.2}FeO_3$ (121) would be accompanied by the decreased intensity of peak Nd_2O_3 (011) as confirmed in Figure 2. It was mean that Nd_2O_3 as a raw material was not perfectly reacted Fe_2O_3 to form $NdFeO_3$ sintered at 950 °C and annealed time varied at 1 h, 2 h, and 3 h. The suggestion it was needed a higher temperature and longer time to heat treatment process to form perfect $Nd_{1.2}FeO_3$ phase. Figure 2 was shown sample was annealed time to 2 h has higher ratio intensity between $Nd_{1.2}FeO_3$ (121) and Nd_2O_3 (011).

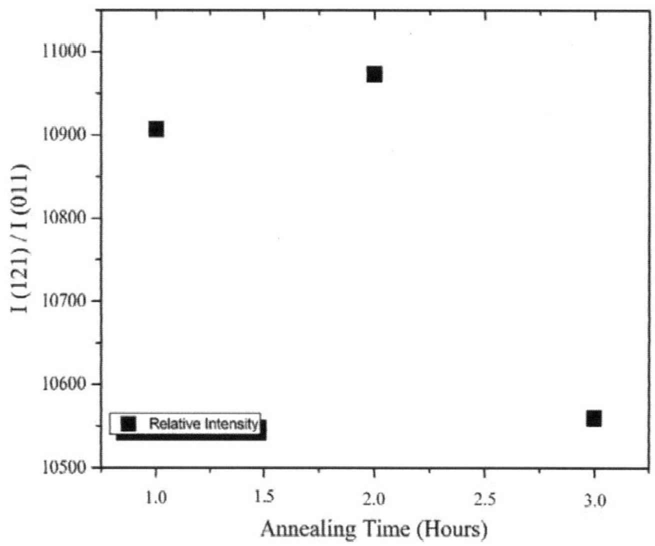

Figure 2. Curve comparison relative intensity of $Nd_{1.2}FeO_3$ (121) and Nd_2O_3 (011) for each variation of annealing time

The crystallite size of oxide material was calculated by Debye Scherer's formula, which is given as:

$$D = \frac{0.94\lambda}{\beta_{hkl}\cos\theta} \tag{1}$$

where D is the crystallite size, β_{hkl} is the full width at half maximum (FWHM) intensity value, θ is the diffraction angle, and λ represents the x-ray wavelength. The crystallite size of $Nd_{1.2}FeO_3$ was shown in

IOP Conf. Series: Materials Science and Engineering **367** (2018) 012037 doi:10.1088/1757-899X/367/1/012037

Table 1. Based on the result as was described in Table 1, phase and crystallite size have similar than previous research [16].

Table 1. Result of XRD analysis for $Nd_{1.2}FeO_3$oxide material

Annealing time (h)	2θ (°)	FWHM (°)	Crystallite Size (nm)
1	32.56	0.22	$\vert 393.08 \pm 0.02 \vert$
2	32.58	0.22	$\vert 393.10 \pm 0.02 \vert$
3	32.58	0.22	$\vert 393.10 \pm 0.02 \vert$

The surface morphology, structure and particle size of the samples were investigated by SEM. Figure 3 was shown the SEM micrograph of the samples at 5000 magnification. SEM micrograph showed that morphology of samples was homogeneous grain orientation and grain size with an estimated size was 0.2 μm. It can be seen, a variation of annealing time as a parameter process has obtained nanostructure with uniform in both morphology and particle size. This result confirms that annealing time at 2 h has more homogeneous morphology [17].

Figure 3. SEM micrograph of $Nd_{1.2}FeO_3$ as variation of annealing time t of (a) 1 h, (b) 2 h, and (c) 3 h, respectively.

EDS analysis results are given in Table 2. The results showed the existence of iron and neodymium as a dominant element. In addition, it can be seen the percentage value the presence of another minor oxide in each sample due to the sample holder.

Table 2. EDS analysis of $Nd_{1.2}FeO_3$ oxide material as variation of annealing time

Compound Norm.	Comp. C [wt%]		
	t = 1 h	t = 2 h	t = 3 h
Nd	71.44	67.95	70.77
Fe	26.86	26.36	26.90

Based on the XRD and SEM analysis has been describing above, the variation of annealing time did not significantly affect on crystalline quality and morphology of $Nd_{1.2}FeO_3$ oxide material. Annealing was not a process to form a new phase, but to maintain the oxygen ion of an oxide material. Therefore annealing treatment not affect significantly the crystal structure and sample morphology, but will affect the electrical properties of the material. This phenomenon was similar to the case of $YBa_2Cu_3O_{7-\delta}$ oxide material which is annealing did not affect the crystal structure and morphology of material. Annealing affects the conductivity of the sample as corresponding to oxygen composition [14, 19, 20]. This also occurred in case of TiO_2 material that annealing variation affects the resistivity of the material [21].

4. Conclusions

$Nd_{1.2}FeO_3$ oxide material has been synthesized by solid-state reaction method with a variation of annealing time. XRD result shown the oxide has a crystal structure with the orthorhombic phase of perovskite-type with the highest peak of *hkl* (121) at 2θ is 32.5°. There were $Nd_{1.2}FeO_3$ and Nd_2O_3 phases contain in each sample. The calculation using Debye Scherer's formula given crystallite size is 393 nm with FWHM 0.22°. SEM micrograph of the samples shown morphology was uniform grain size around 0.2 µm. Based on XRD and SEM results, a sample with annealing time for 2 h has ratio intensity of *hkl* (121) more dominant as an indication of the best sample in this research.

Acknowledgements

This research was funded by Directorate research and Community Services, Ministry of Research, Technology and Higher Education, Republic of Indonesia, under research scheme of *Hibah Kompetensi* fiscal year 2017. The authors acknowledge this financial support.

References

[1] Pena M A and Fierro J L G 2001 Chemical Structures and Performance of Perovskite Oxides *Chemical Reviews* **101** 1981 – 2018

[2] Ho T G *et al* 2011 Nanosized perovskite oxide $NdFeO_3$ as material for a carbon-monoxide catalytic gas sensor *Advanced in Natural Science: Nanoscience and Nanotechnology* **2** 15012 –15015

[3] Taneja S *et al* 1993 Nuclear Instruments and Methods in Physics Research Section B: Beam Interactions with Materials and Atoms *Nuclear Instruments and Methods in Physics Research* **76** 127 – 129

[4] Opuchovic O *et al* 2015 Sol-gel synthesis, characterization and application of selected sub-microsized lanthanide (Ce, Pr, Nd, Tb) ferrites *Dyes and Pigment.* **118** 176 – 182

[5] Tongyun C *et al* 2012 $NdFeO_3$ as anode material for S/O_2 solid oxide fuel cells *Journal of Rare Earths.* **30** 1138 – 1141

[6] Xinshu N *et al* 2003 *Journal of Rare Earths.* **21** 550 – 553

[7] Ru Z *et al* 2010 Electrical and CO-sensing properties of $NdFe_{1-x}Co_xO_3$perovskite system *Journal of Rare Earths.* **28** 591 – 595

[8] Motlagh M *et al* 2013 Chemical Synthesis and Characterization of Perovskite NdfeO3 Nanocrystals via a Co-Precipitation Method *International Journal of Nanoscience and Nanotechnology.* **9** 7 – 14

[9] Song G *et al* 2015 Spin reorientation transition process in single crystal $NdFeO_3$ *Solid State Communication.* **211** 47 – 51
[10] Chanda S, Saha S, and Sinha T 2013 Raman spectroscopy and dielectric properties of nanoceramic $NdFeO_3$ *Material Research Bulletin.* **48** 1688 – 1693
[11] Pikula T *et al* 2014 Structure and hyperfine interactions in $Bi_{1-x}Nd_xFeO_3$ solid solutions prepared by solid-state sintering *Journal of Alloys and Compounds.* **606** 1 – 6
[12] Belessi V C *et al* 1999 Structure and catalytic activity of $La_{1-x}FeO_3$ system (x=0.00, 0.05, 0.10, 0.15, 0.20, 0.25, 0.35) for the NO+CO reaction *Applied Catalysis A : General.* **177** 53 – 68
[13] Eko H S 2008 *Paduan oksida logam Nd1(Fe)XBa$_2$-XCu$_3$OY dan metode pembuatannya* I. D. Patent No. P00200800471 (in Indonesia)
[14] Eko H S *et al* 2001 Crystal Structure and Morphology Analysis of $Nd_{1+x}Ba_{2-x}Cu_3O_7$ Oxide Alloy Surface Developed by Solid State Reaction Method *Physica Status Solidi (A) Applied Research.* **187** 471 – 479
[15] Bhagavannarayana G *et al* 2005 A study of the effect of annealing on Fe-doped $LiNbO_3$ by HRXRD, XRT and FT–IR *Journal of Applied Crystallography.* **38** 768 – 771
[16] Zharvan V *et al* 2017 The Effect of Molar Ratio on Crystal Structure and Morphology of $Nd_{1+X}FeO_3$ (X=0.1, 0.2, and 0.3) Oxide Alloy Material Synthesized by Solid State Reaction Method *IOP Conference: Materials Science and Engineering.* **202** 012072
[17] Singh S *et al* 2013 Fabrication of nanobeads structured perovskite type neodymium iron oxide film: Its structural, optical, electrical and LPG sensing investigations *Sensors and Actuators B: Chemical.* **177** 730 – 739
[18] Tighineanu A *et al* 2010 Conductivity of TiO_2 nanotubes: Influence of annealing time and temperature *Chemical Physics Letters.* **494** 260 – 263
[19] Eko H S 2007 Optimization of Surface Smoothness of Nd 1.12 Ba 1.88 $Cu_3O_{7-\delta}$ Thin Films *Indonesian Journal of Physics.* **18** 107 – 110
[20] Eko H S, Arifin P, and Barmawi M 2002 $YBa_2Cu_3O_{7-\delta}$ thin films deposited by a vertical MOCVD reactor *Materials Chemistry and Physics.***73** 47 – 50
[21] Shanker J, Suresh M B, and Babu D S 2015 Synthesis, Characterization and Electrical Properties of $NdXO_3$(X=Cr, Fe) Nanoparticles *Materials Today: Proceedings.* **3** 2091 – 2100

Porous Fe_2O_3 Microspheres as Anode for Lithium-Ion Batteries

L. Noerochim[1], M. A. T. Indra[1], H. Purwaningsih[1], and A. Subhan[2]

[1]Department of Materials and Metallurgical Engineering, Sepuluh Nopember Institute of Technology, Surabaya 60111, Indonesia
[2]Research Center of Physics, Indonesian Institute of Science – Physics (LIPI – Fisika), Puspiptek, Serpong 15314, Indonesia

E-mail: lukman@mat-eng.its.ac.id

Abstract. In this work, Fe_2O_3 was successfully synthesized by the hydrothermal process at low temperature. $FeCl_3.6H_2O$ as precursor and variation of lysine as hydrolyzing agent were used to preparing Fe_2O_3. SEM images show that the morphology of Fe_2O_3 is porous microsphere with sizes in the range of (1 to 5) μm in diameter. The as-prepared Fe_2O_3 with the 2 M of lysine exhibits excellent cycling performance when used as the anode for lithium ion batteries, obtaining reversible discharge capacity of 172.33 $mA{\cdot}h{\cdot}g^{-1}$ at 0.5 C after 50 cycles. It is attributed to the unique structure of porous microspheres providing a large surface area which maintains good electronic contact between particles during charge-discharge process. This result demonstrates that Fe_2O_3 porous microsphere has a high potential as anode material for application of lithium-ion battery.

Keywords. α-Fe_2O_3, hydrothermal, lithium-ion battery, lysine, and microsphere.

1. Introduction

The lithium-ion battery is the most popular energy storage and energy source today. Lithium-ion batteries are used for various applications such as energy sources in hybrid cars as well as some clean energy storage applications [1]. Research on lithium-ion batteries is still continuously developed. In 2000, the concept of a conversion reaction in the electrochemical-reversible lithium reaction began to be developed on some oxide transition metals, such as FeO, CoO, and NiO [2]. The oxide-transition metal is a good material for the lithium-ion battery anodes because of the metal's ability to deliver a much larger specific capacity than the theoretical capacity of graphite, i.e. 372 $mA{\cdot}h{\cdot}g^{-1}$ [3]. From several existing oxide transition metals, Fe_2O_3 (hematite) is selected for a promising anode material of lithium-ion batteries. This material is selected on the basis of its abundant presence in nature, environmentally friendly, and theoretically specific capacity of 1007 $mA{\cdot}h{\cdot}g^{-1}$ [4]. In the application of Fe_2O_3 as an anode on lithium-ion batteries still have a weakness, such as easy loss of owned capacity and low battery cycling rate [5]. Several previous studies have provided a solution to the problem of synthesizing Fe_2O_3 in nanostructures with varying morphology, such as nanotubes, nanotubes, nanorods and nano coils [6]. The nanostructure will facilitate the transfer of electrons and Li^+ ions by reducing the diffusion paths. In addition, the nanostructures are also capable of improving the kinetic intercalation process by providing larger area contacts [1].

Content from this work may be used under the terms of the Creative Commons Attribution 3.0 licence. Any further distribution of this work must maintain attribution to the author(s) and the title of the work, journal citation and DOI.
Published under licence by IOP Publishing Ltd

From the results of previous studies, the Fe_2O_3 material nanostructure shows a significant increase in electrochemical performance. The problem is low volumetric energy density and low conductivity. One effective strategy to overcome the problem is to prepare the material on a nanoscale. One of the methods to synthesize nanostructured materials is by hydrothermal method. The hydrothermal method has high energy efficient and easy to control the size and morphology of the materials [4].

In this work, the Fe_2O_3 material is prepared and synthesized by hydrothermal method with the addition of lysine. Lysine is used as a hydrolysis agent in Fe_2O_3 synthesis [6]. Therefore, the effect of lysine on the morphology and electrochemical performance of Fe_2O_3 will be studied and written systematically to find the optimum concentration of lysine for application Fe_2O_3 as anode materials for a lithium-ion battery.

2. Materials and methods

The Fe_2O_3 microsphere is synthesized by hydrothermal method with $FeCl_3.6H_2O$ as precursor and variations of lysine concentration. Lysine variation of 0.5 M, 1 M and 2 M are mixed with 1 M $FeCl_3.6H_2O$ in 30 mL of di-water and stirred with magnetic stirring for 30 min. The solution then is transferred to 120 mL Teflon-autoclave and heated in a muffle furnace with a temperature of 180 °C for 12 h. The precipitate is separated by centrifuge for approximately 45 min. Then it is obtained a thin deposit at the bottom of the centrifuge tube and washed with di-water for several times. The remaining sediment is dried in the oven at temperature 90 °C for 24 h. To prepare the electrode, the prepared Fe_2O_3 microsphere is mixed with carbon black, and polyvinylidene fluoride (PVDF) according to mass ratio of 80:15:5. After the mixing process is complete, then N-Methyl-2-Pyrolidone is added to obtain a paste and coated with Cu foil with area 1 cm^2 using doctor blade method. The electrodes are heated into an oven at a temperature of 100 °C for 12 h. Fe_2O_3 microsphere was analyzed with XRD (PANalytical, Cu Kα radiation) to identify the phase and crystal structure. Morphology was carried out using SEM (FEI Inspect 250). Electrochemical performance was tested with CR2032 coin cells between voltage ranges (0.1 to 3.0) V. The coin cells were fabricated with Fe_2O_3 microsphere as working electrode, metallic lithium as counter electrode, 1 M $LiPF_6$ in mixed diethyl carbonate/ethylene carbonate (1:1 in volume) solution as the electrolyte, and Celgard polypropylene as the separator inside glove box system (Vigor) under Ar flow with adjustable rate. Cyclic voltammetry was measured by an electrochemistry workstation (WonATech WBCS 2000) with sweep scan rate 0.1 $mV \cdot s^{-1}$ and potential range between (0.1 to 3.0) V. The coin cells were tested galvanostatic charged – discharged on a recycler battery system (WonATeCH WBCS 2000) with various C – rates from 0.3 C to 3 C with 5 cycles for each various rates and then back to rate 0.3 C. The impedance of coin cells was tested with electrochemical impedance measurement (AutoLab) using (0.1 to 20000) Hz and potential 3 V.

3. Results and discussion

Figure 1 shows XRD patterns of Fe_2O_3 microsphere at various concentration of lysine. It shows that the all of samples is successfully indexed as hematite (JCPDS No. 33-0664) with no impurity peaks detected.

It is clearly observed that the intensity peak of Fe_2O_3 2 M lysine has the highest intensity compare to other samples. It could be at the highest concentration of lysine enhancing the crystallinity of Fe_2O_3. The higher crystallinity and smaller particle size are due to the role of lysine as an inhibitor in the nucleation process by decreasing the rate of crystallization. The addition of lysine concentration plays a significant role in decreasing the size of the crystal by surface adsorption mechanism [7]. The smaller crystal size will decrease the diffusion path, increase the kinetic intercalation and enlarge the contact area of the lithium ion which will improve the electrochemical performance of the material [6].

Figure 2 shows that the all samples of Fe_2O_3 have the morphology microsphere with the size of (0.15 to 1.5) μm. In the three samples show that the increased concentration of lysine reduces the size of the microsphere. The mechanism of microsphere formation starts with the occurrence of nucleation or nucleation; then there is aggregation where the interconnected particle formed to bind each other

IOP Conf. Series: Materials Science and Engineering **367** (2018) 012038 doi:10.1088/1757-899X/367/1/012038

towards the center [6]. Fe_2O_3 material with 2 M of lysine has microsphere size smaller than the other two samples. This is due to the addition of lysine concentration triggering less aggregation during hydrothermal process.

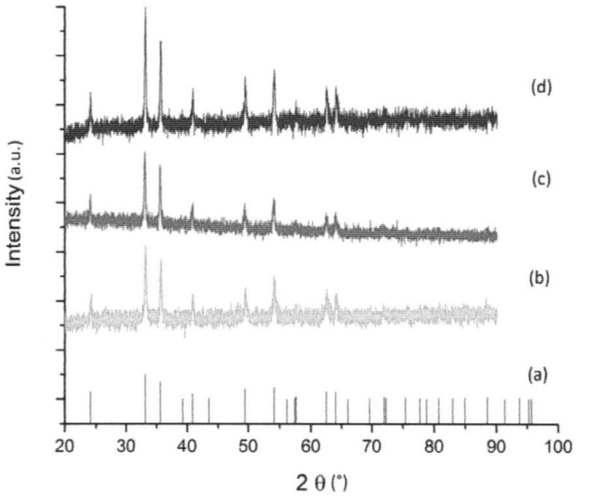

Figure 1. XRD patterns of Fe_2O_3 microsphere with various concentration of lysine (a) JCPDS 33-0664 of hematite (b) 0.5 M (c) 1 M (d) 2 M of lysine

Figure 2. SEM images of Fe_2O_3 with variation (a) 0.5 M (b) 1 M (c) 2 M of lysine

Figure 3 shows the cyclic voltammetry curve of three first cycles of all Fe_2O_3 samples with two reduction peaks and single oxidation peak. The reduction peak is occurred below 1.0 V indicating the reaction process with electrolyte and carbon contained in the anode sample. In this peak, a solid electrolyte interface (SEI) is formed between the surface of the Fe_2O_3 anode and the $LiPF_6$ electrolyte [2]. The absence of the reduction peak below 1.0 V in the second cycle and the third cycle, it could be the SEI film formed is too thick. During the charging process in the half-cell, oxidation reactions occur in Fe_2O_3 that satisfy the reaction Equation (1).

$$3Li_2O + 2Fe \leftrightarrow Fe_2O_3 + 6Li^+ + 6^{e-} \quad (1)$$

The discharge occurs the reduction reaction with Equation (2).

$$Fe_2O_3 + 6Li^+ + 6e^- \leftrightarrow 3Li_2O + 2Fe \quad (2)$$

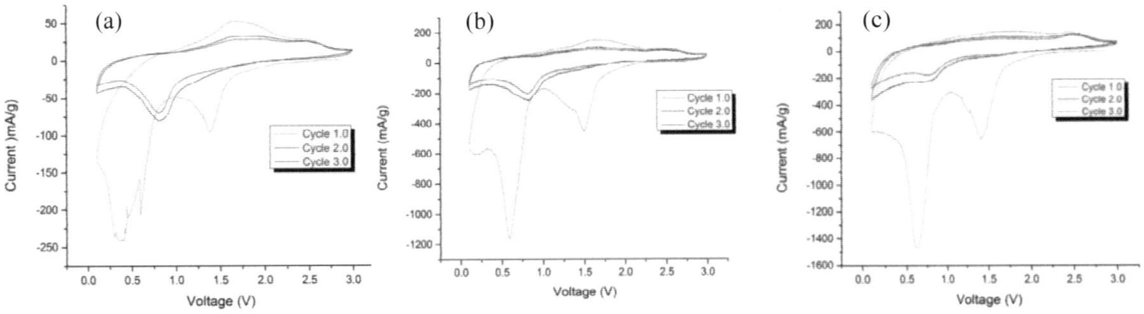

Figure 3. Cyclic voltammetry curves of Fe_2O_3 with variation (a) 0.5 M (b) 1 M (c) 2 M lysine

In the sample with a concentration of a 2-mole lysine has the highest intensity of the reduction peak indicating the amount of Li^+ which is capable of being inserted. Figure 3 (c) also shows that the distance polarization of reduction and oxidation peak is smaller than other samples. It is demonstrated that Fe_2O_3 with a 2-mole lysine has faster intercalation and de-intercalation Li^+ ions. It appears that the greater the addition of lysine, the greater the intensity current. The greater intensity will affect the transfer rate of electrons in the Fe_2O_3 material. Thus, from the CV test, it is found that Fe_2O_3 of a 2-mole lysine has better electron transfer capability than other variations.

It could be the smallest size of particle that it is clearly observed in SEM images of Fe_2O_3 of a 2-mole lysine. The Fe_2O_3 material of a 2-mole lysine has the smallest range of particle sizes, which increases the kinetic intercalation of lithium ions to give a better ion transfer efficiency compared to other samples.

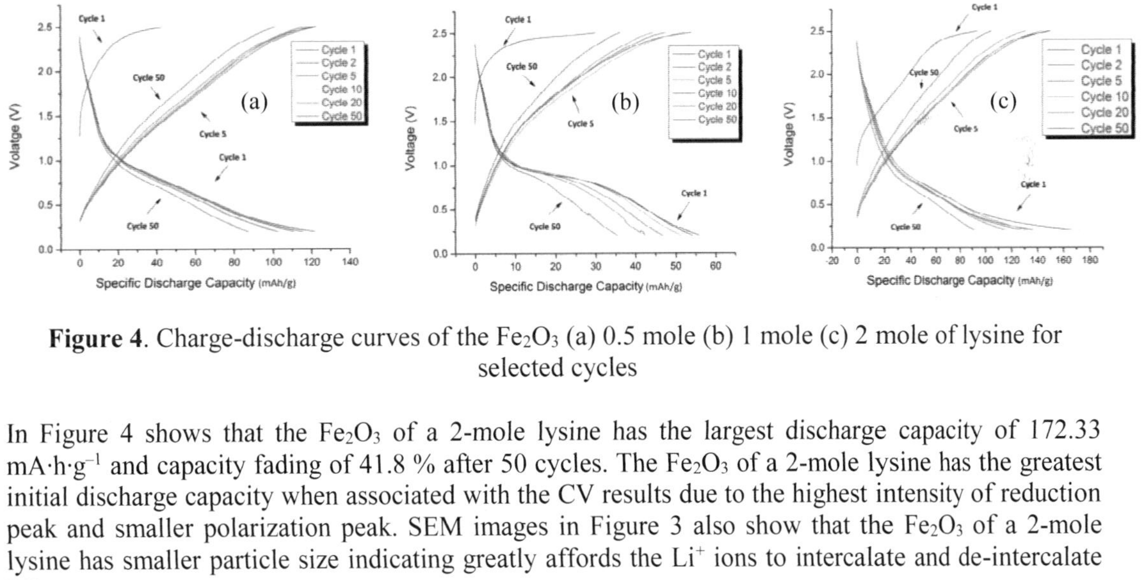

Figure 4. Charge-discharge curves of the Fe_2O_3 (a) 0.5 mole (b) 1 mole (c) 2 mole of lysine for selected cycles

In Figure 4 shows that the Fe_2O_3 of a 2-mole lysine has the largest discharge capacity of 172.33 $mA{\cdot}h{\cdot}g^{-1}$ and capacity fading of 41.8 % after 50 cycles. The Fe_2O_3 of a 2-mole lysine has the greatest initial discharge capacity when associated with the CV results due to the highest intensity of reduction peak and smaller polarization peak. SEM images in Figure 3 also show that the Fe_2O_3 of a 2-mole lysine has smaller particle size indicating greatly affords the Li^+ ions to intercalate and de-intercalate [6].

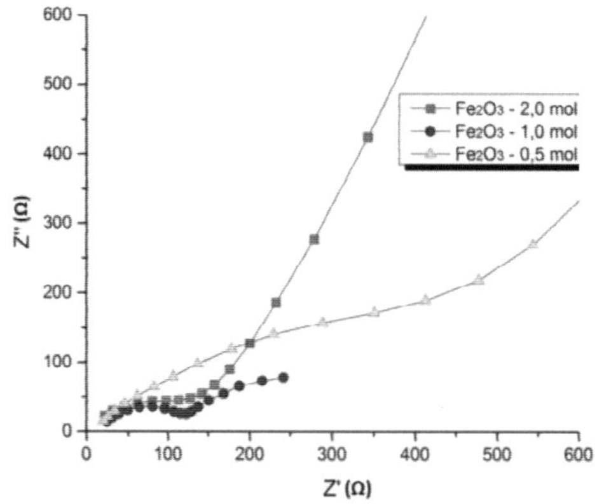

Figure 5. Nyquist plots of Fe_2O_3 with variation 0.5 M, 1 M and 2 M lysine

Figure 5 shows the Nyquist plots of all samples. The Rs value determines the electron transfer capability of the electrolyte of a sample. Here, it is seen that Fe_2O_3 of 2-mole lysine has the lower Rs than other samples indicating that the electrolyte of a 2-mole lysine sample has the highest electron transfer capability. The Act value indicates the impedance of the sample where the smaller impedance will enhance the insertion and extraction of Li^+ ion. It can be seen that the smallest Rct value is owned by a sample with a variation of a 2-mole lysine facilitating fast ion exchange in this sample. It is attributing to the higher crystallinity, and smaller size of Fe_2O_3 of a 2-mole lysine compare to other samples.

4. Conclusions

In this work shows that the Fe_2O_3 of a 2-mole lysine has the best electrochemical performance. Where the sample produces a sharp diffraction peak with the highest intensity indicates more crystalline phase. The Fe_2O_3 of a 2-mole lysine shows the morphological form of the microsphere with the smallest diameter. The Fe_2O_3 of a 2-mole lysine has the highest specific discharge capacity of 172.33 $mA \cdot h \cdot g^{-1}$ and capacity fading of 41.8 % indicating that the anode material can withstand a redox reaction for 50 cycles. The charge transfer resistance of the Fe_2O_3 2-mole lysine is lower than other samples. Therefore, the Fe_2O_3 of 2-mole lysine has great potential as anode material for a lithium-ion battery.

References

[1] Arico, A. S., Bruce, P., Scrosati, B., Tarascon, J. M., & Van Schalkwijk, W. 2005. Nanostructured materials for advanced energy conversion and storage devices. *Nature materials*, 4(5), 366.

[2] Mo, Z., Gou, H., He, J., Yang, P., Feng, C., & Guo, R. 2012. Controllable synthesis of functional nanocomposites: Covalently functionalize graphene sheets with biocompatible L-lysine. *Applied Surface Science*, 258(22), 8623-8628.

[3] Penki, T. R., Shivakumara, S., Minakshi, M., & Munichandraiah, N. (2015). Porous flower-like α-Fe2O3 nanostructure: A high performance anode material for Lithium-ion batteries. *Electrochimica Acta*, 167, 330-339.

[4] Sun, Y., Zhang, J., Huang, T., Liu, Z., & Yu, A. (2013). Fe_2O_3/CNTs composites as anode materials for lithium-ion batteries. *Int. J. Electrochem. Sci*, *8*(2), 2919-2931.

[5] Zhang, J., Sun, Y., Yao, Y., Huang, T., & Yu, A. (2013). Lysine-assisted hydrothermal synthesis of hierarchically porous Fe_2O_3 microspheres as anode materials for lithium-ion batteries. *Journal of Power Sources*, *222*, 59-65.

[6] Song, M. K., Park, S., Alamgir, F. M., Cho, J., & Liu, M. (2011). Nanostructured electrodes for

lithium-ion and lithium-air batteries: the latest developments, challenges, and perspectives. *Materials Science and Engineering: R: Reports*, *72*(11), 203-252.
[7] Nazar L F *et al* 2011 *Inorganic Materials* **3** 191 – 200

IOP Conf. Series: Materials Science and Engineering **367** (2018) 012039 doi:10.1088/1757-899X/367/1/012039

Analysis of Porosity Defects in Aluminum as Part Handle Motor Vehicle Lever Processed by High-pressure Die Casting

L. Anggraini[1] and Sugeng[2]

[1]Faculty of Engineering, President University, Jalan Ki Hajar Dewantara, Jababeka Education Park, Cikarang Baru, Bekasi 17550, Indonesia.
[2]Undergraduate Study Program of Mechanical Engineering, President University, Jalan Ki Hajar Dewantara, Jababeka Education Park, Cikarang Baru, Bekasi 17550, Indonesia.

E-mail: lydia.anggra@president.ac.id

Abstract. This research discusses the testing and analysis of cracking Aluminum (Al) material. Al as a handle lever was used for a braking device on a motor vehicle. Cracking of handle lever due to the part content of porosity from hydrogen gas. The existence of the H_2 can be caused by the casting process and dies design that is less perfect, especially at the gate or brisket and overflow. This research is to optimize the process of making Al part handle lever, and the construction dies by following the standard. The results of these improvements were re-evaluated through the chemical and mechanical testing properties stages, such as density test and tensile test on the workpiece as part handle lever. The loads on the tensile test are 25 kg and 35 kg, and the tensile test result has met the standard set by the motor vehicle company. The optimization result has the porosity defect can be reduced by 99 %. Therefore the best part handle lever can be produced.

Keywords. Aluminum, manufacturing, microstructure, part handle lever, and properties.

1. Introduction

Aluminum (Al) is a light metal that has properties against corrosion resistance and good electrical conductivity [1]. Future Al usage is expected to be wide open as the main material and supporting material, with the availability of the abundant Al seeds on earth. Al can be used for household appliances, aircraft materials, automotive, ships, constructions and others [2–5]. The aluminum products can be produced through the casting and forming process.

Al alloys have been used in many applications and have been widely used in industry because they have superior properties [6–12]. These properties make aluminum alloys into metals that are very suitable and economical for various applications and have made aluminum alloy as the most widely used metal after steel [13]. Al is commonly used for the manufacturing part produced in the motor vehicle companies [14–15]. Handle lever is a part that serves as braking or reduction of production rate in a motor vehicle.

Al is a common element found in the earth's crust and is the third most abundant element after oxygen (O) and silicon (Si) [16]. The excess of Al compared to other metals is that Al has a low specific gravity which makes this material light and has good corrosion resistance due to the 20

Content from this work may be used under the terms of the Creative Commons Attribution 3.0 licence. Any further distribution of this work must maintain attribution to the author(s) and the title of the work, journal citation and DOI.
Published under licence by IOP Publishing Ltd

IOP Conf. Series: Materials Science and Engineering **367** (2018) 012039 doi:10.1088/1757-899X/367/1/012039

passivation phenomenon [1, 17–20]. In order to improve the mechanical properties, other elements were added into Al. On the other hand, Al also has some shortcomings, lack of Al among others, such as easy to mix with gases; hydrogen in liquid conditions that can cause porosity [21–23].

In any manufacturing or production process, various problems that can affect the quality of production are generally found. Those are because the production process is not perfect and the design needs to be an improvement. The manufacturing process of handle lever in an automotive company, especially handles lever part was found porous during the initial trial of production. Therefore, to overcome that problem, our study is to examine the cause of the occurrence of porosity made from Al as part handle lever produced by high pressure die casting.

2. Materials and methods

The purpose of simulation performed on the Al handle lever is to estimate the use of parts in the application, as an example: the part is used for braking at the speed of the motor vehicle. In the work system to clarify the problem on the Al part handle lever, can be seen in Figure 1.

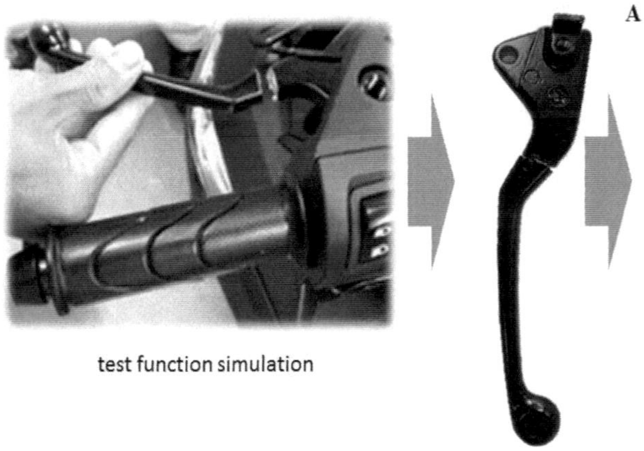

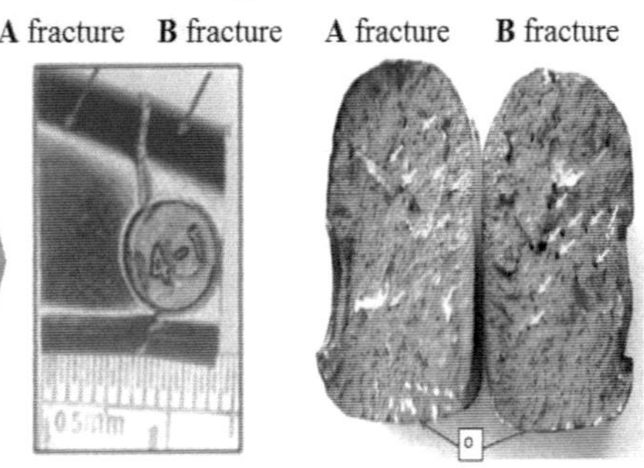

Figure 1. The braking system on a motor vehicle.

Figure 2. Optical microscope image of Al part handle lever.

Analysis performed on the cracked or broken part is done based on five aspects: user, working method, materials, machine, and environment. Before the parts are manufactured, testing should be made to ensure that the parts fit into the quality standard. The first procedure of part handle lever from the casting process is done by simulation test in function. The second step is to determine the mass of the type of time on the object by density testing. The density examination resulted from the outcome of the weighing of the specimen, it is theoretically performed, namely *Archimedes* test. Discovery of crack or broken cases found in the production of Al part handle lever, part analysis problem was carried out by Scanning Electron Microscope with Energy Dispersive X-ray Spectroscopy (SEM-EDS) microstructure image analysis.

3. Results and discussion

In advance of the parts are being manufactured, the first test has been made to ensure that the parts fit into the quality standard. In this simulation test, the result of Al part handle lever is NG (No Good), that part has a crack (broken) obtained from the casting process. The density result which was performed by Archimedes test has 92.105 % density or 7.894 % porosity.

By microstructure analysis, the crack propagation was obtained from the O point. However, the initial crack was not found. Figure 2 was obtained by an optical microscope (OM). The SEM images clearly show the fracture surface of Al part handle lever in different magnifications. Figure 3 shows the crack obtained by a casting process and the morphology of H_2 inserted to the Al part. In the observations with EDS, the impurities with element contents of carbon and oxygen were found more clearly. The SEM-EDS results show that the defects were not in the direction of fault propagation. In addition, the cause means of the fracture is not due to the impurities as shown in Figure 4.

The summary of the above analysis is: the 'cause' of the crack case in the Al part handle lever is the porosity of the casting process, gas porosity or shrinkage porosity.

From the material composition analysis results obtained by the spectrometer observation, the composition contained in the part handle lever corresponds to the composition of the material specification. Figure 5 shows the test location of material and Table 1 shows the composition results of materials by the spectrometer.

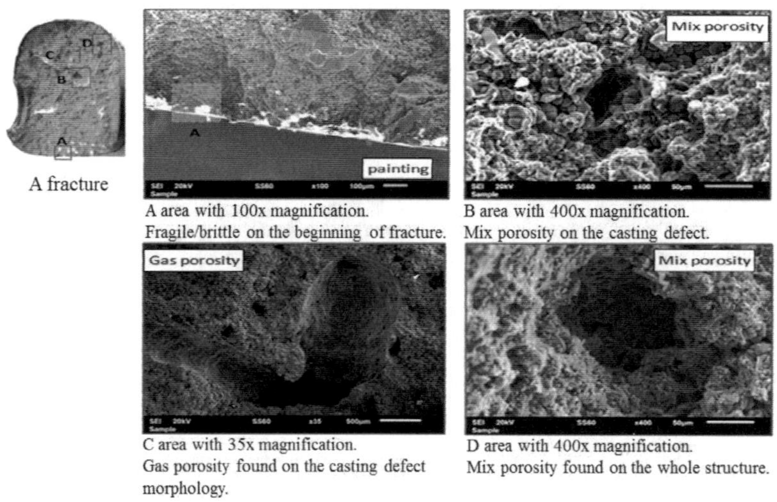

Figure 3. SEM fracture images of Al part handle lever with different area and various magnification.

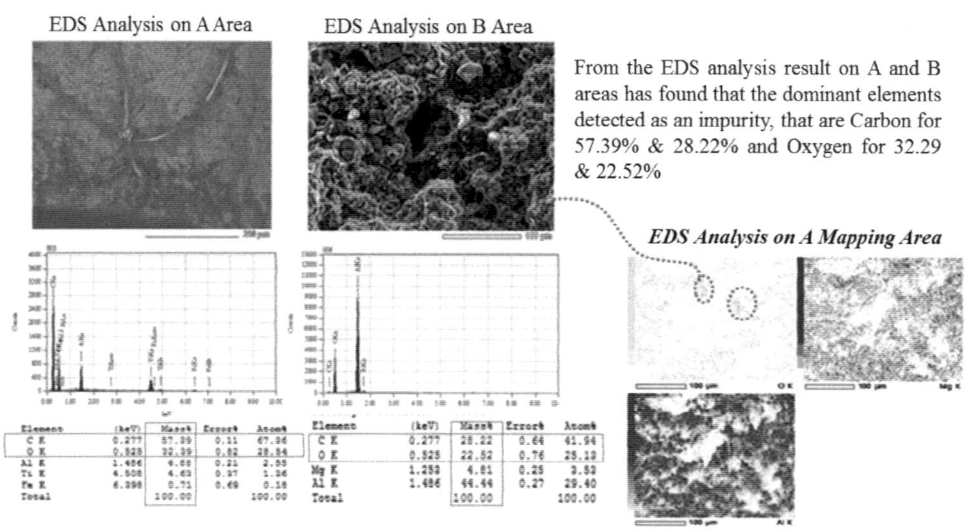

Figure 4. EDS analysis results of Al part handle lever and fractography with impurity.

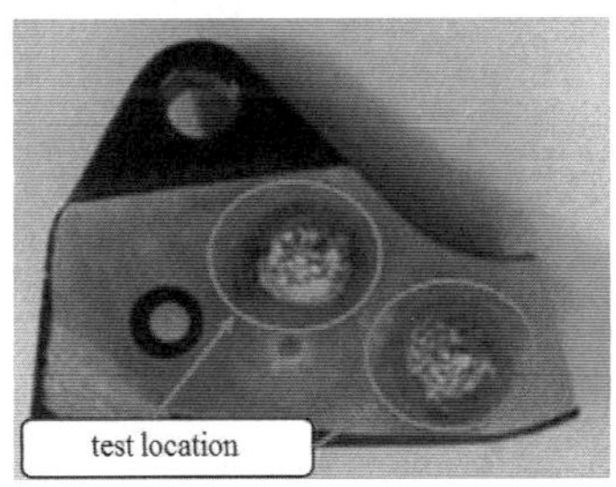

Figure 5. Material test location.

Table 1. The composition result of materials by the spectrometer

Items	Composition of Elements (%)							Result
	Si	Fe	Cu	Mn	Mg	Zn	Ni	
Standard HD4 HES C-101-03	0.50 – 1.10	0.50 – 0.80	0.10 Max	0.50 – 0.80	4.00 – 5.50	0.10 – Max	0.10 – Max	OK
Spark 1	0.89	0.72	0.05	0.60	4.63	0.03	0.01	
Spark 2	0.87	0.70	0.04	0.59	4.61	0.03	0.01	

The following tensile test simulation results on the Al part handle lever based on updated parameters. The illustrations and results of the tensile test can be seen as following Figure 6 to10, respectively.

Based on the standard that has been set by the company on the strength of the part handle lever, it can be concluded that the production of Handle Lever parts is fine and safe for the use of mass production. Compared with the production of other parts with the similar material Al alloy, and the similar process die-casting shows increasing mechanical properties, which has been proofed by Avalle *et al.* described the influence of casting defects on static and fatigue strength is investigated for a high pressure die cast aluminum alloy. In this case, defects count for their size and location, while quality control often takes no account of component working conditions [24]. Besides, other researchers Nakata et al. have proven improvement in the mechanical properties was accomplished due to the microstructural modification of an aluminum die casting alloy by multi-pass friction stir processing (MP-FSP), which is a solid-state microstructural modification technique using a frictional heat and stirring action. The hardness of the MP-FSP sample is about 20 Hv higher than that of the base metal.

The tensile strengths of the MP-FSPed specimens were significantly increased to about 1.7 times if compare of the base metal [25]. Thus, the application of the high pressure die-casting is a very effective method for the mechanical improvement of aluminum alloys.

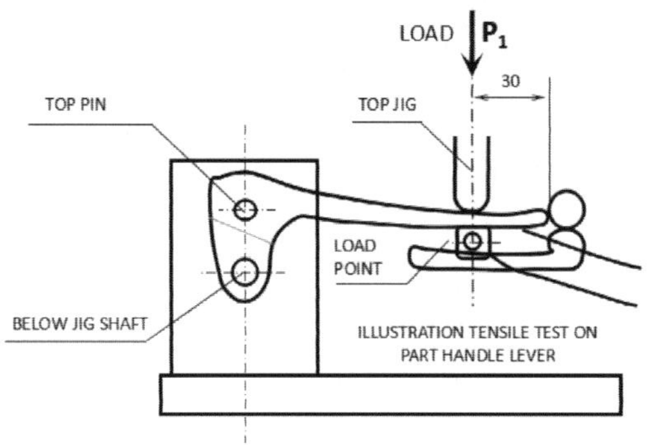

Figure 6. Illustration of tensile test workpiece

Compressive extention at preset point (load 20 kgf) mm	*Compressive extention at preset point (load 35 kgf) mm*	*Maximum load (kgf)*
2,80856	3,82955	122,44880

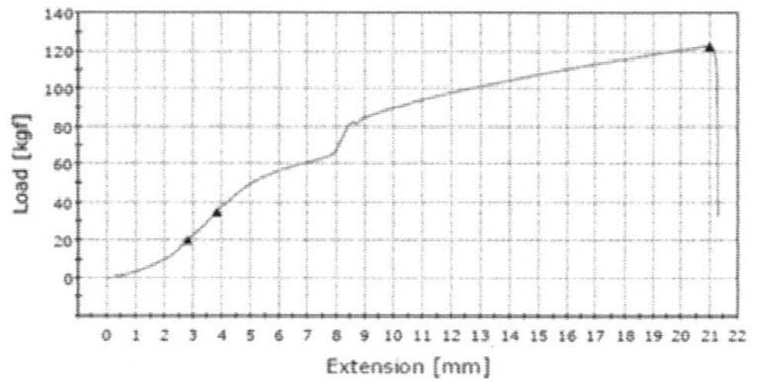

Figure 7. The values of compressive extension test 1 at preset point is2.80856 mm

Test Load	Standard		Actual		Judgement
			Sample No.	Result	
20 Kgf	Max Displacement	Max 5 mm	1	2,81 mm	OK
			2	3,66 mm	OK
35 kgf	Max Displacement	Max 5 mm	1	3,83 mm	OK
			2	4,91 mm	OK
Max Load	Min Load	60 kgf	1	122,45 kgf	OK
			2	114,23 kgf	OK

Figure 8. Results of compression testing

Compressive extention at preset point (load 20 kgf) mm	*Compressive extention at preset point (load 35 kgf) mm*	*Maximum load (kgf)*
3,66180	4,91224	114,23241

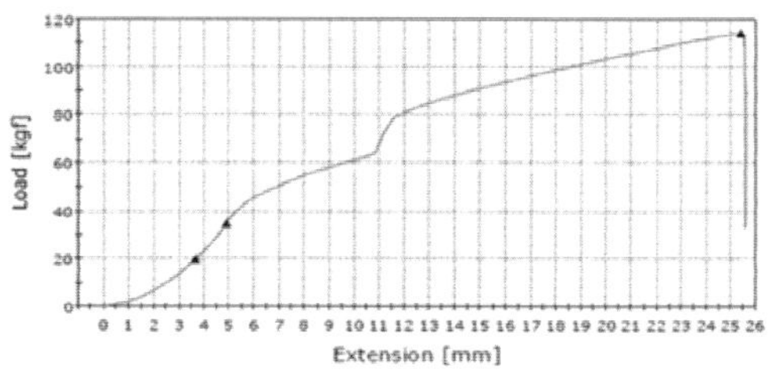

Figure 9. The values of compressive extension test 2 at preset point is3.66180 mm

4. Conclusions

Analysis of porosity defects in Al as the part handle motor vehicle lever has been processed by high pressure die casting, by a standard operational procedure such as the fluxing process frequency standard and the optimization of special particle lever parameter on the casting process has been well determined. The optimum dies design with the addition of two runners is accompanied by the change of gate and overflow. From the results clearly, show the improvements in the process have been successfully made, and the mechanical properties and safety have met the standards. Therefore, the application of the high pressure die-casting is proven as a very effective method for the mechanical improvement of aluminum alloys.

References

[1] Davis J R 1999 *Aluminum and Aluminum Alloys* (ASM International)
[2] Starke E A and Staley J T 1996 Application of modern aluminum alloys to aircraft *Progress in Aerospace Sciences* **32** 131 – 172
[3] Hu Y, Bakker M C M, and De Heij P G 2011 Recovery and distribution of incinerated aluminum packaging waste *Waste Management* **31** 2422 – 2430
[4] Das S.. *J. Minerals, Metals, and Materials Society* **52** 41 – 44

[5] Minghua Q 2012 *Construction Technology* **6** 011
[6] Rao S K et al 2005 Reasons for superior mechanical and corrosion properties of 2219 aluminum alloy electron beam welds *Materials Characterization* **55** 345 – 354
[7] Polmear I J and Couper M J 1988 Design and development of an experimental wrought aluminum alloy for use at elevated temperatures *Metallurgical and Materials Transactions A* **19** 1027 – 1035
[8] Sankaran K K and Grant N J 1980 The structure and properties of splat-quenched aluminum alloy 2024 containing lithium additions *Materials Science and Engineering* **44** 213 – 227
[9] Yar A A *et al* 2009 Microstructure and mechanical properties of aluminum alloy matrix composite reinforced with nano-particle MgO *Alloys and Compounds* **484** 400 – 404
[10] Elangovan K and Balasubramanian V 2008 Influences of post-weld heat treatment on tensile properties of friction stir-welded AA6061 aluminum alloy joints *Materials Characterization* **59** 1168 – 1177
[11] Ferrasse S *et al* 1997 Microstructure and properties of copper and aluminum alloy 3003 heavily worked by equal channel angular extrusion. *Metallurgical and Materials Transactions A* **28** 1047 – 1057
[12] Wang W T et al 2010 Influences of Ce addition on the microstructures and mechanical properties of 2519A aluminum alloy plate *J. Alloys and Compounds* **491** 366 – 371
[13] Barnes T A and Pashby I R 2000 Joining techniques for aluminium spaceframes used in automobiles: Part I—solid and liquid phase welding. *J. Materials Processing Technology* **99** 62 – 71
[14] Miracle D B 2005 Metal matrix composites–from science to technological significance *Composites Science and Technology* **65** 2526 – 2540
[15] Watarai H 2006 *NISTEP Science & Technology Foresight Center*
[16] Exley C and Darwin 2009 Darwin, natural selection and the biological essentiality of aluminium and silicon *Trends in Biochemical Sciences* **34** 589 – 593
[17] Schweitzer P A 2005 *Paint and Coatings: Applications and Corrosion Resistance* (CRC Press)
[18] Schwartz M 2002 *Encyclopedia of Materials, Parts, and Finishes* (CRC Press)
[19] DeGarmo E P *et al* 1997 *Materials and Process in Manufacturing* (Prentice Hall)
[20] Seah K H W et al 2002 Effects of temperature and reinforcement content on corrosion characteristics of LM13/albite composites *Corrosion Science* **44** 761 – 772
[21] Zhao H, White D R and DebRoy T 1999 Current issues and problems in laser welding of automotive aluminium alloys *International Materials Reviews* **44** 238 – 266
[22] Chen J K and Huang I S 2013 Thermal properties of aluminum–graphite composites by powder metallurgy *Composites Part B: Engineering* **44** 698 – 703
[23] Tsui T Y, Oliver W C, and Pharr G M *J. Materials Research* **11** 752 – 759
[24] Avalle M et al 2002 Casting defects and fatigue strength of a die cast aluminium alloy: a comparison between standard specimens and production components *International Journal of Fatigue* **24** 1 – 9
[25] Nakata K et al 2006 Improvement of mechanical properties of aluminum die casting alloy by multi-pass friction stir processing *Materials Science and Engineering: A* **437** 274 – 280

Synthesis and Characterization of Barium-Hexaferrite-Based Nanocomposite on X-Band Microwave

Y. E. Gunanto[1], M. P. Izaak[1], S. S. Silaban[1], and W. A. Adi[2]

[1]Department of Physics Education, Faculty of Education, University of Pelita Harapan, Karawaci, Tangerang 15811, Indonesia
[2]Center for Science and Technology of Advanced Materials, BATAN, Tangerang Selatan 15314, Indonesia

E-mail: yohanes.gunanto@uph.edu

Abstract. In this research, synthesis and characterization of barium-hexaferrite-based nanocomposites were conducted with silicone rubber as a microwave absorber in x-band frequency. Barium-hexaferrite nanoparticle powder is prepared with high energy milling for 75 h. Nanocomposites were prepared by weight percentage of barium-hexaferrite to silicon rubber with a variation of 20 % to 50 %. X-ray diffraction (XRD) is used for structural analysis and phase in still-powder conditions. Scanning electron microscopy (SEM) and transmission electron microscopy (TEM) are used for surface morphology analysis. The vibration of sample magnetometer (VSM) and vector network analyzer are respectively used for static magnetic analysis and microwave absorption properties. The characterization results show that the sample has a single phase with a hexagonal crystal structure. The surface morphology is almost homogeneous with an almost spherical grain shape and a nanoparticle size of 20 nm to 30 nm. Magnetic saturation (Ms) reached 52.0 $emu \cdot g^{-1}$, magnetic remanent (Mr) 33.0 $emu \cdot g^{-1}$ and coercivity (Hc) 0.32 T. Ability to absorb microwaves expressed by reflection loss (RL) reached the maximum for the sample with 20 % composition, i.e. about –15 dB at a frequency of 10.8 GHz. While for the composition of 30 % and 50 % the ability to absorb microwaves in a row is –7 dB (10.8 GHz) and –5.2 dB (10.6 GHz).

Keywords. Barium-hexaferrite, microwave absorber, nanocomposite.

1. Introduction

Technological developments, especially in microwave absorption technology, attracted the attention of many researchers. The type of material that can be used as a microwave absorber has a magnetic, electrical and dielectric loss. Among the magnetic materials that can be used as a microwave absorber is barium hexaferrite which has a hexagonal structure, high magnetic saturation and chemical stability.To further enhance its ability to absorb microwaves, it is first processed through doping procedures in the position of Ba, either in the Fe position or both at once [1–7].The work already carried out refers to variations in magnetic permeability, dielectric permittivity at GHz, and absorption properties with composition and sample-making procedures but has not much been studied about the composite of the compound.

Content from this work may be used under the terms of the Creative Commons Attribution 3.0 licence. Any further distribution of this work must maintain attribution to the author(s) and the title of the work, journal citation and DOI.
Published under licence by IOP Publishing Ltd

Polymers have several advantages, including ease of processing and formation, thermomechanical stability, and low cost. Epoxy resin is commonly used because it is extensively used for the production of high tech composite in the fields of automotive, aerospace and electronics industries, mostly due to their high stiffness [8, 9]. The barium ferrite composite as the content with the epoxy resin will have dielectric permittivity, magnetization and magnetic saturation increased with the increased barium ferrite content [10]. Particles of nanosize in carbon-epoxy resin composites will produce better microwave absorption [11]. Conductivity, saturation magnetization and coercivity decrease with the incorporation of MnO_2 contents [12]. We hope that the thinner thickness of silicon rubber and barium-strontium hexaferrite can improve the ability to absorb microwaves.

In this paper, the nanocomposite is formed from silicone rubber and barium-strontium hexaferrite nanoparticles with a content variation of 20 % to 50 % by weight. Nanocomposite printed sheets with a thickness of 1 mm. Analysis of structures and phases in the form of powder used X-ray diffraction (XRD) and for surface morphology analysis used scanning electron microscopy (SEM) and transmission electron microscopy TEM, while the analysis of the ability to absorb microwaves used vector network analyzer (VNA).

2. Materials and methods

The $Ba_{0.6}Sr_{0.4}Fe_{10}MnTiO_{19}$ sample with certain composition was prepared using a conventional milling technique through stoichiometric quantities of analytical-grade $BaCO_3$, $SrCO_3$, Fe_2O_3, $MnCO_3$, and TiO_2 precursors with high purity were mixed and milled using high-energy milling (HEM) type Spex 8000 for up to 5 h. The samples were characterized by powder X-ray diffraction (XRD) using Pan Analytical diffractometer with CuKα radiation (α = 1.5406 Å). The powders were then compacted into pellets and sintered in the electric chamber furnace Thermolyne at 1000°C for 5 h to obtain crystalline materials. The $Ba_{0.6}Sr_{0.4}Fe_{10}MnTiO_{19}$ sample was re-milled for 75 h to obtain nanoparticles.

The absorber nanocomposite was made from silicon rubber and $Ba_{0.6}Sr_{0.4}Fe_{10}MnTiO_{19}$ nanoparticle. The matrix of nanocomposite absorber used was methyl vinyl silicone rubber and toluene as vulcanized assistants. The silicon rubber and $Ba_{0.6}Sr_{0.4}Fe_{10}MnTiO_{19}$ nanoparticle were mixed and guaranty a better dispersion. The diameter of nanocomposite absorber was about 15 cm and the thickness was about 2 mm. The various volume fractions of the $Ba_{0.6}Sr_{0.4}Fe_{10}MnTiO_{19}$ nanoparticle were added into the silicone rubber with the weight ratio was 20 wt%, 30 wt% and 50 wt% then were mixed in each composite. Nanostructural of the samples were observed using the scanning electron microscope (SEM) and transmission electron microscope (TEM). The magnetic properties were evaluated using an Oxford Instrument Vibrating Sample Magnetometer (VSM). Meanwhile the reflection loss of microwave was measured using the vector network analyzer (VNA) with a frequency range of 8 GHz to 12 GHz.

3. Results and discussion

Figure 1a shows the refinement pattern of X-ray diffraction sample $Ba_{0.6}Sr_{0.4}Fe_{10}MnTiO_{19}$ with milling time of 75 h. The result of this refinement produces excellent quality fittings with a very small R factor as well. The R factor is a very small criterion of fit and χ2 (chi-squared) (1.036).The complete refinement results can be seen in Table 1.

Table 1. Lattice parameters, crystal structure, R-factor and χ^2

Lattice parameters, crystal structure, R-factor and χ^2 Phase $BaFe_{12}O_{19}$			
space group : P 63/m m c (194), crystal structure : Hexagonal Lattice parameters: $a = b = 5.8984(3)$ Å and $c = 23.171(1)$ Å, $\alpha = \beta = 90^o$ and $\gamma = 120^o$, V = 698.1(1) Å^3 and $\rho = 5.614$ g.cm^{-3}			100.00 %
R-factor	*w*Rp = 3.10	Rp = 2.25	$\chi^2 = 1.036$

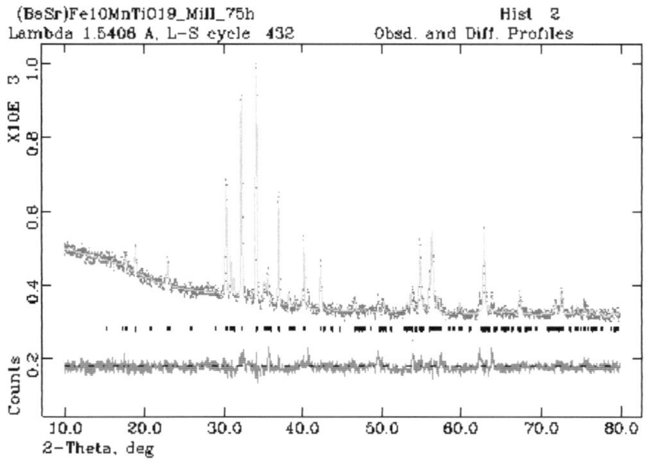

Figure 1a. Refinement of x-ray diffraction pattern of $Ba_{0.6}Sr_{0.4}Fe_{10}MnTiO_{19}$

Figure 1b. Cell unit of Barium Hexaferrite [13]

Figure 1b shows crystal structure of M-type barium hexaferrite $BaFe_{12}O_{19}$ [13]. This hexagonal structure is constructed into 4 sub-unit blocks called the sub-units S and S* blocks with the chemical formula $(Fe_6^{3+}O_8^{2-})^{2+}$ and sub-unit blocks R and R* with the chemical formula $(Ba^{2+}Fe_6^{3+}O_{11}^{2-})^{2-}$. The (*) indicates that the sub-unit relationship is rotated 180° around the hexagonal axis.The sub-unit blocks S and S* are spinels with 2 layers O binding to 6 Fe^{3+} ions (4 Fe^{3+} ions form the octahedral and 2 Fe^{3+} ions form the tetrahedral). While the sub-unit blocks R and R* consisting of 3 layers of O, 1 layer O bind to the Ba^{2+} ions and the other 2 layers O bond with 6 Fe^{3+} ions to form octahedral. The colors of spheres; red, pink, gray, blue and yellow represent iron atoms in 2a,4f$_1$,12k,4f$_2$, and 2b sites respectively. Small green spheres represent oxygen atoms, while larger cyan spheres represent barium atoms [13].

The single phase $Ba_{0.6}Sr_{0.4}Fe_{10}MnTiO_{19}$ has a homogeneous microstructure of surface morphological observations as shown in Figure 2. Surface morphological observation results show that the sample has a shape and a particle size distribution of about 20 nm to 30 nm. This result is very much in accordance with the characterization results using TEM which can be seen in Figure 3. S.S.S.Afghahi *et al* obtained an average particle size of about 250 nm to 300 nm [14].

Figure 2. Surface morphology of $Ba_{0.6}Sr_{0.4}Fe_{10}MnTiO_{19}$ using SEM

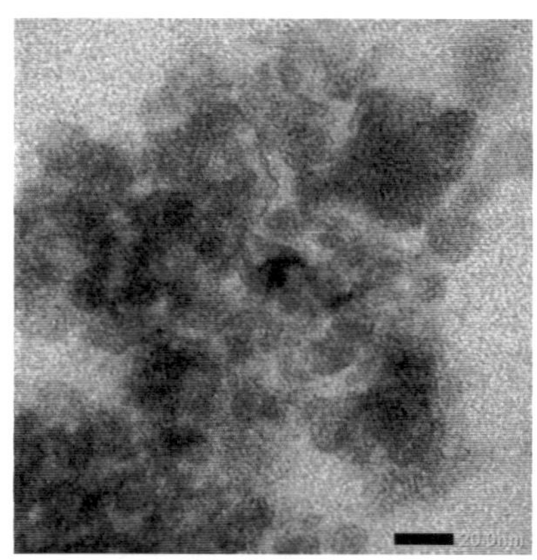

Figure 3. Characterization of sample particles $Ba_{0.6}Sr_{0.4}Fe_{10}MnTiO_{19}$ using TEM.

The result of characterization of magnetic properties of $Ba_{0.6}Sr_{0.4}Fe_{10}MnTiO_{19}$ particle can be seen in Figure 4.The loop of the hysteresis curve consists of the intrinsic saturation of Ms, remanence Mr, and coercivity Hc. This saturation is an area where the material is unable to absorb strong magnetic fields such as magnetization strengthening and does not produce significant changes in magnetic flux density. Remnant Mr for the samples shows that the magnetization left in the media after the outside magnetic field has been removed and the coercivity indicating the coercive power of the material is the same as the demagnetizing force required to reduce residual induction to zero in the magnetic field after the magnetization is saturated. The complete result of magnetization can be seen in Table 2.

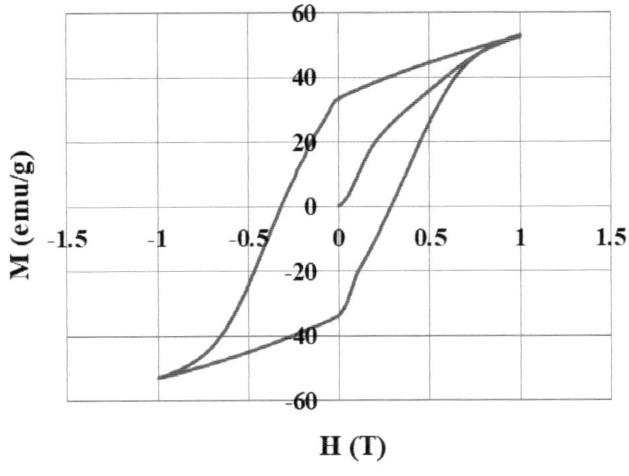

Figure 4. Hysteresis curve of $Ba_{0.6}Sr_{0.4}Fe_{10}MnTiO_{19}$ using VSM

Tabel 2. Summary of magnetic properties data$Ba_{0.6}Sr_{0.4}Fe_{10}MnTiO_{19}$

t (h)	Coercivity, Hc (T)	Saturation, Ms (emu·g^{-1})	Remanence, Mr (emu·g^{-1})
75	0.32	52.0	33.0

Ability to absorb microwaves is done using VNA. For composites with variations of the filler composition $Ba_{0.6}Sr_{0.4}Fe_{10}MnTiO_{19}$ 20 % to 50 % can be seen in Figure 5. It appears that the 20 % composition actually shows as the best absorbent compared to the larger composition. The maximum value of Reflection Loss (RL) is about –15 dB at a frequency of 10.8 GHz. S.S.S.Afghahi *et al* found that the maximum value of reflection loss was about –2.4 dB [14].

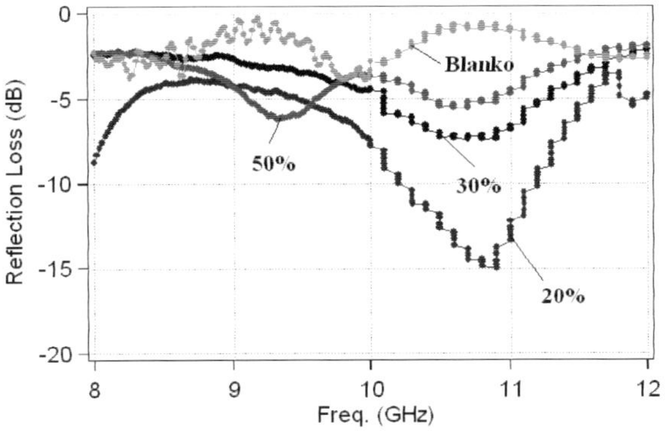

Figure 5. Reflection loss curve (RL) $Ba_{0.6}Sr_{0.4}Fe_{10}MnTiO_{19}$ with filler variation between 20 % to 50 %.

4. Conclusions

Nanocomposite has been successfully manufactured with barium-strontium hexaferrite nanoparticles as fill and silicon rubber with a content variation of 20 wt% to 50 wt%. The characterization of the ability to absorb microwaves using a vector network analyzer (VNA) found that the higher the content, the weaker the ability to absorb microwaves. The 20 % composition has a maximum reflection loss value (RL) of about –15 dB at a frequency of 10.8 GHz with a bandwidth of 1.0 GHz.

Acknowledgements

This work was supported by University of Pelita Harapan Grants 2017 No. **P-005-FIP/I/2017**, Development of Anti-Radar Nanoparticle Material Based Barium-Strontium Hexaferrite

References

[1] Gunanto Y E, Jobiliong E, and Wisnu A A 2016 Composition and phase analysis of nanocrystalline $Ba_xSr_{1-x}Fe_{12}O_{19}$ (x = 10; 06; and 04) by using general structure analysis system *AIP Conference Proceedings* **1719** p 030019

[2] Singh J *et al* 2016 Tunable Microwave Absorption in Co/Al Substituted M–Type Ba/Sr Hexagonal Ferrite *MaterDes.* **110** 749–761

[3] Shams M H *et al* 2016 Effect of Mg2+ and Ti4+ Dopants on the Structural Magnetic and High–Frequency Ferromagnetic Properties of Barium Hexaferrite *J. Magn. Magn. Mater.* **399** 10–18

[4] Trukhanov A V *et al* 2016 Crystal Structure Magnetic and Microwave Properties of Solid Solutions $BaFe_{12-x}Ga_xO_{19}$ (01 ≤ x ≤12) *Phys. Solid–State* **58** 1792–1797

[5] Gunanto Y E, Cahyadi L, and Wisnu A A 2016 Effect of Mn and Ti substitution on the reflection loss characteristic of $Ba_{06}Sr_{04}Fe_{11-z}MnTi_zO_{19}$ (z = 0 1 2 and 3) *AIP Conference Proceedings* **1725** p 020023

[6] Yohanes E, G Eric J and Wisnu A A 2016 Microwave Absorbing Properties of $Ba_{06}Sr_{04}Fe_{12-z}Mn_zO_{19}$ (z = 0 – 3) Materials in XBand Frequencies *J. Math. FundSci.* **48** 55–65

[7] Liu J *et al* 2016 Magnetic Properties of a Highly Textured BariumHexaferrite Quasi–Single Crystal and Its Application in Low–Field Biased Circulators *J. Electr. Mater.* **45** 5069–5073

[8] Asimakopoulos I A, Psarras G C, and Zoumpoulakis L 2014 Barium titanate/polyester resin nanocomposites: Development structure–properties relationship and energy storage capability *eXPRESS Polymer Letters* **8** 698–707

[9] Tomara G N *et al* 2015 Dielectric response and energy storage efficiency of low content TiO_2–polymer matrix nanocomposites *Composites Part A: Applied Science and Manufacturing* **71** 204–211

[10] Kanapitsas A, Tsonos C, Psarras G C, and Kripotou S 2016 Barium ferrite/epoxy resin nanocomposite system : Fabrication dielectric magnetic and hydration studies *eXPRESS Polymer Letters* **10** 227–236

[11] Micheli D, Apollo A, Pastore R, and Marchetti M 2010 X–band microwave characterization of carbon–based nanocomposite material absorption capability comparison and RAS design simulation *Composite Science and Technology* **70** 400–409

[12] Ali K *et al* 2017 Synthesis structural dielectric and magnetic properties of $CuFe_2O_4/MnO_2$nanocomposites *Journal of Magnetism and Magnetic Materials* **434** 30–36

[13] Moitra, A., Kim, S., Kim, S. G., Erwin, S. C., Hong, Y. K., & Park, J. 2014. Defect formation energy and magnetic properties of aluminum-substituted M-type barium hexaferrite. *Computational Condensed Matter*, *1*, 45-50.

[14] Afghahi S S S, Jafarian M, and Atassi Y 2016 Microstructural and magnetic studies on $BaMg_xZn_xX_{2x}Fe_{12-4x}O_{19}$ (x = Zr Ce Sn) prepared via mechanical activation method to act as a microwave absorber in X–band *Journal of Magnetism and Magnetic Materials* **406** 184–191

Semiconductor Ceramic $Mn_{0.5}Fe_{1.5}O_3$-Fe_2O_3 from Natural Minerals as Ethanol Gas Sensors

H. Aliah[1], D. G. Syarif[2], R. N. Iman[1], A. Sawitri[1], M. Sanjaya WS[1], M. Nurul Subkhi[1], and P. Pitriana[3]

[1]Department of Physics, UIN Sunan Gunung Djati Bandung, Jl. A. H. Nasution 105, Bandung 40614, Indonesia
[2]PTNBR-BATAN, Jl.Tamansari 71, Bandung, Indonesia
[3]Department of Physics Educations, UIN Sunan Gunung Djati Bandung, Jl. A. H. Nasution 105, Bandung 40614, Indonesia

E-mail: hasniahaliah@uinsgd.ac.id

Abstract. In this research, Mn and Fe-based ceramic gas sensing were fabricated and characterized. This research used natural mineral which is widely available in Indonesia and intended to observe the characteristics of Mn and Fe-based semiconducting material. Fabricating process of the thick films started by synthesizing the ceramic powder of $Fe(OH)_3$ and Mn oxide material using the precipitation method. The deposition from precipitation method previously was calcined at a temperature of 800 °C to produce nanoparticle powder. Nanoparticle powder that contains Mn and Fe oxide was mixed with an organic vehicle (OV) to produce a paste. Then, the paste was layered on the alumina substrate by using the screen printing method. XRD method was utilized to characterize the thick film crystal structure that has been produced. XRD spectra showed that the ceramic layer was formed from the solid $Mn_{0.5}Fe_{1.5}O_3$ (bixbyite) and Fe_2O_3. In addition, the electrical properties (resistance) examination was held in the room that contains air and ethanol to determine the sensor sensitivity of ethanol gas. The sensor resistance decreases as the ethanol gas was added, showing that the sensor was sensitive to ethanol gas and an n-type semiconductor. Gas sensor exhibit sensitive characterization of ethanol gas on the concentration of (100 to 300) ppm at a temperature of (150 to 200) °C. This showed that the $Mn_{0.5}Fe_{1.5}O_3$-Fe_2O_3 ceramic semiconductor could be utilized as the ethanol gas detector.

Keywords. Bixbyite, gas sensor, jarosite minerals, nanoparticle, semiconductor ceramic, XRD.

1. Introduction

Jarosite mineral is one of the minerals that widely available in Indonesia. Jarosite mineral that contains Fe_2O_3, can be used as the main component of oxide ceramic-based semiconductor manufacturing. Several studies have been conducted to utilize the Fe_2O_3 in Jarosite mineral, including in fabrication of $CuFe_2O_4$ ceramics [1, 2], Fe_2O_3- NiO [3], Fe_2TiO_5 [4] and Fe_2TiO_5/Nb_2O_5 [5]. The gas sensor is an example of metal oxide ceramic application, especially on detection of flammable gases and hydrocarbons [6, 7]. Nowadays, alcohol gas sensor is one of the most popular necessaries. Alcohol gas sensor can be applied to detect the alcohol vapor on human's breath, foods, beverages, or alcohol vapor in the air. The oxide ceramic-based semiconductor material is sensitive to certain gases. MnO_2 is

Content from this work may be used under the terms of the Creative Commons Attribution 3.0 licence. Any further distribution of this work must maintain attribution to the author(s) and the title of the work, journal citation and DOI.
Published under licence by IOP Publishing Ltd

a useful dopant because of the multivalency characteristic of Mn ion [8] and can improve the sensor quality with the thermistor constant value [4]. In the present study, Fe_2O_3 ceramic is suitable for alcohol sensor application [9]. However, the sensing performance of the sensor has not been qualified for sensitivity, operating temperature and response rates. The doping technique is one way that could develop the sensor performance as a function of Mn dopant to qualify insensitivity, sensor operating temperature and response rates. Furthermore, Mn addition could produce the bixbyite solid solution that contains different particle size than Fe_2O_3 phase in composite [10–12].

2. Materials and methods

2.1. Synthesis of $Fe(OH)_3$

$Fe(OH)_3$ was synthesized by precipitation methods using Jarosite minerals from PD Kerta Pertambangan as precursors. An amount of 20 g Jarosite minerals was dissolved in HCl solution and precipitated using NH_4OH solution that was slowly dropped until the pH value 10. The solution was filtered to separate the precipitate. It was heated in an oven at 110 °C for 31 h. Finally, it was crushed to get nanopowder.

2.2. Fabrication of Semiconductor Ceramic $Mn_{0.5}Fe_{1.5}O_3$-Fe_2O_3

Ceramic powder for producing Mn and Fe oxide-based sensors was synthesized using Fe_2O_3 derived from Jarosite mineral. $Fe(OH)_3$ powder (1.391753 g) was dissolved in 30 mL distilled water and MnO powder from Aldrich with purity of 99 % (0.922777 g) was dissolved in 20 mL HCl and both were mixed until homogeneous. The mixture was added with NH_4OH until a precipitate was formed and the precipitate was calcined at a temperature of 800 °C to get nanoparticle powder that contains Mn and Fe oxides. The powders mixed with *Organic Vehicle (*composed of ethyl cellulose and terpineol alpha) in a ratio of 74.1 %: 25.9 %. The mixture was homogeneously stirred to form a paste (Figure 1).

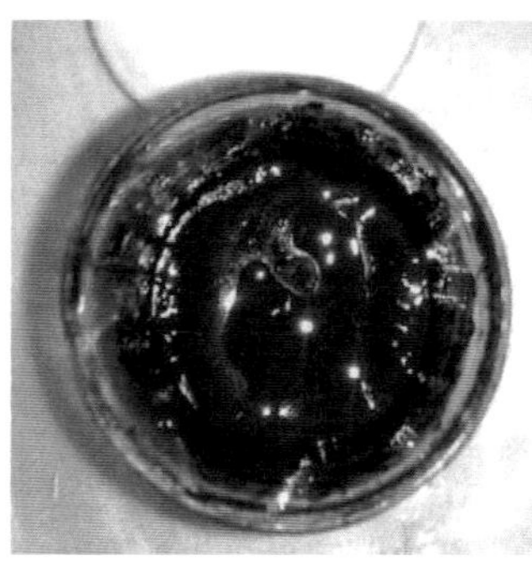

Figure 1. $Mn_{0.5}Fe_{1.5}O_3 – Fe_2O_3$ paste

The paste was printed on a silver-coated alumina substrate using a screen. The film was heated at 800 °C for 2 h in the air.

2.3. Characterization of Semiconductor Ceramic $Mn_{0.5}Fe_{1.5}O_3$-Fe_2O_3

The crystal structure of semiconductor ceramic $Mn_{0.5}Fe_{1.5}O_3$-Fe_2O_3 was analyzed using X-ray diffraction (XRD) from Philips X-ray PW 1710 BASED type. This instrument was equipped with Cu target, voltage source 40 kV, current source 35 mA and wavelength of 1.54056 Å. The electrical resistance of the thick film was measured at various temperatures in the air and in the atmosphere of ethanol gas. The principle of the instrument for measuring the electrical resistance is shown in Figure 2.

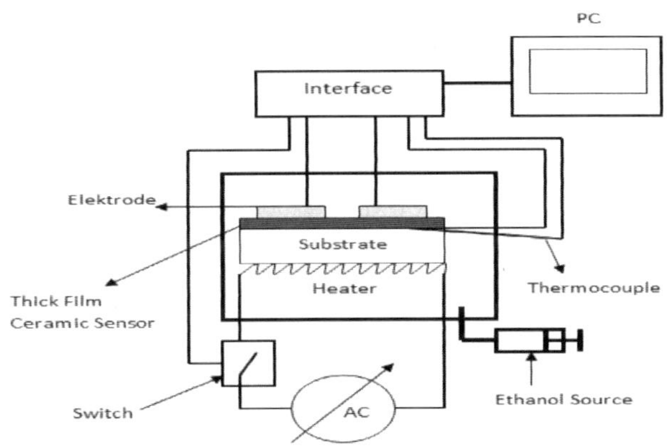

Figure 2. The instrument principle for measuring electrical resistance.

3. Results and discussion

Thick films of $Mn_{0.5}Fe_{1.5}O_3$-Fe_2O_3 were successfully fabricated. A sample as a representative is shown in Figure 3. The thick films of $Mn_{0.5}Fe_{1.5}O_3$-Fe_2O_3 are smooth without cracks on the ceramics surface. The XRD pattern of the $Mn_{0.5}Fe_{1.5}O_3$-Fe_2O_3 thick film is depicted in Figure 4.

Figure 3. Representation of thick film Mn and Fe-based oxide.

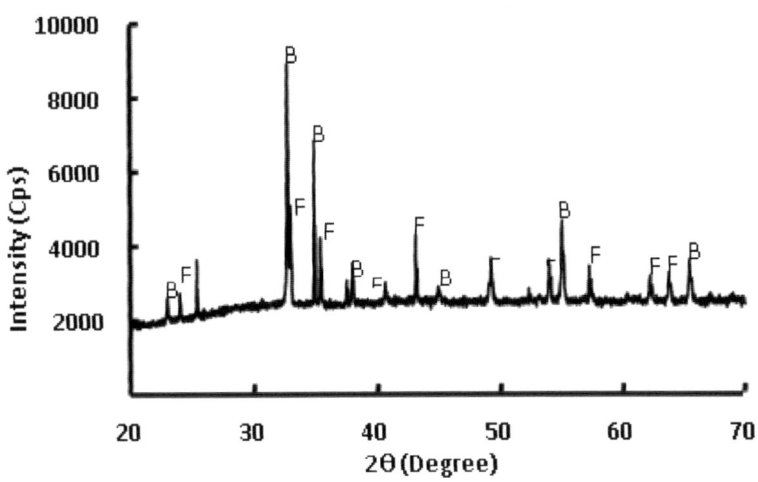

Figure 4. XRD pattern of semiconductor ceramic $Mn_{0.5}Fe_{1.5}O_3$-Fe_2O_3. B is from bixbite, and F is from Fe_2O_3.

Figure 4 shows the XRD profile of $Mn_{0.5}Fe_{1.5}O_3$-Fe_2O_3 thick film ceramics. The analysis was done by comparing the XRD pattern of $Mn_{0.5}Fe_{1.5}O_3$-Fe_2O_3 thick film ceramics to the XRD pattern of JCPDS standard No. 41-1442 for Mn_2O_3 and JCPDS No. 33-0664 for Fe_2O_3. From the analysis it was known that the fabricated ceramic formed a ceramic composite that consists of $Mn_xFe_{2-x}O_3$-Fe_2O_3 solid solution ceramic (indicated by B) and Fe_2O_3 (indicated by F). Here x is 0.5.

The electrical resistance of the thick film ceramics of $Mn_{0.5}Fe_{1.5}O_3$-Fe_2O_3 as a function of temperature is shown in Figure 5. Whereas Figure 6 shows the sensitivity characteristics of the film. As can be seen in Figure 5, the resistance decreases with temperature. This indicates that thick film $Mn_xFe_{2-x}O_3$-Fe_2O_3 material has a semiconducting property with n-type [13, 14]. The resistance of the samples in the ethanol-containing atmosphere (100 ppm) is smaller than that in the air. And when the ethanol concentration is increased from 100 ppm to 200 ppm and from 200 ppm to 300 ppm, the resistance decrease. This data shows that the thick film is sensitive to ethanol gas, and means that the thick film ceramic of $Mn_{0.5}Fe_{1.5}O_3$-Fe_2O_3 has the potentiality to be ethanol gas sensor.

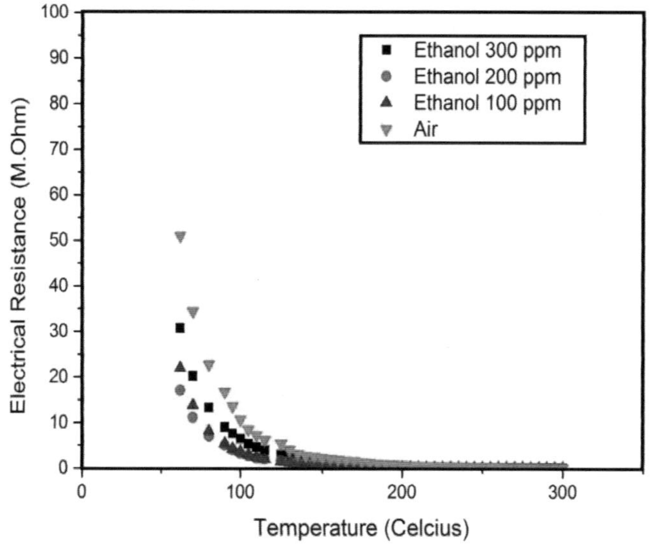

Figure 5. The electrical resistance characteristic of thick film $Mn_{0.5}Fe_{1.5}O_3$-Fe_2O_3 to temperature changes.

Figure 6 shows that the thick film of $Mn_{0.5}Fe_{1.5}O_3$-Fe_2O_3 ceramic is sensitive to ethanol gas. Sensitivity value is calculated with Equation 1. The highest sensitivity is for ethanol gas concentration of 100 ppm. The optimum operating temperature of the sensor is between (150 to 200) °C.

$$S = \left[\frac{R_O - R_G}{R_O} \right] \times 100\% \quad (1)$$

Where S is the sensitivity of thick film ceramics (%), R_O is the electrical resistance of thick film ceramic in the air environment and R_G is the electrical resistance of thick film ceramic in the ethanol gas environment.

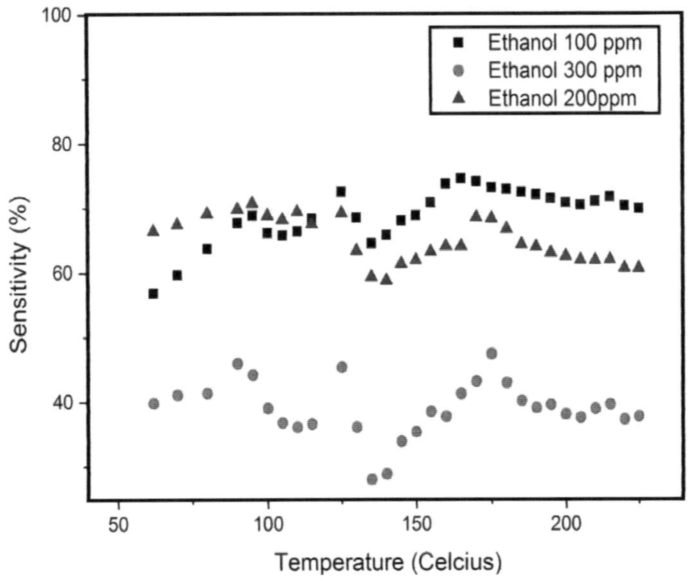

Figure 6. Interpretation of thick Film $Mn_{0.5}Fe_{1.5}O_3$-Fe_2O_3 sensitivity

The decreasing of the resistance of $Mn_{0.5}Fe_{1.5}O_3$-Fe_2O_3 thick film electricity is caused by several factors. One is that when the thick film is in the air environment, the thick film surface absorbs oxygen (O_2) ions and electrons in the conduction band move to oxygen, the barrier increases. When the thick film surface absorbs ethanol gas, a reaction between ethanol gas and adsorbed oxygen is formed. Due to the reaction, barrier resistance decreases. At this time the resistance decreases, increasing the conduction [15].

4. Conclusions

Ethanol gas sensor can be fabricated by using thick film technology. A thick film of $M_{0.5}Fe_{1.5}O_3$-Fe_2O_3 ceramics that is sensitive to ethanol gas has been fabricated. The sensor ceramic has n-type semiconductor characteristic. The resistance decreases with ethanol gas concentration.

Acknowledgments

This research was funded by LP2M UIN Sunan Gunung Djati Bandung and the Ministry of Religious Affairs, Republic of Indonesia.

References

[1] Wiendartun W *et al* 2008 *JUSAMI* **43** 195 – 200
[2] Wiendartun W and Syarif D G 2012 *J. Mater Res.* **1** 70 – 75
[3] Hasanah L, Suhendi E, and Syarif D G 2011 *JUSAMI* **13** 27 – 29
[4] Wiendartun W and Syarif D G 2013 Effect of MnO_2 Addition on Characteristics of Fe_2TiO_5 Ceramics for NTC Thermistor Utilizing Commercial and Local Iron Oxide *J. Aust. Ceram. Soc.* **49** 141 – 147
[5] Wiendartun W, Risdiana R, Fitrilawati F, and Siregar R E 2015 The effect of sintering temperature on electrical characteristics of Fe_2TiO_5/Nb_2O_5 ceramics for NTC thermistor *Mater Sci. Forum* **827** 262 – 265
[6] Ahmed M 2014 *Metal oxide semiconductors as humidity and NOx sensors for environmental monitoring* (PhD thesis Politecnico, Torino)
[7] Wang C, Yin L, Zhang L, Xiang D, and Gao R 2010 Metal oxide gas sensors: sensitivity and influencing factors *Sensors* **10** 2088 – 2106

[8] Tanwar V G and K Sreenivas 2010 Effect of Mn Doping on the Dielectric and Electrical Conductivity of CaBi4Ti4O15 Ceramics *Ferroelectrics* **404** 19 – 26
[9] Rao B 2000 Zinc oxide ceramic semi-conductor gas sensor for ethanol vapour *Mater Chem Phys* **64** 62 – 65
[10] Patrice R, Dupont L, Aldon L, Jumas J C, Wang E, and Tarascon J M 2004 *Chem Mater* **16** 2772 – 2782
[11] Sorensen B, Gaal S, Tangstad M, Ringdalen E, Kononov R, and Ostrovski O 2010 *Properties of manganese ores and their change in the process of calcination* (Ferromanganese – Pretreatment for Smelting, The Twelfth International Ferroalloys Congress Sustainable Future, Helsinki Finland) pp 439 – 448
[12] Kivinen V, Krogerus H, and Daavittila J 2010 *Upgrading of Mn / Fe ratio of low–grade manganese ore for ferromanganese production* (Ferromanganese-Pretreatment for Smelting, The Twelfth International Ferroalloys Congress Sustainable Future, Helsinki Finland) pp 467 – 476
[13] Patil L A 2009 Fe_2O_3-ZnO Based Gas Sensors *Sensors & Transducers Journal* **104** 68 – 75
[14] Williams, D. E. (1999). Semiconducting oxides as gas-sensitive resistors. *Sensors and Actuators B: Chemical*, *57*(1-3), 1-16.
[15] Ansari S G *et al* 1997 Grain size effects on H_2 gas sensitivity of thick film resistor using SnO_2 nanoparticles *Thin Solid Films* **295** 271 – 276

The Effects of Calcination Temperatures on Crystal Structures and Morphologies of $Nd_{1.2}FeO_3$ Synthesized by Solid-State Reaction

E. H. Sujiono[1], M. Y. Dahlan[1], R. A. Imran[1], A.C. M. Said[1], and S. Samnur[1]

[1]Laboratory of Materials Physics Department of Physics, Universitas Negeri Makassar, Makassar 90224, Indonesia

E-mailr: e.h.sujiono@unm.ac.id

Abstract. $NdFeO_3$ is one of the oxide alloys that can be used as a raw material for gas sensor. The $NdFeO_3$ have been synthesized using solid state reaction method by varying calcination temperatures of 750°C, 850°C, and 950°C for 6 h. All of the $Nd_{1.2}FeO_3$ samples were characterized using scanning electron microscope (SEM) and x-ray diffraction (XRD) to identify their morphologies and phases. The results show that all of the samples formed major phase is $NdFeO_3$ and minor phase of Nd_2O_3 and have homogenous morphology with estimating grain size is 0,2 µm for all samples. The value of FWHM and the crystal size of $Nd_{1.2}FeO_3$ was obtained for each sample is 0.22° and 372 nm. The orthorhombic phase with a dominant peak at *hkl* (121) is an indication that material has potential application as a gas sensor.

Keywords. Crystal structure, morphology, calcination, $NdFeO_3$, and solid state method.

1. Introduction

As increasing awareness of environmental issues and the development of industrial rapidly that affects pollutant gas emissions makes the demand for sensors increases. The active material in the gas sensor can be metal, metal oxide, composite polymer and conductive polymer but now also developed active material on gas sensor derived from oxide alloy material. In recent years, $NdFeO_3$ perovskite structure has been investigated its usefulness in a wide variety of applications such as in oxide fuel cells [1], gas sensors [2], the photocatalysis and catalytic converter [3]. $NdFeO_3$ has a perovskite-type orthorhombic structure [4]. The preparation of $NdFeO_3$ has been successfully investigated by many methods, such as combustion [5], hydrothermal [6], sol-gel citrate method [7], precipitation [8], sonication assisted precipitation [9], and solid state reaction [10] are used. Solid state reaction is the most widely used for the synthesis of inorganic materials because it is easy and inexpensive by involving the heating components at a high temperature for a relatively long period. We have experiences in fabrication of such an oxide material, e.g., YBa2Cu3Oy, NdBaCuO (off-stoichiometric), and $NdFe_xBa_{2-x}Cu_3O_y$, the results have reported [11-13].

In this article, we reported our current results in the development of $Nd_{1.2}FeO_3$ oxide alloy material as one potential candidate for sensor application. $Nd_{1.2}FeO_3$ oxide have been synthesized using solid-state reaction method with two stages of heat treatment process and varying the calcination

Content from this work may be used under the terms of the Creative Commons Attribution 3.0 licence. Any further distribution of this work must maintain attribution to the author(s) and the title of the work, journal citation and DOI.
Published under licence by IOP Publishing Ltd

temperature. Characterization of material has been done by X-ray Diffraction (XRD) and Scanning Electron Microscope (SEM).

2. Materials and methods

$Nd_{1.2}FeO_3$ oxide alloy has been synthesized using solid-state reaction method [14]. The raw material Nd_2O_3 99.99 % (Strem Chemicals) and Fe_2O_3 99.99 % (Sigma Aldrich) were mixed and grinded together for 3 h then calcined for 6 h at temperature 700 °C. The mixed powder then grinded for 5 h then sintered for 6 h at temperature 950 °C. The synthesis process and the heating are then repeated to obtain a better sample homogeneity [15]. The mixed powder was grinded for 3 h and calcined at temperature 750 °C, this process was repeated for temperature 850 °C and 950 °C. All of the powders were grinded for 5 h and sintered at temperature 950 °C for 6 h.

$Nd_{1.2}FeO_3$ powder characterized by X-ray diffractometer [Rigaku Mini Flex II, $2\theta = 20° - 65°$ (CuKα, $\lambda = 0.154$ nm)] to determine the crystal structure which includes the value of FWHM (Full Width at Half Maximum) and peak height. The analysis of surface morphology and elemental of the powder were investigated using Scanning Electron Microscope and Energy Dispersive Spectroscopy (SEM-EDS) [Tescan Vega3SB] with a magnification of 5000 times.

3. Results and discussion

XRD diffraction patterns of oxide material $Nd_{1.2}FeO_3$ powder were synthesized by using the solid-state reaction method with variations of calcination temperature at temperatures of 750 °C, 850 °C and 950 °C are shown in Figure 1.

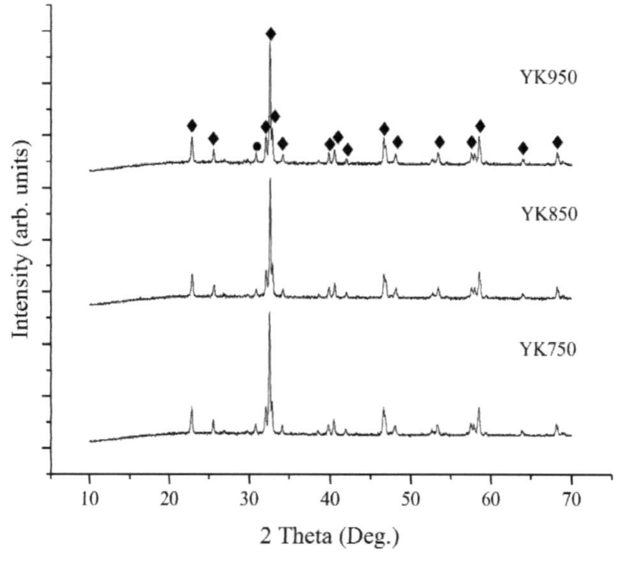

Figure 1. XRD patterns of $Nd_{1.2}FeO_3$ as variation of calcination temperatures (♦ = $NdFeO_3$, • = Nd_2O_3)

Figure 1 shows the peak of $NdFeO_3$ and Nd_2O_3 phase have been identified based on data adjustment using the Match! Software. This crystallographic curve shows that Nd_2O_3 and Fe_2O_3 raw materials have been formed of the new phase of $NdFeO_3$ which crystallizes in the orthorhombic system. The existence of minor phase formation of Nd_2O_3 is an indication that $Nd_{1.2}FeO_3$ raw material does not produce perfect phase. The reaction of imperfection is suspected due to the adjustment of calcination temperature and the heating time is less than optimal. On the other hand, Niu Xinshu et al.

also successfully synthesized $NdFeO_3$ with a temperature of 950 °C [16] and Yabin Wang et al. with a temperature of 1000 °C [17]. The results are similar to the current study with an indication of the dominant phase formation of $NdFeO_3$ located at $2\theta = 32.56°$ corresponding to the *hkl* value (121). The dominant phase intensity *hkl* (121) increases when the heating temperature is increased [18].

The crystal size can be estimated by using Debye-Scherer equation as described in Equation 1:

$$D = \frac{0.9\lambda}{\beta \cos\theta} \quad (1)$$

Where λ is the wavelength of the radiation Cu Kα ($\lambda = 0.154$ nm), θ is the angle Bragg (°), and β = FWHM at the peak of *hkl* (121) is association 2θ of 32.56° [19]. The calculation results of crystal size and FWHM can be seen in Table 1.

Table 1. Data Position (2θ), intensity, FWHM value and crystal size of $Nd_{1.2}FeO_3$ phase

Samples	2θ (°)	Intensity (Counts)	FWHM (°)	Crystal Size (nm)
YK 750	32.56	13063.33	0.22	372.17 ± 0.02
YK 850	32.56	12686.67	0.22	372.22 ± 0.02
YK 950	32.56	13050.00	0.22	372.17 ± 0.02

Based on the Table 1, it can be seen that the FWHM values for each sample are same in order of 0.22°. Full-width at half maximum (FWHM) is still an effective method to confirm the quality of crystal structure [17]. FWHM value was influenced by the intensity of each crystal plane. The higher intensity is resulting in smaller FWHM value which indicating the good crystallinity of the samples.

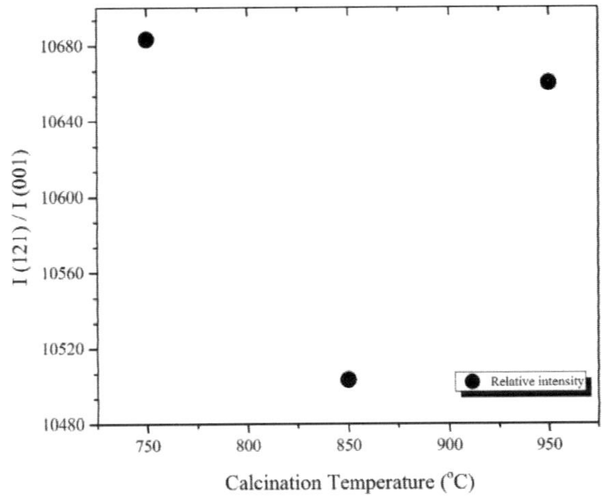

Figure 2. The comparison of the relative peak intensity of $Nd_{1.2}FeO_3$ samples with the variations of calcination temperature

Figure 2 shows the calculation result of relative intensities curve for each variation of calcination temperature. These results found that the variation of calcination temperature did not a significant change of crystal size of the sample. In fact, the existences of the atom due to the Nd_2O_3 phase will reduce the diffraction intensity of each sample. Sample with calcination temperature of 850 °C at peak *hkl*(121) is more dominant than other peaks. Thus, the $Nd_{1.2}FeO_3$ oxide material with the parameters process as has explained above will be useful for the application as gas sensors as has been reported elsewhere [2, 9, 16].

IOP Conf. Series: Materials Science and Engineering **367** (2018) 012042 doi:10.1088/1757-899X/367/1/012042

The morphology, structure and particle size of samples $Nd_{1.2}FeO_3$ as a variation of calcination temperature were investigated by SEM. Figure 3 shows the SEM micrograph of the samples.

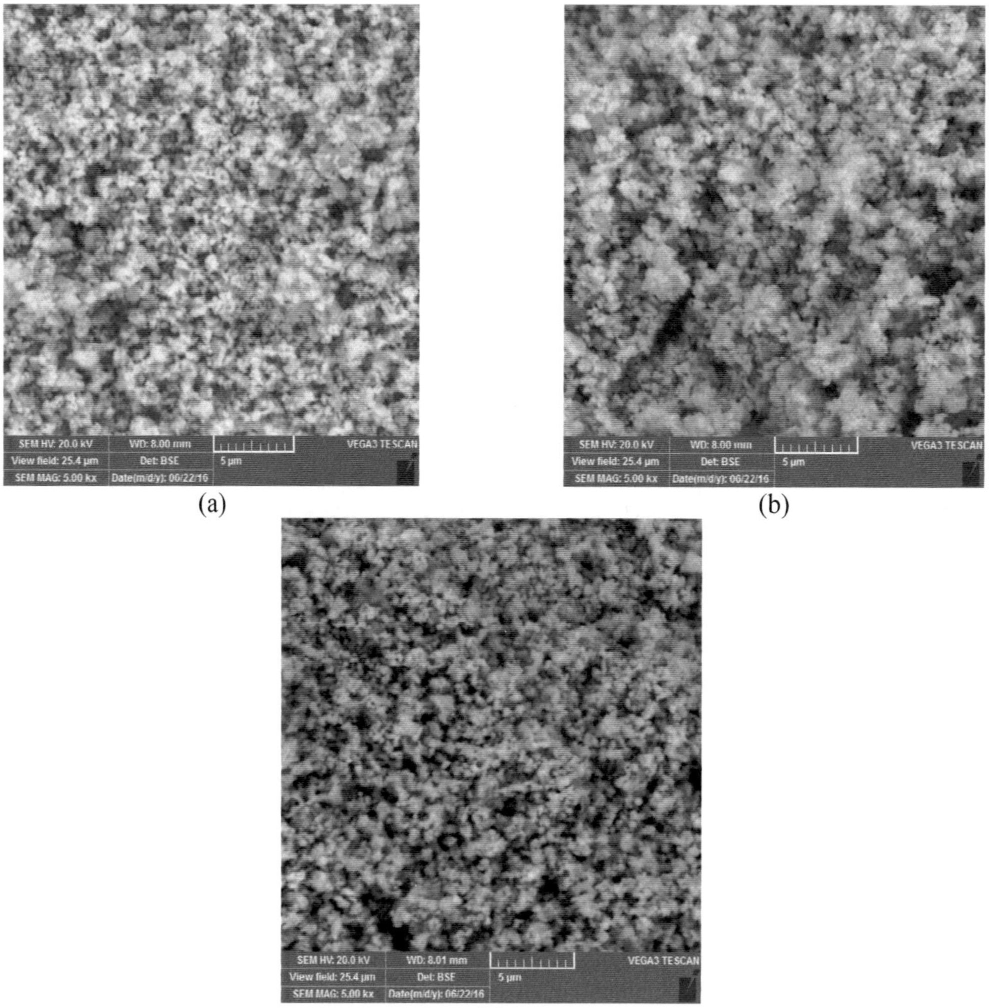

(a) (b)

(c)

Figure 3. Morphology of sample $Nd_{1.2}FeO_3$ as a variation of calcination temperature (a) YK750, (b) YK850 and (c) YK950, respectively

Table 2. Data composition element of $Nd_{1.2}FeO_3$ samples using EDS

Element	Norm. C [wt%]			Error (3 Sigma) [wt%]		
	YK750	YK850	YK950	YK750	YK850	YK950
Sodium	1.30	1.04	1.54	0.40	0.35	0.42
Magnesium	0.61	0.31	0.76	0.22	0.16	0.24
Aluminium	0.34	0.39	0.56	0.16	0.17	0.19
Silicon	0.17	0.46	-	0.12	0.16	-
Oxygen	17.20	17.33	17.19	6.63	6.83	5.77
Potassium	0.12	0.10	0.04	0.10	0.10	0.09
Titanium	0.17	0.21	0.14	0.11	0.12	0.11
Iron	20.38	20.16	20.09	1.74	1.75	1.49
Copper	0.35	0.15	0.74	0.17	0.13	0.22

Neodymium	59.36	59.83	58.94	4.76	4.91	4.02
Total :	100	100	100			

In Figure 3, it can be observed that all samples have high homogeneity indicated by the morphology of the sample forming small uniform granules, while the estimated grain size of each sample is 0.2 μm. This powder has high porosity, and this is one of the benefits to improve the characteristics of the $NdFeO_3$ oxide alloy material as a gas sensor application, as disclosed by Ho et al. [2].

The EDS results showed that $Nd_{1.2}FeO_3$ samples of YK750, YK850, and YK950 has contained Fe (20.38 wt%), Fe (20.16 wt%), Fe (20.09 wt%) and Nd (59.36 wt%), Nd (59.83 wt%), Nd (58.94 wt%), respectively and also contains a minor phase as shown in Table 2. It can be seen; there is no significant effect on the constituent elements of each sample. That existing of minor phase as indication due to the sample holder preparation process.

4. Conclusions

The $Nd_{1.2}FeO_3$ powders as a variation of calcination temperature of 750 °C, 850 °C, and 950 °C have been successfully synthesized using solid state reaction method. The results of X-ray diffraction analysis showed $NdFeO_3$ and Nd_2O_3 phase, in which the crystal structure of the phase $NdFe_{1.2}O_3$ is orthorhombic to the space group Pnma. Variation of calcination temperature higher than 700 °C did not the significant influence of diffraction intensity, FWHM, and crystallite size.

All of the samples have homogeneous morphology and high porosity with an estimated grain size of 0.2 μm. This study has been obtained compound $NdFe_{1.2}O_3$ oxide alloy with the dominant peak of *hkl* (121) which indicated that the sample is a good candidate for a gas sensor material as has been reported elsewhere.

Acknowledgements

This research was funded by Directorate Research and Community Services, Directorate General of Research and Development, Ministry of Research, Technology, and Higher Education, Republic of Indonesia, under research scheme of *Hibah Kompetensi* fiscal year 2017.

References

[1] Chen T *et al* 2012 $NdFeO_3$ as anode material for S/O_2 solid oxide fuel cells *Journal of Rare Earths* **30** 1138 – 1141
[2] Ho T G *et al* 2011 *Advanced in Natural Science: Nanoscience and Nanotechnology* **2** 015012 – 15021
[3] Niu X, Li H, and Liu G 2005 Preparation, characterization and photocatalytic properties of REFeO3 (RE= Sm, Eu, Gd) *Journal of Molecular Catalysis A: Chemical* **232** 89 – 93
[4] Chen L *et al* 2012 The role of 4f-electron on spin reorientation transition of $NdFeO_3$: A first principle study *Journal of Applied Physics* **111** 103905
[5] Wu A *et al* 2009 Preparation of $ReFeO_3$ nanocrystalline powders by auto-combustion of citric acid gel *Asia-Pacific Journal of Chemical Engineering* **45** 518 – 521
[6] Zheng W *et al* 2000 Hydrothermal synthesis of $LaFeO_3$ under carbonate-containing medium *Materials Letters* **43** 19 – 22
[7] Cui H, Zayat M, and Levy D 2006 Epoxide assisted sol–gel synthesis of perovskite-type $LaMxFe_{1-x}O_3$ (M= Ni, Co) nanoparticles *Journal of Non-Crystalline Solids* **352** 3035 – 3040
[8] Khorasani-Motlagh M et al 2013 *International Journal of Nanoscience and Nanotechnology* **9** 17 – 14
[9] Singh S *et al* 2013 Fabrication of nanobeads structured perovskite type neodymium iron oxide film: Its structural, optical, electrical and LPG sensing investigations *Sensors and Actuators B: Chemical* **177** 730 – 739
[10] Smart L and Moore E 2005 *Solid State Chemistry* (CRC Press, Boca Raton)
[11] Sujiono E H *et al* 2001 Crystal Structure and Morphology Analysis of $Nd_{1+x}Ba_{2-x}Cu_3O_7$ Oxide

IOP Conf. Series: Materials Science and Engineering **367** (2018) 012042 doi:10.1088/1757-899X/367/1/012042

Alloy Surface Developed by Solid State Reaction Method *Physica Status Solidi (A) Applied Research* **187** 471 – 479

[12] Sujiono E H, Arifin P, and Barmawi M. 2002 $YBa_2Cu_3O_{7-\delta}$ thin films deposited by a vertical MOCVD reactor *Materials Chemistry and Physics* **73** 47 – 50

[13] Sujiono E H 2017 Paduan oksida logam Nd1(Fe)XBa$_2$-XCu$_3$OY dan metode pembuatannya ID Patent No. P00200800471

[14] Zharvan V et al 2017 The Effect of Molar Ratio on Crystal Structure and Morphology of $Nd_{1+X}FeO_3$ (X=0.1, 0.2, and 0.3) Oxide Alloy Material Synthesized by Solid State Reaction Method *IOP Conference: Materials Science and Engineering* **202** 012072

[15] Mir S A, Ikram M, and Asokan K 2014 *Optik - International Journal for Light and Electron Optics* **125** 6903 – 6908

[16] Niu X *et al* 2003 *Journal of Rare Earths* **21** 630

[17] Wang Y *et al* 2010 Growth rate dependence of the $NdFeO_3$ single crystal grown by float-zone technique *Journal of Crystal Growth* **318** 927 – 931

[18] Aono H *et al* 1998 Characterizations of $NdFe_{0.5}Co_{0.5}O_3$ Trimetallic Oxide Prepared by Thermal Decomposition of Heteronuclear Complex , $Nd[Fe_{0.5}Co_{0.5}(CN)_6]_4H_2O$ *Journal Ceramic Society of Japan* **106** 10

[19] Chanda S *et al* 2013 Raman spectroscopy and dielectric properties of nanoceramic $NdFeO_3$ *Materials Research Bulletin* **48** 1688 – 1693

IOP Conf. Series: Materials Science and Engineering **367** (2018) 012043 doi:10.1088/1757-899X/367/1/012043

Synthesis and Characterization of Monodisperse Core-shell Lanthanide Upconversion Nanoparticles $NaYF_4$: Yb,Tm/SiO_2

R. V. Manurung[1], G. Wiranto[1], and I. D. P. Hermida[1]

[1]Research Center for Electronics & Telecommunication – Indonesian Institute of Science Jl. Cisitu No.21/154D, Komplek LIPI Sangkuriang, Bandung, Jawa Barat - 40135, Indonesia

E-mail : rvmanurung@gmail.com

Abstract. Lanthanide up-converting luminescent nanoparticles (UCNPs) are exciting and promising materials for optical bioimaging, biosensor and theranostic due to their unique and advantageous optical and chemical properties. The UCNPs absorb low energy near-infrared (NIR) light and emit high-energy shorter wavelength photons (visible light). Their unique features allow them to overcome various problems associated with conventional imaging probes such as photostability, lack of toxicity, and to provide versatility for creating nanoplatforms with both imaging and therapeutic modalities. This paper reports synthesis and characterization of core-shell structured of $NaYF_4$:Yb,Tm/SiO_2 microspheres. The synthesis of lanthanide upconversion nanoparticles $NaYF_4$:Yb,Tm was prepared by thermal decomposition process which involves dissolving organic precursors in high-boiling-point solvents oleic acid (OA) and octadecene (ODE). After that, the $NaYF_4$:Yb,Tm phosphors was coated by silica via reverse microemulsion process to obtain core-shell structured $NaYF_4$:Yb,Tm/SiO_2. Scanning electron microscopy, transmission electron microscopy, specific area electron diffraction, and photoluminescence were applied to characterize these samples. The obtained core-shell structured $NaYF_4$:Yb,Tm/SiO_2 phosphors exhibit a perfect cubic morphology with narrow size distribution and smooth surface. Upon IR excitation at 980 nm, the $NaYF_4$:Yb,Tm/SiO_2 samples exhibit whitish blue upconversion (UC) luminescence, respectively. These phosphors show potential applications in the displaying on biological fields and biosensing.

Keywords. Core-shell, lanthanide, photoluminescence, and upconversion.

1. Introduction

Organic dyes and fluorescent proteins have been used for fluorescent probes in biological detections including bioimaging and biosensing in last decades. It is well known that organic fluorescent dyes have several advantages such as small size, excellent biocompatibility, easy surface modifications for covalent conjugations, and relatively high fluorescent intensity [1]. However, some drawbacks have been found such as short detection times due to high photobleaching efficiency and chemical degradation [2]. Also, organic dyes have narrow absorption and broad emission spectra with long tailing which limits their detection [1–2]. Semiconductor quantum dot (QD) has advantages compared to conventional organic dyes such as size-tunable emission, bright photoluminescence, good photostability, broad ultraviolet (UV) excitation and narrow emission have emerged as alternative and promising fluorescent labels agent [3]. However, QD has several drawbacks such as cytotoxicity issue,

Content from this work may be used under the terms of the Creative Commons Attribution 3.0 licence. Any further distribution of this work must maintain attribution to the author(s) and the title of the work, journal citation and DOI.
Published under licence by IOP Publishing Ltd

and chemical instability[4] have limited application for long-term usage. Moreover, UV or short wavelength excitation usually required by traditional biolabels such as organic dyes, fluorescent proteins, and QDs. The short wavelength may result in severe limitations such as low light-penetration depth, possible damage or even death of biomolecules caused by long-term irradiation, and low signal-to-noise ratio due to significant autofluorescence (background) from biological samples in the UV short wavelength regions.

Nowadays, non-linear optical processes phenomena known as upconversion (UC) where the conversion of near-infrared radiations into visible light by combining two or more low energy photons to generate a single high-energy photon by an anti-Stokes process. The applications of upconversion nanoparticles (UCNPs) have become very prominent in bioimaging and biosensor field. In particular, UCNPs serve as an alternative and excellent substitute for traditional fluorescent labels. In contrast to conventional luminescent probes, UCNPs exhibit a great photostability, sharp multi-peak line emission, and continuous emission capability. Furthermore, light scattering by biological tissues is substantially reduced with near-infrared (NIR) excitation. By absorb NIR excitation result in the reduction in light scattering, and it implies in depth of penetration in tissue much more significant than that obtained under ultraviolet or visible excitation [5]. UCNPs typically comprises three key components: a host matrix, a sensitizer (absorber) and an activator [6]. The host matrix is one of the essential elements of UCNPs due to their essential and unique UC optical properties, including luminescence efficiency and emission profile. The sensitizer can be efficiently excited from specific light source and then transfers the energy to the activator where radiation can be emitted. Thus, the activator is the luminescence center in UCNPs, and the sensitizer enhances the UC luminescence efficiency. The dopants, sensitizer, and activator are added to the host lattice in relatively low concentrations (~20 mol% for the sensitizer and ~2 mol% for the activator). The trivalent Yb^{3+} ion, having a straightforward energy level scheme, therefore it is suitable as a sensitizer. Er^{3+}, Tm^{3+}, and Ho^{3+} ions have energy levels in the form of a ladder and hence are most frequently used as UC activators. Among the available types of UC host materials, fluorides have proven to be ideal host candidates for UC because of their very low phonon energies and high chemical stability.

In this paper, we report the synthesis of UCNPs of $NaYF_4$:Yb,Tm in oleic acid (OA) and octadecene (OD) by thermal decomposition method and followed by a SiO_2 coating by using microemulsion for water dispersibility. Transmission electron microscopy, scanning electron microscopy, specific area electron diffraction, and photoluminescence emission were applied to characterize these samples.

2. Materials and methods

2.1. Synthesis of $NaYF_4$:Yb,Tm nanocrystals

All the chemicals used in this work were purchased from Sigma-Aldrich without further purification. $NaYF_4$:25%Yb, 0.3%Tm nanocrystals were synthesized as follows [7]: 0.75 mmol YCl_3 (Yttrium trichloride, 99.99 %), 0.25 mmol $YbCl_3$ (ytterbium(III)chloride, 99.99 %) and 0.003 mmol $TmCl_3$ (thulium(III)chloride) were mixed with 6 mL oleic acid (OA) and 15 mL octadecene (ODE) in a 50 mL flask. The solution was heated to 160 °C to form a homogeneous solution and then cooled down to room temperature. Then, 10 mL methanol solution containing NaOH (2.5 mmol) and NH_4F (4 mmol) was slowly added to the mini schlenk line flask then stirred for 30 min. The solution was slowly heated to remove methanol, degassed at 100 °C for 10 min, and followed by heating at 300 °C and maintained for 1 h under inert gas protection. After the solution was cooled naturally, nanocrystals were precipitated from the solution with ethanol and washed with ethanol/water (1:1 v/v) for three times.

2.2. Synthesis of the SiO_2 shell on the $NaYF_4$:Yb,Tm nanoparticles

The hydrophobic $NaYF_4$:Yb,Tm nanocrystals, in hexane, surfactants, and ammonia were added to form a water-in-oil reverse microemulsion. A relatively high nanocrystal concentration was used, and the emulsion was sonicated to ensure that all the nanocrystals were encapsulated in a microemulsion pool. Firstly we prepared a mixture of 0.1 mL of CO-520 (IGEPAL), 6 mL of hexane, and 4 mL of 10 mM $NaYF_4$:Yb,Tm nanoparticles in hexane solution. The solution was stirred for 10 min; 0.4 mL of CO-520 and then add 0.08 mL of 30 wt% ammonia followed by sealing the container. The container was then sonicated for 20 min until a transparent emulsion was formed. Then, 0.04 mL of TEOS was added to the solution. The solution was then stirred for 48 h at 600 rpm. $NaYF_4$:Yb,Tm/SiO_2 core/shell nanoparticles were precipitated by adding acetone and washed twice with ethanol/water (1:1 v/v), and then re-dispersed in water.

2.3. Characterization

SEM and TEM images of the nanocrystals were obtained using a FESEM 6700F, JEOL, Japan and FEI Tecnai G2 20 S-Twin transmission electron microscope equipped with energy dispersive spectroscopy (EDS) and selective area diffraction (SAD), respectively. Upconversion fluorescence spectra were obtained on a Maya2000 Pro spectrometer from Ocean Optics with an external 980 nm laser (5 W, continuous wave Shanghai Dream laser) as the excitation source.

3. Results and discussion

3.1. Morphology and structure of the nanocrystals

Figure 1.a exhibited the morphology of $NaYF_4$:Yb,Tm under SEM investigation while Figure 1.b showed the TEM images of the as-prepared core $NaYF_4$:Yb,Tm and the core-shell $NaYF_4$:Yb,Tm/SiO_2 nanoparticles. Under TEM investigation, UCNPs with spherical morphology were observed, which indicated that the adsorption of the hydrophobic capping ligand of oleic acid on the crystal surface prevented the further aggregation of $NaYF_4$ toward macroscopic crystals, leading to the formation of the nanocrystal. UCNPs also possessed narrow size distribution with an average diameter of 10 nm. High magnification TEM image on inset indicated that the core $NaYF_4$:Yb,Tm were single crystal, with an interplanar distance of 0.3 nm (Figure 1.b). The problem of thermal decomposition is that it typically produces nanocrystals with hydrophobic surfaces. The ideal luminescence nanoparticles used for biology should meet several requirements, such as being homogeneously dispersed and colloidally stable in aqueous solvents, maintaining a high quantum yield and showing high specific binding to biological components. The insolubility of such nanocrystals in aqueous solution thus greatly limits their bio-applications and silica coating as one of the solutions to overcome the problem. Silica layer on the core UCNPs surface has been succeeding to growth by reverse microemulsion as a common technique of SiO_2 coating along with Stober method [8]. The dense amorphous SiO_2 shell thickness can be tuned by changing the concentration of tetraethyl orthosilicate (TEOS). However, coating single nanoparticles with uniform thin shells are very challenging [9–10]. The inset of Figure 1.c exhibit ± 7 nm thickness of SiO_2 shell on the surface of $NaYF_4$:Yb,Tm.

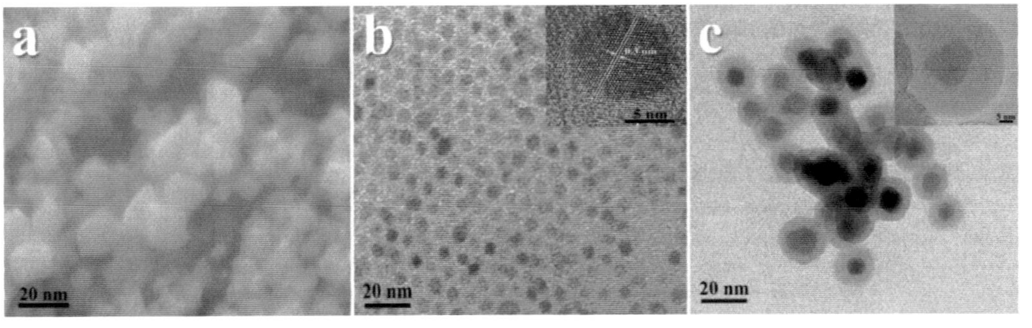

Figure 1. (a) SEM, and TEM images of (b) the core $NaYF_4$:Yb,Tm (c) the core-shell $NaYF_4$:Yb,Tm/SiO_2 nanoparticles. The inset in b, and c refer to high magnification TEM images of the core and core-shell, respectively.

IOP Conf. Series: Materials Science and Engineering **367** (2018) 012043 doi:10.1088/1757-899X/367/1/012043

3.2. Upconversion Fluorescence and elemental properties of upconversion nanoparticles

Fluorescence spectra of transparent colloidal solutions of $NaYF_4$:Yb,Tm nanospheres in hexane is shown in Figure 2.a. The emission peaks of $NaYF_4$:Yb,Tm nanospheres has been observed at 360, 450, 478, 649, 679, 700 and 800 nm were due to the electrons transitions from the energy levels $^1D_2 \rightarrow {}^3H_6$; $^1D_2 \rightarrow {}^3F_4$; $^1G_4 \rightarrow {}^3H_6$; $^1G_4 \rightarrow {}^3F_4$; $^3H_5 \rightarrow {}^3H_6$, $^3F_2 \rightarrow {}^3H_6$ and $^3H_4 \rightarrow {}^3H_6$ [11]. The inset of figure 2(a) exhibited an optical photograph of quartz cuvette showing blue luminescence from core $NaYF_4$:Yb,Tm nanocrystals and it is showed that the color obtained from highly radiative relaxation from the electron which is populated on the blue region (478 nm). The luminescence and radiant transition result of our core $NaYF_4$:Yb,Tm nanocrystals indicate the similar result with previous publication reported by Sudheendra *et al* [12]. However, our radiant transition and fluorescence spectra exhibit more additional peaks compare to the results reported by Sudheendra *et al* [12]. Moreover, the core of UCNPs displayed higher optical absorption around 977 nm which indicate a match with excitation wavelength as shown in Figure 2.b due to utilizing ion pairs (sensitizer/activator) of Yb^{3+}/Tm^{3+} for enhanced excitation at ~975 nm. The result indicates Yb^{3+} works perfectly as a sensitizer, as it has a sufficiently large absorption cross-section in the NIR region at ~975 nm [13]. The energy dispersive spectroscopy indicated their chemical symbol identifies different elemental peaks. The EDS result confirmed the presence of Yb and Tm in the upconversion nanoparticles, and the amount of Y, Yb, and Tm was quantified using this technique. Figure 2(d) shows the selected area diffraction (SAD) pattern of the $NaYF_4$:Yb,Tm nanocrystals. The positions of the polycrystalline diffraction ring corresponding to the (111), (200), (220), (311), (222), (400), (331) and (420) planes. These identified planes agreed with the single-phase cubic or alpha (α) $NaYF_4$ crystals (JCPDS: 06-0342), and no peaks of impurities were detected [11, 12]. The result shows that thermal decomposition method is an appropriate method for preparing doped $NaYF_4$ nanocrystals with a purely cubic phase. The SAED result also suggests that $NaYF_4$:Yb,Tm successfully crystallize on the surface amorphous SiO_2 as an outer coating layer. No other phase related with the doped Yb^{3+} and Tm^{3+} can be observed, indicating the high purity of the sample.

Figure 3 showed the luminescence spectra of $NaYF_4$:Yb,Tm at a various excitation power of 980 nm laser source start from 1 W to 5 W. Each of signature peak of $NaYF_4$:Yb,Tm indicated upconversion process was observed at (360, 450, 478, 649, 679, 700 and 800) nm. These peaks were associated to the electrons transitions from the energy levels $^1D_2 \rightarrow {}^3H_6$; $^1D_2 \rightarrow {}^3F_4$; $^1G_4 \rightarrow {}^3H_6$; $^1G_4 \rightarrow {}^3F_4$; $^3H_5 \rightarrow {}^3H_6$, $^3F_2 \rightarrow {}^3H_6$ and $^3H_4 \rightarrow {}^3H_6$ [11] and plot as shown in Figure 3 (a to d). The result indicates either core and core-shell of $NaYF_4$:Yb,Tm/SiO_2 has consistent respond on each of the signature peaks of luminescence spectra to different excitation power. The result also provides understanding that the more incoming photon from 980 lasers by increasing power excitation. The more electron transition transfer within the electronic state of lanthanide trivalent (Yb^{3+} and Tm^{3+}) produced upconversion luminescence as mentioned as well by Sudheendra *et al* [12] and Chen *et al* [13]. The up-conversion photoluminescence intensity (I_{pl}) has a dependence on the power (P) of the excitation light given by [13]

$$I_{pl} = KP^n \tag{1}$$

where ‘n’ is an integer that determines the number of photons utilized in the photoluminescence process, and K is a material-related coefficient.

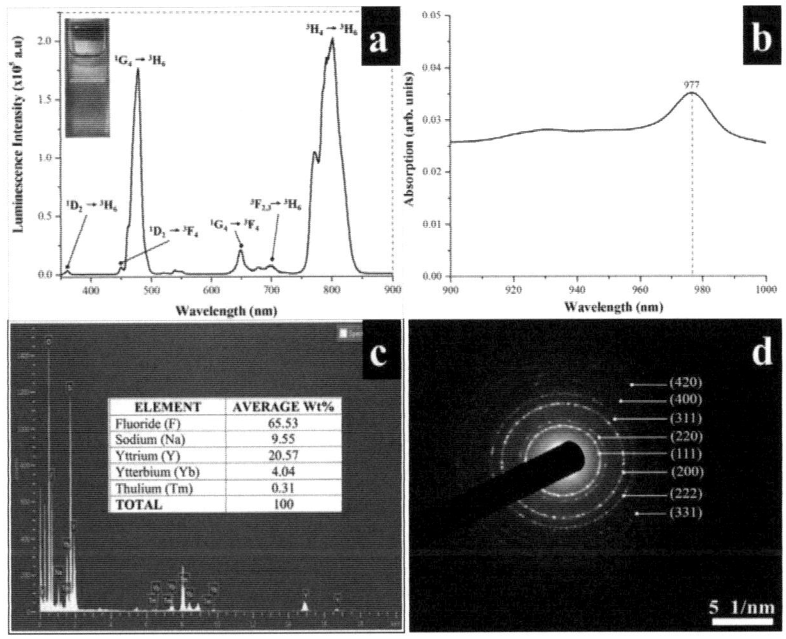

Figure 2. (a) The luminescence spectra of $NaYF_4$:Yb,Tm emission in hexane, under excitation power 5 W of 980 nm, (b) absorption spectra, (c) energy dispersive spectra (EDS), and (d) selected area diffraction (SAD). Their chemical symbols identify the elemental peaks in the EDS (c), and diffraction rings are distinguished (by the arrows and Miller indices) according to JCPDS No. 06-0342 (d).

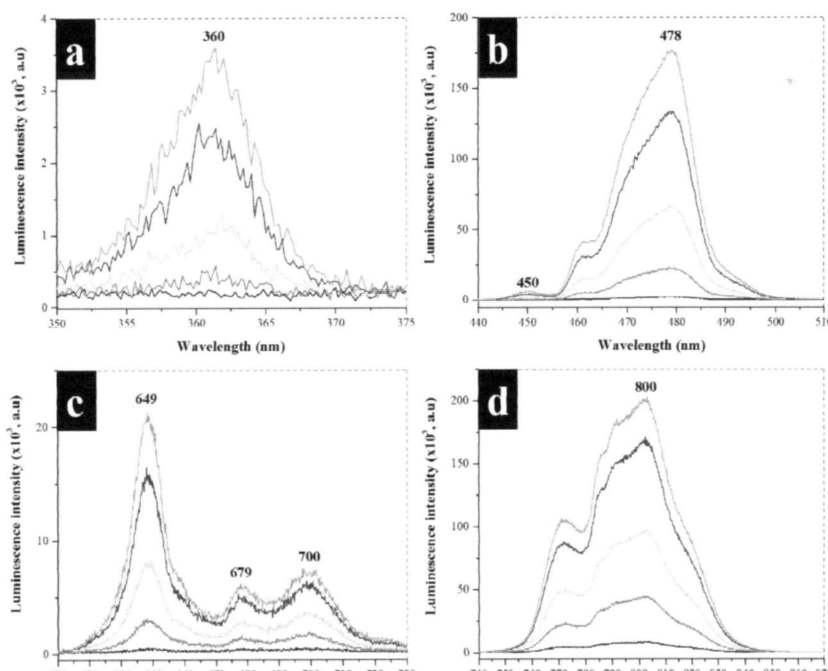

Figure 3. The luminescence spectra of the (a) 360 (b) 450, 478 (c) 649, 679, 700 and (d) 800 nm peak emission of $NaYF_4$:Yb, Tm in hexane, under different excitation power 1-5 W of 980 nm.

4. Conclusions

A monodisperse and purely cubic phase of $NaYF_4$:Yb,Tm nanocrystals were prepared with oleic acid and octadecene as a chelating agent and stabilizer have obtained by thermal decomposition method. The nanocrystals have average diameter ±10 nm with lattice spacing 0.3 nm. SiO_2 coating has been succeeding by reverse microemulsion to increase solubility, and the thickness was around 7 nm and can be adjusted by controlling the concentration of TEOS during hydrolysis. Under 980-nm excitation, the core-shell $NaYF_4$:Yb,Tm/SiO_2 upconversion nanoparticles emit strong upconversion fluorescence. The further application of UCNPs $NaYF_4$:Yb,Tm has open new opportunities in fields such as bioimaging, biosensor and photodynamic therapy.

Acknowledgments

This study was supported by the Ministry of Research, Technology and Higher Education of the Republic of Indonesia under INSINAS Program 2017. The authors thank Functional Nanomaterial Laboratory – National Yang-Ming University for EDS and SAED measurements.

References

[1] Shen J *et al* 2008 Luminescent rare earth nanomaterials for bioprobe applications *Dalton Trans.* **42** pp 5687–5697
[2] Smith A M *et al* 2004 Quantum dot nanocrystals for in vivo molecular and cellular imaging *Photochem Photobiol.* **80** pp 377–385
[3] Chan W C W and Nie S M 1998 Quantum dot bioconjugates for ultrasensitive nonisotopic detection *Science* **281** pp 2016–2018
[4] Chang E *et al* 2006 Evaluation of Quantum Dot Cytotoxicity Based on Intracellular Uptake *Small* **2** pp 1412–1417
[5] Zhang Y *et al* 2014 *Chem. Soc.* **136** pp 4893–4896
[6] Wang F and Liu X G 2009 Recent advances in the chemistry of lanthanide-doped upconversion nanocrystals *Chem. Soc. Rev.* **38** pp 976–989
[7] Manurung R V *et al* 2016 A plasmon-tuned 'gold sandwich'for metal enhanced fluorescence in silica coated $NaYF_4$: Yb, Er upconversion nanoparticles *RSC Adv* **6** pp 87088–87095
[8] Zhang H *et al* 2011 Highly spectral dependent enhancement of upconversion emission with sputtered gold island films *Chem. Commun.* **47** pp 979–981
[9] Eichailbum M and Rademann K 2009 Plasmonic Enhancement or Energy Transfer On the Luminescence of Gold-, Silver-, and Lanthanide-Doped Silicate Glasses and Its Potential for Light-Emitting Devices *Adv. Funct. Mater.* **19** pp 2045–2052
[10] Verhagen E, Kuipers L and Polman A 2009 Field enhancement in metallic subwavelength aperture arrays probed by erbium upconversion luminescence *Opt. Express* **17** pp 14586–14598
[11] Yin A X *et al* 2010 Colloidal synthesis and blue based multicolor upconversion emissions of size and composition controlled monodisperse hexagonal NaYF 4: Yb, Tm nanocrystals *Nanoscale* **2** pp 953–959
[12] Sudheendra L *et al* 2011 *Chem. Mater.* **23** pp 2987–2993
[13] Chen G *et al* 2014 *Chem. Rev.* **114** 5161–5214

IOP Conf. Series: Materials Science and Engineering **367** (2018) 012044 doi:10.1088/1757-899X/367/1/012044

Fabrication of Polyvinylpyrrolidone Fibers by Means of Rotary Forcespinning Method

D. Andjani[1,2], I. Sriyanti[1,2], A. Fauzi[1], D. Edikresnha[1,2], M. M. Munir[1,2] and Khairurrijal[1,2]

[1]Department of Physics, Faculty of Mathematics and Natural Sciences, Institut Teknologi Bandung, Bandung 40132, Indonesia
[2]Bioscience and Biotechnology Research Center, Institute for Research and Community Services, Institut Teknologi Bandung, Bandung 40132, Indonesia

E-mail: miftah@fi.itb.ac.id

Abstract. Fibers made from polymer materials have been widely developed as a carrier medium of active ingredients in drug delivery systems. In this research, PVP polymer was chosen because of its wide and safe use in the medical field. The purpose of this study was to produce PVP fibers that can later be applied as a carrier of active ingredients in drug delivery systems. The rotary forcespinning (RFS) method was chosen to shorten the time of production and to overcome the limitations of electrospinning method such as the use of high voltage and dielectric solutions. The PVP solution was varied in several concentrations (8 wt%, 10 wt%, 12 wt%, 14 wt%, 16 wt%, and 18 wt%) to achieve the best fibers morphology. The morphology and the diameter of fibers were analyzed using a digital microscope. From the microscope images, it can be shown that beaded fibers were formed when the concentration of polymer in the precursor solution was low. The number of beads decreased as the concentration of polymer increased. Beads-free fibers were fully formed at above certain polymer concentration.

Keywords. Fibers, polyvinylpyrrolidone and rotary forcespinning.

1. Introduction

Fibers, at a size ranging from micro-to-nanometer, have been developed intensively in engineering material for a wide range of applications. One of the broadest applications of micro/nanofibers is in the medical field. They can be used as a drug delivery system, body tissue scaffolding, membrane filter, transparent conductive fibers, biocatalysts and wound dressing [1]. The fibers at this size have many advantages such as it has a very high ratio of surface area to unit volume, low density, high porosity, and high flexibility [2–3]. These characteristics are very much needed if the fiber is used as a carrier for the active agent in drug delivery systems. By possessing these characteristics, the solubility, the dissolution rate, and the biological availability of the active agent can be increased significantly although it previously has low solubility when being swallowed directly in its original form.

Since the building block of fibers is a polymer, a precursor solution with a certain concentration of polymer must be first prepared before the fiber is being synthesized. Some commonly used polymers are poly(ethylene glycol) (PEG), polyvinylpyrrolidone (PVP), polyvinyl alcohol (PVA), polyacrylic

Content from this work may be used under the terms of the Creative Commons Attribution 3.0 licence. Any further distribution of this work must maintain attribution to the author(s) and the title of the work, journal citation and DOI.
Published under licence by IOP Publishing Ltd

acid (PAA), polyacrylamides, polylactic acid (PLA), cellulose acetate, and others [4]. In this study, PVP powder was used since it has been proven by the US Food and Drug Administration (FDA) for health applications, especially for food and medicine. Therefore, it can be used as a carrier of active agents in drug delivery systems [5]. Also, PVP can be dissolved very easily in ethanol which often being used as the solvent in medical application[6]. Additionally, there were reports that PVP powder can be successfully synthesized into nanofibers with the conventional electrospinning method [1,7]. Electrospinning is a technique to produce fibers by using a high electric field to form fibers [8]. Unfortunately, this method requires a very long time [1]. Therefore, a better method must be developed to reduce the duration needed.

Rotary forcespinning (RFS) is a method that can eliminate and/or minimize the limitations encountered in electrospinning. The high rotational speed of RFS ensures that the solution will run out rapidly compared to the time-consuming electrospinning [4,9]. Therefore, the use of RFS technique can improve the selection of material, increase the production rate, and lower the cost of fibers synthesis through environmentally friendly processes [10–11]. In our previous experiment, we used the RFS technique directly to encapsulate natural extract into PVP fibers [12]. In this research, we varied the concentration of PVP solution to observe the change of morphology of the fibers. To the best of our knowledge, there is no study that relates the concentration of PVP to the diameter of rotary forcespun fibers, although the similar study has been conducted using PEG [11].

2. Materials and methods

2.1. Synthesis of precursor solution

PVP powder K85-95 (1,300 kg·mol^{-1}) was purchased from Sigma Aldrich and ethanol, as the solvent, with 96 % purity was purchased from Sakura Pharmacy Store. The solution was synthesized at 40 °C, stirred by a magnetic stirrer at ± 400 rpm for ± 60 min until a clear homogeneous solution was achieved.The concentration of PVP in the solution was varied at 8 wt%, 10 wt%, 12 wt%, 14 wt%, 16 wt%, and 18 wt%.

2.2. The production of fibers using rotary forcespinning method

In the RFS technique, the solution is placed inside a rotating container, in which the solution is then ejected out from the reservoir through a hole. The electric field used to attract fibers in the electrospinning process is replaced by the centrifugal force of motor rotation. As known from elementary physics, the centrifugal force (F) depends on mass (m), rotational speed (ω) and radius (R) and they are related according to $F= m\omega^2R$. The centrifugal force becomes the main factor in the production of fibers, and the stretch of this force causes the solution to turn into fibers. There are three main systems of the RFS technique: a motor system, a collector, and a heating system that controls the temperature and humidity around the motor [4,10]. Some steps of the RFS method in producing fibers are as follows (also described in Figure 1):

1. Jet initiation to stimulate the flow of the polymer solution through the hole,
2. Jet extension to increase the surface area of the polymer solution flow driven out from the reservoir through the hole (elongation process),
3. Solvent evaporation and solidification processes [13].

By using centrifugal force to stretch the fiber and by adjusting the parameters involved such as the inner diameter of the needle, the distance between the needle tip, and the collector, the concentration of the solution, the flow rate, and the surface tension, the RFS technique is a proficient method to produce fibers.

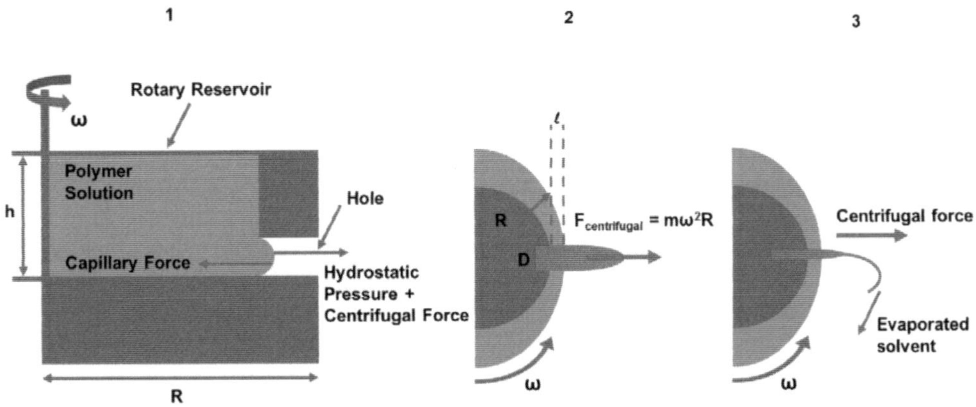

Figure 1. The mechanism of fibers formation using the RFS technique[13]

The homogeneous PVP solution was placed in a single nozzle syringe, which was connected to a silicon hose with an inner diameter of 0.45 cm. The collector was coated with an aluminum foil in which the fiber was attracted onto. The hole diameter was 0.6 cm, the distance between the hole and collector was 10 cm, the rotational speed was set to be 15,000 rpm. The flow rate of the solution in the syringe was 30 mL/h, and the relative humidity inside the chamber was maintained at 80 %. The spinning process was carried out at 23 °C.

2.3. Characterization of morphology and diameter of fibers

The characterization of the morphology of the fiber was performed using a digital microscope (National, DC3-163). The magnification of digital microscope used was varied from 100×, 200×, 400× to 1000×.To calculate the diameter of the composite fiber, we used ImageMIF software developed in the Laboratory of Electronics and Instrumentation in Department of Physics, Institut Teknologi Bandung.

3. Results and discussion

The synthesis of PVP fibers was initiated by preparing precursor solutions at six different concentrations, which were 8 wt%, 10 wt%, 12 wt%, 14 wt%, 16 wt%, and 18 wt%. The concentration of the solution was varied to observe the effect of concentration on the morphology of the fibers. Other than the concentration, all other parameters were kept constant. The resulting fibers were then characterized using the digital microscope (Figure 2).

As illustrated in Figure 2, the variation of the polymer concentration has a significant effect on the formation of beads and the diameter of the fibers formed. The fiber synthesized from precursor solution with the lowest concentration (8 wt%) produced beaded fibers with a large number of beads. As the concentration of the solution increased, the number of beads decreased. The fibers made from a precursor solution with a concentration of 16 wt% were beads-free. However, when the concentration of the polymer solution was increased further to 18 wt%, beaded fibers were re-formed although at a smaller number compared to the beads at made from the precursor solution with a concentration lower than 16 wt%. Therefore, it can be concluded that most uniform morphology of fibers was achieved when the concentration of polymer was 16 wt%

To explain the formation of fiber beads, one must refer to the surface tension and viscosity of the precursor solution. In this experiment, the concentration of 14 wt% can be called as the critical concentration since it produced the very small amount of beads. The term critical concentration is a specific concentration in which the morphology of the fibers starts to be continuous and also free of beads. When the concentration is less than the critical concentration, the viscosity of the solution is low. At low viscosity, the polymer chain has low binding energy, which is then not strong enough to

IOP Conf. Series: Materials Science and Engineering **367** (2018) 012044 doi:10.1088/1757-899X/367/1/012044

form fibers. On the other hand, high surface tension causes the jet to become unstable, resulting in the decrease of surface area per unit of mass and the jet tends to form spherical shapes. Therefore, spherical droplets will appear on the fibers and the viscoelastic force in the polymer solution will withstand sudden changes of shape (spheres) which will then produce fibers with beads. As the concentration of polymer becomes higher, the viscosity will be larger as well. This will cause the polymer chain to have a stronger bond as the concentration goes near to the critical concentration. At this situation, beads-free fibers are then formed. Even if there is still some small amount of beads, the slight increase in polymer concentration will reduce the number of beads although the diameter of beads will increase [11].

However, as shown in Figure 2, there were beads of fibers made from the precursor solution at a concentration of 18 wt%, which should not exist based on the argument in the previous paragraph. This shows that there are other factors that also trigger the formation of beads other than the viscosity and surface tension. The factors that may trigger the formation of beads in the fiber are the rotational speed and the centrifugal force generated by the ac motor. Just as there is critical concentration, there also exists critical rotational speed, which is the critical rotational speed for each concentration of polymer in the precursor solution to form fibers [14]. The greater the concentration of the solution, the greater the rotational speed is required.

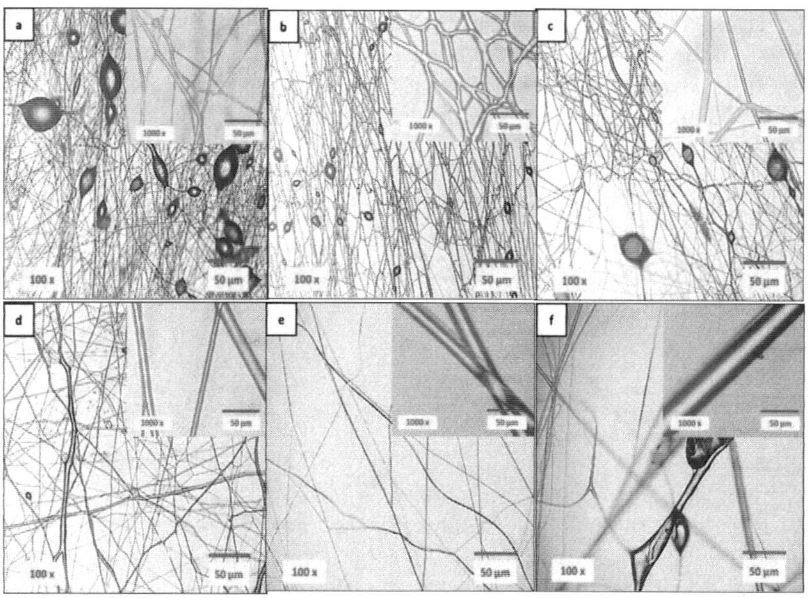

Figure 2. The morphology of PVP fibers under a digital microscope with a magnification of 100× for PVP concentrations of (a) 8 wt%, (b) 10 wt%, (c) 12 wt%, (d) 14 wt%, (e) 16 wt% and (f) 18 wt%

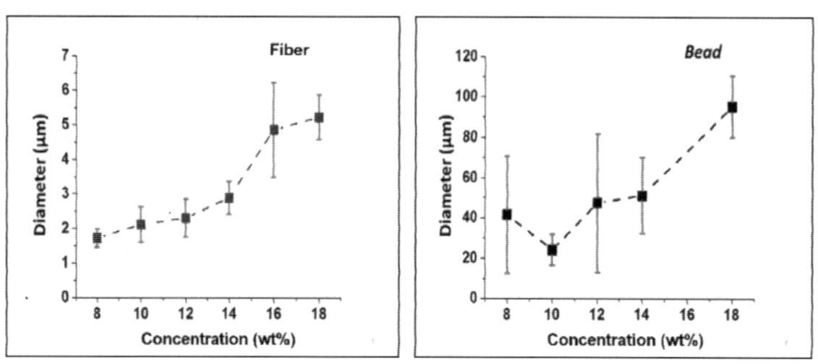

Figure 3. The effect of increasing the concentration to the average diameter of fibers and beads.

IOP Conf. Series: Materials Science and Engineering **367** (2018) 012044 doi:10.1088/1757-899X/367/1/012044

In this experiment, the rotational speed was kept constant for all concentrations of the precursor solution. At the concentration of 18 wt%, the rotational speed of 15000 rpm was not sufficient to produce a centrifugal force capable of attracting fibers. Thus, the fibers formed were not perfectly cylinder, and some spherical droplets appeared. Due to its large concentration, the resulting fibers had a large diameter and large beads. The exact relationship between the concentration and diameter is illustrated in Figure 3. It can be seen that for a higher concentration, the average diameter of fibers is greater as well.The fibers made from the solution concentration of 18 wt% had the largest average diameter, which was 5.22μm [4,11,15]. Lastly, Figure 4 depicts the distributions of the resulting fiber diameter. To decide whether a group of fibers is homogenous, a ratio called as the coefficient of variation (CV) is used, which is simply a ratio between the standard deviation of fibers diameter over the average fiber diameter. For CV larger than 0.3, the fibers are non-uniform, otherwise, the fibers are uniform.

As shown in Table 1, all CVs have a value less than 0.3. However, it should be taken into note that the beaded fibers were fully achieved when the precursor solution was PVP 14 and 16 wt%. Since both sets of fibers have CVs below 0.3, the resulted fibers can be categorized as homogenous [15].

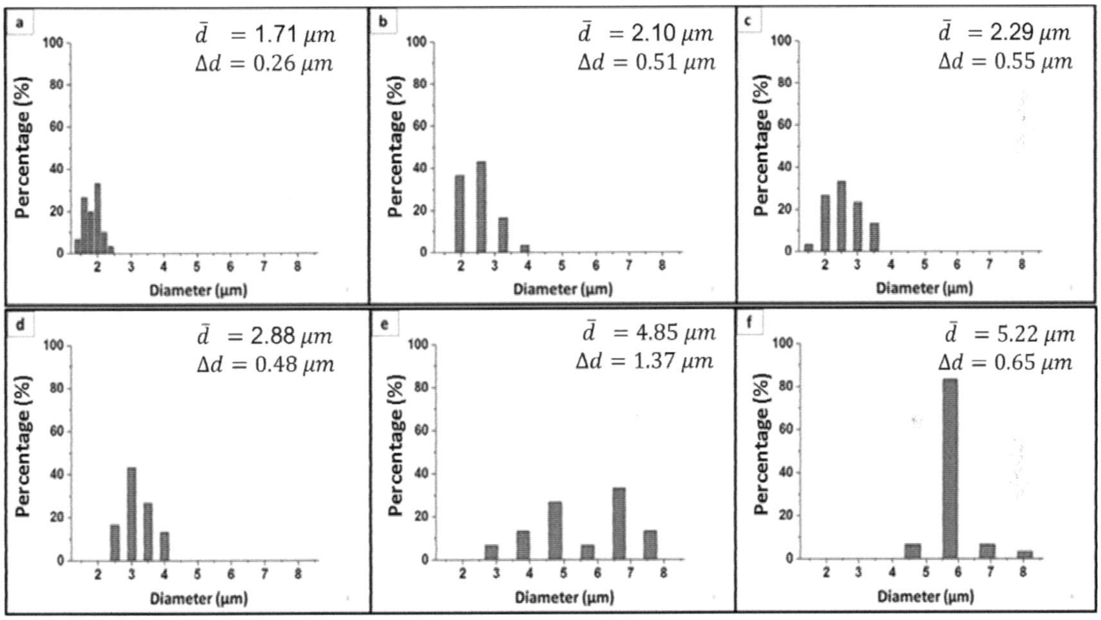

Figure 4. The distribution of PVP fibers diameter produced from the precursor solution concentrations of (a) 8 wt%, (b) 10 wt%, (c) 12 wt%, (d) 14 wt%, (e) 16 wt%, and (f) 18 wt%

Table 1. The physical characteristics of rotary forcespun fibers

Solution concentration (wt%)	Average diameter of fiber [μm]	Coefficient of variation
8	1.71±0.26	0.15
10	2.10 ± 0.51	0.24
12	2.29 ± 0.55	0.24
14	2.88 ± 0.48	0.17
16	4.85 ± 1.37	0.28
18	5.22 ± 0.65	0.12

4. Conclusions

The fibers built up from PVP have been successfully synthesized using rotary forcespinning method although the fibers contained some beads. The morphology and diameter of fibers are affected by several parameters, such as solution concentration, viscosity, surface tension, and rotational speed. The greater the concentration of the solution, the viscosity will be greater thus resulting in the greater average diameter of fibers, and vice versa. The value of viscosity also indicates the binding strength of between polymer chain, which is a very important to determine whether fibers will be formed. The large surface tension leads to the formation of beads on the fibers. A solution with low viscosity indicates that the bonds in the polymer chain are not strong enough to be attracted and to form fibers.

Acknowledgments

This research was financially supported by Directorate of Research and Community Engagement of Ministry of Research, Technology and Higher Education, the Republic of Indonesia, under the University's Excellence Research (PUPT) Grant in the fiscal years 2016-2017.

References

[1] Munir M M *et al* 2009 High performance electrospinning system for fabricating highly uniform polymer nanofibers *Rev.Sci.Instrum.* **80** p 026106
[2] Rathinamoorthy R 2012 Nanofiber For Drug Delivery System – Principle And Application *Pakistan Textile J.* **61** pp 45–48
[3] Yu D 2009 *Health* (Irvine. Calif) **1** pp 67–75
[4] Kadajji V G and Betageri G V 2011 Water soluble polymers for pharmaceutica applications *Polymers* **3** pp 1972–2009
[5] Saffoon N *et al* 2011 Enhancement of Oral Bioavailability and Solid Dispersion: A Review *J. Appl. Pharm. Sci.* **1** pp 13–20
[6] Teodorescu M and Bercea M 2015 Poly(vinylpyrrolidone) – A versatile polymer for biomedical and beyond medical applications *Poly. Plastic Tech. Eng.* **54** pp 923–943
[7] Munir M M *et al* 2009 Scaling law on particle-to-fiber formation during electrospinning *Polymer* **50** pp 4935–4943
[8] Munir M M *et al* 2008 A constant-current electrospinning system for production of high quality nanofibers *Rev. Sci. Instrum.* **79**
[9] Padron S *et al* 2013 Experimental study of nanofiber production through forcespinning *J. Appl. Phys.* **113**
[10] Sarkar K *et al* 2010 Electrospinning to Forcespinning™ *Mater. Today* **13** pp 12–14
[11] Andjani D *Engineering Physics International Conference, EPIC 2016 in Procedia Engineering* **170** (edited by H. Kasai et al. (Elsevier Ltd., 2017) pp 14–18
[12] Zhang X and Lu Y 2014 Centrifugal spinning: an alternative approach to fabricate nanofibers at high speed and low cost *Poly. Rev.* **54** pp 677–701
[13] Fauzi A *et al* 2015 Synthesis of Styrofoam Fibers Using Rotary Forcespinning Technique *Mater. Sci. Forum* **827** pp 279–284
[14] Li Z and Wang C 2013 Effects of working parameters on electrospinning *One-Dimensional nanostructures* pp 15–28
[15] Matulevicius J *et al* 2016 The comparative study of aerosol filtration by electrospun polyamide, polyvinyl acetate, polyacrylonitrile and cellulose acetate nanofiber media *J. Aerosol Sci.* **92** pp 27–37

A paper-based Colorimetric Indicator Label using Natural Dye for Monitoring Shrimp Spoilage

A. Listyarini[1], W. Sholihah[1], C. Imawan[1]

[1]Department of Physics, FMIPA Universitas Indonesia, Kampus Depok, Indonesia 16124

E-mail: cuk.imawan@sci.ui.ac.id

Abstract. Shrimp is a type of perishable food. This study developed a simple indicator label using colorimetric method for monitoring shrimp freshness. This indicator label was made from natural dye extract of *Ruellia simplex* flowers which immobilized on cellulose paper by dip coating method. The indicator labels were used for examining freshness of shrimp. In this experiment, shrimp were stored in sealed bottles that have been labeled using the indicator and stored at 13 °C, 25 °C, and 40 °C for a certain range of time. Color changes of the indicator labels were observed using digital photography after shrimp storage for 0 h, 2 h, 17 h and 24 h. The color changes that occur were quantified and analyzed using the ImageJ program. The color of the indicator label when detecting the fresh shrimp was pink and after the shrimp spoilage began, the color of the label changed to purple and then became yellow when the shrimp is badly spoilage. The color change rates of label indicator increases as the shrimp storage temperature increased. These results indicate that this label indicator can be used as an indicator of the freshness of shrimps and it is not toxic and safe for food.

Keyword. Colorimetric, dip coating, *Ruellia simplex* and shrimp spoilage.

1. Introduction

Shrimp freshness is the most basic indication in determining the quality of the shrimp, whether it can be consumed or not. Therefore, various methods have been developed by researchers to be able to determine the quality of shrimp. Methods to determine shrimp freshness include sensory methods, determination of microbial levels in shrimp, and determination of levels of Total Viable Count (TVC) or Total Volatile Basic-Nitrogen (TVB-N) levels [1]. Determination of quality of shrimp with the sensory method is the simplest way but become an obstacle when shrimp covered in a plastic package. Determining the amount of contamination through microbiological tests, such as measurement of TVC levels and TVB-N levels, takes a long time and should be done in the laboratory [2]. Therefore, it is necessary to develop a method that is simple, practical, and can determine the freshness of shrimp directly and integrated into shrimp packaging. Recently, a method to detect the freshness of perishable food using colorimetric indicator label has been developed. Research on this indicator has been done by other researchers by developing dye-colored labels, either using synthetic or natural dyes [3]. Research on this label by using environmentally friendly materials is very interesting to be developed further. Utilization of natural dyes as an indicator of the freshness of food such as the utilization of red cabbage extract [4], spinach [5], turmeric [1], and black rice [6] have been studied. Natural colorants

Content from this work may be used under the terms of the Creative Commons Attribution 3.0 licence. Any further distribution of this work must maintain attribution to the author(s) and the title of the work, journal citation and DOI.
Published under licence by IOP Publishing Ltd

that can be used for the indicator should have enough stability but can also be affected by factors such as acid-base to change its color, one of the famous natural dyes is anthocyanin [7].

The aim of this research is to develop an indicator label by using *Ruellia simplex* flower extract which is found in Indonesia as an ornamental plant. As the matrix of this dye extract is cellulose paper, all materials used in making this label are environmentally friendly. The indicator label is then tested to detect the freshness of the shrimp. The working principle of the label in the detection of the freshness of food used in this experiment is the colorimetric method.

2. Materials and methods

2.1. Materials
The chemical materials used in this experiment were Ethanol absolute purchased from JT Baker and Sodium Hydroxide and Chloride Acid which are both purchased from Merck with purity 99.99 %.

2.2. Method

2.2.1. Preparation of Anthocyanin Extract from Ruellia Simplex flowers
The *Ruellia simplex* flower used in this experiment were collected from the plant and then washed with clean water and then drained. Flowers weighed as much as 80 g then put into a blender. After that the solution of ethanol/aquadest (7:3) as much as 100 mL mixed into the blender and then the flowers were chopped together. The solution mixture was then dropped with a solution of hydrochloric acid (HCl) until the pH of the solution becomes 2. The solution mixture then stored in a dark bottle for 24 h at 5 °C. After 24 h, the solution mixture was filtered using Whatman no. 41 paper. The resulting supernatant was stored in a dark bottle and used as a natural dye extract from the *Ruellia Simplex* flower. Part of the solution of the *Ruellia simplex* flower extract was further investigated its absorbance spectra in the UV-visible region using a UV-Vis Genesys 10S Thermo Scientific spectrophotometer.

2.2.2. Indicator Label Preparation
The prepared indicator labels were composed of the *Ruellia simplex* dye extract and the matrix made of cellulose-based paper. The cellulose paper used in this research was Whatman no 1 cat.1001. The indicator label was made by cutting the cellulose paper with the size of 10 cm × 1 cm, then dipped into the solution of flower extract of *Ruellia simplex* with pH 2 for 30 s. The indicator label then dried at room temperature and ready for use.

2.2.3. Shrimp freshness detection test
In this study, the shrimp species used for freshness testing were fresh *Fenneropenaeus merguiensis* obtained from the local market. The shrimp after purchased directly transported to the laboratory and immediately prepared for the observation. For the preparation of the test, the shrimp weighed as much as 10 g and then put into a transparent glass bottle. The indicator label was hung on the bottle cap containing the shrimp. Label indicator was placed not in direct contact with the shrimp. The distance between the label and the shrimp was approximately 5 cm. The bottle was then tightly closed. Similar samples were prepared like this and then the bottles were tested for observation in various conditions of shrimp storage temperature 13 °C (in refrigerator) temperature, 25 °C (room temperature), and temperature of 40 °C (in the oven). Color changes that occur on the indicator label due to the decay process experienced by the shrimp were observed after a storage time of 2 h, 17 h, and 24 h at each storage temperature condition. The color change of the label indicator was photographed using the Canon EOS 750D camera and for subsequent color changes was analyzed using the ImageJ program.

3. Results and discussion

The *Ruellia simplex* flower extract produced in this study has a red color at pH 2 as shown in Figure 1.a. The UV-visible spectrum of this extract is shown in Figure 1.b. The red color of this extract shows that under conditions of pH 2 the dominant compound present in anthocyanin is in the form of stable flavylium cation [8]. In the visible spectrum, the *Ruellia simplex* flower extract has an absorbance peak at a wavelength of about 535 nm. This wavelength value is similar to the wavelength value when the maximum absorbance is found in the *Bauhinia blakeana dunn* flower extract [9] or also grape skin extract [10]. The UV-visible spectrum for the *Ruellia simplex* flower extract on base condition (pH value of 11) is also measured (Figure 1.b). This characterization is used as a preliminary detection to determine the color change of the extract. The color change of the extract from the acidic pH 2 condition to the base condition is likely to occur when the anthocyanin detects a rotten shrimp. When the shrimp decay, the condition is base. In the base condition, the color of the extract is yellow and the wavelength at the maximum absorbance value shifts to the right to about 625 nm.

(a)

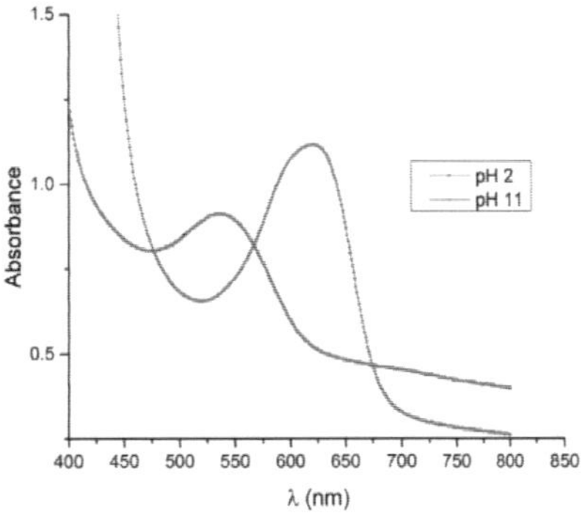

(b)

Figure 1. The color of the Ruellia simplex flower extract conditioned on the pH value of 2 (a), the UV-visible spectrum of the *Ruellia Simplex* flower extract in buffer pH value 2 and pH value 11 (b).

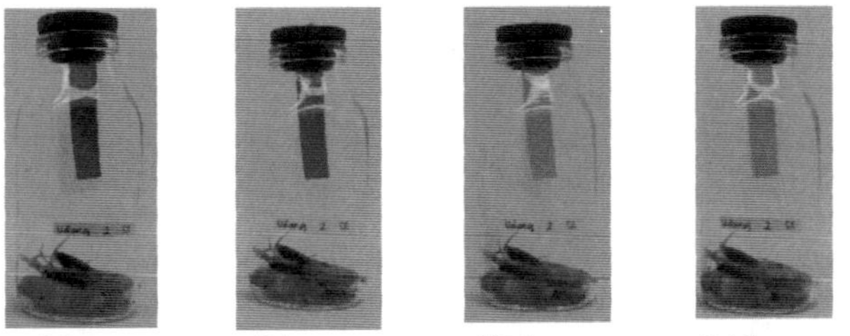

0 hour 2 hours 17 hours 24 hours

Figure 2. Indicator label's color changes upon detecting gas emitted by the shrimp when spoilage process occurs. Measurement is performed at 25 °C.

The indicator label made of cellulose paper that has been dyed into the *Ruellia simplex* flower extract produces a purplish pink color. This color has not changed significantly and is still similar to the color of the extract solution. This is also found in the Zhai *et al* 2017 study that uses anthocyanins from roselle flower extract using starch / polyvinyl alcohol matrix, the color of the early label is purplish pink, in accordance with the color of roselle flower extract at pH 2 [11].

Testing the indicator label to detect shrimp freshness is done by hanging the label inside the shrimp filled bottle, as shown in Figure 2. The bottle containing the indicator and shrimp label is then stored in three different storage temperature conditions, at 13 °C storage temperature, room temperature at 25 °C, and extreme heat conditions with a temperature of 40 °C . Observations of color changes that occur on the label indicator performed simultaneously with an organoleptic test of shrimp. The grading for the shrimp organoleptic conditions is based on the SNI 01-2728.1-2006 standard on fresh shrimp part 1-specification. The results of an organoleptic test of the shrimp are shown in Table 1.

Based on the standards used, shrimp come into the fresh category and can be consumed if it has a value of 7 to 9 [12]. Thus, based on organoleptic test results, storage at 13 °C for up to 24 h indicates that shrimp can still be consumed even though the quality has decreased. While at 25 °C, after being stored at 17 h or more, the shrimp is not worth consuming. At a temperature of 40 °C when the shrimp has been stored for 17 h or more, the shrimp should not be consumed or have been very rotten.

The color map of the indicator label for the observation of the color change in the three storage temperature conditions and the effect of the storage time given in Figure 3.

Table 1. The results of shrimp organoleptic test at 13 °C, 25 °C and 40 °C storage temperature as a function of storage time.

Storage temperature	Results of organoleptic test (score)			
	0 h	2 h	17 h	24 h
13 °C	The shape is unchanged, the color of the flesh is clear luminescent, smells very fresh, very elastic texture, compact and solid (9)	The color of the flesh is less clear, fresh smell, elastic texture (8)	The clear color of the flesh is somewhat lost, specific odor, less elastic texture (7)	
24 °C			The color of the flesh is rather pink, arising ammoniac smell, not elastic texture (5)	The color of the flesh becomes red, smell of strong ammonia, soft texture (3)
40 °C		The clear color of the flesh is somewhat lost, specific odor, less elastic texture (7)	The flesh become very dull red color, dark spots, strong ammonia smell and foul odor, soft and watery texture (1)	

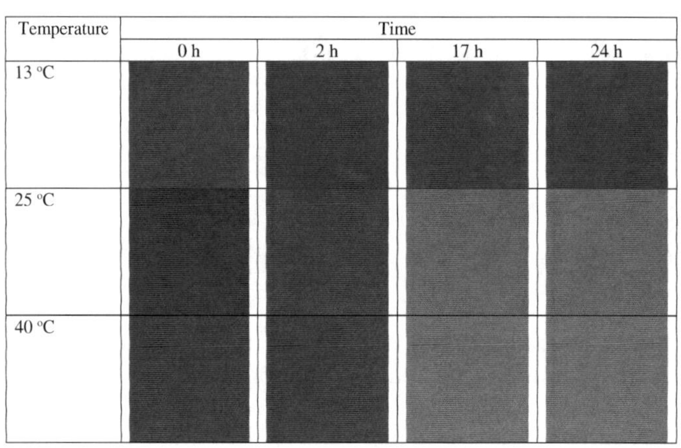

Figure 3. Map of color change of the indicator labels on shrimp storage temperature of 13 °C, 25 °C and 40 °C as a function of time.

After 2 h of storage time, the indicator label for all observed storage temperature conditions changes its color from purplish pink to purplish blue. At a storage temperature of 13 °C, the color of the indicator label with increasing storage time up to 24 h has not undergone significant changes in purplish blue color. On the other hand, the color of the indicator label when measuring shrimp freshness at storage temperatures of 25 °C and 40 °C for storage time up to 17 h, changes further from purplish-blue to greenish gray. The more prolonged storage time of up to 24 h in both conditions of this temperature causes the color of the label to change further into a yellowish gray.

Based on the observations for the color change of the indicator label shown on the color map in Figure 3 and the organoleptic observation results of the shrimp shown in Table 1 it can be concluded that under conditions of storage temperature of 25 °C and 40 °C for 17 h storage time has made the shrimp become rotten and should not be consumed. The color change of this indicator label occurs due to ammonia released during the decomposition process of the shrimp [2]. Ammonia vapor will shift the equilibrium of the anthocyanin form contained in the dye extract in the indicator label. The shift in anthocyanin equilibrium when affected by ammonia corresponds to a shift of the cation of flavylium having a red color to a quinoidal base having a blue color. So the color of the indicator label that originally purplish pink color change to purplish blue. With increasing time, the ammonia produced from the decomposition process will increase steadily so that the anthocyanin that has changed to the quinoidal base will continue to change to a calcon form that has a yellow color [8]. The phenomenon of shifting from this form of anthocyanin is observed when the shrimp become very rotten, the color of the indicator label becomes yellowish gray. The change to yellowish color is also found when the Ruellia simplex flower extract is mixed in the buffer base solution with a pH value of 11 as shown in Figure 1.b. The results of this experiment indicate that the color change of the indicator label due to the changing environmental conditions that became the base for this case due to the ammonia released from the process of shrimp decomposition.

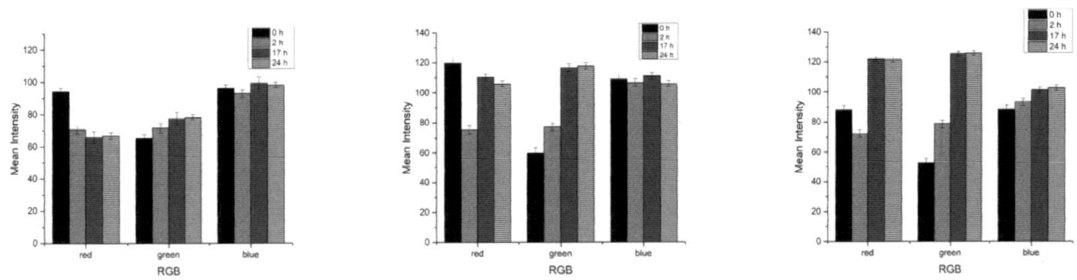

(a) (b) (c)

Figure 4. The RGB histogram of the indicator label as a function of shrimp storage time at the temperature of (a) 13 °C, (b) 25 °C, and (c) 40 °C

The color of the indicator label is further analyzed to quantify the color change using the ImageJ program. The results of the analysis are histograms of RGB as shown in Figure 4. RGB values at various storage temperatures show different trends for changes in the value of each color component. A more regular pattern of changes in color values is indicated by the green color component. At a storage temperature of 13 °C the value of the green color component becomes larger as the length of storage time increases. The trend of increasing values of the green color component is also found for storage temperatures of 25 °C and 40 °C. Based on Figure 4, the green color component can be used to indicate a marker of increased shrimp decomposition process. Figure 5 shows the relationship between the intensity of the green color component as a function of storage time for various storage temperatures.

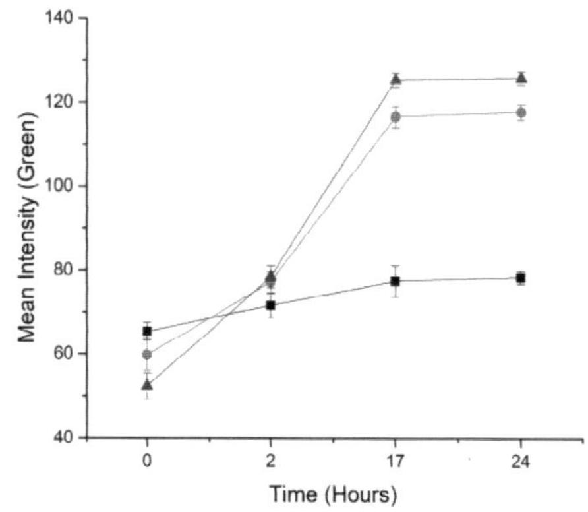

Figure 5. Intensity of green color components as a function of time for various storage temperature (—■— for 13 °C , —●— for 25 °C, and —▲— for 40 °C)

Figure 6. Recommended color for label indicator used to detect shrimp freshness

The intensity of green color components at storage temperatures of 13 °C did not increase significantly compared to the storage temperatures of 25 °C and 40 °C. Storage conditions at cold temperatures can slow down the shrimp process so that the ammonia released by the shrimps is small as a result the label of the indicator does not change color dramatically. When the storage temperature is high, the shrimp become more rapidly decomposed so that the amount of ammonia vapor released increases and the production is faster. This phenomenon causes the labels to change color significantly. The storage time also has an effect on the speed of the indicator color change. The longer of shrimp storage time, the shrimp freshness condition decreased causing the amount of ammonia released increased and caused the label to change color.

Based on the results of the observations described earlier the recommended indicator label is applied to the consumer along with an explanation of the color change can be seen in Figure 6. In Figure 6 for the real application can be given four types of color indicators to monitor the shrimp freshness condition: when the shrimp is in fresh condition, the shrimp can still be consumed even though the quality has decreased, the shrimp is not proper to be consumed, and the shrimp is very rotten. This simple indicator label that simply looks at the color change when it detects the shrimp and compares the color with the given color explanation can help the consumer to quickly identify the freshness quality of the shrimp, whether the shrimp is still fresh or rotten. A similar color change phenomenon was observed in a study of Zhang *et al* 2014 using colorimetric labels from Bauhinia and chitosan flowers extracts to observe the freshness of the fish. Colors of colorimetric labels changed from pink to brown after contact with fish for 12 h. Then change from brown to green after 16 h at 25 °C. The label color turned green after the fish sample was stored over 20 h [9].

IOP Conf. Series: Materials Science and Engineering **367** (2018) 012045 doi:10.1088/1757-899X/367/1/012045

4. Conclusions

A simple method of detecting shrimp freshness has been developed using indicator labels using colored cellulosic paper with *Ruellia simplex* flower extract. The color change of the indicator label from the purplish pink color to the purplish blue then turns further into greenish gray and becomes yellowish gray can be used to express the quality of the shrimp from the fresh to the already rotten sequentially. Quantitatively the quality of shrimp freshness can also be expressed by measuring changes in the intensity of the green color component of the RGB analysis of the indicator label color. Practically this indicator label can be further developed by determining four color indicators that are given color information describing the condition of the shrimp.

References

[1] Olafsdottir G *et al* 2004 Multisensor for fish qualitydetermination *Trends in Food Science and Technology* **15** 86–93

[2] Kuswandi B *et al* 2012 Real-Time Monitoring of Shrimp Spoilage Using On-Package Sticker Sensor Based on Natural Dye of Curcumin *Food Anal. Methods* **5** 881–889

[3] Vanderroost M *et al* 2014 Intelligent food packaging: The next generation *Trends in Food Science and Technology* **39** 47–62

[4] Chen X and Gu Z 2013 Absorption-type optical pH sensitive film based on immobilized purple cabbage pigment *Sensors and Actuators B* **178** 207 – 211

[5] Veiga-Santos P, Dicthfield C, and Tadini C C 2011 Development and evaluation of a novel ph indicator biodegradable film based on cassava starch *Applied Polymer Science* **120** 1069–1079

[6] Huang and Xiao-Wei *et al* 2014 Determination of pork spoilage by colorimetric gas sensor array based on natural pigments *Food Chem.* **145** 549 – 554

[7] Reyes L F and Zevallos L C 2007 Degradation kinetics and colour of anthocyanins in aqueous extracts of purple- and red-flesh potatoes (*Solanum tuberosum* L.) *Food Chem.* **100** 885 2007

[8] Rodriguez-Amaya D B 2016 Natural food pigments and colorants *Current Opinion in Food Science* **7** 20 – 26

[9] Zhang X and Chen Xi 2014 A visual ph sensing film using natural dyes from Bauinia blakeana Dunn *Sensors and Actuators B* **198** 268–273

[10] Ma Q and Wang L 2016 Preparation of a visual pH-sensing film based on tara gum incorporating cellulose and extracts from grape skins *Sensors and Actuators B* **235** 401–407

[11] Zhai X *et al* 2017 Novel colorimetric films based on starch/polyvinyl alcohol incorporated with roselle anthocyanins for fish freshness monitoring *Food Hydrocolloids* **69** 308 – 317

[12] SNI 2016 *Udang segar bagian 1-spesifikasi* (SNI 01-27281-2006) (in Indonesia)

Kinetics and Equilibrium of Fe^{3+} Ions Adsorption on Carbon Nanofibers

Alimin[1], La Agusu[2], L. O. Ahmad[1], L. O. Kadidae[1], L. Ramadhan[1], M. Nurdin[1], N. Isdayanti[1], Asria[1], P Aprilia M[1] and Hasrudin[1]

[1]Department of Chemistry, Universitas Halu Oleo, Kampus Hijau Bumi Tridharma, Anduonohu-Kendari 93232, Indonesia.
[2]Department of Physics, Universitas Halu Oleo, Kampus Hijau Bumi Tridharma, Anduonohu-Kendari 93232, Indonesia.

E-mail: alimin@uho.ac.id

Abstract. Generally, the interaction between metal ions and adsorbent is governed by many factors including; concentration of metal ions, interaction time and solution pH. In this work, we applied liquid phase adsorption for studying the interaction between Fe^{3+} ions and Carbon Nanofibers (CNFs) irradiated by ultrasonic waves. Kinetics and isotherms model of the Fe^{3+} ion adsorption was investigated by varying contact time and pH. We found that the Fe^{3+} ions were efficiently adsorbed on CNFs for 0.5 h in acidic pH of around 5. In order to obtain the best-fitted isotherms model, Langmuir and Freundlich's isotherms were used in this work. The adsorption equilibrium Fe^{3+} metal ions on CNFs tend to follow Langmuir. Adsorption kinetics of Fe^{3+} ions on CNFs were investigated by using both pseudo-first and pseudo-second orders. The adsorption kinetics coincided well with the pseudo-second-order.

Keywords. Carbon nanofibers (CNFs), Fe^{3+} ions, kinetics and nanofiber.

1. Introduction

Study of kinetics and equilibrium of metal ions adsorption on adsorbents such as activated carbon [1–5], modified activated carbon [6, 7] and nanocarbon [8, 9] has been intensively studied for a wide variety of purposes. Kinetics and equilibrium for iron ion removal by activated carbon have been intensively studied [1–5]. Adsorption behavior of Fe^{2+} ion from aqueous solution on activated carbon has been reported [1–3]. Kinetics and thermodynamics of Fe^{3+} ions adsorption onto activated carbon from biomass have been investigated by Mahmoud *et al* [4]. Adsorption of Fe^{3+} on *Acanthaceae* activated carbon has been investigated [5]. In addition, modified activated carbons have been studied [6, 7]. Removal of iron from simulated contaminated groundwater using Fe^{3+} impregnated activated carbon has been reported by Mondal *et al* [6]. Vassileva *et al* [7] have studied the adsorption of Fe^{3+} onto ignited-based activated carbons modified by oxidation.

In the last decades, carbon nanotubes (CNTs) and carbon nanofibers (CNFs) as members of carbon family have attracted the attention of researchers because they have the large surface area, hence they are projected to be one of the promising adsorbents for adsorbing metal ions [8, 9]. Adsorption of the iron ion on the nanocarbon has been intensively reported [8, 9]. Ozcan *et al* [8] have applied multi-wall carbon nanotubes to remove Fe^{3+} from the water. Furthermore, Abdullah *et al.* [9] have reported

Content from this work may be used under the terms of the Creative Commons Attribution 3.0 licence. Any further distribution of this work must maintain attribution to the author(s) and the title of the work, journal citation and DOI.
Published under licence by IOP Publishing Ltd

Fe^{3+} ions removal using CNFs. However, most of those studies were addressed to remove metal ions from wastewater [1–9].

In our study, prior to producing Fe nanoparticles on CNFs through liquid phase adsorption, it was crucial to study of kinetics and equilibrium of Fe^{3+} metal ions adsorption on CNFs because the metal ions adsorption on adsorbent strongly depends on parameters such as contact time, the initial pH of the solution and the initial dose of metal ions. Based on literature review, to the best of our knowledge, not so many researchers focusing the effect of the ultrasonic wave on Fe^{3+} adsorption. Ultrasonic energy should aid adsorption of Fe^{3+} ions onto CNF during the interaction period. Therefore, in the present work, adsorption of Fe^{3+} metal ions on CNFs using liquid phase adsorption [11] was done in various sonication time and initial pH of the solution [11]. These parameters are two crucial parameters of kinetics and thermodynamics allowing for maximizing the amount of adsorbed metals. A significant amount of Fe^{3+} ions adsorbed onto CNFs is useful for producing Fe nanoparticles onto CNFs.

2. Materials and methods

The effect of contact time on the metal ions adsorption was studied by doping 5 mg of CNFs with 50 mL of 25 mg/L solution of Fe^{3+} assisted by an ultrasonic wave of 53 Hz in various times sonication (5 – 210) min. The effect of initial pH on the metal ions was studied by means doping 5 mg of CNFs with 50 mL of 25 mg/L solution of Fe^{3+} within initial pH ranging from 2 to 10 assisted by the ultrasonic energy for 30 min. All pH measurements were conducted using a pH meter (370 pH Meter Jenway) in which the initial pH levels of experimental solution were adjusted using 0.1 M solution of HCl or NaOH. The effect of initial concentration the metal ions on the adsorption was studied by means each 5 mg of CNFs was dipped ultrasonically for 30 min in different concentration of 50 mL solution of Fe^{3+} with initial pH around 5. After doping treatment, solid and filtration were separated by filtration using a 0.45 pore size Whatman membrane filter paper. The final and the initial concentrations of the metal ions infiltrate were determined by Atomic absorption Spectrophotometer (Hitachi Z-2000). The concentration of the metal ions adsorbed in CNFs was obtained from the difference between the initial and the final metal ions concentrations. Here, all the calculated parameters of adsorption equilibrium of the metal ions such as the amount of the ions, the percentage of the adsorbed ions, and adsorption isotherms refer to work in [2].

The amount of the metal ions adsorbed into CNFs at equilibrium, q_e (mg/g) is given by:

$$q_e = \frac{(C_o - C_e) \times V}{M} \tag{1}$$

where C_o and C_e (mg/L) are the liquid phase concentrations of the metal ions at the initial and at equilibrium states, respectively; V (L) is the volume of solution; and M (g) is the mass of dried CNFs used.

The percentage of metal ions that was adsorbed in CNFs was calculated by the following Equation 2:

$$\text{Percentage of Adsorption } (\%) = \frac{(C_o - C_e)}{C_o} \times 100 \tag{2}$$

where C_o and C_e (mg/L) are the liquid phase concentrations of the metal ions at the initial and at equilibrium states, respectively.

The procedure of contact time effect as adsorption kinetics aspect in this study was similar to that of batch equilibrium. However, the amount of the metal ions adsorbed in CNFs at equilibrium, q_e (mg/g) is obtained by calculating the metal ions uptake at selected time intervals. The amount of the adsorbed metal ions at any time, q_t (mg/g) was calculated by:

$$q_t = \frac{(C_o - C_e) \times V}{M} \tag{3}$$

IOP Conf. Series: Materials Science and Engineering **367** (2018) 012046 doi:10.1088/1757-899X/367/1/012046

where C_t (mg/L) is the liquid phase concentration of the metal ions at any time (min). Langmuir and Freundlich's models were employed for studying adsorption isotherms. Langmuir adsorption isotherm is given by:

$$\frac{C_e}{q_e} = \frac{1}{q_m K_L} + \frac{C_o}{q_m} \quad (4)$$

where C_e (mg/L) is the equilibrium liquid phase concentration of metal, q_e (mg/g) is the number of metals adsorbed at equilibrium, q_m (mg/g) is the maximum adsorption capacity and K_L is Langmuir equilibrium constant. R_L associated with a characteristic of Langmuir isotherm is determined by using:

$$RL = \frac{1}{(1 + q_m K_L)} \quad (5)$$

Freundlich adsorption isotherm is given by:

$$\log q_e = \log q_m + \frac{1}{n} \log C_e \quad (6)$$

where C_e (mg/L) is the equilibrium liquid phase concentration of metal, q_e (mg/g) is the number of metals adsorbed at equilibrium, q_m (mg.g^{-1})(L.g^{-1})$^{1/n}$ is Freundlich constant associated with adsorption capacity. 1/n (mg.L^{-1}) is Freundlich constant related to sorption intensity of the sorbent.

3. Results and discussion

The effects of sonication time and initial pH on the amount of adsorbed Fe^{3+} on CNFs are presented in Figure 1 and Figure 2, respectively.

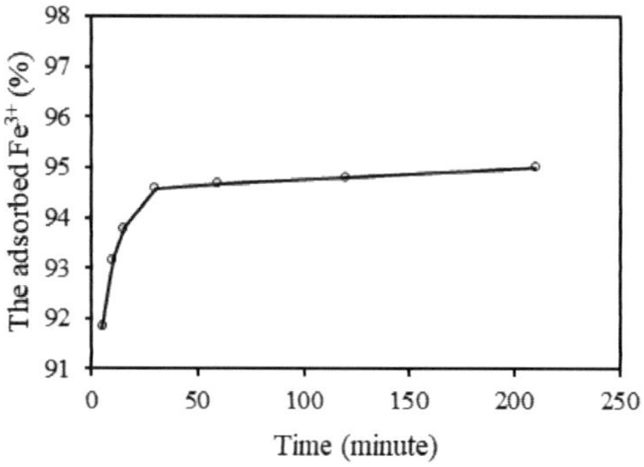

Figure 1. The effect of sonication time on the amount of adsorbed Fe^{3+} on CNFs

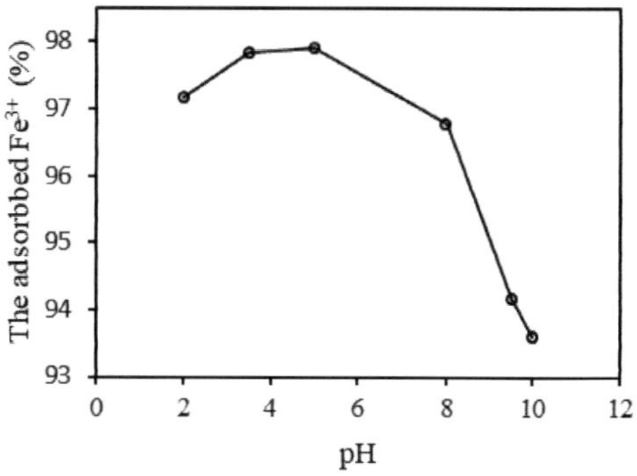

Figure 2. The effect of initial pH on the amount of adsorbed Fe^{3+} on CNFs

Effect of sonication time presented in Figure 1 exhibits that the metal ions adsorptions goes quickly from the initial stage of 5 min until 20 min. At the second stage with range time of around 20 min to 30 min, the metal ions uptake becomes slower until it reaches a constant time of sonication at about 30 min. After sonication for 30 min, the metal ions adsorption tends to be constant. The fast uptake of metal ions at the initial stage might be caused by a large number of surface sites available on CNFs for adsorption of the metal ions. The next stage, at sonication time above 30 min the metal ions sorption is nearly getting constant and the rate of the adsorption is getting slower with the time due to the sites were almost entirely occupied by the adsorbate. It is possible that the bulk Fe^{3+} ions adsorbed on the surface sites and Fe^{3+} ions in solution will repulse each other. Consequently, surface sites of CNFs would not be able to adsorb Fe^{3+} metal ions anymore. Therefore, sonication time of 30 min was selected for all the subsequent experiments.

The effect of initial pH on the amount of adsorbed Fe^{3+} on CNFs is presented in Figure 2. It is well known that acidic solution contains high proton concentration with high mobility property. The H^+

ions should compete with Fe^{3+} metal ions on active sites of the adsorbent. Due to higher abundance and higher mobility of H^+ ions in solution, Fe^{3+} ion should not be able to compete with H^+ ions. As a result, the amount of the adsorbed Fe^{3+} ion on CNFs is relatively small [13]. It is similar to high pH of a solution having OH^- high abundance. Commonly, in high pH region (above pH 7), the adsorption of metal ions decreases dramatically due to the formation of metal hydroxide.

In our study, however, we found the unusual phenomenon of the adsorption in which the metal ions uptake was relatively high even though at low pH (pH < 3.5) i.e 97.17 %. At pH range around 3.5 until 5, the adsorption of metal ions tends to increase with the increase of solution pH. The highest metal ions uptake was reached at pH 5 (98 %). At pH region above 5 the adsorption of the metal ions indeed decreases with the increase of solution pH, even though the amount of Fe^{3+} uptake is still relatively high. The amount of Fe^{3+} uptake at pH range of 8 to 10 is more than 93 %. The relatively high amount of Fe^{3+} uptake at those pH regions might result from ultrasonic energy. Probably, the ultrasonic wave has facilitated the electrostatic attraction between the metal ions and active sites of CNFs, hence reduction of Fe^{3+} to be Fe^{o} on the active sites take place easily. The role of the ultrasonic wave on metal ions reduction on nanocarbon has been reported by the author [11]. Another possibility is the high adsorption because of the large surface area of CNFs.

In this work, Langmuir and Freundlich's models were used for studying the adsorption isotherm. Figure 3 and Figure 4 shows curves of Langmuir and Freundlich adsorption isotherms, respectively. The straight line in Figure 3 was obtained by plotting C_e/q_e versus C_e using Langmuir adsorption isotherm (Equation 4). The q_m value was determined through the slope of the straight line (Figure 3). In this case, the slope equal to $1/q_m$ and $1/q_mK_L$ is defined as intercept. The characteristic of Langmuir isotherm can be evaluated from R_l value as Equation 5. The equilibrium data were fitted to Langmuir isotherm. The q_m, K_L, R^2 and R_L values as Langmuir isotherm parameters were then obtained. The values of those parameters are q_m (347.222 mg/g), K_L (9.931 L/mg), R^2 (0.981) and R_L (0.00029). The value of R_L which is less than 1 indicates that Langmuir adsorption isotherm is favorable [2].

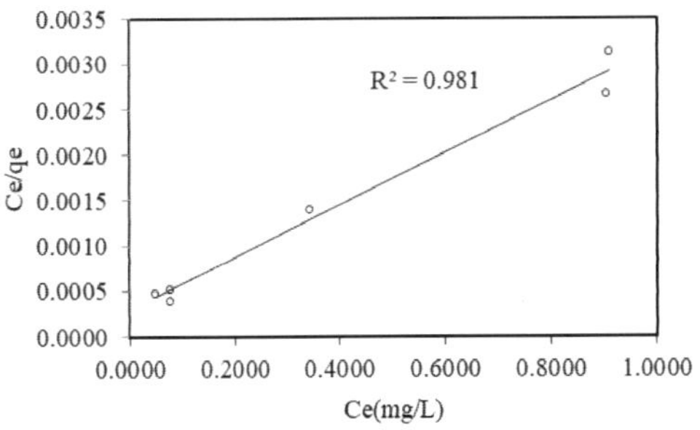

Figure 3. Langmuir adsorption isotherm

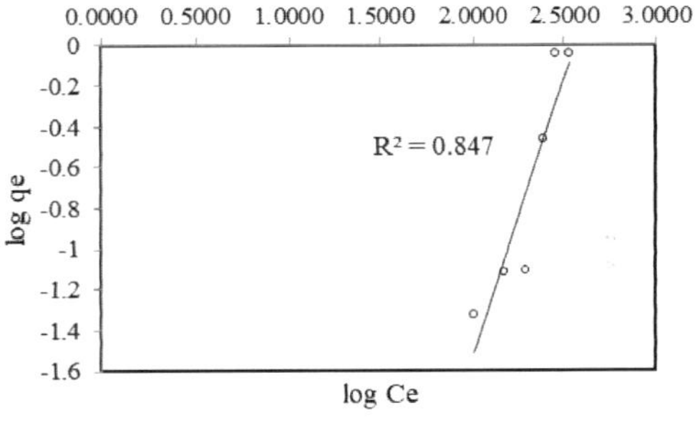

Figure 4. Freundlich adsorption isotherm

Figure 4 exhibits plot of log q_e against log C_e. The plot gives a straight line was obtained from Freundlich adsorption isotherm (Equation 6) with a slope of 1/n and intercept of log q_m. The 1/n is Freundlich constant related to sorbent intensity, whereas q_m is Freundlich constant associated with adsorption capacity. The Freundlich isotherm parameters are n (0.374mg/L), q_m ((2.034 mg.g^{-1}) (L.g^{-1})$^{1/n}$), and R^2 (0.847). The larger value of q_m means the higher adsorption capacity of CNFs. The adsorption of Fe^{3+} metal ions on CNFs is realistically described by Langmuir and Freundlich isotherms. Based on the R^2 values, however, Langmuir isotherm provided the better fit compared to Freundlich isotherm. It suggests that maximum adsorption of Fe^{3+} metal ions occurred was covered by a monolayer of adsorbate.

In this work, two common models, namely, pseudo-first order and pseudo-second-order models employed for the modeling of the kinetics of adsorption of Fe^{3+} metal ions are given by Equation 7 and Equation 8, respectively:

$$\log(q_e - q_t) = \log q_e - \frac{k_{1t}}{2.303}t \tag{7}$$

$$\frac{t}{q_t} = \frac{1}{k_2 q_e^2} + \frac{t}{q_e} \tag{8}$$

where q_e and q_t (mg/g) are the amounts of Fe^{3+} ions uptake at equilibrium and at any time, t (h), respectively, k_1 (min^{-1}) and k_2' ($g{\cdot}mg^{-1}{\cdot}min^{-1}$) are the adsorption rate constants for the pseudo first-order and pseudo second-order models, respectively. The plot of log ($q_e - q_t$) versus t gives a straight line with a slope of k_1 and intercept of log q_e as shown in Figure 5. The values of k_1 and R^2 obtained from the plots for adsorption of the metal ions on the CNFs are 0.1221 min^{-1} and 0.9983, respectively. The plot of $\frac{t}{q_t}$ against t provides a straight line with slope and intercept of $\frac{t}{q_e}$ and $\frac{1}{k_2 q_e^2}$, respectively as presented in Figure 6.

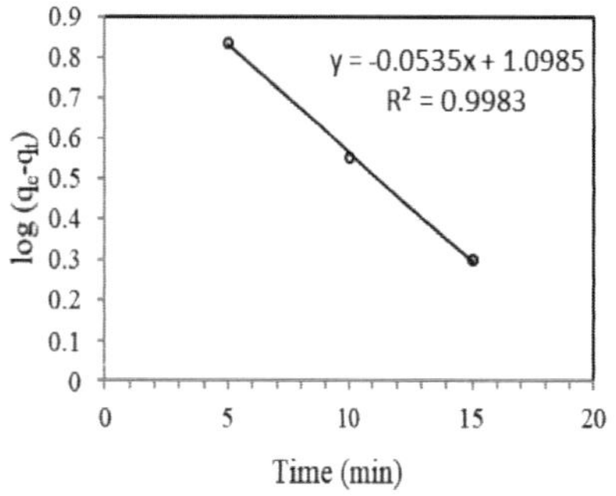

Figure 5. Pseudo-first order kinetic model for Fe^{3+} metal ions adsorption

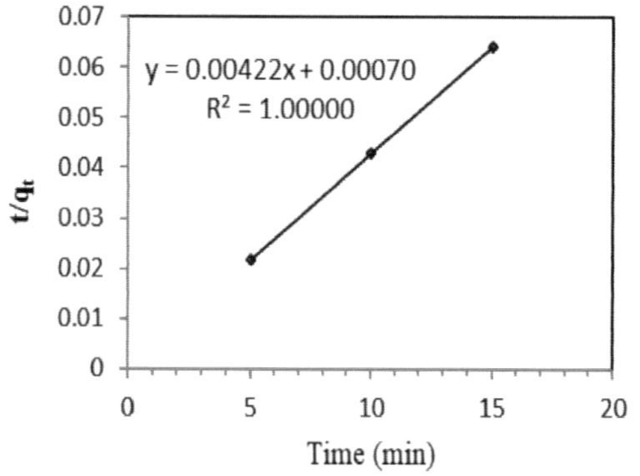

Figure 6. Pseudo-second order kinetic model for Fe^{3+} metal ions adsorption

The values of k_2 and R^2 obtained from the plots for adsorption of the metal ions on the CNFs are 0.025441 $g{\cdot}mg^{-1}{\cdot}min^{-1}$ and 1.0, respectively. In comparison, an R^2 value obtained from the pseudo-first-order slightly lower than that of the pseudo-second-order model indicating that the adsorption of Fe^{3+} metal ions on CNFs tends to follow the pseudo-second-order kinetics model. Since, our experimental data is in a good agreement with pseudo-second-order model, then the q_e value should be calculated from the slope of the graph in Figure 6. The calculated q_e is 236.967 mg/g. This value is similar to experimental q_e value i.e., 236.470 mg/g [2]. The q_e value was obtained by involving C_o = 25 mg/L, M = 0.05g, C_e = 1.3530 mg/L and V = 0.05 L in Equation 1. It is important to be noted that the experimental q_e value associates with equilibrium adsorption capacity at optimum time (q_e). As a comparison, the calculated q_e value obtained from the intercept of pseudo-first-order is 0.0408 mg/g which is significantly different from the experimental value.

4. Conclusions

We investigated that the Fe^{3+} ions were efficiently adsorbed on CNFs for 0.5 h in acidic pH of around 5 i.e., 98 %. However, we found the unusual phenomenon of the adsorption in which the metal ions uptake was still relatively high even at low pH (pH < 3.5) i.e. 97.17 %. The finding is beneficial for producing Fe nanoparticles onto CNFs. Adsorption equilibrium of Fe^{3+} metal ions on CNFs was fitted to Langmuir while the adsorption kinetic tends to follow the pseudo-second-order.

Acknowledgments
The authors thank Directorate of Research and Community Service-The Ministry of Research, Technology and Higher Education of the Republic of Indonesia for funding this work through the Research Grant for Applied Product #065/ADD/SP2H/LT/DRPM/VIII/2017. The authors sincerely thank Professor Katsumi Kaneko of Shinshu University-Japan for supplying CNFs.

References
[1] Kouakou U, Ello A S, Yapo J A, and Trokourey A 2013 Adsorption of iron and zinc on commercial activated carbon *J. Environ. Chem. Ecotoxicol* **5(6)** 168–171
[2] Alslaibi T M, Abustan I, Ahmad M A, and Foul A A 2013 *Desalination and Water Treatment*
[3] Thilagam G, Mani N, Arivoi S, and Hema M 2016 Adsorption behavior of Fe (II) ion from aqueous solution onto nano carbon *International Journal of Chemical Studies*, **4(1)** 96–102
[4] Mahmoud M A, Gawad E A, Hamoda E A, and Haggag E A 2015 Kinetics and thermodynamic of Fe(III) adsorption type onto acti- vated carbon from biomass: kinetics and thermodynamicsstudies *ESAIJ* **11(4)** 128–136
[5] Hussain A A, Mohammed S R, Nallu M, and Arivoli S 2012 Adsorption of Fe(III) from aqueous solution by acanthaceae activated carbon *J. Chem. Pharm. Res.* **4(4)** 2325–2336
[6] Mondal P, Balomajumder C, and Mohanty B 2007 A laboratory study for the treatment of arsenic, iron, and manganese bearing ground water using Fe(3+) impregnated activated carbon: effects of shaking time, pH and temperature *Journal of Hazardous Materials* **144** 420–426
[7] Vassileva P St and Detcheva A K 2010 Adsorption of some transition metal irons [Cu(II), Fe(III), Cr(III) and Au (III)] onto Lignite-Based Activated Carbons Modified by Oxidation *Adsorption Science & Technology* **28(3)**
[8] Ozcan S G, Satiroglu N, and Soylak M 2010 Column solid phase extraction of iron(III), copper(II), manganese(II) and lead(II) ions food and water samples on multi-walled carbon nanotubes *Food and Chemical Toxicology* **48** 2401–2406
[9] Abdullah N, Rinaldi A, and Hamid S B A 2015 Synthesis and Adsorption Performance of Carbon Materials for the Removal of Iron (III) from Aqueous Solution *Applied Mechanics and Materials* **699** 988–993
[10] Huang F, Xu Y, Liao S, Yang D, Hsieh Y L, and Wei Q 2013 Preparation of Amidoxime Polyacrylonitrile Chelating Nanofibers and Their Application for Adsorption of Metal Ions *Materials* **6** 969–980
[11] Alimin A, Narsito N, Kartini I, and Santosa S J 2015 Production of Silver Nanoparticle Chains inside Single Wall Carbon Nanotube with a Simple Liquid Phase Adsorption *Bulletin of Chemical Reaction Engineering & Catalysis* **10(3)** 266–274
[12] Hameed B, and M El–Khaiary 2008 Equilibrium, kinetics and mechanism of malachite green adsorption on activated carbon prepared from bamboo by K2CO3 activation and subsequent gasification with CO2 *J. Hazard. Mater.* **157** 344–351
[13] Bozˇic´D, Stankovic´V, Gorgievski M, and Bogdanovic´G 2009 Adsorption of heavy metal ions by sawdust of deciduous trees *J. Hazard. Mater.* **171** 684–692

Effect of Molar Ratio on Crystal Structure and Surface Morphology of $Nd_1(Fe)_xBa_{2-x}Cu_3O_7$ Oxide Alloy by Solid-State Reaction Method

N. A. Humairah[1], D. Sartika[1], Muris[2] and E. H. Sujiono[2]

[1]Department of Physics, *Universitas Sulawesi Barat*, Majene 91413, Indonesia
[2]Laboratory of Materials Physics, Department of Physics, *Universitas Negeri Makassar*, Makassar 90224, Indonesia[2)]

E-mail: e.h.sujiono@unm.ac.id

Abstract. Processing of materials $Nd_1(Fe)_xBa_{2-x}Cu_3O_7$ oxide alloy by varying the molar ratio of Fe doping, which x varied from +0.1 to +0.3 has been developed by solid-state reaction method. This research has shown that the *hkl* plane of the sample has *c*-axis oriented, i.e., (005), (006), (007), (008) and (009), respectively. All of the samples have the most dominant peak of (006) with the crystal structure is tetragonal. The molar ratio of the samples having a composition of Nd:Ba:Fe:Cu = 1:(2 – x):x:3.1. Other results, by SEM characterization, indicate that variation of Fe doping influence of the grain size and grain orientation of the various samples. The result of XRD and SEM characterization analysis indicates that Fe doping has an effect on the crystal structure, composition, and surface morphology of Nd(Fe)BCO oxide alloy.

Keywords. Crystal structure, molar ratio, morphology, NdBaCuO, SEM, solid state reaction, andXRD

1. Introduction

Material science and engineering are fundamental to the development of modern technology, both for structural applications, electronics, thermal, electrochemical, environmental and biomedical. One of the oldest applied sciences has defined as process metallurgy. Its history can be traced back to 6000 BC. Currently, there are 86 known metals, and only 24 species were found during the 19th century. Each era is characterized by the discovery of new materials. The mechanical properties of the material are mainly influenced by the process of synthesis and material composition [1, 2, 3].

The highest technology application of superconductor materials not only at very low temperatures but also can be applied to high-temperature superconductors. It will require energy, i.e., an HTS (High-Temperature Superconductor) second-order sigma-delta modulator, and a pulse stretcher. It used as an interface between the modulator and the first semiconducting amplifier stage. Moreover, based on HTS (High-Temperature Superconductor) technology are very attractive since they can operate under the considerably relaxed cooling effort, which is one of the main problems with LTS (Low-Temperature Superconductor) circuits [4, 5].

With advantages of understanding physical properties of high-T_c superconductors, development of practical materials of bulks, tapes, and thin films are now in rapid progress. The present concern is

Content from this work may be used under the terms of the Creative Commons Attribution 3.0 licence. Any further distribution of this work must maintain attribution to the author(s) and the title of the work, journal citation and DOI.
Published under licence by IOP Publishing Ltd

now focused on $YBa_2Cu_3O_{7-x}$ while $Rba_2Cu_3O_{7-x}$ (R = light rare earth such as Nd, Sm, Eu, Gd) [5, 6, 7] is also promising especially for high field applications. The reason is that, e.g., $NdBa_2Cu_3O_{7-x}$ has higher T_c (~96 K) and possible higher irreversible field, H_{irr} (>10 T at 77 K) compared with those of YBaCuO [1, 7]. One of the properties of metal oxide alloys is superconducting at critical temperatures, for example, YBCO oxide alloys having 87 K and 87.4 K, deposited at 680 °C and 700 °C [8] and are magnetic at room temperature [9].

In addition, the development of NdBCO material has higher chemical stability than the YBCO system. The NdBCO system has the same structure as the YBCO system [8, 9, 10], but NdBCO is known to be higher and has stronger pinning flux. Thus it is possible to hold impedance doping variation [9]. The solid-state reaction method is a method for the preparation of polycrystalline solids from a mixture of solid starting materials [5]. Other oxide material processing techniques are making sol-gel [10], hydrothermal [11], Chemical Vapor Deposition (CVD) [8] and etc. There are advantages of preparing using solid state reaction because this method is cheap and easy to complete. Another thing that the product of the reaction has high purity and excellent crystallinity [2].

In this paper, we present the influence of the molar ratio of Fe doping on NdBCO with x = (0.1, 0.2 and 0.3) on crystal structure and surface morphology. The effects of doping determine by analyzing the data results using XRD and SEM characterization. The results describe the crystal quality presenting the value of FWHM, the lattice constant, and crystal orientation, also the surface morphology indicating at grain size, and grain orientation.

2. Materials and methods

$Nd_1(Fe)_xBa_{2-x}Cu_3O_7$ alloys have been synthesized by solid reaction method. The sintering process has been done with the mixing stage of the Nd_2O_3 99.99 %; CuO 99.99 %; $BaCO_3$ 99.99 % and Fe_2O_3 99.9 % powder by grinding for 3 h until the precursor's powder has homogeneous. Then the precursor's powder was calcined for 20 h at a temperature of 950 °C, then continuing with the pelletization process which was 0.5 cm in diameter and 0.25 cm. Samples bulk form had sintered for 10 h at a temperature of 950 °C and followed by annealing process for 20 h at a temperature of 450 °C, as shown in Figure 1.

The same procedure was used for all sample has a variation of x=0.1 to x = 0.3, respectively. Characterization process was conducted to identify the phases using X-ray diffraction with λ = 1.54439 Å [12, 13], and analyzed the surface morphology using SEM (Scanning Electron Microscope) [14].

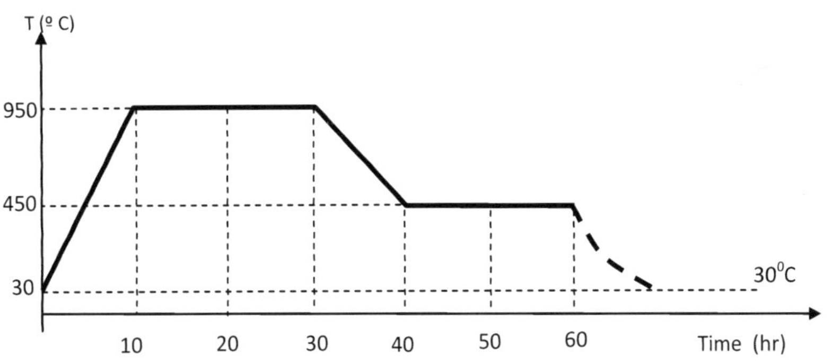

Figure 1. Scheme process of sintering and annealing temperature as a function of the time processing.

3. Results dan Discussion

XRD diffraction patterns of oxide material Nd(Fe)BCO powder were synthesized by using the solid-state reaction method with variations of the molar ratio at a x = 0.1, x = 0.2 and x = 0.3 are shown in Figure 2.

In Figure 2 shows the *hkl* plane of the sample indicate of all peaks oriented to the *c*-axis orientation, i.e., peak I (005); peak II (006), peak III (007), peak IV (008), and peak V (009), where the most dominant peak is the peak II (006) for all sample with variation of x is 0.1, 0.2 and 0.3, respectively.

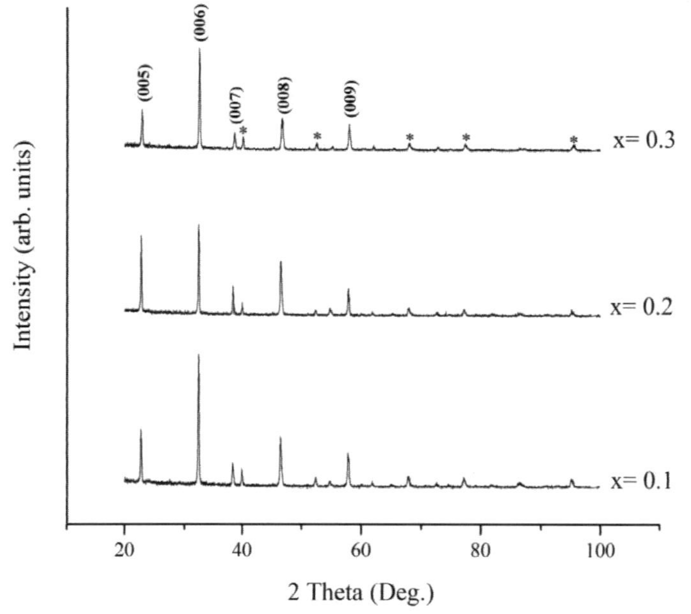

Figure 2. XRD pattern of sample Nd(Fe)BCO for variation of x to 0.1, 0.2, and 0.3, respectively.

Table 1. Position (2θ), intensity and FWHM value of Nd(Fe)BCO phase as a variation of molar ratio

Molar Ratio	2θ(°)	Intensity (counts)	FWHM (°)
x = 0.1	32.40 ± 0.12	900 ± 2.5	0.12
x = 0.2	32.38 ± 0.04	900 ± 2.5	0.04
x = 0.3	32.50 ± 0.03	1195 ± 1.0	0.03

Table 1 shows the influence of molar ratio on the position of 2θ, intensity, and the value of FWHM (Full Width at Half Maximum) of Nd(Fe)BCO samples. As illustrated, the increasing molar ratio of Nd content gradually shifted the XRD peaks towards a lower angle or higher *d* value due [2, 3] to larger ionic radii of Nd than Fe [15]. Peak position value of 2θ at 32.50° had the highest intensity compare with other peaks, and this peak was related to the *hkl* (006). Relating to the intensity, the molar ratio has influenced the intensity of the sample [1]. It can be seen in Table 1, that the sample with a molar ratio of x = 0.3 had the narrowest and highest intensity which indicated the growth of Nd(Fe)BCO crystallite sample has excellent crystallinity and conductivity. This result is in accordance with that most orderly arrangement and homogeneous of crystals in the correlation of the smaller of the FWHM value and tip width of the peak as has reported E. H. Sujiono *et al* [2]. The sample with a molar ratio of x = 0.3 has lowest FWHM value among others and has a similar crystal structure; this also affects the electrical properties of the sample [16]. Samples having an FWHM value lower than

0.5 will select a good crystal quality level, these results are similar with has reported Wickenden *et al* [17].

To determine the lattice parameter has used a formula derived from Bragg law as described in Equation 1 [18].

$$2\ \sin\theta = \lambda\sqrt{\frac{h^2}{a^2}+\frac{k^2}{b^2}+\frac{l^2}{c^2}} \tag{1}$$

with λ is the wavelength of the x-rays, *a, b,* and *c* are lattice constants of the crystal structure, and θ is the diffraction angle.

Based on Table 2, the characterization results, it was found that the Nd(Fe)BCO sample in this study has a tetragonal crystal structure with the lattice parameters obtained for all peaks ranging from (11.2828 to 11.7518) Å. It can be assumed that variation of molar ratio x = 0.1 to x = 0.3 has a lattice constant value similar as was reported elsewhere [2, 4, 6]. These results indicate that oxide material produced in this study have excellent quality as superconducting material properties. The crystalline properties as has been describing, i.e., *c*-axis oriented with shaped the structure of crystals and small value FWHM as an indication of high crystal quality [8, 10]. The difference in height intensity and width of the peak as representing at the value of FWHM indicating of the effect variation of molar ratio x. The highest intensity and smallest of FWHM means that *hkl* plane in which orientation has a form more dominant. These phenomena are similar as has reported elsewhere [1, 8].

Table 2. The lattice parameters for each variation of molar ratio

Peaks (*hkl*)	Samples	Lattice parameter (Å)
	$Nd_1(Fe)_{0.1}Ba_{1.9}Cu_3O_7$	11.7518
(006)	$Nd_1(Fe)_{0.2}Ba_{1.8}Cu_3O_7$	11.7518
	$Nd_1(Fe)_{0.3}Ba_{1.7}Cu_3O_7$	11.7518
	$Nd_1(Fe)_{0.1}Ba_{1.9}Cu_3O_7$	11.2614
(005)	$Nd_1(Fe)_{0.2}Ba_{1.8}Cu_3O_7$	11.2708
	$Nd_1(Fe)_{0.3}Ba_{1.7}Cu_3O_7$	11.2828

Figure 3 shows the characterization result of SEM image of molar ratio variation (a) x = 0.1, (b) x = 0.2, and (b) x = 0.3, it can be seen that increasing of Fe ratio on NBCO oxide alloy reducing the number of grain boundaries. The surface morphology image is appearing of many variations of grain size as shown in Figure 3.a, the number grain size is reduced at molar ratio Fe of 0.2 and almost disappears at ratio Fe of 0.3. The analysis of samples showing the effect of Fe doping on of Nd(Fe)BCO alloy caused granular changes of the elongated rectangular pattern, into a short rectangular pattern [4]. In case of Fe doping of +0.3, it appears that a short rectangular granular group is identified as Fe_2O_3 formation as shown in Figure 3. The substitution of Fe_2O_3 which magnetic properties may lead to changes the surface morphology of NBCO compound as explained. Nd(Fe)BCO as a new alloy can improve the magnetism properties of NBCO at room temperature and as the center of pinning at superconducting state [9]. The developed of oxide alloy not only produce new superconducting material but also found in magnetic material that is very useful in the application of spintronics devices. The surface morphology results indicate that variation of Fe molar ratio changed the grain size and reduced the grain boundaries of the alloys caused improving the crystal quality.

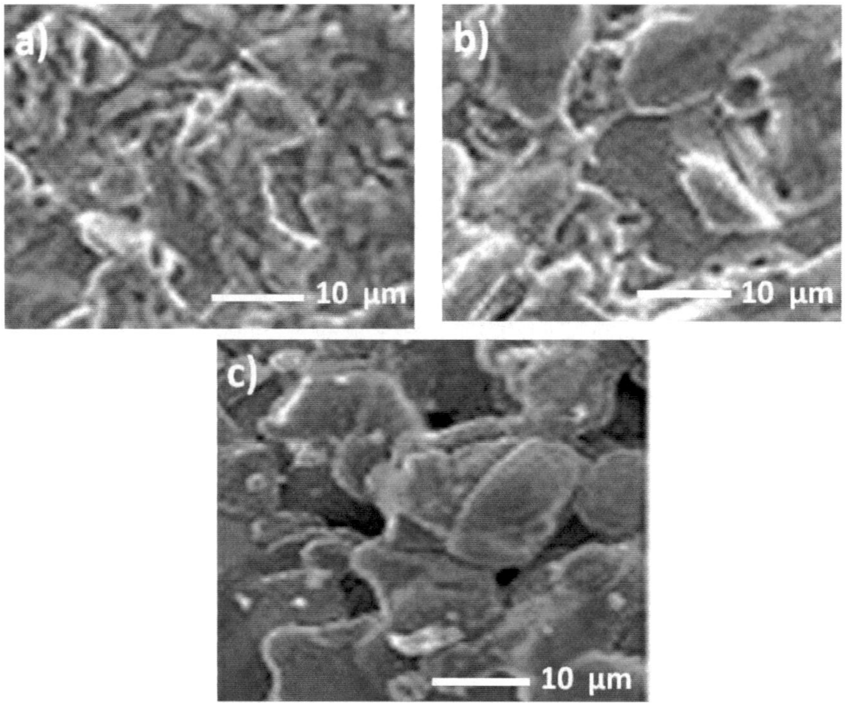

Figure 3. The SEM micrograph of Nd(Fe)BCO oxide alloy as variation molar ratio Fe of (a) x = 0.1 (b) x = 0.2 dan (c) x = 0.3, respectively.

4. Conclusions

The Nd(Fe)BCO alloy with the variations of molar ratio x of 0.1, 0.2 and 0.3 have been successfully synthesized using solid-state reaction method. The variation of molar ratio has a significant effect on the intensity, FWHM, and the crystallite size of the samples. It was found that the Nd(Fe)BCO sample in this study has a tetragonal crystal structure with the lattice parameters obtained for all peaks ranging from (11.2828 to 11.7518) Å. The surface morphology also changes with the variation of Fe doping indicating with reduced of grain boundaries and caused granular changes of the elongated rectangular pattern.

Acknowledgments

This research has fully funded by the Ministry of Research, Technology, and Higher Education, Republic of Indonesia, under the research scheme of *Pekerti* fiscal year 2017.

References

[1] Zharvan V, Kamaruddin Y N I S, Samnur, and Sujiono E H 2017 *IOP Conference: Materials Science and Engineering* **202** 012072

[2] Sujiono E H , H Ahmad, F Rera, S Samnur 2011 *Jurnal Fisika dan Aplikasinya* **7(1)** 1 –3

[3] Cramb A 1996 *A Short History of Metals* (Carnegia-Mellon University, USA)

[4] Roussy A, Karthikeyan S, Oomen I, Ortlepp Th, Sujiono E H, Brinkman A, and Rogalla H 2005 *IEEE Transactions on Applied Superconductivity* **15** 457 – 460

[5] Monot-Laffez I, Tancret F, Guilmeau E, Desgardin G and Raveau B 2000 *Superconductor Sci. Technol.* **13** 655 – 659

[6] Sujiono E H 2007 *Indonesian Journal of Physics* **18** 107 – 110

[7] Koshizuka N, Pradhan A K, Shibata S, Feng Y, Machi T, and Nakao K 2001 *Physica C* **364 – 365** 320 – 325

IOP Conf. Series: Materials Science and Engineering **367** (2018) 012047 doi:10.1088/1757-899X/367/1/012047

[8] Sujiono E H, Arifin P, and Barmawi M 2002 *Materials Chemistry and Physics* **73** 47 – 50
[9] Sujiono E H 2017 *Paduan Oksida Logam* $Nd_1(Fe)_xBa_{2-x}Cu_3O_y$ *dan Metode Pembuatannya* (ID Paten 44179 P00200800471) (in Indonesia)
[10] Sujiono E H, Sani R A, Saragi T, Arifin P, and Barmawi M 2001 *Physica Status Solidi (A) Applied Research* **187** 471 – 479
[11] Septia E C, Nuryadin B W, Iskandar F, Abdullah M , and Khairrurijal 2014 *Jurnal Matematika dan Sains* **18** 99 –102 (in Indonesia)
[12] Osman M O 2016 *Metallurgical And Material Transactions* 1 –8
[13] Gremenok V F, Reddy K T R, Titanov M S, and Patryn A 2017 *Przeglad Elektrotechnizny* **93** 88 –91
[14] Wibowo K M, Zahdan M Z, Asmah M T, Saim H, Adriyanto F, Suyitno, Hadi S 2017 *International Research and Innovation Summit* **226** 1 –6
[15] Prabeer B, Recham N, Chotard J N, Jebbal K, Walker W, Armand M, and Tarascom J M 2010 *Journal of Material Chemistry* 1 – 8
[16] Husain H, Risal M, Sujiono E H 2015 *Jurnal Sainsmat* **4** 1 –6 (in Indonesia)
[17] Wickenden A E, Wickenden D K, and Kistenmacher T J 1994 *Journal of Applied Physics* **75** 5367
[18] Zettli N 2009 *Quantum Mechanics* (Library of Congress Cataloging, Publication Data)

Photocurrent Responses of Metanil Yellow and Remazol Red B Organic Dyes by Using TiO_2/Ti Electrode

M. Nurdin [1], T. Azis[1], M. Maulidiyah[1], A. Aladin[2], N. A. Hafid[1], L. O. A. Salim[1], and D. Wibowo[1]

[1]Department of Chemistry, Faculty of Mathematics and Natural Sciences, Universitas Halu Oleo, Jl. HEA Mokodompit Kampus Baru, Kendari 93232, Indonesia
[2]Department of Chemical Engineering, Universitas Muslim Indonesia, Jln. Urip Sumoharjo, KM 5. Makassar 90231, Indonesia.

E-mail: mnurdin06@yahoo.com

Abstract. Studies on organic dyes profile were performed by using TiO_2/Ti electrode photoelectrochemical system. The TiO_2/Ti electrode was prepared by an anodizing technique. These photoelectrochemical profiles of metanil yellow and remazol red B organic dyes were investigated by Linear Sweep Voltammetry (LSV), Cyclic Voltammetry (CV), Chronoamperometry (CA), and Potentiometric (V-t) methods. The profile tests of these organic dyes in concentrations of 2.0 ppm, 4.0 ppm, 6.0 ppm, 8.0 ppm, and 10 ppm were applied by Multi Pulse Amperometry (MPA). The highest of photocurrent responses of metanil yellow and remazol red B with the addition of electrolyte solution of 0.1 M $NaNO_3$ were 0.827 µA and 0.475 µA, respectively, meanwhile in non-electrolyte solution were 0.56 µA and 0.213 µA, respectively. The electrolyte solution influenced pH of the metanil yellow and remazol red B dyes in which their pH values were 6.31, 6.26, 6.18, 6.10, 6.07, 6.27, 6.25, 6.22, 6.20, and 6.15, for the concentrations of 2.0 ppm, 4.0 ppm, 6.0 ppm, 8.0 ppm, and 10 ppm, respectively.

Keywords. Electrode, organic compounds, pH, photoelectrochemical, and TiO_2/Ti.

1. Introduction

The textile industry is one of the fastest growing industries in Indonesia. The increasing number of the textile industry has a negative impact on the environment, especially dyestuff wastewater [1], [2]. Dyestuff waste resulting from textile dyeing processes, are usually difficult to decompose, causing environmental pollution, especially for aquatic environment [1], [3]. Organic dyes are widely used in the textile industry namely metanil yellow and remazol Red B. The existence of textile dyestuff waste has an effect on aquatic organisms due to the decrease of water quality so as it interferes with human health which can cause skin irritation, eye irritation and cancer [4].

Various methods have been done to overcome the dyestuff waste, including the adsorption method using activated carbon, filtration, coagulation and ozonation [2]. However, each of these methods has limited use and is less efficient. Coagulation method has excellent efficiency in waste treatment but generates new waste namely coagulant that cannot be reused. The purpose of activated carbon for dye degradation also requires high cost because the price of activated carbon is relatively high. One way to deal with dye waste more efficiently is to employ photocatalysis using semiconductors [5], [6].

Content from this work may be used under the terms of the Creative Commons Attribution 3.0 licence. Any further distribution of this work must maintain attribution to the author(s) and the title of the work, journal citation and DOI.
Published under licence by IOP Publishing Ltd

Titanium dioxide (TiO_2) is one of the most widely used semiconductors for photocatalysis to treat organic wastes because they are inert, stable and strong oxidizers [2], [7]. The utilization of TiO_2 as a photocatalyst to degrade organic compounds is converting them into inorganic products that are environment-friendly in the form of CO_2 and H_2O [8]. Through the mechanism of photocatalysis the recombination occurs between electron and hole, so this process does not optimize 95 % of absorbed light energy. Improvement of semiconductor photocatalysis performance can be made through the photoelectrochemical process using semiconductor coating on conductor substrate as photoelectrode. This process is called photoelectrocatalysis, i.e the process of merging photocatalysis with the provision of potential [3].

Several previous studies on photoelectrocatalysis, for example Maulidiyah *et al* (2015), have performed photoelectrocatalytic response test using SnO-F as the conductor substrate on Innert Wall Conductive Glass Tube (IWCGT) against glucose, potassium hydrogen phthalate (PHP) and benzoic acid compounds [1]. Nurdin *et al* (2017) have also degraded the organic compounds using TiO_2/Ti electrodes [7]. However, these studies still use simple compounds, so it is necessary to develop an electrochemical profile system of TiO_2/Ti electrodes using more complex compounds [3], [9].

Based on literature search, no one has reviewed the electrochemical profile system of the TiO_2/Ti electrode in organic dyes (metanil yellow and remazol Red B) as organic wastes that can pollute the environment. In this research there will be tests of electrode activity of TiO_2/Ti that have been prepared by an anodizing method using Linear Sweep Voltammetry (LSV), Cyclic Voltammetry (CV), Chronoamperometry (CA) and Potentiometry methods. Subsequently, the photocurrent response profile test was performed by Multi Pulse Amperference (MPA) from organic compounds. These studies are expected to be used as a model for the development of TiO_2 sensors electrochemically from various organic compounds.

2. Materials and methods

2.1. Fabrication of TiO_2/Ti electrode

Preparation of Ti plate was carried out by cutting Ti plate with a purity of 99.7 % and thickness of 0.5 mm with the size of 4 cm × 1 cm. Further furnished using sandpaper with size of 1200 cm^3 until the surface was clean and shiny. The Ti plate was then washed using detergent, water, and aquadest. Then, the Ti plate was etched using a mixture of HF, HNO_3, and aquadest with a ratio of 1:3:6 for 2 min.

The prepared Ti plate was inserted in the probe which has been filled with an electrolyte solution of 0.27 M NH_4F and aquadest in 98 % glycerol. The anodic oxidation process was carried out by placing the Ti plate as anode and Cu as a cathode, and providing a potential bias of 25 V from the power supply. This anodic oxidation process was applied for 4 h. The final step was to calcine the Ti plate for 1.5 h with a temperature of 500 °C to evaporate the remaining electrolyte solution on the surface of Ti plate.

2.2. Photoelectrochemical activity test

The activity test of TiO_2/Ti electrode was performed by Linear Sweep Voltammetry (LSV), Cyclic Voltametry (CV), Chronoamperometry (CA) and Potentiometric (V-t) methods using 0.1 M $NaNO_3$ solution. The LSV test was conducted from potentials of -1 V to 1 V with a scan rate of (1.0×10^{-4}) V/s. The CV was performed from potentials of -1.5 V to 1.0 V with a scan rate of 0.05 V/s. The CA was conducted from potentials of -1 to 1 V with a scan rate of (1.0×10^{-3}) V/s, and the potentiometric test was done from potentials of -2 V to 2 V. Testing of the TiO_2/Ti electrode in each method was carried out by giving a variation of UV light intensity.

2.3. Photoelectrochemical test based on organic compounds

Measurements of dyes (metanil yellow and remazol Red B) photocurrent responses were performed using the TiO_2/Ti catalyst with Multi Pulse Ampermmetry (MPA) technique at a potential bias of 0.5

V for 60 s. The concentration variations used were 2 ppm, 4 ppm, 6 ppm, 8 ppm and 10 ppm with the addition of 0.1 M $NaNO_3$.

2.4. Determining pH of organic compounds
A pH meter was used to measure pH of the dyes. Measurements were made for the concentrations of 2 ppm, 4 ppm, 6 ppm, 8 ppm and 10 ppm with $NaNO_3$ and without $NaNO_3$. Then, the relationship chart between pH and analytical concentration was made.

3. Results and discussion

3.1. Photoelectrochemical activity test
Measurement of photocurrent activity as a potential function was derived from electron transfer during the analyte oxidation-reduction process of 0.1 M $NaNO_3$ using Pt counter electrode, TiO_2/Ti working electrode and Ag/AgCl comparison electrode. This measurement was performed using a potentiostat with LSV, CV, CA, and V-t methods (Figure 1).

The LSV technique was used for qualitative test of TiO_2 crystal photoactivity. Using this method, we can know TiO_2 activity could be observed by the increase of photocurrent on the LSV curve after the electrode is irradiated with UV lamps. The LSV activity test of TiO_2/Ti electrode showed the highest activity during UV lamp irradiation (Figure 1.a). This indicates the presence of photons absorbed by TiO_2 so that the electrons are excited from valence band to conduction band. TiO_2 has photoelectrocatalysis activity on UV irradiation because of the TiO_2 is only active at a wavelength less than 388 nm related to the bandgap energy of 3.2 eV. While in the dark condition, there is no current increase (zero-generated current) so that no electron from the conduction band to the valence band.

Measurement of CV aims to determine the activity of anodizing TiO_2/Ti plate. In this method, current and potential changes were observed. Figure 1.b shows that in the dark condition, the current is generated at the zero region, while during UV lamp was illuminated the current increases. This indicates that without lamp illumination, the TiO_2/Ti electrode does not exhibit photoelectrochemical activity because there is no generation of electron-hole pairs on the electrode surface.

The measurement of CA is an electrochemical method that uses a potential bias. Figure 1.c shows differences in current generated. The current increases when the electrode is subjected to UV light, while in the dark condition the obtained current value is lower. According to Adan *et al* (2016), when the electrode is subjected to UV light the increase in anodic spike occurs due to the separation of the electron-hole pairs, then a decrease in current due to the recombination of electron-hole pairs [10], [11].

The method of potentiometric analysis is based on the measurement of electric photocurrent as the function of potential electrical bias. Figure 1.d shows that the potential produced by the UV illuminated TiO_2/Ti electrode is lower than potential in the dark. According to Cheshideh *et al* (2017), when the TiO_2 electrode in the electrolyte solution is irradiated, the electrons and holes are formed. Furthermore their potential will switch to a negative value because the holes are trapped on the surface and switches to the solution, therefore the electrons are accumulated in the electrode.

IOP Conf. Series: Materials Science and Engineering **367** (2018) 012048 doi:10.1088/1757-899X/367/1/012048

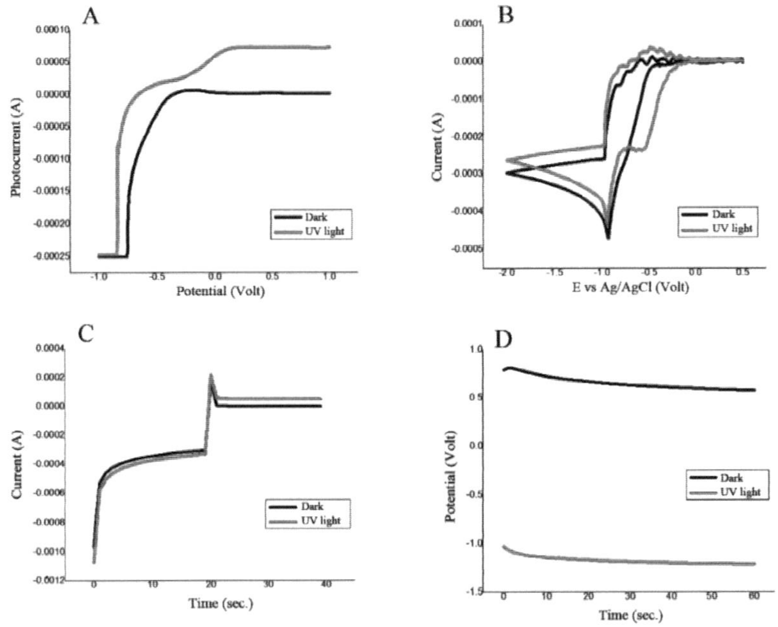

Figure 1. Voltammogram Graph. (a) Linear Sweep Voltammetry (LSV), (b) Cyclic Voltammetry (CV), (c) Cronoamperometry (CA), and (d) Potentiometric (V-t)

3.2. Photoelectrochemical test based on organic compounds

The measurement of photocurrent response generated from the photoelectrocatalytic system was developed from a test solution containing several organic compounds. Photocurrent measurements were made using a potentiostat under the MPA technique. The potential bias used was 200 mV, carried out based on the previous optimization results at a potential bias of 200 mV to 500 mV in capturing light and preventing the direct electrochemical process.

Metanil yellow is azo aromatic group dye with a molecular formula of $C_{18}H_{14}N_3O_3SNa$. Measurement of photocurrent response with a variation of metanil yellow concentration of 2.0 ppm, 4.0 ppm, 6.0 ppm, 8.0 ppm, and 10 ppm was carried out using TiO_2/Ti electrode with UV light.

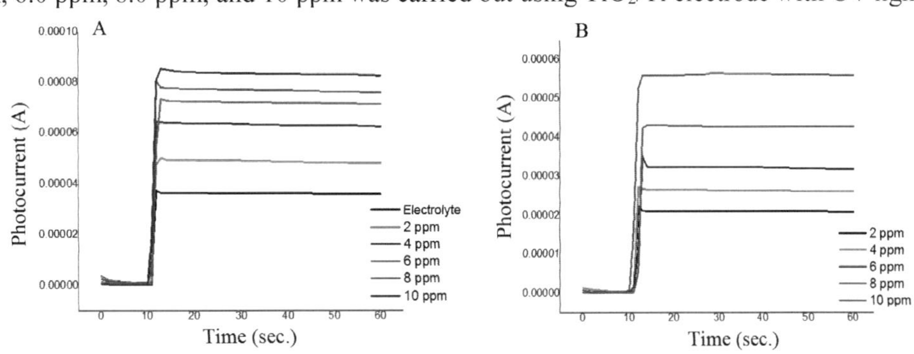

Figure 2. Photocurrent response based on metanil yellow, (a) by adding the electrolyte solution 0.1M $NaNO_3$ (b) without the electrolyte solution

Figure 2.a shows that the higher the concentration of a compound the higher the resulting photocurrent value. The photocurrent response was measured horizontally, the possibility of this compound difficult to degrade by TiO_2 photocatalyst so that the photocurrent decrease is not significant [10]. In Figure 2.a it is also seen that the current does not coincide with current blank of $NaNO_3$ electrolyte which indicates that metanil yellow was not degraded perfectly. This may be due to metanil yellow having aromatic groups with strong aromatic bonding which is difficult to degrade. The photocurrent response without the electrolyte was lower when compared to the photocurrent response in metanil yellow due to the electrolyte solution affecting photocurrent as can be seen in Figure 2.b.

Remazol Red B is one of mono-azo synthetic dye because it has one double bond of N=N [11]. Measurement of photocurrent response was performed using the TiO_2/Ti electrode under UV light with a variation of Remazol Red B concentration of 2.0 ppm, 4.0 ppm, 6.0 ppm, 8.0 ppm, and 10 ppm.

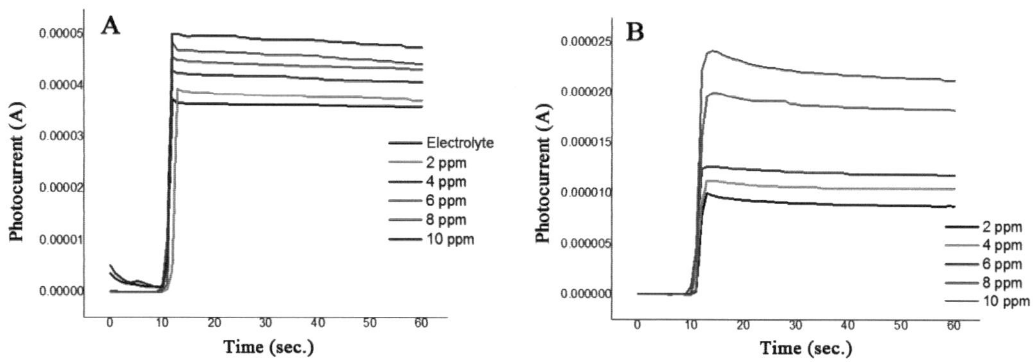

Figure 3. Photocurrent response based on remazol red B, (A) by adding the electrolyte solution 0.1M NaNO3 (B) without the electrolyte solution

Figure 3.a shows the increase of photocurrent response as the concentration of the test solution increases. The photocurrent response indicates that this compound was not degraded well enough by photoelectrocatalysis. The photocurrent of remazol red B does not coincide with a blank current of the electrolyte at a concentration more than 4 ppm. This observation indicates that remazol red B cannot be degraded perfectly within the observed time range. It is also possible that the electrode has been saturated by continuous measurements on the same electrode. However, the initial photocurrent values tend to increase in proportion to concentrations.

Figure 3.b shows differences in photocurrent responses of electrolyte and non-electrolyte test compounds. The photocurrent response to the electrolyte test compound is higher than that of the non-electrolyte test compound. This is because of the electrolyte solution which acts as a current conductor can increase the photocurrent in the test compound.

The linear regression equation applied to the results must have intercept values not significantly different from zero. If the non-zero significant intercept is obtained, it must be proved that this has no effect on the accuracy of the method [12].

Figure 4.a shows the linearity of test compounds by electrolytes. Metanil yellow and remazol red B have increased photocurrent along with increasing concentrations. However, there is a change of slope value on each test compound. It can be seen that the increased linearity of the current due to the increase of metanil yellow and remazol red B concentrations have R^2 = 0.9539 and R^2 = 0.9714, respectively. This shows the effect of the photodegradation process of each test compound. According to Nurdin *et al* (2009), at low concentration the photodegradation process is only influenced by photo hole capture process on catalyst surface while at high concentration compound structures affect the photodegradation process of the compounds [5], [13].

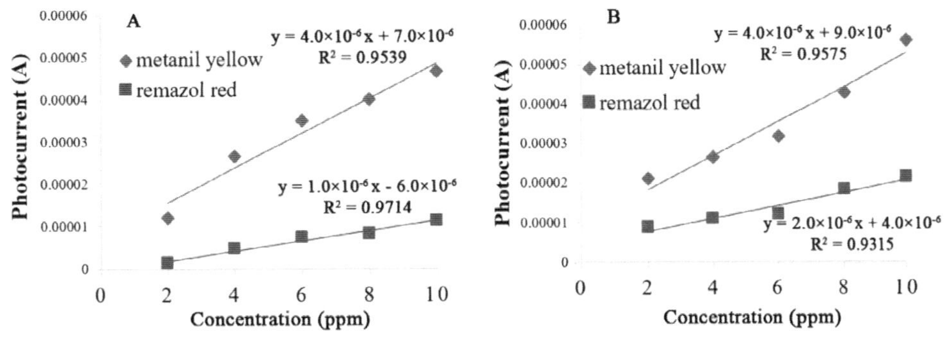

Figure 4. Linear graph, (a) by using electrolyte 0.1M $NaNO_3$ solution, (b) without the electrolyte solution

Based on Figure 4.b it can be seen that the metanil yellow and remazol red B compounds have increased the current along with increasing concentrations. It can be seen that the increase linearity of the current due to the increase of metanil yellow and remazol red B concentrations that have values of R^2 = 0.9575 and R^2 = 0.9315, respectively. However, the resulting linearity values using non-electrolyte test solution are lower than using solution with electrolyte. This is because the electrolyte solution can increase the photocurrent to the test compound.

3.3. Determination of pH in organic compounds

The pH value has an important role in determining waste characteristics and the formation of hydroxyl radicals. Photocatalysis of dye may occur by reaction mechanism, i.e., hydroxyl radical attack and direct oxidation by holes, depending on conditions of substrate base and pH [8]. The effect of $NaNO_3$ electrolyte solution on pH can be determined by making the correlation between analyte concentration and pH on metanil yellow and remazol red B test compounds as shown in Figure 5.

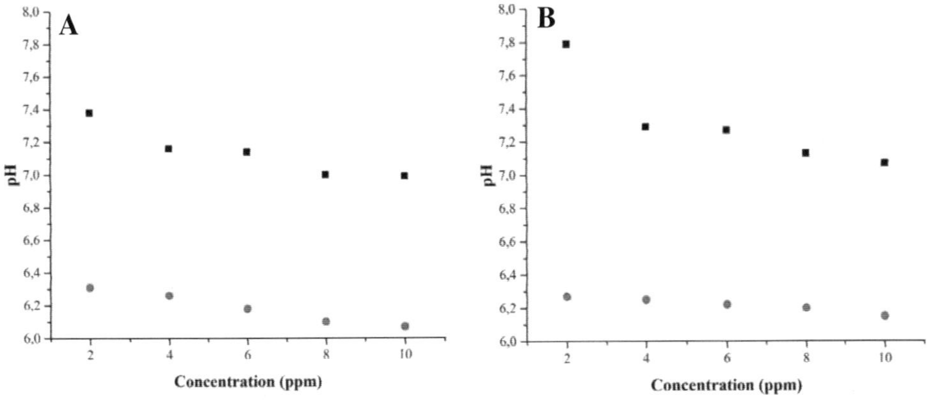

Figure 5. Correlation between concentration and pH, dot square: by adding electrolyte; dot cycle: without electrolyte (a) metanil yellow (b) remazol red B

The correlation between concentration and pH on electrolyte and a non-electrolyte metanil yellow test compound is the decrease in pH values at high concentrations (Figure 5.a). Each test compound has a maximum pH at a concentration of 2 ppm i.e. 6.31 and 7.38. Non-electrolyte metanil yellow exhibited alkaline properties, while electrolyte metanil yellow exhibited acid properties. This is due to the structure of the metanil yellow which shows the basic properties of the amine group and the acidic properties of sulphonate group. Figure 5.b also shows the correlation between concentration and pH on test compound where there is a decrease in the pH value at high concentration. Each test compound has a maximum pH at concentration of 2 ppm i.e. 6.27 and 7.29. Non-electrolyte remazol red B exhibited alkaline properties, whereas electrolytes Remazol red B exhibited acid properties. This is due to the structure of remazol red B which can show the alkaline properties of a hydroxyl group and the acidic properties of the sulphonate group.

4. Conclusions

It can be concluded that the photocurrent responses produced by the metanil yellow and remazol red B test compounds are higher along with the increase of test solution concentrations. Metanil yellow and remazol red B compounds with the addition of electrolyte ($NaNO_3$) at 10 ppm concentration yielded photocurrent values of 0.827 μA and 0.475 μA, respectively. For metanil yellow and remazol red B test compounds without electrolyte produced photocurrent values of 0.56 μA and 0.213 μA, respectively. The results of metanil yellow and remazol red B test profile showed good photocurrent response. The photocurrent response of metanil yellow and remazol red B test compounds was influenced by pH, the more acidic pH of the test solution the higher the photocurrent response values.

Acknowledgments

We acknowledge the financial support of the DRPM - Ministry of Research, Technology and Higher Education, the Republic of Indonesia.

References

[1] R. H. Maulidiyah, R. Salamba, D. Wibowo, and M. Nurdin, "Organic compound rhodamine B degradation by TiO_2/Ti electrode in a new portable reactor," *Int J ChemTech Res*, vol. 8, no. 6, pp. 645–653, 2015.

[2] R. H. Maulidiyah, R. Salamba, D. Wibowo, and M. Nurdin, "Organic compound rhodamine B degradation by TiO_2/Ti electrode in a new portable reactor," *Int J ChemTech Res*, vol. 8, no. 6, pp. 645–653, 2015.

[3] Maulidiyah, D. S. Tribawono, D. Wibowo, and M. Nurdin, "Electrochemical profile degradation of amino acid by flow system using TiO_2/Ti nanotubes Electrode," *Anal. Bioanal. Electrochem.*, vol. 8, no. 6, pp. 761–776, 2016.

[4] M. Maulidiyah, M. Natsir, F. Fitrianingsih, Z. Arham, D. Wibowo, and M. Nurdin, "Lignin Degradation of Oil Palm Empty Fruit Bunches using TiO_2 Photocatalyst as Antifungal of Fusarium Oxysporum," *Orient. J. Chem.*, vol. 33, no. 6, pp. 3101–3106, 2017.

[5] L. Mursalim *et al.*, "Synthesis and Photoelectrocatalytic Performance of Mn-N-TiO_2/Ti Electrode for Electrochemical Sensor," in *IOP Conference Series: Materials Science and Engineering*, 2017, vol. 267, p. 012006.

[6] M. Maulidiyah, M. Natsir, F. Fitrianingsih, Z. Arham, D. Wibowo, and M. Nurdin, "Lignin Degradation of Oil Palm Empty Fruit Bunches using TiO_2 Photocatalyst as Antifungal of Fusarium Oxysporum," *Orient. J. Chem.*, vol. 33, no. 6, pp. 3101–3106, 2017.

[7] M. Nurdin, M. Z. Muzakkar, M. Maulidiyah, N. Maulidiyah, and D. Wibowo, "Plasmonic Silver—N/TiO_2 Effect on Photoelectrocatalytic Oxidation Reaction," *J Mater Env. Sci*, vol. 7, no. 9, pp. 3334–3343, 2016.

[8] M. Nurdin, A. H. W. Maulidiyah, N. Abdillah, and D. Wibowo, "Development of extraction method and characterization of TiO_2 mineral from ilmenite," *Int. J. ChemTech Res.*, vol. 9, no. 4, pp. 483–491, 2016.

[9] Z. Arham, M. Nurdin, and B. Buchari, "Photoelectrocatalysis Performance of La_2O_3 doped TiO $_2$/Ti Electrode in Degradation of Rhodamine B Organic Compound," *Int J ChemTech Res*, vol. 9, no. 11, pp. 113–120, 2016.

[10] M. Morozová, P. Kluson, J. Krysa, M. Vesely, P. Dzik, and O. Solcova, "Electrochemical properties of TiO_2 electrode prepared by various methods," *Procedia Eng.*, vol. 42, pp. 573–580, 2012.

[11] R. G. Saratale *et al.*, "Decolorization and detoxification of sulfonated azo dye CI Remazol Red and textile effluent by isolated Lysinibacillus sp. RGS," *J. Biosci. Bioeng.*, vol. 115, no. 6, pp. 658–667, 2013.

[12] R. Veerasamy, H. Rajak, A. Jain, S. Sivadasan, C. P. Varghese, and R. K. Agrawal, "Validation of QSAR models-strategies and importance," *Int. J. Drug Des. Discov.*, vol. 3, pp. 511–519, 2011.

[13] D. Wibowo, M. Nurdin, and others, "Determination of COD based on Photoelectrocatalysis of $FeTiO_3$. TiO_2/Ti Electrode," in *IOP Conference Series: Materials Science and Engineering*, 2017, vol. 267, p. 012007.

Immobilization of Chromium from Liquid Waste of Electroplating Home-Industries by Fly Ash Geopolymerization

H. Fansuri[1], E. Erviana[1], M. Rosyidah[1], R. M. Iqbal[1], W. P. Utomo[1)] and Nurlina[2]

[1]Department of Chemistry, Institut Teknologi Sepuluh Nopember, Kampus ITS Sukolilo, Surabaya, Indonesia
[2]Department of Chemistry, Universitas Tanjungpura, Jl. Prof. Dr. H. Hadari Nawawi, Pontianak, Indonesia

E-mail: h.fansuri@chem.its.ac.id

Abstract. Immobilization is one of methods that can be used to overcome the problem of water pollution by heavy metals, such as chromium, which is possibly produced in an electroplating home-industry. One method that can be used to immobilize heavy metals is by using the waste water as a reactant in a geopolymerization process. The resulting geopolymer will contain such heavy metals and keep them from being dispersed into the environment. In this research, the immobilization was carried out by geopolymerization of fly ash from PT. IPMOMI (International Power Mitsui Operation and Maintenance) in Probolinggo, Indonesia, which pass through 60 (FA 60), 100 (FA 100), and 200 (FA 200) mesh sieve and using wastewater from electroplating home-industries in Ngingas Village, Sidoarjo-Indonesia, or a solution of $Cr(NO_3)_3.9H_2O$ as one of its reactants. The composition of the geopolymerization follows the results of previous research, i.e using the ratio of Solid to Liquid (S/L) = 3.59, Na_2O/SiO_2 = 0.67, H_2O/Na_2O = 3.33 and SiO_2/Al_2O_3 = 6.46. Experimental results show that the compressive strength of geopolymers without heavy metals addition increases when the particle size decreases. Geopolymer made of FA 100 has the highest compressive strength i.e 35.05 MPa. The addition of a chromium-containing wastewater in FA 100 geopolymerization produce a geopolymer with a lower compressive strength i.e 26.42 MPa while the addition of $Cr(NO_3)_3\cdot 9H_2O$ solution produce geopolymer with compressive strength of 24.47 MPa. The SEM-EDX photo shows that chrome was evenly dispersed in the geopolymer. Leaching test with TCLP method (Toxicity Characteristic Leaching Procedure) on the geopolymer product showed that no chrome content in the leachate was found. This suggests that chromium is immobilized by geopolymers very well.

Keywords. geopolymer, immobilization, chrome waste, and fly ash.

1. Introduction

Ngingas village in Sidoarjo, East Java, Indonesia is well known as a center for metal-plating home industry. The metal plating is started with cleaning the metal using acid solutions and followed by plating process. Chrome electroplating is one of important metal-plating processes in the area to produce various chrome plated metals [1].

Content from this work may be used under the terms of the Creative Commons Attribution 3.0 licence. Any further distribution of this work must maintain attribution to the author(s) and the title of the work, journal citation and DOI.
Published under licence by IOP Publishing Ltd

Chromium, especially in the form of Cr(VI), is very toxic and it is an agent that can cause cancer, cell mutation, and malformation [2]. It also can cause cancer of skin and respiratory system [3]. Therefore, although Cr(VI) concentration in a normal chrome plating waste is small, it release to the environment must be controlled cautiously.

Immobilization of chrome by geopolymerization is a promising method to control chromium release to the environment. Immobilization of Cr^{3+} by geopolymerization using fly ash of PT. IPMOMI in Probolinggo, Indonesia can hold almost 100 % Cr^{3+} in the geopolymer when tested by TCLP method [4]. Other researcher reported that 96 % of chromium in the form of Cr^{6+} ion remains in the geopolymer after leaching [5].

Although it has been reported that chromium can be excellently immoblized by geopolymerization, the chromium ion was introduced to the geopolymer mixture in the form of its salt [5]. The effect migh be different if the source of chromium is a homogenous solution of chromium salts (such as $Cr(NO_3)_6 \cdot 9H_2O$) or from the waste water of a chrome-electroplating home industry because it might also contain other components which my alter the immobilization of chromium ion by geopolymerization process. Therefore, it is interesting to study the immobilization of chromium ion in the waste water of chrome-electroplating home industry. This paper reports our effort to immobilize chromium in waste water of chrome-electroplating home industry in Ngingas village, Sidoarjo, Indonesia by using geopolymerization of fly ash of PT. IPMOMI.

2. Materials and methods

The chemicals that were used in this research are fly ash, demineralized water, technical grade waterglass ($2SiO_2 \cdot Na_2O$) and p.a. grade sodium hydroxide (NaOH), aluminum hydroxide ($Al(OH)_3$), and acetic acid glacial. The ashes were taken from PT. IPMOMI power plant in Probolinggo, East Java, Indonesia.

The chemical composition of the fly ash was characterized by XRF and the result is shown in Table 1. The composition of waterglass has been reported elsewhere [6] and it was comprised of 19.16 wt.% SiO_2, 37.99 wt% Na_2O, and 28.07 wt% H_2O.

Table 1. Elemental composition of PT. IPMOMI fly ash

Element (reported as its most stable oxide)	% mass	Element (reported as its most stable oxide)	% mass
SiO_2	50.67	CaO	12.7
Al_2O_3	13.76	MgO	0.19
Na_2O	0.19	K_2O	1.28
TiO_2	0.76	Fe_2O_3	11.99
MnO_2	0.14	SO_2	0.63
Se_2O_3	0.12	P_2O_5	0.08
LOI	1.42		

2.1. Preparation of $Al(OH)_3$ and Activating Solution

The activating solution is a mixture of NaOH solution and waterglass. The NaOH solution was prepared by dissolving 28 g of NaOH in demineralized water so that the total volume was 50 mL. After leaving the resulted solution at room temperature for 24 h, 80 g waterglass was added and stirred well and then left at room temperature for 6 h. Meanwhile, $Al(OH)_3$ solution was prepared by dissolving 3.4 g $Al(OH)_3$ in 15 g of demineralized water.

2.2. Preparation of Geopolymer

The geopolymer was prepared by mixing 260 g of FA60, FA100 or FA 200 with the activating solution and then stirred well using a mixer for 40 s. After that, $Al(OH)_3$ solution was added and the mixture was stirred again for another 2.5 min. The resulted paste was then poured into a 2 cm height and 4 cm tall PVC cylindrical mold and left at room temperature for 24 h. The geopolymer was then taken out of the mold and put into a plastic bag and sealed before cured in an oven at 55 °C for 24 h. The compressive strength of the geopolymer was measured in the 7th day of curing. The mixture had solid/liquid (S/L) = 3.59; Na_2O/SiO_2 = 0.67; Na_2O/Al_2O_3 = 4.31; and SiO_2/Al_2O_3 = 6.46. The geopolymer is called chromium free-geopolymer.

2.3. Preparation of Geopolymer to Immobilize Chromium in the Waste Water
The preparation of geopolymer to immobilize chromium in the waste water was carried out by the same procedure as the preparation of the strongest chromium free-geopolymer. However, the concentration of sodium hydroxide was increased to reduce its volume while maintaining the same amount of sodium ion in the geopolymer paste mixture. To avoid the precipitation of chrome due to high alkaline concentration, the addition of waste water was carried out just after the addition of $Al(OH)_3$ solution. The geopolymer is denoted as chrome containing-geopolymer.

2.4. Leaching Test
Leaching test was carried out for the chrome containing-geopolymer. The geopolymer paste was immersed in 2.5% acetic acid solution and stirred by a magnetic stirrer for 24 h. The ratio of volume of acetic acid solution to the mass of geopolymer paste was 1:20. At 1, 2, 4, 8, 16 and 24 h, 10 mL of the immersing solution was taken out to be analyzed by Atomic Absorption Analyzer (AAS) to determine the concentration of leached cromium from the gropolymer.

2.5. Characterization
Compressive strength of all geopolymer pastes were measured using Universal Testing Machine (UTM). SEM-EDX analyses were carried out using Zeiss EVO MA 10 on Au-Pd coated polished cross-section of the geopolymer samples.

3. Results and discussion

3.1. Characteristics of the Waste Water and Fly Ash
Concentration of total chromium content in the waste water, which was analyzed by an AAS, is shown in Table 2. Sample L1 and L2, which were taken from the unprocessed waste water, contain relatively high chromium concentration. The other two (L3 and L4), which were taken from waste water treatment plants, contains no chromium.

The maximum alowable chromium content, in the form of Cr^{6+}, in surface water is 0.1 mg/L (0.1 ppm) [7]. Although the analysis method (AAS) could not differentiate the chromium species whether it is in the form of Cr^{3+} or Cr^{6+}, the chromium content in L1 and L2 are very high (6.0 and 28.8 ppm, respectively). Therefore, the waste water is suitable for the study of immobilization by geopolymer and the highest one (L2) was used in the geopolymerization experiments.

The mineral composition of the fly ash was determined by XRD and the diffractogram is shown in Figure 1. The diffractogram shows peaks of quartz at 2θ = 20.58°, 26.61° and 50.04°, mullite ($3Al_2O_3\cdot 2SiO_2$or $2Al_2O_3\cdot SiO_2$) at 2θ= 33.44° and 42.95°, and magnetite at 2θ = 35.56° and 62.40°. The fly ash also contains amorphous phase of aluminosilicate which is shown by a hump at 2θ between 20° and 40° [8]. The diffractogram also shows that major elements as in Table 1 (Si, Al and Fe) are distributed in crystalline and might also be found in the amorphous phases in the fly ash.

Table 2. Concentration of chromium in the waste water

Sample ID	Cr_{total} (ppm)
L1	6.0
L2	28.8
L3	0.0
L4	0.0

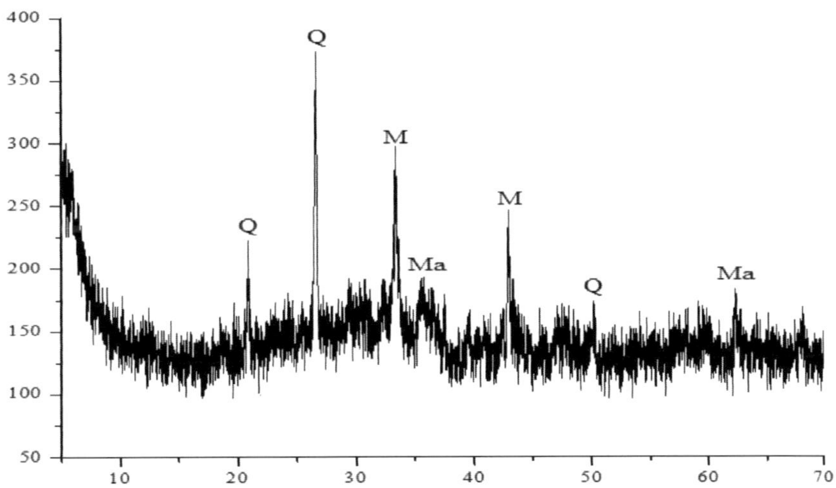

Figure 1. X-ray diffractogram of fly ash of PT. IPMOMI. Q = quartz, M = Mullite and Ma = Magnetite.

Acording to Supriadi, *et al.* [6], the amorphous phase is the most reactive component of fly ash and reacted completely in geopolymerization process while the crystalline phase only partially reacted. However, Atmaja, *et al.* [7] reported that although the crystalline phases are only partially reacted, their existence do not have significant effect to the overall strength of geopolymer. Therefore, it is assumed that all silicon and aluminum content in the fly ash will be involved in the geopolymerization process and counted in the calculation of Si/Al ratio.

Before the preparation of the geopolymer, fly ash of PT. IPMOMI was sieved through 60 (FA 60), 100 (FA 100) and 200 (FA 200) mesh sieve to get a uniform particle size and to study the effect of particle size to the compressive strength of the resulting gepolymer. The sieve opening of each mesh is 0.250 mm, 0.149 mm and 0.074 mm respectively. According to Wang, *et al.* [8], particle size affected the reactivity of the ash to alkaline activator during the geopolimerzation process. Fly ash with smaller particle size is more reactive in the geopolymerization process than fly ash with larger particle size.

Table 3 shows the compressive strength of chromium free-geopolymer. The test was carried out in the day 7 of curing. The table shows that the strongest geopolymer was made of FA 100, which is the intermediate particle size of fly ash. The result seems contradict to Wang et al. [8] who reported that the strength of geopolymer was inversely proportional to the geopolmer strength.

In our case, particle size of FA 100 was larger but more uniform than FA 200. The uniformity of fly ash comes from the sieving process where fly ash particles smaller than 0.149 mm but larger than 0.074 mm will be stay in 100 mesh sieve while the smaller particles will pass through the 200 mesh sieve. Therefore, although smaller particle will produce stronger and better geopolymer, there is a possibility that the uniformity, in combination with small particle size of the fly ash, also has an important role in the geopolymerization process. Based on the results in Table 3, FA 100 was then used to make chromium containing-geopolymer.

Table 3. Compressive strength of chromium free-geopolymer

Trial No	Compressive Strength (MPa)		
	FA 60	FA 100	FA 200
1	25.00	33.95	31.45
2	21.45	36.18	33.00
3	22.55	35.00	32.24
Average	23.00	35.05	32.20

3.2. Immobilization of Chromium by the Geopolymer

Table 4 shows the compressive strength of chromium containing-geopolymer made of FA 100. It is shown that the geopolymer strength was reduced from 35.05 MPa to 24.47 MPa – 26.42 MPa when it contains chromium. The results seem contradict to a report by Fatmawati [4] where compressive strength of fly ash geopolymer increased at increasing chromium content up to 6,860 ppm (the concentration of chromium is relative to the mass of fly ash used in the synthesis of geopolymer) before it drastically decreased at higher chromium content. However, in Fatmawati's reports the strength of chromium containing-geopolymer at low chromium content was about 29 MPa, which was close to our results. Therefore, the strength of chromium containing geopolymer in this research is in agreement with low chromium content geopolymer in Fatmawati's report [4].

Figure 2 shows SEM-EDX images of cross section of chromium containing-geopolymer. Although concentration of chromium was very low, the SEM-EDX image shows the existence of the metal and its distribution across the cross section.

Table 4. Compressive strength (in MPa) of chromium containing-geopolymer made of FA 100

Chromium from waste water	Chromium from $Cr(NO_3)_6.9H_2O$ Solution
26.42	24.47

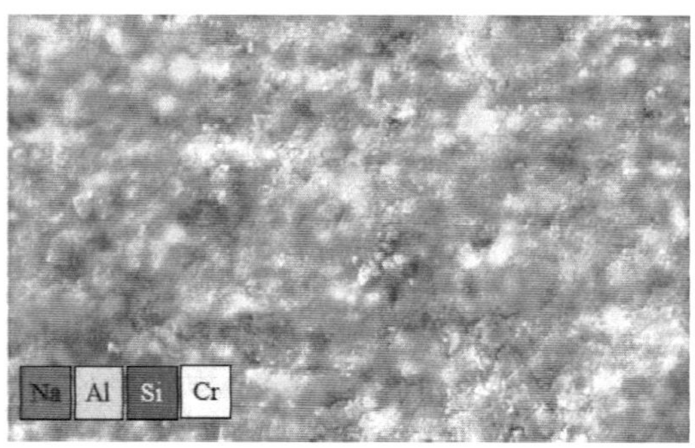

Figure 2. SEM-EDX photo of the cross section of chromium containing geopolymer

Leaching test results of chromium containing geopolymer are listed in Table 5. In the table, the name of "with waste water" means that the geopolymer was prepared by using the waste water while number in "Geo-2000" means that the geopolymer was prepared with the addition of 2000 ppm chromium, relative to total fly ash used in the preparation. The table shows that no chromium was leached out from geopolymers that were prepared with waste water.

When the geopolymer was made with chromium $Cr(NO_3)_6 \cdot 9H_2O$ solution, a small amount of chromium (1 % for Geo-2000 to 3 % Geo-8000) was leached out Its means that more than 97%

IOP Conf. Series: Materials Science and Engineering **367** (2018) 012049 doi:10.1088/1757-899X/367/1/012049

chromium was immobilized by the geopolymer. The percentage of immobilized chromium is slightly lower than the result that was reported by Zhang, *et al.* [9]. However, Zhang, *et al.* [9] converted all chromium ion to Cr^{3+} before the geopolymer preparation, which was not done in this research.

The immobilization of chromium by the geopolymer can also be seen from SEM-EDX results in Figure 3. The figure shows that the chromium content stayed in the geopolymers after leaching process.

Table 5. Leaching test results of chromium containing geopolymer

No	Geopolymer	$t_{Leaching}$ (h)	$[Cr]_{leached}$ (%)
1	with waste water	1	0
2	with waste water	2	0
3	with waste water	4	0
4	with waste water	8	0
5	with waste water	16	0
6	with waste water	24	0
7	Geo-2000*	24	1
8	Geo-4000*	24	1
9	Geo-8000*	24	3

*Geo-2000: chromium content = 2000 ppm relative to total fly ash used in the geopolymer preparation

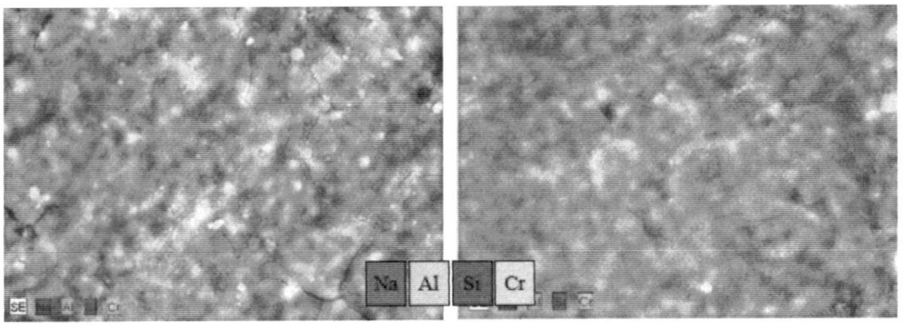

Figure 3. SEM-EDX photo of the cross section of chromium containing geopolymer: (a) before and (b) after leaching

4. Conclusions

Experimental results showed that the highest compressive strength of chromium free-geopolymer (35.05 MPa) was given by the geopolymer made of PT. IPMOMI fly ash with particle size between 0.074 mm and 0.149 mm (FA 100). The results indicated that uniformity of the particle size of fly ash may also play an important role in the geopolymerization process. The addition of a chromium-containing waste water and $Cr(NO_3)_3 \cdot 9H_2O$ solution reduced compressive strength of the resulted geopolymer to 26.42 MPa and 24.47 MPa, respectively. SEM-EDX images showed that chromium was evenly dispersed in the geopolymer. Leaching test result showed that no chromium was found in the leachate when the source of chromium was waste water from Ngingas village with the highest chromium concentration. When using higher chromium solution, which was made of $Cr(NO_3)_3 \cdot 9H_2O$, up to 3% of chromium was leached out of the chromium containing-geopolymer. This suggests that chromium is immobilized by geopolymers very well and geopolymerization can immobilize cromium in the waste water from electroplating home-industry in Ngingas village, Sidoarjo.

IOP Conf. Series: Materials Science and Engineering **367** (2018) 012049 doi:10.1088/1757-899X/367/1/012049

Acknowledgments
The reseach was funded by Indonesian Ministry of Research, Technology and Higher Education under research contract No. 078/SP2H/LT/DRPM/II/2016 and also by LPPM ITS under contract No. 820/PKS/ITS/2017. The authors thank to Dr. Subaer of Departement of Physics, Universitas Negeri Makassar, for fruitfull discussion, accommodation and other supports that was given to us during our visit to his laboratory.

References
[1] Martak F, Candrawati I, and Cahyo Y I, *J. Pure App. Chem. Res*, vol. 6, pp. 117–126, 2017.
[2] J. Vera-Candioti, S. Soloneski, and M. L. Larramendy, "Acute toxicity of chromium on Cnesterodon decemmaculatus (Pisces: Poeciliidae)," *Theoria*, vol. 20, no. 1, 2011.
[3] J. Deja, "Immobilization of Cr^{6+}, Cd^{2+}, Zn^{2+} and Pb^{2+} in alkali-activated slag binders," *Cem. Concr. Res.*, vol. 32, no. 12, pp. 1971–1979, 2002.
[4] Fatmawati A, "Immobilization of heavy metal cation Cr3+ in geopolymer based on fly ash of PT. IPMOMI," Institut Teknologi Sepuuh Nopember, Surabaya, skripsi, 2014.
[5] Y. L. Galiano *et al.*, "Behaviour of Fly Ash-Based Geopolymer Panels Under Fire," *Waste Biomass Valorization*, vol. 8, no. 7, pp. 2485–2494, 2017.
[6] W. Supriadi, R. Bayuaji, R. Burhan, H. Fansuri, and others, "The Effect of Pb2+ and Cd2+ Addition to Mechanical Properties of Fly Ash Geopolymer Paste," in *Materials Science Forum*, 2016, vol. 841, pp. 178–185.
[7] L. Atmaja, H. Fansuri, and A. Maharani, "Crystalline Phase Reactivity In The Synthesis Of Fly Ash-Based Geopolymer," *Indones. J. Chem.*, vol. 11, no. 1, pp. 90–95, 2011.
[8] H. Wang, H. Li, and F. Yan, "Synthesis and mechanical properties of metakaolinite-based geopolymer," *Colloids Surf. Physicochem. Eng. Asp.*, vol. 268, no. 1–3, pp. 1–6, 2005.
[9] Zhang B, Kang F, Tarascon J M, and Kim J M, *J.K. RRecent. Prog. Mater. Sci*, vol. 76, pp. 319–380, 2015.

IOP Conf. Series: Materials Science and Engineering **367** (2018) 012050 doi:10.1088/1757-899X/367/1/012050

Preparation, Structural and Dielectric Behaviors of $Co_xMn_{1-x}Mn_2O_4$ ($0 \leq x \leq 1$) Nanoparticles

A. Taufiq[1], M. Muzammil[1], A. Fuad[1], N. Hidayat[1], S. Sunaryono[1], N. Mufti[1], A. Hidayat[1], M. Diantoro[1] and M. Munasir[2]

[1]Department of Physics, Faculty of Mathematics and Natural Sciences, Universitas Negeri Malang Jl. Semarang 5, Malang 65145, Indonesia
[2]Department of Physics, Faculty of Mathematics and Natural Sciences, Universitas Negeri Surabaya, Jl. Ketintang Surabaya 60231, Indonesia

E-mail: ahmad.taufiq.fmipa@um.ac.id

Abstract: Cobalt-manganese oxide nanoparticles become remarkable metal oxides due to their physical characters, for example, their electrical properties. In this paper, we report the synthesis of $Co_xMn_{1-x}Mn_2O_4$ ($0 \leq x \leq 1$) nanoparticles via a precipitation method. The structural and dielectric properties were investigated by means of X-ray diffractometer (XRD) and LCR-meter. From the data analysis, it was found that the $Co_xMn_{1-x}Mn_2O_4$ particles structured spinel cubic with the particle size ranging from 22.7 to 28.6 nm. Increasing Co^{2+} ions led to declining lattice and crystal volume of the $Co_xMn_{1-x}Mn_2O_4$ nanoparticles. Such phenomenon was originated from the substitution process of Co^{2+} ions that change the metal construction both at the octahedral and tetrahedral sites. Furthermore, the Co^{2+} ion substitution resulted in an increase in the dielectric properties of the $Co_xMn_{1-x}Mn_2O_4$ nanoparticles regarding with dipole moment. Interestingly, the increase was also influenced by reducing the particle size and crystal volume of the $Co_xMn_{1-x}Mn_2O_4$ nanoparticles.

Keywords: $Co_xMn_{1-x}Mn_2O_4$, dielectricity, nanoparticles, and nanostructure.

1. Introduction

Over recent decades, investigations on nanomaterials, especially on metal oxides have been widely performed by many researchers and engineers regarding preparations, characterizations, and applications of the metal oxides. One of the various metal oxides that play an essential role in sophisticated technology is hausmannite (Mn_3O_4) nanoparticles. Mn_3O_4 nanoparticles constructed in a spinel system with tetragonal structure have been fabricated in various sizes, forms, and morphologies to enhance their application performance. Some literature showed that the hausmannite had been utilized for many applications as supercapacitor [1, 2], superior anode materials for Li-ion batteries [3], highly efficient hydrogen evolution [4], electro-oxidation of ethanol and propanol [5], nonenzymatic glucose sensor [6], superior sunlight photocatalysts [7], and so forth.

In order to develop the sophisticated applications of the Mn_3O_4 nanoparticles, many researchers have made various efforts, especially which are generally started by developing nanoparticle synthesis methods. The methods have been conducted to produce the Mn_3O_4 nanoparticles in various shapes or geometry, sizes, and distributions. Several papers presented that the Mn_3O_4 nanoparticles have been successfully prepared by ultrasonic irradiation assisted by co-precipitation method [8], auto-

Content from this work may be used under the terms of the Creative Commons Attribution 3.0 licence. Any further distribution of this work must maintain attribution to the author(s) and the title of the work, journal citation and DOI.
Published under licence by IOP Publishing Ltd

combustion method [9], one-pot method [10], ultrasonic method [2], co-precipitation method [11], and others. In this experiment, based on the methods mentioned above, we apply the co-precipitation method due to its simplicity and effectiveness in producing the nanomaterials. By using such method, the Mn_3O_4 nanoparticle can be prepared without a high temperature in relatively short time.

One of the exciting properties of the Mn_3O_4 nanoparticles is related to their electrical properties such as dielectric property. In nanometric size, the nanomaterials have entirely different electrical properties from those of their bulk counterparts [11]. Furthermore, in order to enhance the dielectric property of the Mn_3O_4 nanoparticles, it is also essential to substitute several metal ions in the Mn_3O_4 particles. Theoretically, the substitution can be conducted by inserting some metals such as Fe, Zn, Mn, Cr, Co, and other metals. Therefore, in this work, we propose the substitution of Co ion in Mn_3O_4 to form $Co_xMn_{1-x}Mn_2O_4$ ($0 \leq x \leq 1$) nanoparticles. The range of x value is determined to keep the samples in spinel system with a tetragonal structure. In this paper, we report the preparation of the $Co_xMn_{1-x}Mn_2O_4$ nanoparticles using a simple coprecipitation route. Moreover, the structural and dielectric properties of the samples are also presented.

2. Materials and methods

The materials used in this experiment were $MnCl_2.4H_2O$, $CoCl_2.6H_2O$, NH_4OH, H_2O, and alcohol. In the first stage, the $MnCl_2.4H_2O$ and $CoCl_2.6H_2O$ were weighed based on the stoichiometric composition (x value) to form $Co_xMn_{1-x}Mn_2O_4$ ($0 \leq x \leq 1$) particles in 5 samples, i.e., x = 0; 0.4; 0.6; 0.8; and 1.0. The $MnCl_2.4H_2O$ and $CoCl_2.6H_2O$ were then diluted in H_2O (90 mL) using a hot plate magnetic stirrer at 70 °C and followed by wisely dropping NH_4OH (30 mL) for 60 min to obtain sediment. The sediment was then filtered and washed with H_2O and alcohol for several times until pH condition was obtained. The process was followed by a calcination at 100 °C for 120 min to achieve the $Co_xMn_{1-x}Mn_2O_4$ ($0 \leq x \leq 1$) particles in powders.

The prepared samples were characterized via an X-ray diffractometer (XRD) using Cu-Kα for 2-theta in the range from 20° to 70° with a step size of 0.02°. The X-ray diffraction data were then analyzed using a Rietica program by employing Rietveld method. To investigate the dielectric property as well as a dielectric constant, an LCR-meter was utilized at ambient temperature by varying the frequency in the range of 1 kHz to 200 kHz. The dielectric constant data was then analyzed by comparing the data by x content and frequency with the crystal structure and crystallite size.

3. Results and discussion

X-ray diffraction patterns of the $Co_xMn_{1-x}Mn_2O_4$ particles are presented in Figure 1. It is clear from Figure 1a that a higher value of x corresponds to the increasing Co content and decreasing Mn content. The peaks of the XRD data from small 2-theta present (*hkl*) planes of respective (112), (200), (103), (211), (004), (220), (105), (312), (303), (321), (224), and (400) which coincides with JCPDS No. 18-0803. Moreover, the figure also indicates a peak is broadening along with the increasing Co content. It could prove that the Co^{2+} ions have been successfully injected into the Mn_3O_4 without producing new phases or impurities. The wider peak broadening also becomes evident that the increasing Co content resulted in decreasing the particle size of the samples. Moreover, from Figure 1b, it can be seen that the Co^{2+} ions occupied the position of Mn^{2+} ions at the tetrahedral site. In accordance with this work, Salek *et al*. reported that the Co^{2+} ions mainly occupied the tetrahedral site [12]. Such site is occupied by the metal ions surrounded by four oxygen atoms. Meanwhile, the octahedral site was occupied by the metal ions surrounded by six oxygen atoms [13]. In order to study the crystal structure comprehensively, we analyzed the XRD data using a Rietica program by employing a Rietveld analysis. The output of quantitative data analysis, as well as crystallite size and crystal volume, are presented in Figure 2.

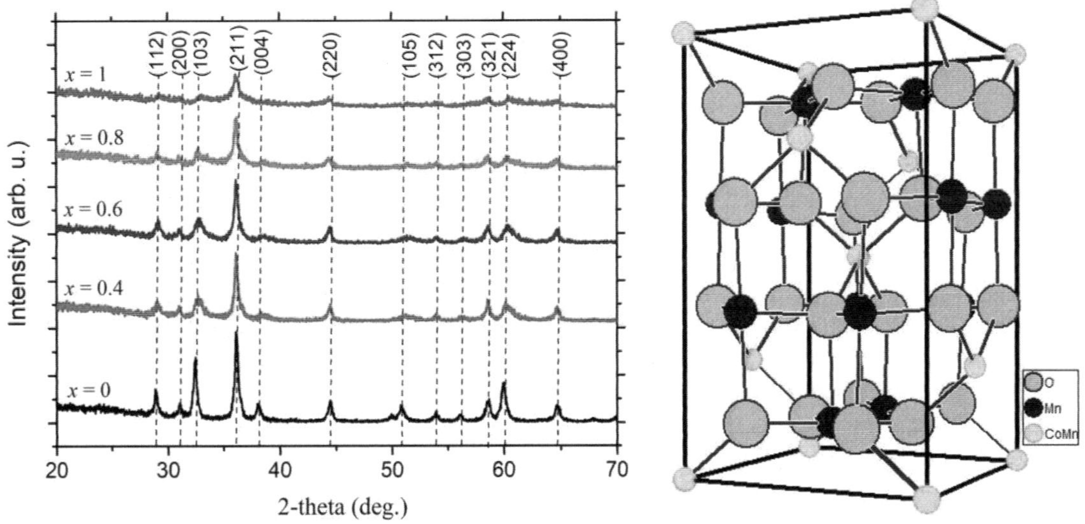

(a) (b)

Figure 1. (a) XRD patterns of the $Co_xMn_{1-x}Mn_2O_4$ ($0 \leq x \leq 1$) particles and (b) the crystal structure for a certain x composition

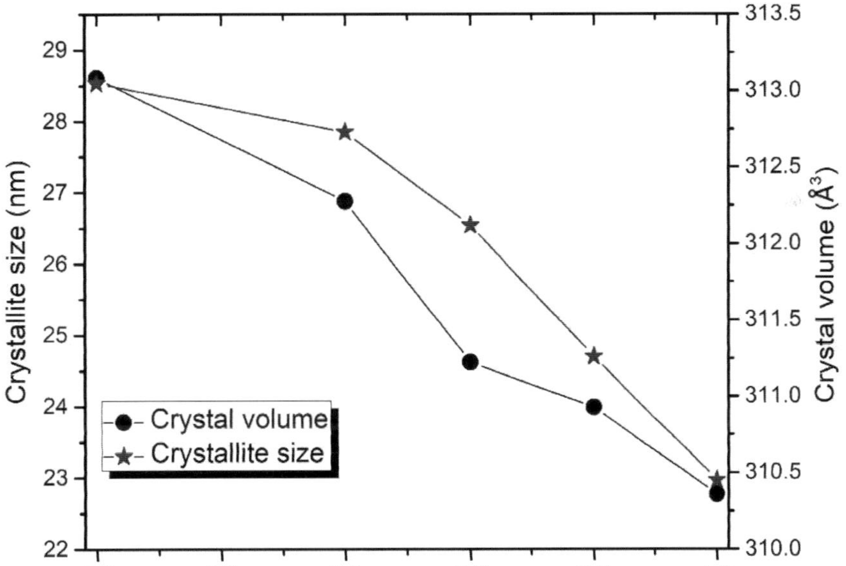

Figure 2. Crystallite size and crystal volume of the $Co_xMn_{1-x}Mn_2O_4$ ($0 \leq x \leq 1$) particles

The increasing crystal volume as shown in Figure 2 originating from the lattice parameters of the $Co_xMn_{1-x}Mn_2O_4$ particles is strongly affected by the incorporation of Co^{2+} ions. These ions have been successfully inserted into the tetrahedral site that was initially occupied by Mn^{2+} ions. Because the ionic radii of Co^{2+} ions (0.74 Å) were lower than those of Mn^{2+} ions (0.80 Å) [14], the lattice parameters (a, b and c) of the system became smaller and therefore directly decreased its crystal volume. The results of data analysis also presented that there is not a significant connection between the *c/a* ratio and the degree of *c-a/a* distortion. The ratio of the five samples was around 1.63 and 1.64 that became another evidence that the Co^{2+} ions successfully substituted the Mn^{2+} ions and not the Mn^{3+} ions. Thus, it only disrupted the tetrahedral site while the octahedral site occupied by the Mn^{3+} ions was not disrupted although the *c/a* ratio and the degree of *c-a/a* distortion were strongly influenced by the Jahn-Teller effect, which was purely occupied by Mn^{3+} ions at the octahedral site. In principle, the crystal field theory could explain the tetrahedral distortion caused by the Jahn-Teller effect [15]. Such distortion caused a lower system of symmetry and degeneration level. If the two ligands in *z*-axis move away from the core atom in the same directions with the $-z$ and $+z$ axes, the repulsion between the electrons and ligands and the central atom that occupies *d* orbitals containing *z* components such as d_{xz}, d_{yz}, d_z^2 would decrease. Consequently, the orbitals would stabilize or experience a decrease in energy level. On the contrary, the *d* orbitals that do not possess any *z* components, namely d_{xy} and d_x^2-d_y^2, would be unstable or would have an increasing level of energy.

The smallest crystallite size of 22.77 nm was obtained on $x = 1$ sample, and the largest was 28.61 nm on $x = 0$ sample. Meanwhile, the $x = 0.4$ sample had 26.88 nm crystallite size, the $x = 0.6$ sample had 24.62 nm crystallite size, and the $x = 0.8$ sample had 23.99 nm crystallite size. Figure 2 shows such information in detail. The results are in good agreement with the initial prediction about the peak area of the X-ray diffraction pattern in a nanometric size. From such explanation, it could be concluded that the $Mn_{3-x}Co_xO_4$ nanoparticles have been successfully prepared. The crystallite size of the $Mn_{3-x}Co_xO_4$ nanoparticles for $x = 1$ in this experiment is relatively similar to that of the sample prepared by a newly adopted hydrothermal route [16]. The data analysis via Scherer's equation obtained a continuously decreasing crystallite size along with the increasing composition of the Co^{2+} ions. It related to the concept of the compound formation when Mn_3O_4 is treated at a certain temperature while simultaneously added with a composition of another material, in this case, a Co^{2+} dopant. Therefore, it requires a longer period to form a new compound when the dopant composition increased. As a result, the formation of the new compound of $Co_xMn_{1-x}Mn_2O_4$ would be imperfect, which resulted in a smaller crystallite size as a higher amount of Co^{2+} dopant was added.

Figure 3 presents the dielectric constant of $Co_xMn_{1-x}Mn_2O_4$ nanoparticles for $0 \le x \le 1$ as a function of frequency. The figure shows the increasing value of dielectric constant along with the increasing Co^{2+} content. It means that the Co^{2+} incorporation into the tetrahedral site in the system contributed significantly to enhancing the dielectric constant of the $Co_xMn_{1-x}Mn_2O_4$ nanoparticles. Theoretically, it can be explained that the Co^{2+} ions with the smaller ionic radii than those of Mn^{2+} ions contributed to reducing valence electron of the $Co_xMn_{1-x}Mn_2O_4$ nanoparticles and followed by increasing their dipole moments. The increasing dipole moments could be reflected by lowering the crystal volume of the $Co_xMn_{1-x}Mn_2O_4$ nanoparticles that directly increased the dielectric constant. The decreasing crystal volume of the $Co_xMn_{1-x}Mn_2O_4$ nanoparticles was started by a declining crystal volume of the tetragonal structure.

The interesting phenomenon in this work is a decreased dielectric constant along with the increasing frequency. Figure 3 illustrates the dielectric constant of the same sample, i.e., for similar *x* composition, the constant dielectric lead to a decrease in frequency. The phenomenon can be explained by the physical concept that relates to the correlation between capacitance and frequency as shown in Equation 1 [17].

$$C = C_g \frac{s\tau}{1+\tau^2\omega^2} \tag{1}$$

Where *C* is capacitance, C_g is geometrical capacitance, *S* is conductance, τ is time relaxation for a dipole, and ω is angular frequency. The equation guides us that the capacitance decreases by an increase in frequency. Implicitly, it could be explained by the polarization mechanism of a charge carrier in the $Co_xMn_{1-x}Mn_2O_4$ nanoparticles by increasing the frequency. Therefore the polarization would decrease following the decreasing dielectric constant [17]. Rathore and co-workers reported that dielectric polarization happens as ionic, electronic, dipolar, or interfacial polarization. The dipole will no longer be able to rotate sufficiently fast when the frequency is increased. Furthermore, the dipole will be unable to follow the field when the frequency is further increased, and the orientation polarization will stop. Therefore, at higher frequencies, the dielectric permittivity of the samples decreases and tends to be constant. Such phenomenon is believed originating from the vanished space charge or interfacial polarization [18].

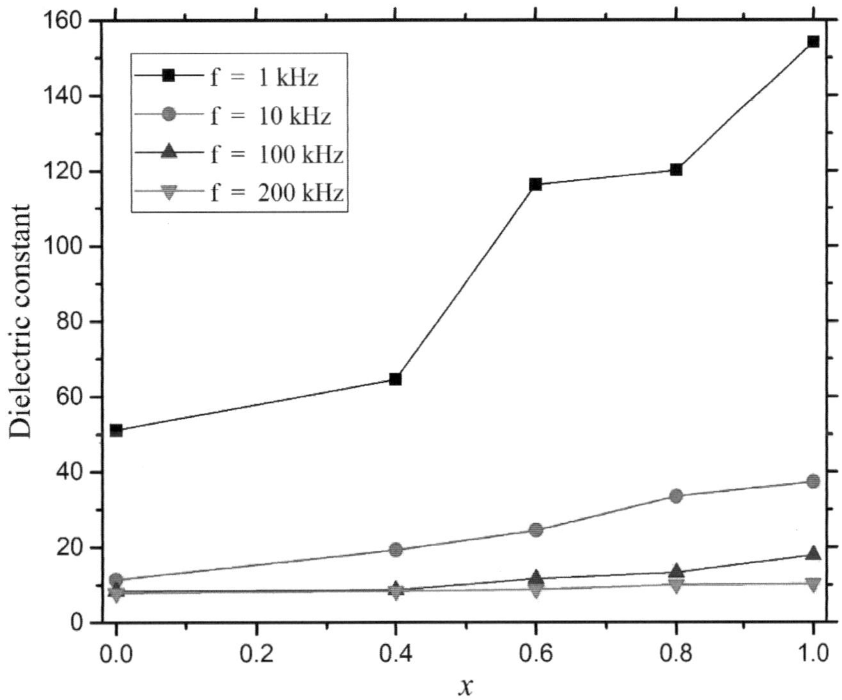

Figure 3. Dielectric constant of the $Co_xMn_{1-x}Mn_2O_4$ ($0 \leq x \leq 1$) nanoparticles as a function of frequency

Another work reported that the effect of increasing frequency could be expressed by considering samples composing of well-conducting grains separated by poorly conducting grain boundaries [19]. Furthermore, the charge displacement can explain the microscopic mechanism in the material. The electric dipole movement caused by the electrical field in the material that is still in the range of short transition since the valency electrons could not move freely due to the addition of Co^{2+} ions to the $Co_xMn_{1-x}Mn_2O_4$ nanoparticles. Therefore, it strengthens the core bond towards the outermost electron resulting in a larger dipole moment. As a consequence, the dielectricity of the material will increase. Besides, the growing dielectric constant also could be seen from the comparison between the polarization and crystal volume based on the Equation 2.

$$P = \frac{\sum p_i}{V} \qquad (2)$$

Where $\boldsymbol{P}$ is the polarization, $\boldsymbol{p}_i$ is the dipole moment, and V is the volume [20]. The equation shows that the volume is inversely proportional to the polarization so that the increasing amount of Co^{2+} ions in the $Co_xMn_{1-x}Mn_2O_4$ nanoparticles will cause its volume to shrink resulting in a growing polarization and dielectric constant. Moreover, polarization at the tetrahedral complex in the spinel structure of the $Co_xMn_{1-x}Mn_2O_4$ nanoparticles also contributes to the increasing dielectricity. The polarization at such tetrahedral complex has a particular value (not zero) even though the sample has not obtained the electric field yet.

From Figure 2 and 3, it is shown that the declining crystal volume is proportional to the increase of dielectric constant. It shows that the Co^{2+} ions contributing in the tetrahedral site that increases the dielectricity of the material. A dielectric constant of nanoparticles could be influenced by the technique of synthesis, particle size, and cation distribution. It is known in this research that besides the ionic radii of a dopant that is smaller than the parent ion, the dielectric constant of the $Co_xMn_{1-x}Mn_2O_4$ nanoparticles is also affected by its crystallite size. It is in good agreement with the statement of Raghavender and Jadhav that the smaller the size of a crystal, the higher the dielectric constant value of it [21]. Based on Figure 2 and 3, the crystallite size tends to decline and result in an increase of the dielectricity of the samples. Such increase in the value of the dielectric constant at a certain frequency was found to be significant, namely 126 %. The samples with $x = 0.8$ and $x = 1.0$ had an almost similar percentage of the dielectric constant increase, i.e., 134 %. The percentage of such increase is affected by the declining crystallite size. Such facts are in good agreement with others work that a small crystallite size would have a high value of dielectric constant. The increase of the dielectric constant as the decrease of the crystallite size can be described by the correlated barrier hopping model. In principle, the ions or charge carriers be supposed to be separated by anisotropy barrier or potential well and to jump from one to another position over the potential well [18]. In this condition, when the particle size becomes smaller as a single-domain in nanometric size, a weak field is necessary to overcome the potential well and then finally to increase the dielectric permittivity of the samples. Furthermore, the Coulombic interaction mechanism could be stronger due to the shrinking crystal volume that tends to contributes more in the dielectric constant value rather than in the development of the crystallite size, which is relatively small. The comprehensive explanation of the size effect on dielectric constant can be found in the literature [19,22]. As a final remark, in order to investigate the hierarchical nanostructure in more detail of the $Co_xMn_{1-x}Mn_2O_4$ nanoparticles in 3-dimensions, it is necessary to conduct a further experiment by employing small angle scattering using neutron or synchrotron sources.

4. Conclusions

Based on the data analysis, it can be concluded that the $Co_xMn_{1-x}Mn_2O_4$ particles have been successfully prepared via a precipitation route. The $Co_xMn_{1-x}Mn_2O_4$ particles constructed spinel cubic structures with the particle size ranging from 22.7 nm to 28.6 nm. Increasing Co^{2+} ions with ionic radii larger than those of Mn^{2+} ions caused declining lattice parameters and crystal volume of the $Co_xMn_{1-x}Mn_2O_4$ nanoparticles. Interestingly, the Co^{2+} ion substitution results in an increasing dielectric property of the $Co_xMn_{1-x}Mn_2O_4$ nanoparticles regarding dipole moment. Moreover, the increase was also influenced by the shrinking particle size and crystal volume of the $Co_xMn_{1-x}Mn_2O_4$ nanoparticles. Finally, the dielectric constant was also affected by the variation of frequency during characterization.

Acknowledgements

The authors appreciate KEMENRISTEKDIKTI for giving research award in the scheme of "PDUPT" for AT in 2017-2018.

References

[1] Bui P T M, Song J-H, Li Z-Y, Akhtar M S and Yang O-B 2017 Low temperature solution processed Mn_3O_4 nanoparticles: Enhanced performance of electrochemical supercapacitors *J. Alloys Compd.* **694** 560–7

[2] Raj B G S, Ramprasad R N R, Asiri A M, Wu J J and Anandan S 2015 Ultrasound assisted synthesis of Mn_3O_4 nanoparticles anchored graphene nanosheets for supercapacitor applications *Electrochimica Acta* **156** 127–37

[3] Zhao Y, Ma C and Li Y 2017 One-step microwave preparation of a Mn_3O_4 nanoparticles/exfoliated graphite composite as superior anode materials for Li-ion batteries *Chem. Phys. Lett.* **673** 19–23

[4] Yan D, Li F, Xu Y, Liu C, Wang Y, Tang Y, Yang L and Wang L 2017 Three-dimensional reduced graphene oxide– Mn_3O_4 nanosheet hybrid decorated with palladium nanoparticles for highly efficient hydrogen evolution *Int. J. Hydrog. Energy*

[5] Zhang Y, Yi Q, Chu H and Nie H 2017 Catalytic activity of Pd-Ag nanoparticles supported on carbon nanotubes for the electro-oxidation of ethanol and propanol *J. Fuel Chem. Technol.* **45** 475–83
[6] Yang S, Li G, Wang G, Zhao J, Gao X and Qu L 2015 Synthesis of Mn_3O_4 nanoparticles/nitrogen-doped graphene hybrid composite for nonenzymatic glucose sensor *Sens. Actuators B Chem.* **221** 172–8
[7] Sheikhshoaie I, Ramezanpour S and Khatamian M 2017 Synthesis and characterization of thallium doped Mn_3O_4 as superior sunlight photocatalysts *J. Mol. Liq.* **238** 248–53
[8] Tholkappiyan R, Naveen A N, Vishista K and Hamed F 2017 Investigation on electrochemical performance of Hausmannite Mn_3O_4 nanoparticles by ultrasonic irradiation assisted co-precipitation method for supercapacitor electrodes *J. Taibah Univ. Sci.*
[9] Chen B, Rao G, Wang S, Lan Y, Pan L and Zhang X 2015 Facile synthesis and characterization of Mn_3O_4 nanoparticles by auto-combustion method *Mater. Lett.* **154** 160–2
[10] Yang S, Liu L, Wang G, Li G, Deng D and Qu L 2015 One-pot synthesis of Mn_3O_4 nanoparticles decorated with nitrogen-doped reduced graphene oxide for sensitive nonenzymatic glucose sensing *J. Electroanal. Chem.* **755** 15–21
[11] Dhaouadi H, Ghodbane O, Hosni F and Touati F 2012 Mn_3O_4 Nanoparticles: Synthesis, Characterization, and Dielectric Properties *ISRN Spectrosc.* **2012** 1–8
[12] Salek G, Dufour P, Guillemet-Fritsch S and Tenailleau C 2015 Sustainable low temperature preparation of $Mn_{3-x}Co_xO_4$ ($0 \leq x < 3$) spinel oxide colloidal dispersions used for solar absorber thin films *Mater. Chem. Phys.* **162** 252–62
[13] Cullity B D and Graham C D 2008 *Introduction to Magnetic Materials* (Hoboken, N.J: Wiley-IEEE Press)
[14] Dong R, Ye Q, Kuang L, Lu X, Zhang Y, Zhang X, Tan G, Wen Y and Wang F 2013 Enhanced Supercapacitor Performance of Mn_3O_4 Nanocrystals by Doping Transition-Metal Ions *ACS Appl. Mater. Interfaces* **5** 9508–16
[15] Effendy E 2007 *Perspektif Baru Kimia Koordinasi Jilid I* (Malang: Bayumedia Publishing)
[16] Vigneshwaran P, Kandiban M, Senthil Kumar N, Venkatachalam V, Jayavel R and Vetha Potheher I 2016 A study on the synthesis and characterization of $CoMn_2O_4$ electrode material for supercapacitor applications *J. Mater. Sci. Mater. Electron.* **27** 4653–8
[17] Yadav V S, Sahu D K, Singh Y, Kumar M and Dhubkarya D C 2010 Frequency and Temperature Dependence of Dielectric Properties of Pure Poly Vinylidene Fluoride (PVDF) Thin Films American Institute of Physics Conference Series vol 1285 pp 267–78
[18] Rathore D, Kurchania R and Pandey R K 2014 Influence of particle size and temperature on the dielectric properties of $CoFe_2O_4$ nanoparticles *Int. J. Miner. Metall. Mater.* **21** 408–14
[19] Kumar E R, Kamzin A S and Prakash T 2015 Effect of particle size on structural, magnetic and dielectric properties of manganese substituted nickel ferrite nanoparticles *J. Magn. Magn. Mater.* **378** 389–96
[20] Griffiths D J 1999 *Introduction to electrodynamics* (Upper Saddle River, N.J: Prentice Hall)
[21] Raghavender A T and Jadhav K M 2009 Dielectric properties of Al-substituted Co ferrite nanoparticles *Bull. Mater. Sci.* **32** 575–8
[22] Ian M, Li M and Li J C 2011 Effect of size on dielectric constant for low dimension materials *Phys. B Condens. Matter* **406** 541–4

A First-principles Investigation of The Adsorption of CO and NO Molecules on Germanene

M. R. Al Fauzan[1], W. D. Astuti[1], G. Al Fauzan[1], and Sholihun[2]

[1]Department of Electrical Engineering and Informatics, Vocational Collage, Universitas Gadjah Mada, Bulaksumur Yogyakarta 55281, Indonesia.
[2]Department of Physics, Faculty of Mathematics and Natural Sciences, Universitas Gadjah Mada, Bulaksumur Yogyakarta 55281, Indonesia.

E-mail: alfauzan.rifqi@ugm.ac.id

Abstract. The adsorption of common pollutant gases on germanene is investigated using first-principles calculation by means of density functional theory. In this work, CO and NO molecules are used as pollutant gases. Our calculation results show that CO molecule is physisorbed on germanene, while NO molecule is chemisorbed since it opens band gap of germanene at the Dirac point. To investigate the mechanism of the adsorption process, we calculate the transfer of charges between germanene and gas molecules. In order to obtain understanding into the transformation of electron density upon the adsorption processes, Mulliken analysis is used to evaluate atomic charge of each atom. We found that electron charge of NO molecule obviously increased due to adsorption process, while CO molecule slightly increased.

Keywords. CO molecule, electron density, germanene, and NO molecule.

1. Introduction

Carbon monoxide (CO) and nitric oxide (NO) are poisonous gases that are extremely dangerous for human health. They can cause a wide range of health effect form dizziness to brain damage or even death [1], [2]. Therefore, in order to reduce the amount of pollutant in the atmosphere, it is necessary to find material that has the ability to adsorb molecules especially pollutant gases. One of the most promising candidates is germanene (germanium analog to graphene). Graphene has a high sensitivity to the chemical environment [3]. Hence, germanene is expected to give a similar result. Experimentally, germanene has successfully grown on Pt (111) [4] and Au (111) [5] surface. In the previous studies, germanene has used as an adsorbent material for transition metal atoms[6], group IV elements [6], and organic molecules [7], [8]. Xia *et al* were also reported the ability of germanene to adsorb various gas molecules including CO and NO [9].In our study, we did further calculation in order to understand the mechanism of adsorption of CO and NO on germanene. We investigated the charge difference distribution and electron charge transfer in each system.

We carried out the first-principles electronic-structure calculations to investigate adsorption processes and its mechanism of CO and NO molecules on germanene. We use a $4 \times 4 \times 1$ supercell of germanene which contains 32 germanium atoms with a concentration of CO and NO about 6.25 %.

Content from this work may be used under the terms of the Creative Commons Attribution 3.0 licence. Any further distribution of this work must maintain attribution to the author(s) and the title of the work, journal citation and DOI.
Published under licence by IOP Publishing Ltd

2. Materials and methods

Unlike the fully planar honeycomb lattice of graphene, germanene is predicted to be buckled [10]. This buckled lattice is composed of two sub-lattice with different z coordinates (see Figure 1). Furthermore, we introduce CO and NO molecule on germanene as shown in Figure 2. To obtain the most stable geometry and electronic properties of the material we use density functional theory as implemented in Open Source Package for Material Explorer (OpenMX) code [6]. It has an efficient implementation of first-principles calculations by means of density functional theory that allow us to investigate electronic, magnetic, and geometrical structures of a wide variety of materials. The OpenMX code can be freely obtained through its official website.

OpenMX code is based on norm-conserving pseudopotentials [10] and the wave functions were expanded by a linear combination of multiple pseudo-atomic orbitals (LCPAO) [6], [10], [11]. All calculations were performed with Generalized-Gradient Approximation (GGA) with Perdew-Burke-Ernzerhof (PBE) [12] as the exchange-correlation form. The vacuum space in the z-direction is set to 20 Å to avoid interactions between the neighboring layers. We use a $10 \times 10 \times 1$ k-points in Brillouin zone for electronic structure calculation with energy convergence criteria is determined at 10^{-6} Hartree.

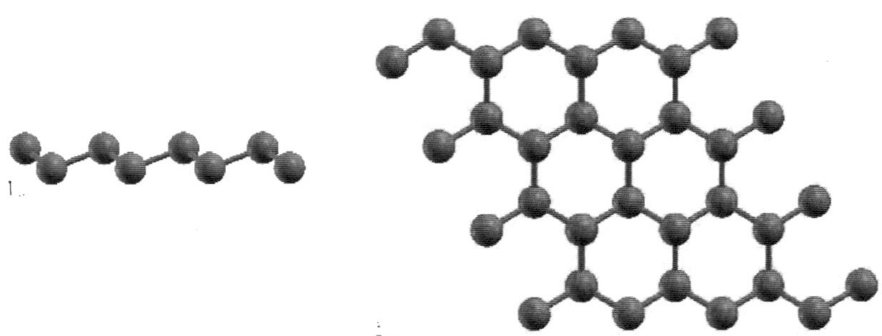

(a) (b)

Figure 1. Buckled honeycomb lattice of germanene viewed from (a) y-z plane and (b) x-y plane. Purple balls represent germanium atoms.

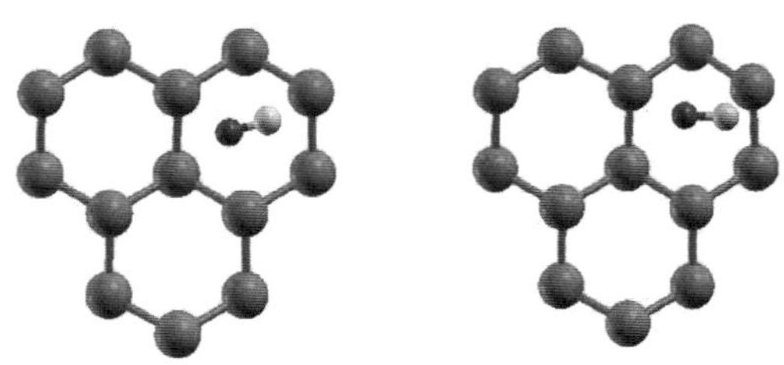

(a) (b)

Figure 2. The most stable positions of (a) CO and (b) NO molecules on germanene. The purple, red, yellow, and green balls are germanium, oxygen, carbon, and nitrogen atoms respectively.

To evaluate the adsorption of gas molecules on germanene, the adsorption energy (Eads) is defined as:

$$E_{ads} = E_{molecule/germanene} - \left(E_{germanene} + E_{molecule}\right) \quad (1)$$

Where $E_{molecule}$, $E_{germanene}$, and $E_{molecule/germanene}$ are the total energy of the gas molecule, pristine germanene, and gas molecule adsorbed on germanene, respectively.

3. Results and discussion

Our structural calculations begin by finding lattice constant of germanene. We have a varied lattice constant value in a certain range to get the lowest energy system. Our calculation indicated that 4.053 Å provided the lowest ground state energy of germanene structure (see Figure 3). This result is in a good agreement with other work reported by Acun *et al* [13]. We also found that buckling height of honeycomb germanene about 0.692 Å, it also seems in a good agreement with other result reported by Scalise *et al* [14].

The most stable adsorption configurations of gas molecules on germanene are shown in Figure 2. By employing Equation 1, we can evaluate adsorption energy of each gas molecule. The corresponding results are listed in Table 1. Both CO and NO adsorption are spontaneous reactions. It means no external energy needed to perform the adsorption processes.Charges of the total density of states (DOS) were analyzed, as plotted in Figure 4, to investigate the adsorption type of each model. Our results showed that the electronic properties of germanene are semi-metal with zero-gap at the Dirac point. This is consistent with other calculation result reported by Ni *et al* [15]. Adsorption of CO molecule does not change the electronic properties of germanene. Therefore, CO molecule is physisorbed on germanene. However, we found that NO molecule is chemisorbed since it changes the electronic properties of germanene. These results are in a good agreement with previous calculation conducted by Xia *et al* [9]. The existence of NO molecule on germanene opened band gap at Dirac point about 0.18 eV.

The mechanism of adsorption process can be evaluated by investigating the interaction between gas molecules and atoms of germanene. We explore this interaction by considering charge transfer which can be verified through calculation of charge difference distribution ($\Delta\rho$) as defined in Equation 2.

$$\Delta\rho = \rho_{molecule/germanene} - \left(\rho_{germanene} + \rho_{molecule}\right) \tag{2}$$

Where $\rho_{molecule}$, $\rho_{germanene}$, $\rho_{molecule/germanene}$ are the charge density of the gas molecule, pristine germanene, and gas molecule adsorbed on germanene, respectively. The results are illustrated in Figure 5. The electron accumulation is depicted by the red region, while the blue region shows the electron loss. Electron flowed from blue region to red region. This result proves the presence of charge transfer between germanene and gas molecules.

To gain a better understanding of the mechanism of adsorption of gas molecules, we investigate the effective charge around each atom using Mulliken charge analysis. Its results are presented in Table 2.

Table 1. Energy and type of adsorption of CO/germanene and NO/germanene models.

Model	E_{ads} (eV)	Type of Adsorption
CO/germanene	-0.558	Physisorption
NO/germanene	-1.141	Chemisorption

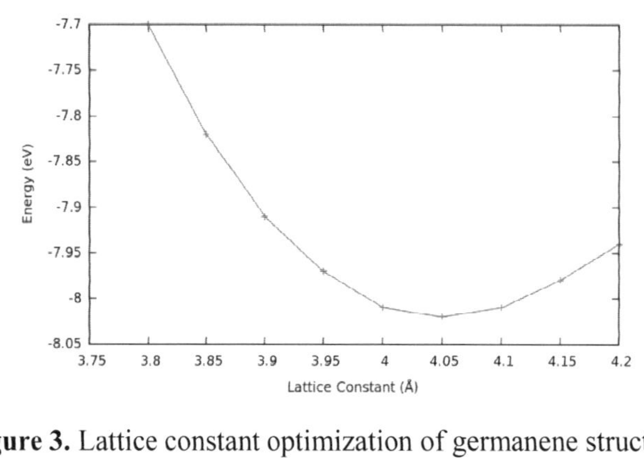

Figure 3. Lattice constant optimization of germanene structure.

IOP Conf. Series: Materials Science and Engineering **367** (2018) 012051 doi:10.1088/1757-899X/367/1/012051

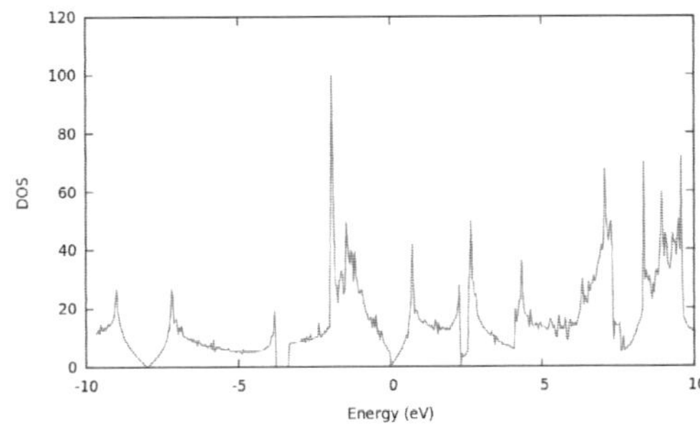

(a)

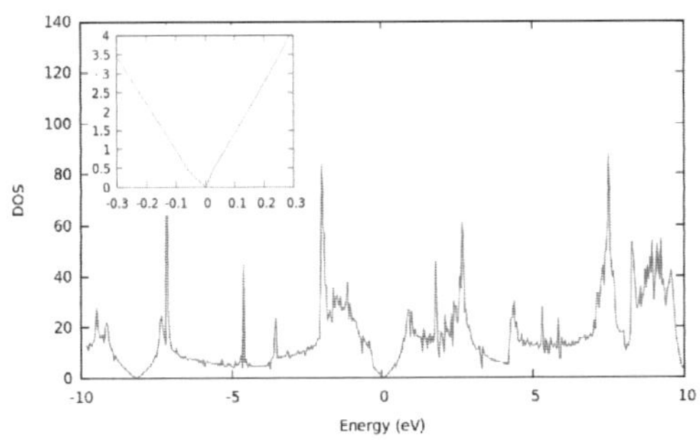

(b)

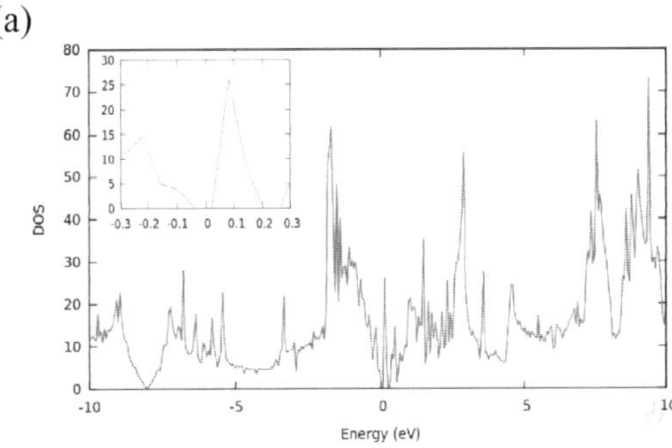

(c)

Figure 4. Total density of states of (a) germanene (b) CO/Germanene (c) NO/Germanene.

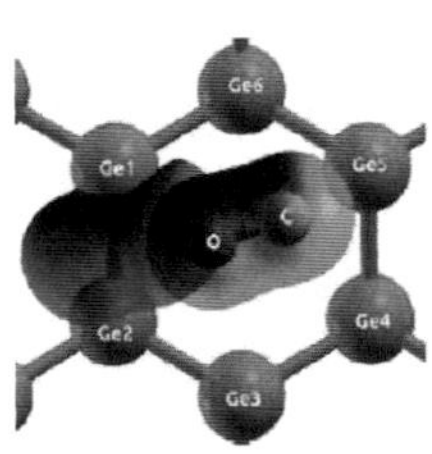

(a)

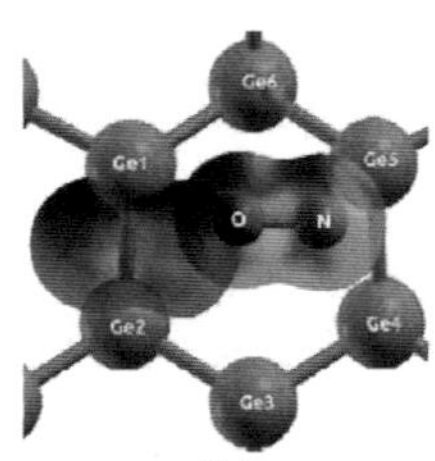

(b)

Figure 5. Charge difference distribution of (a) CO/germanene and (b) NO/germanene models. The red and blue regions indicate the positive and negative values. Only gas molecules and its neighboring germanium atoms are shown.

Table 2. Atomic charges (*e*) of gas molecules and neighboring atoms of germanene based on Mulliken charge analysis

Atom	Before adsorption	After adsorption	
		CO/germanene	NO/germanene
Ge1	4.000	3.961	3.922
Ge2	4.000	4.004	4.011
Ge3	4.000	3.981	3.952
Ge4	4.000	4.026	3.987
Ge5	4.000	3.981	3.848
Ge6	4.000	4.030	4.043
O_{CO}	6.341	6.370	-
C	3.659	3.640	-
O_{NO}	6.211	-	6.268
N	4.790	-	5.004

Since each gas molecule is adsorbed to germanene, the accumulated charge can be observed around gas molecules. It causes an increase in the electron charge of CO and NO molecules about 0.010 *e* and 0.271 *e*, respectively.

4. Conclusions

In summary, with first-principles calculations, the adsorption of CO and NO molecules on the germanene sheet was successfully investigated. Our results showed the adsorption processes of both CO and NO molecules are a spontaneous reaction but with different type of adsorption. CO molecule is physisorbed on the germanene, while NO molecule is chemisorbed. Mulliken charges analysis result indicated that significant charge transformation ensued between germanene and NO molecule but much smaller in CO molecule.

Acknowledgments

This work was funded by*Bantuan Pendanaan Perguruan Tinggi Badan Hukum Direktorat Penelitian Universitas Gadjah Mada* under contract No. 376/DIT.LIT/2017. The calculations in this research were performed using the high-performance computing facilities (DSSDI) at Universitas Gadjah Mada.

References

[1] H. Ischiropoulos, M. F. Beers, S. T. Ohnishi, D. Fisher, S. E. Garner, and S. R. Thom, "Nitric oxide production and perivascular nitration in brain after carbon monoxide poisoning in the rat.," *J. Clin. Invest.*, vol. 97, no. 10, pp. 2260–2267, 1996.

[2] J. A. Raub, M. Mathieu-Nolf, N. B. Hampson, and S. R. Thom, "Carbon monoxide poisoning—a public health perspective," *Toxicology*, vol. 145, no. 1, pp. 1–14, 2000.

[3] F. Schedin *et al.*, "Detection of individual gas molecules adsorbed on graphene," *Nat. Mater.*, vol. 6, no. 9, p. 652, 2007.

[4] X. Li and C. Li, "Improved CEEMDAN and PSO-SVR Modeling for Near-Infrared Noninvasive Glucose Detection," *Comput. Math. Methods Med.*, vol. 2016, 2016.

[5] R. P. F. F. da Silva, T. A. P. Rocha-Santos, and A. C. Duarte, "Supercritical fluid extraction of bioactive compounds," *TrAC Trends Anal. Chem.*, vol. 76, pp. 40–51, Feb. 2016.

[6] T. Ozaki, "Variationally optimized atomic orbitals for large-scale electronic structures," *Phys. Rev. B*, vol. 67, no. 15, p. 155108, 2003.

[7] P. Rubio-Pereda and N. Takeuchi, "Adsorption of organic molecules on the hydrogenated germanene: a DFT study," *J. Phys. Chem. C*, vol. 119, no. 50, pp. 27995–28004, 2015.

[8] T. P. Kaloni, "Tuning the structural, electronic, and magnetic properties of germanene by the adsorption of 3d transition metal atoms," *J. Phys. Chem. C*, vol. 118, no. 43, pp. 25200–25208, 2014.

[9] W. Xia, W. Hu, Z. Li, and J. Yang, "A first-principles study of gas adsorption on germanene," *Phys. Chem. Chem. Phys.*, vol. 16, no. 41, pp. 22495–22498, 2014.

[10] N. Troullier and J. L. Martins, "Efficient pseudopotentials for plane-wave calculations," *Phys. Rev. B*, vol. 43, no. 3, p. 1993, 1991.

[11] S. Suzuki *et al.*, "Nitroxide-substituted nitronyl nitroxide and iminonitroxide," *J. Am. Chem. Soc.*, vol. 132, no. 45, pp. 15908–15910, 2010.

[12] C. G. Pereira and M. A. A. Meireles, "Supercritical fluid extraction of bioactive compounds: fundamentals, applications and economic perspectives," *Food Bioprocess Technol.*, vol. 3, no. 3, pp. 340–372, 2010.

[13] A. Acun *et al.*, "Germanene: the germanium analogue of graphene," *J. Phys. Condens. Matter*, vol. 27, no. 44, p. 443002, 2015.

[14] E. Scalise, "Vibrational properties of silicene and germanene," in *Vibrational Properties of Defective Oxides and 2D Nanolattices*, Springer, 2014, pp. 61–93.

[15] A. Kara *et al.*, "A review on silicene—new candidate for electronics," *Surf. Sci. Rep.*, vol. 67, no. 1, pp. 1–18, 2012.

Synthesis and Characterization Hierarchical Three-Dimensional TiO_2 Structure via Hydrothermal Method

N. Syuhada[1], B Yuliarto[1], and Nugraha[1,2]

[1]Advance Functional Materials Laboratory, Department of Engineering Physics, Fakultas Teknologi Industri Institut Teknologi Bandung, 40132, Indonesia
[2]Research Center for Nanoscience and Nanotechnology (NRCN), Institut Teknologi Bandung, Fakultas Teknologi Industri, 40132, Indonesia

E-mail : brian@tf.itb.ac.id

Abstract. TiO_2 is one of the most potential candidates due to its fascinating properties for multi-discipline fields. One dimensional nanostructure TiO_2 such as nanotube and nanorods has been widely used for many devices technology. Compare with one-dimensional nanostructure TiO_2; the hierarchical TiO2 has not been widely applied. Three dimensional TiO_2 play a promising role for application in many different fields such as photovoltaics, photocatalytic and a gas sensor. Herein, we report that the hierarchically structures TiO_2 have been successfully obtained by the one-pot Hydrothermal process. The growth mechanism of Titania was controlled by Titanium (IV) isopropoxide (TTIP). Ethylene glycol (EG). Hydrochloric acid (HCl). Hexadecyltrimethylammonium bromide (CTAB) molar ratio. TTIP was used as titanium source and CTAB as a soft template. The molar ratio of TTIP. EG. HCl. CTAB was 0.1:0.2:0.4:0.001. Those samples were synthesized using the hydrothermal method at 180 °C for 20 h. The purpose of this work was focused on investigating morphology, crystallite size, crystalline phase, and particle size. The properties of these materials were characterized by X-Ray Diffraction, Energy Dispersive Spectroscopy and Scanning Electron Microscope. It was found all particles exhibited unique spherical morphology which arranged by nanorods and good distribution nanoparticle. The Average size of the sphere has range 1 μm to 3 μm with diameter nanorods 60 nm to 100 nm. The TiO_2 spheres were constructed of interconnected nanorods and formed a three dimensional (3D) porous framework. XRD analysis confirmed that sample consisted of pure rutile crystal structure with crystallite size was 50 nm, and EDS revealed an elemental content of Ti 61.03 % and O 38.97 %.

Keywords. Hierarchical Structure, Three Dimensional TiO_2, Hydrothermal Synthesis.

1. Introduction

Nanostructured materials have got great interest as catalysts and other application because of their unique texture and structural characteristics. Much effort has concentrated on the important metal oxide such as TiO_2, SnO_2. VO_2 and ZnO. Titanium dioxide (TiO_2) had been used in fields such as industrial manufacturing, aerospace, ocean exploring, environmental protection, resource development, gas sensing, photocatalysts, photoelectrodes for photo splitting water, ceramic material, cosmetics, medical diagnose, solar cells and even as a food coloring agent and in toothpastes [1]. Titania has been extensively investigated because of the stability of its chemical, physical, optical, and

Content from this work may be used under the terms of the Creative Commons Attribution 3.0 licence. Any further distribution of this work must maintain attribution to the author(s) and the title of the work, journal citation and DOI.
Published under licence by IOP Publishing Ltd

electronic properties and have an advantage such as biocompatibility, non-toxic, photo corrosion-free, and cost-effective. In those applications, it is a vital factor to control morphology, particle size distribution, phase composition and porosity of TiO_2.

TiO_2 exists in three mineral forms: Anatase, Rutile, and Brookite [2]. Anatase type has a tetragonal crystalline structure (dipyramidal habit) and mainly used as a photocatalysts under-irradiation. Rutile type also has a tetragonal crystal structure (prismatic habit) and mainly used as white pigment in paint. Brookite type has an orthorhombic crystalline structure. TiO_2 morphologies have many shapes such as nanotubes, nanowires, nanorods and mesoporous structures [1].

There are numerous methods for synthesis TiO_2 such as chemical precipitation [3]–[5], the sol-gel method [5] hydrothermal [6] and solvothermal process [7], Wang et al. [8] reported the synthesis of anatase TiO_2 by sol-gel method using titanium tetra-isopropoxide with ethylene glycol had been obtained the TiO_2 particle with high surface area, well crystallized and small crystallite size. The method of Kim et al.[7] successfully investigated the TiO_2 obtained from Titanium isopropoxide as a precursor in toluene solution with a solvothermal synthesis produced a uniform anatase structure with mid-size particles on the nanoscale level. Among those reports, the sol-gel technique is the most frequently applied, but there are some problems like the precipitates derived from the process are amorphous, and the photocatalytic and photoelectrical conversion efficiency of those TiO_2 are not high enough for the industrial purpose. To improve TiO_2 characteristics, several methods had presented like formed high surface area, improve interfacial charge separation with developing the defect structures and modification the TiO_2 with metal or other semiconductors. Hydrothermal synthesis has become one of the important methods to use in nanomaterial and nanotechnology production. Hydrothermal is the reaction that conducted in steel pressure vessels called autoclaves under controlled temperature or pressure. The temperature can be elevated above the boiling point of water, reaching the pressure of vapor saturation. This method is widely used for the production of small particles in the ceramic industry [1]. Occurring high temperature and pressure condition, the benefit of this method allows to make materials have monodispersed and highly homogeneous nanoparticles.

The performance of TiO_2 was influenced by the particle size of the TiO_2 structure. Some authors report that the specifics surface area and surface to volume ratio increase as the size of particle size decrease. The TiO_2 material is expected become an important target in helping solve many serious problems on environmental, pollution and energy crisis based on the use of solar energy on photovoltaic device. In this works, we apply a hydrothermal method to synthesize TiO_2 nanoparticles by using CTAB as a surfactant of the forming agent. Hydrothermal treatment is helpful to crystallize nanoparticle and stabilize the sphere structure.

2. Materials and methods

All chemical reagents used in present experiments were obtained from commercial sources and used without purification. The reagent used for the synthesis of titanium dioxide were titanium (IV) isopropoxide ($C_{12}H_{28})_4Ti$) (Aldrich, 97 % purity), Ethylene Glycol ($C_2H_6O_2$) (Merck, 99 %), Cetyltrimethylammonium bromide ($C_{19}H_{42}BrN$) (Merck, 97 %), hydrochloric acid (HCl) (Merck 37 %). The molar ratio of TTIP.EG.HCl.CTAB is 0.1:0.2:0.4:0.001. Firstly, 3 mL titanium isopropoxide (TTIP) was added into hydrochloric acid (HCl) solution during vigorous stirring (bottle A) at room temperature, then ethylene glycol (EG) were dropped into bottle A under stirring. Secondly, 0.5 g cetyltrimethylammonium bromide was dissolved in 2.5 mL distilled water to form CTAB solution (Bottle B). finally, after stirring for 15 min, bottle B was added into bottle A and stirring for 30 min to form TTIP solution. After stirring the solution was transferred to an autoclave and heated at 180 °C for 20 h, after that the autoclave was cooled naturally. The obtained samples were washed with distilled water several times, to collected the resultants and removed the organic solvent is using the centrifuge. Then, dried at 100 °C for 3 h and calcination at 550 °C for 6 h. The obtained TiO_2 powders were ready for further characterization.

The phase and crystal structure were studied by X-ray diffraction (XRD) Rigaku with the standard Cu-Kα radiation source, λ = 0.15418 nm) operated at 40 kV and 40 mA at room temperature in the 2Θ range from 15° to 65° at a scanning rate of 0.01° s^{-1}. The size and morphology of samples were characterized using scanning electron microscopy (SEM, Hitachi S-4800 microscope) and to analyzed elemental content of samples were using Energy Dispersive Spectroscopy.

3. Results and discussion

The one-pot hyhdrothermal method has been used to fabricate the TiO_2 hierarchical microspheres. The key point of this synthesis is to control the hydrolysis rate of the TiO_2 precursor. The addition of EG can retard the hydrolysis rate of precursor during the hydrothermal process and reduce the size of the CTAB surfactant micelles. Then some CTAB miczelles would dissolve into individual molecules and adsorb on the surface the TiO2 crystal nucleus and act as a template to form the structure of TiO_2 hierarchical microsphere. Huang *et al.* [9] prepared TiO_2 by the sol-gel process at room temperature using tetra-butyl titanate as the precursor without any templates. Antonneli and Ying synthesized TiO_2 by a modified sol-gel method using surfactants as templates. All of them reported the sample was successfully obtained, but samples using surfactant has the smaller particle size, small agglomeration and good uniform of morphology.

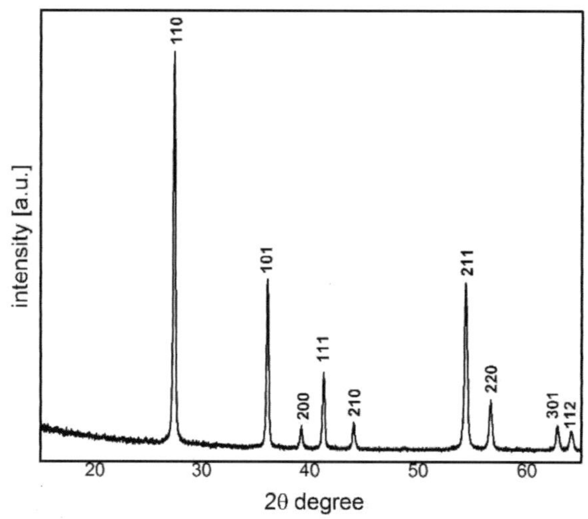

Figure 1. Diffraction pattern of TiO2

The XRD Pattern of the Titanium dioxide that was synthesized at 180 °C and calcined at 550 °C for 6 h are shown in Figure 1. X-ray diffraction (XRD) characterization was used to identify the phase and crystal structure of the sample. As shown in Figure 1 the TiO_2 sample has phase crystal of rutile with the tetragonal crystal structure (a = b = 0.459 nm, c = 0.296 nm, space group P42/mnm). The diffraction peaks at 27.5°, 36.01°, 39.26°, 41.28°, 44.12°, 54.4°, 56.7°, 62.92°, and 64.14° were corresponding to the (110), (101), (200), (111), (210), (211), (220), (301) and (112) planes. The obtained result showed broad rutile peak which is plane 110 has the highest intensity. All diffraction peaks show the complete formation of crystalline rutile phase and were indexed according to the JCPDS card No. 01-073-2224. No peaks related to either the anatase or brookite phases were observed in the XRD pattern. The hydrothermal synthesis route has advantages to obtain phase-pure TiO_2nanoparticles at both lower temperatures and reaction times. From the diffractogram, the widths of the bases of the peaks and its intensity indicate the size of nanocrystals; small crystals could promote more intense spreading due to internal reflections that occur in the system. The peaks were rather sharp, which indicated that the samples had relatively high crystallinity. Titanium powder could be controlled by the molar ratio of starting material, washing process and calcinate temperature.

Crystalline sized calculated from XRD peaks are listed below in table 1. The average nanocrystal size was determined using the Scherrer equation

$$D = \frac{K\lambda}{\beta \sin\theta} \tag{1}$$

Where D is the average crystallite size, K is the Scherrer constant (0.90), λ is the wavelength of the X-ray radiation (0.15425 nm for Cu-Kα), β is the peak position.

Table 1. The crystallite size of microsphere TiO_2

Bragg Planes (hkl)	Crystallite Size (nm)
110	63.02
101	61.22
200	59.40
111	59.50
210	43.66
211	41.64
220	41.21
301	45.34
112	42.41
Average	50.82

The SEM images showed a spherical hierarchical structure with average uniform size about 1.5 µm (Figure 2). It can be seen that the TiO_2 material is made up of large quantities of microsphere architectures and composed of many nanorods with average diameter distribution in about 62 nm, these nanorods covered the sphere. Those structures indicated that the hierarchical TiO_2 has large specific surface. The literature reported the microsphere TiO_2 showed high specific surface large area leading to high photocatalytic activity due to a large contact area. The result of hierarchical microsphere TiO_2 structure that obtained in this work generally similar with previous studies [10], [11] however, there is probably different result that obtained like particle uniform and difference in particle diameter and surface area. This occurs because of reactant concentration of the solution, acid condition [12], reaction time [12], and hydrothermal temperature [11]. Based on the report the usage of acid can control the hydrolysis reaction of titanium while the reaction time and temperature synthesis give significant influence on crystallite size and surface area [11].

Figure 2. Morphology of Microsphere Hierarchal TiO_2

Energy dispersive X-ray spectrometry (EDX) analysis (Figure 3) of TiO_2nanoparticle showed for Ti element and oxygen. There is no trace of any other impurities that could be seen. However, there is

2 peak show which is carbon and gold. The Peak of gold could be seen because of the coating material that had been used.

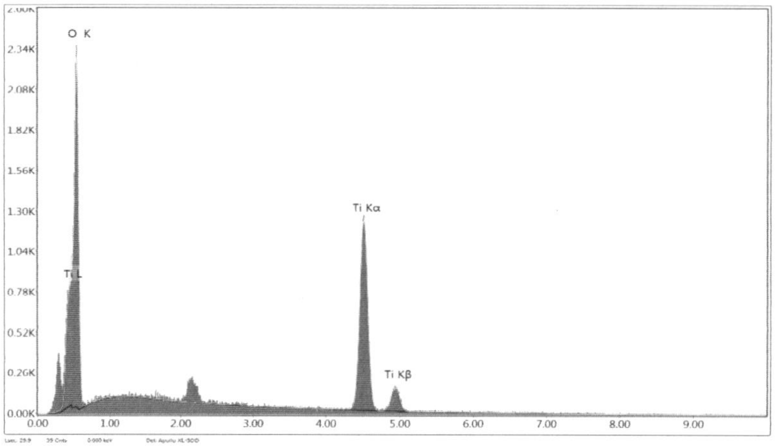

Figure 3. Elemental content of hierarchical structure TiO_2

Table 2. Elemental content of sample

Element	Weight %	Atomic %	Error %	Net Int.	K Ratio	Z	R	A	F
O K	38.97	65.66	10.33	494.57	0.115	1.1743	0.9381	0.2513	1.0001
Ti K	61.03	34.34	4.42	515.71	0.5372	0.8778	1.0239	1.0032	1

4. Conclusions

A rutile phase Titanium dioxide was synthesized by the hydrothermal route at 180 °C for 20 h. SEM images showed the titanium dioxide that was obtained have microsphere hierarchical structured with average uniform size about 1 μm to 3 μm that was arranged by nanorods. The diameter particle of nanorods was 62 nm. A major advantage of this route is the synthesis of titanium dioxide rutile anatase can be performed at moderate temperatures and reaction times. The TiO_2 powder showed good characteristics and good uniform distributed EDS revealed an elemental content of Ti 61.03 % and O 38.97 %. This synthesis method provides a simple route to fabricated nanostructured titanium dioxide that can be applied in various fields.

Acknowledgments

This work was supported by Laboratorium of Advance Functional Material, Department of Engineering Physics Institut Technology Bandung and Indonesia endowment fund for education (LPDP).

References

[1] M. Malekshahi Byranvand, A. Nemati Kharat, L. Fatholahi, and Z. Malekshahi Beiranvand, "A review on synthesis of nano-TiO2 via different methods," *J. Nanostructures*, vol. 3, no. 1, pp. 1–9, 2013.

[2] G. P. Li, A. Bono, D. Krishnaiah, and J. Collin, "Preparation of titanium dioxide photocatalyst loaded onto activated carbon support using chemical vapor deposition: a review paper.," *J. Hazard. Mater.*, vol. 157, no. 2–3, pp. 209–219, 2008.

[3] F. Pedraza and A. Vazquez, "Obtention of TiO2 rutile at room temperature through direct oxidation of TiCl3," *J. Phys. Chem. Solids*, vol. 60, no. 4, pp. 445–448, 1999.

[4] E. Scolan and C. Sanchez, "Synthesis and characterization of surface-protected nanocrystalline titania particles," *Chem. Mater.*, vol. 10, no. 10, pp. 3217–3223, 1998.

[5] M. Gartner *et al.*, "Spectroellipsometric characterization of sol–gel TiO_2–CuO thin coatings," *Thin Solid Films*, vol. 455, pp. 417–421, 2004.
[6] M. Wu *et al.*, "Sol-hydrothermal synthesis and hydrothermally structural evolution of nanocrystal titanium dioxide," *Chem. Mater.*, vol. 14, no. 5, pp. 1974–1980, 2002.
[7] C.-S. Kim, B. K. Moon, J.-H. Park, S. T. Chung, and S.-M. Son, "Synthesis of nanocrystalline TiO2 in toluene by a solvothermal route," *J. Cryst. Growth*, vol. 254, no. 3–4, pp. 405–410, 2003.
[8] G. Wang, "Hydrothermal synthesis and photocatalytic activity of nanocrystalline TiO_2 powders in ethanol–water mixed solutions," *J. Mol. Catal. Chem.*, vol. 274, no. 1–2, pp. 185–191, 2007.
[9] D. Huang, G. Luo, L. Yang, and Y. Wang, "Synthesis of mesoporous TiO_2 materials with high specific area using inorganic acids as catalysts," *China Particuology*, vol. 3, no. 3, pp. 176–180, 2005.
[10] Y. Cai *et al.*, "Hierarchically structured porous TiO_2 spheres constructed by interconnected nanorods as high performance anodes for lithium ion batteries," *Chem. Eng. J.*, vol. 281, pp. 844–851, 2015.
[11] G. Collazzo, S. Jahn, N. Carreño, and E. Foletto, "Temperature and reaction time effects on the structural properties of titanium dioxide nanopowders obtained via the hydrothermal method," *Braz. J. Chem. Eng.*, vol. 28, no. 2, pp. 265–272, 2011.
[12] J. Liu, T. An, G. Li, N. Bao, G. Sheng, and J. Fu, "Preparation and characterization of highly active mesoporous TiO_2 photocatalysts by hydrothermal synthesis under weak acid conditions," *Microporous Mesoporous Mater.*, vol. 124, no. 1–3, pp. 197–203, 2009.

Monofilament Wire Silver Sheathed Bi,Pb-Sr-Ca-Cu-O Prepared by Four-Pass Rolling and Repeated Heating Process

Hendrik[1], P. Sebleku[1], S. Herbirowo[1], S. D. Yudanto[1], B. Siswayanti[1], Lusiana[1], H. Nugraha[1], A. Imaduddin[1] and A. W. Pramono[1]

[1]Research Center for Metallurgy and Materials-Indonesian Institute of Sciences (P2MM-LIPI) Kawasan Puspiptek, Serpong, Tangerang 15314, Indonesia.

E-mail: hend027@lipi.go.id

Abstract. A four-step process was proposed to manufacture monofilament wire Bi,Pb-Sr-Ca-Cu-O using silver (Ag) sheath fabricated by Powder-In-Tube (PIT) method in which the thickness of wire were subjected to rolling from 10 mm Ø to2 mm Ø and heated at temperature of 860 °C in 24 h. The effect of the rolling and heating process on the properties of four samples Ag-sheathed Bi,Pb-Sr-Ca-Cu-O wires in each step were studied. In x-ray diffraction (XRD) patterns of monofilament wire fabricated by PIT method in which the repeated heating process involved in rolling can alter the precursor phase. At a first and second deformation process, samples sintered at 860 °C in 24 h still contained the constituent elements of precursor powders. Third deformation step were shown to have superconducting phase Bi,Pb-Sr-Ca-Cu-O 2212. According to resistivity measurement analysis, rolling and repeated heating process were shown to have a critical effect on superconductivity. First, second and fourth rolling were found to have no superconductivity, while third rolling were found to exhibit superconductivity of which its value of critical temperature monofilament wire ($T_{c,onset}$) was 90 K. Furthermore, according to Scanning Electron Microscopy (SEM) analysis, the rolling and repeated heating process were observed to have important effect on morphology of samples. Wire superconductors on first and second rolling were shown to have crack and high porosity, but after third rolling the crack and porosity were diminished. Overall the deformation process have yielded promising results particularly when the superconducting phase Bi,Pb-Sr-Ca-Cu-O 2212 were found optimum in third deformation process.

Keywords. Monofillament and PIT method.

1. Introduction

Throughout the ten decades, the technological applications of superconductivity such as magnet and electrical power application have been used for long-distance and low-voltage electric grids without transmission loss; fast magnetically levitated trains; superefficient motors and generators where some of them are in demonstration and experimental stage. Therefore, many researchers have devoted their efforts on composites of superconducting materials and metals especially they own better mechanical properties and electrical properties than superconducting materials alone give [1–4]. As one of the promising high temperature superconductors, Bi,Pb-Sr-Ca-Cu-O (BPSCCO) have remarkable properties in the elongated forms such as wire and tapes for higher capacity superconductors cables of which some of them have been produced commercially by several companies around the world [5].

Content from this work may be used under the terms of the Creative Commons Attribution 3.0 licence. Any further distribution of this work must maintain attribution to the author(s) and the title of the work, journal citation and DOI.
Published under licence by IOP Publishing Ltd

However, other superconducting materials such as low temperature superconductors (NbTi, Nb_3Sn, Nb_3Al) is also attractive for this application.

On the one side, as composites of superconducting materials and metals are brittle and difficult to fabricate, many researchers reported how they formulated these composites of superconducting materials and metals into three stages consisting of the precursor powder preparation stage [6–7], composites precursor preparation stage [8] and thermomechanical processing stage [9]. Among these composites, cuprate superconductors is believed to be the best potential candidate for a large scale of the technological application of superconductivity since it is operating in above the boiling point of liquid nitrogen range, of which the use of liquid nitrogen is environmentally clean coolants instead of oil. On the other side, previous researchers reported that bubbles were serious problem during thermomechanical process [10].

In this research, the preparation of thermomechanical processing stage of the composite of BPSCCO and silver(Ag) sheath fabricated by PIT have been modified with four-pass rolling and repeated heating process. It is interesting that four-step deformation process influence the variation of $T_{c,onset}$ value and $T_{c,final}$ value. We observed the superconducting phase BPSCCO 2212 were found optimum in third deformation process. For this purpose, the addition of four-pass rolling and repeated heating process on the quality of superconducting phases was systematically investigated. The quality of superconducting phases was analyzed from the particle size, crystallite structure and impurities.

2. Materials and methods

The preparation of composite of Bi,Pb-Sr-Ca-Cu-O (BPSCCO) using silver (Ag) tube fabricated by traditional PIT (Powder-In-Tube) method through the rolling and repeated heating process were attempted. In Figure 1, the flowchart described a four-step process of making monofilament wire BPSCCO. Based on the stoichiometry in first step, in order to obtain the phase of $Bi_{1.6},Pb_{0.4}Sr_2Ca_2Cu_3O_{6+\alpha}$, the weight ratio of powders Bi_2O_3, PbO,$SrCO_3$, $CaCO_3$,CuO was fixed at 4.17:1:3.3:2.24:1.78. In second step, powders were mixed and inserted into Ag tube with initial thickness of Ø 10 mm. In third step, the Ag tube was rolled into diameter of Ø 8 mm to avoid crack. In fourth step, the composite of BPSCCO and Ag tube were heated in an electric furnace at 860 °C for 24 h under atmospheric air atmosphere. The thickness reductions of all samples during four-pass rolling process were 8 mm, 6 mm, 4 mm and 2 mm, respectively, as illustrated in Table 1.

Table 1. Process of making monofilament BPSCCO with four-step process

Step	Rolling	Heating
1	10 mm to 8 mm Ø	860 °C, 24 h
2	8 mm to 6 mm Ø	860 °C, 24 h
3	6 mm to 4 mm Ø	860 °C, 24 h
4	4 mm to 2 mm Ø	860 °C, 24 h

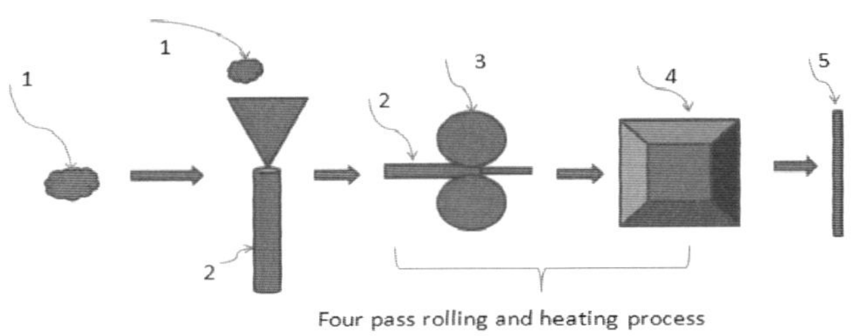

Figure 1. Flow chart of four-step process of making monofilament wire (Bi,Pb)SCCO In order to verify the quality of composite of superconducting BPSCCO and Ag, sample in each pass rolling process were investigated by four-point probe method, X-ray diffraction analysis and scanning electron microscopy (SEM) with an energy dispersive analysis (EDS).

3. Results and discussion

In Figure 2, the XRD patterns of monofilament wire BPSCCO 2212 using Ag sheath were obtained after peeling of the Ag sheath of sample sintering in air at 860 °C for 24 h for third pass rolling process. This is in agreement with the previous report [11]. As it can be seen in Figure 2a and Figure 2b, before third pass rolling, small peaks corresponding with Bi_2O_3 or BPSCCO constituent oxide powders are indicated. On the contrary, as indicated in Figure 2c, small peaks are observed corresponding to BPSCCO 2212, then disappears after fourth pass rolling because it is too small to be indicated.

In Figure 3, from the *R-T* curves showed the resistivity analysis of sample monofilament wire after third pass rolling. The present of the superconducting properties for sample after third pass rolling is quite interesting. The sample simulates the growing of BPSCCO 2212 with the superconducting state transition beginning at 90 K and completing at 70K, as is shown in Figure 2. This result points out the activation for the reaction between powders due to heating after 48 h.

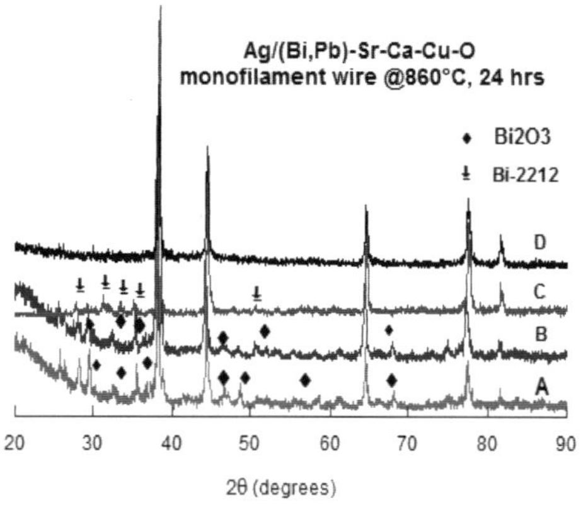

Figure 2. X-ray diffraction patterns of monofilament wire BPSCCO using Ag sheath after sintering at 860 °C for 24 h in (a) First step process, (b) Second step process, (C) Third step process, and (d) Fourth step process

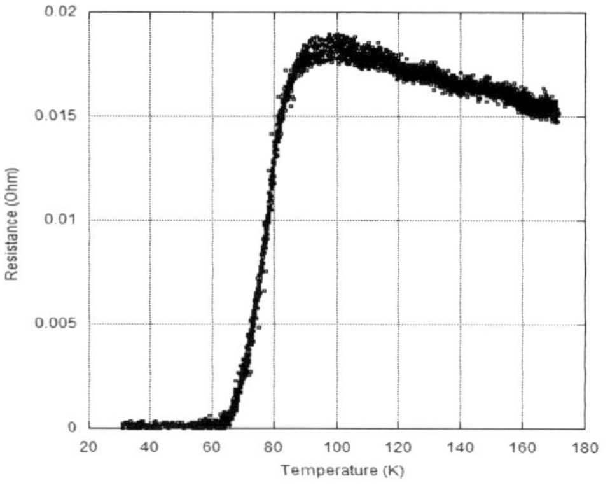

Figure 3. Resistance dependence on temperature for sample with third step

IOP Conf. Series: Materials Science and Engineering **367** (2018) 012053 doi:10.1088/1757-899X/367/1/012053

Figure 4. (a) The surface morphology of sample after first step, (b) The surface morphology of sample after second step, (c) The surface morphology of sample after third step, (d) The surface morphology of sample after fourth step

In Figure 3, the different microstructure and phase content observed in the scanning electron microscopy morphology of samples sintered in air atmosphere at 860 °C for 24 h are reflected in the *R-T* curves. The four microstructures were taken from the sintered samples prepared under different step-rolling process. The sintered contained fine microstructure of deformed-shaped grains. The anisotropic grain shape is considered to originate from the deformed shape of the rolling process. In Figure 4a, the presence of crack and small voids appearing along the thickness of composite is noticed in the matrix. While, in Figure 4b, for sample after rolled at second pass, the small voids and cracks were decreased. In Figure 4c, the sample sintered in third pass rolling has changed the grain structure to BPSCCO 2212 flower-like grain with a length of about 10 µm. In Figure 4d, BPSCCO 2212 flower-like grains structure disappear with less pores both in size and population. The presence of BPSCCO 2212 flower-like grain formation in the sinter after rolled at third pass is quite interesting. This differs from the expectation that both the size and number of residual pores inside the monofilament tape decrease will enhance the textural growth of BPSCCO 2212-plate-like grains with traditional PIT process [10]. The population of the flower-like grain structure in the matrix is fairly uniform implies the larger effective current carrying areas (transport properties) of BPSCCO 2212, as is seen in Figure 2. Obviously, the decrease of both the size and population of residual pores played an important factor in highly textured BPSCCO 2212 and enhance the transport properties of BPSCCO 2212. This suggests that reaction of BPSCCO 2212 with Ag sheath play a crucial role in the suppression of BPSCCO 2212 formation during heat treatment. It is known that intimate contact of between BPSCCO 2212 and Ag sheath can be considered as instantaneous diffusion of which Cu dissolution in Ag at the BPSCCO/Ag interface to activate Ag-Cu-O solid solution is a thermodynamically favorable reaction [12]. Cu was considered as impurity because the purity of Ag used in this study was 92.5 %.

However, Cu in the Ag sheath in filament is considered beneficial to process kilometer length of BPSCCO 2212 superconducting wire. In order to check the reaction of BPSCCO 2212 into Ag sheath, more detailed study are required to provide clear understanding of this phenomenon.

4. Conclusions

Monofilament-wire silver sheathed BPSCCO superconductor has been successfully made. The XRD pattern of sample sintered in the first and second rolling process shows the traceable amounts of constituent oxide powders, while samples sintered in third rolling process shows superconducting phase BPSCCO 2212. The amount of superconducting phase in sample sintered in fourth rolling process is not traceable. Meanwhile, $T_{c,onset}$ and $T_{c,final}$ superconductivity (90 K and 70 K) is found after third pass rolling and heating process. As observed in SEM image, the superconducting BPSCCO 2212 flower-grain like shape is anisotropic due to rolling process. T_c was not found in first and second step due to small void and cleavage along the matrix.

Acknowledgments

This project has been supported by the fund of Indonesian Institute of Sciences (LIPI) in 2016

References

[1] Sotelo A *et al* 2014 *Journal of the European Ceramic Society*
[2] Awana V P S *et al* 2013 *Materials Chemistry and Physics* **139** pp 681–688
[3] Lu T *et al* 2015 *Superconductivity and its applications* **519** pp 24–27
[4] Cicek O, Yetis H and Gencer A 2014 *Cryogenics* **63** pp 143–147
[5] Malozemoff A P 2013 *Physica C*
[6] Darsono N *et al* 2015 *Supercond Nov Magn*
[7] Patel R H, Nabialek A and Niewczas M 2005 *Supercond. Sci. Technol.* **18** pp 317–324
[8] Pramono A W 2000 *Quantitative analysis of the thermomechanical properties of Cu-18wt%Nb in-situ metal matrix composite wire and the Cahn-Hilliard simulation of its thermal phase evolution* (Ph.D. thesis, RWTH Aachen)
[9] Li Q *et al* 2001 *U.S. Patent No. 6,311,386 B1* (6 November 2001)
[10] Zhang S *et al* 2015 *Physics Procedia* **65** pp 57–60
[11] Kadar J *et al* 2016 *Supercond. Sci. Technol* **29**
[12] Weselowski D E *et al* 2005 *Supercond. Sci. Technol.* **18** pp 934–943

Modified Working Electrode by Magnetite Nanocomposite for Electrochemical Sensor Application

R. N. Suhanto[1], R. Rahmawati[1,2], D. A. Setyorini[3], I. Noviandri[3], Suyatman[1], and B Yuliarto[1,4]

[1] Department of Engineering Physics, Faculty of Industrial Technology, Institut Teknologi Bandung, Jl. Ganesha 10, Bandung 40132, Indonesia
[2] Department of Physics, Faculty of Sciences and Technology, UIN Sunan Kalijaga Yogyakarta, Jl. MarsdaAdisucipto No 1 Yogyakarta 55281, Indonesia
[3]Department of Chemistry, Faculty of Mathematics and Natural Sciences, Institut Teknologi Bandung, Jl. Ganesha 10, Bandung 40132, Indonesia
[4]Research Center for Nanoscience and Nanotechnology, Institut Teknologi Bandung, Bandung, Indonesia

E-mail: brian@tf.itb.ac.id

Abstract. In this research, magnetite (Fe_3O_4) has been successfully composite with graphene by a sonochemical method. The formation of the nanocomposite was confirmed by X-ray diffraction (XRD). The XRD pattern showed the magnetic phase of crystalline Fe_3O_4that could be attributed to the cubic structure. However, the graphene peak did not appear in the XRD pattern of the nanocomposite. The crystallite sizes of the Fe_3O_4 nanoparticles were estimated about 10 nm. Furthermore, the morphology of the nanocomposite was investigated using scanning electron microscope (SEM). The SEM image showed the interaction and decoration of graphene sheets with Fe_3O_4.Meanwhile, the synthesized graphene-Fe_3O_4 nanocomposite can be separated by applied an external magnetic field, thus demonstrating the magnetic properties of the nanocomposite. A working electrode was fabricated by modified a surface of carbon paste electrode (CPE) with the graphene-Fe_3O_4 nanocomposite. The modified electrode was characterized using cyclic voltammetry (CV) to evaluate electrochemical properties. By Modifying the surface of the working electrode with graphene-Fe_3O_4 nanocomposite, it can be increased the oxidation and reduction peak current that refer to excellent electron transfer materials. Therefore, the results of this research showed that the use of the graphene-Fe_3O_4 nanocomposite is very feasible for electrochemical sensor.

Keywords. Electrochemical sensor, graphene, magnetite, nanocomposite, and sonochemical method.

1. Introduction

Electrochemical sensors are a device to detect and analyze a specific compound by measured the chemical signals produced. The chemical signals can be measured by analyzing current, potential, or

Content from this work may be used under the terms of the Creative Commons Attribution 3.0 licence. Any further distribution of this work must maintain attribution to the author(s) and the title of the work, journal citation and DOI.
Published under licence by IOP Publishing Ltd

conductance changes [1]. Electrochemical sensors have been applied for many analysis, such as in clinical [2], environmental [3], [4], biological [5], pharmaceutical, and process control in food processing [6]. The attraction of applying electrochemical sensors based on their high sensitivity, good selectivity, inexpensive, the simplicity of operation, simple pretreatment procedure, and fast analytical time.

Nowadays, many researchers are focusing to improve the analytical performance of electrochemical sensors. Highly sensitivity is one of the main keys to obtain more precise detection results. Applying nanomaterials to electrochemical sensors have found able to enhance the effective surface area of electrodes and accelerate electron transfer across the electrode surface, thus improve the sensitivity and the stability of analytical response [2]. Moreover, nanomaterials exhibit fabulous in optical, electrical, mechanical, catalytic, thermal, and magnetic characteristics by comparing with bulk materials.

Magnetite nanoparticles (Fe_3O_4) is one of the nanomaterials that has attracted much research and has been using in various applications, such as magnetic resonance imaging (MRI), drug delivery, cell labeling, hyperthermia, removal of contaminants from water, and biosensing [7]. This is because of their superparamagnetic property, large surface area, biocompatibility, and low toxicity. However, high energy surface of Fe_3O_4 leads it to agglomerate in clusters and Fe_3O_4 have poor conductivity. Therefore, composing Fe_3O_4 with a material with high electrical conductivity is an effort to produce new material that can reduce agglomeration, increase the conductivity, and enhance detection signal. Graphene is one of good material for electrochemical sensors because of its unique structural and catalytic properties, high electrical and thermal conductivity, large surface area, and high loading of biocatalysts [8], [9]. Due to its unique properties, graphene is very competent to develop in biological applications and energy storage, like fuel cells, batteries, and supercapacitors [10]. Accordingly, composing Fe_3O_4 with graphene can integrate the advantages possessed of $Fe_3O_4$and graphene and may offer great application potential in electrochemical sensors.

Magnetite (Fe_3O_4) have been composite with multi-walled carbon nanotube (MWCNT) by using coprecipitation-ultrasonic irradiation [10] and sonochemical method [11]. In this work, a sonochemical method at room temperature was used to synthesize graphene-Fe_3O_4 nanocomposite. The graphene-Fe_3O_4 was prepared using graphene powder, $FeCl_3.6H_2O$, and $FeCl_2.4H_2O$ as starting materials. The characteristic of graphene-Fe_3O_4confirmed by XRD, SEM, and CV. The result displayed that modified the surface of the working electrode with graphene-Fe_3O_4 can enhance of reduction and oxidation currents. It was indicated that the graphene-Fe_3O_4 nanocomposite could provide good material for constructing electrochemical sensors.

2. Materials and methods

2.1. Synthesis of Graphene-Fe_3O_4 Nanocomposite

The synthesis of the graphene-Fe_3O_4 nanocomposite was started by dispersed graphene into ethanol and ultrasonicated for 10 min. Then, $FeCl_3{\cdot}6H_2O$ and $FeCl_2.4H_2O$ with a molar ratio 2:1 completely dissolved in water and added to the graphene suspension. Subsequently, ammonia solution was added slowly until the pH of the resulting solution maintained at 10 under ultrasonic irradiation for 30 min. The nanocomposite collected by magnetic separation and washed with water and ethanol several times. Finally, the produced nanocomposite was dried in an oven at 100 °C to get graphene-Fe_3O_4 nanocomposite powder.

2.2. Fabrication of Working Electrode

The bare carbon paste electrode (CPE) as the basic working electrode was prepared by mixed 70 wt% graphite powder and 30 wt% paraffin oil. The mixture of the carbon paste was packed into one end of a glass tube. A copper wire was inserted establish electrical contact. The CPE surface was smoothed by polishing it on a paper surface. To prepare a modified surface electrode (CPE/graphene-Fe_3O_4), 5 μL of 1 mg.mL^{-1} graphene-Fe_3O_4 nanocomposite solution in ethanol was dropped onto the surface of bare CPE.

2.3. Characterization

The formation of graphene-Fe_3O_4, namely crystalline phase was analyzed using X-ray diffractometer (XRD). The morphology of graphene-Fe_3O_4investigated using scanning electron microscope (SEM).Cyclic voltammetry (CV) was used for measuring the electrochemical performance with three-electrode system consisted of the bare CPE or CPE/graphene-Fe_3O_4as a working electrode, platinum (Pt) wire as a counter electrode and silver/silver chloride (Ag/AgCl) as a reference electrode.

3. Results and discussion

3.1. XRD Analysis

The XRD patterns of the graphene-Fe_3O_4 nanocomposite in Figure 1 was found to be the characteristic peaks at 2θ = (30.14°, 35.86°, 43.32°, 53.96°, 57.40°, and 62.92°) corresponding to (220), (311), (400), (422), (511), (440). These peaks are similar to Fe_3O_4 with spinel inverse cubic structure based on JCPDS card no. 19-0629. The broad peaks suggest that the crystalline size of Fe_3O_4 nanoparticles are very small and the sharp peak indicate the good crystallinity phase of Fe_3O_4. Moreover, the Fe_3O_4 nanoparticles have a fine crystalline quality cause of the absence of other impurity peaks from the XRD patterns. However, the diffraction peak attributed to graphene was no evidence. The XRD patterns only showed a hump between 2θ = (23° to 28°) indicate the formation of graphene [12].

According to the XRD data, the average crystallite size of Fe_3O_4 nanoparticles loaded on graphene nanosheets is estimated using Debye-Scherrer formula at 2θ = 35.86° found to be 10.05 nm. This result was smaller than Shenmin Zhu, and Deosarkar has reported for preparation Fe_3O_4-graphene nanocomposite by sonochemical co-precipitation method, which was 45 nm and 21.22 nm, respectively [13], [14]. The acoustic cavitation arises from ultrasonic irradiation resulting in local high temperature and pressure instantaneously. These extreme conditions cause the formation of smaller sized particles.

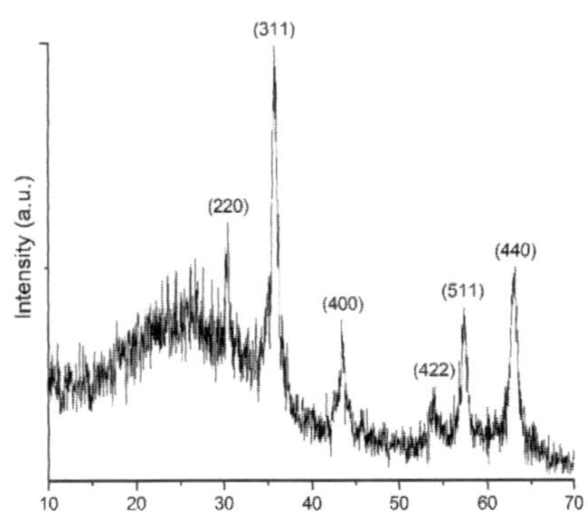

Figure 1. XRD patterns of the graphene-Fe_3O_4 nanocomposite

3.2. Morphological Analysis

Figure 2.a presented the SEM image of the graphene-Fe_3O_4 nanocomposite. The SEM image showed some aggregation of Fe_3O_4because of their surface energy interaction and plenty of Fe_3O_4 nanoparticles attached to the graphene nanosheets. The agglomeration size distribution of the Fe_3O_4 nanoparticles was estimated about 8 nm to 10 nm, as shown in the Figure 2.b. The attachment Fe_3O_4 nanoparticles between the layer graphene nanosheets can reduce the aggregation of Fe_3O_4 nanoparticles. The distribution of Fe_3O_4 nanoparticles on the layer graphene lead to the formation of

IOP Conf. Series: Materials Science and Engineering **367** (2018) 012054 doi:10.1088/1757-899X/367/1/012054

porous nanocomposite with a large amount of void space. That porous nanocomposite with several void spaces can enhance the cyclic performance [14].

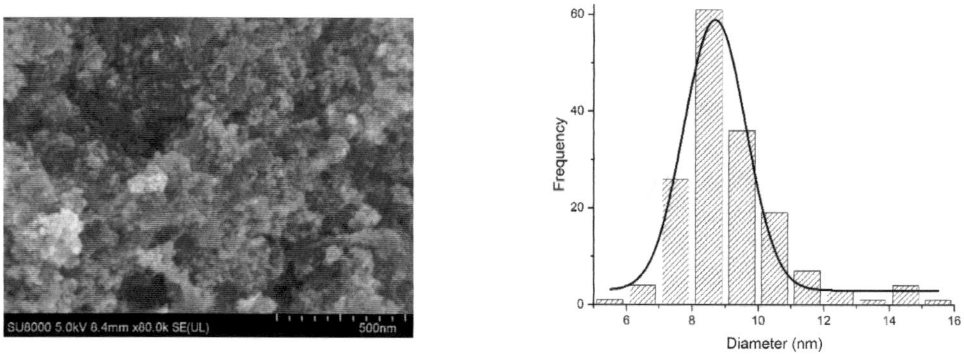

(a) (b)

Figure 2. (a) SEM image of the graphene-Fe_3O_4 nanocomposite and (b) the agglomeration size distribution of the Fe_3O_4 nanoparticles

3.3. Magnetic Properties

The magnetic properties of graphene-Fe_3O_4wasinvestigated by looking at the attraction when placing near a permanent magnet. For the experiment, the graphene-$Fe_3O_4$was dispersed into ethanol (1 mg/mL^{-1}) via ultrasonication. Then, this suspension attracted using a magnet for 5 min. The experiment resulted that graphene-Fe_3O_4can be uniformly and dispersed stably in ethanol. After 5 min, graphene-Fe_3O_4 attracted to the vial wall near to magnet and separated from the solvent. Although the Fe_3O_4 particles attached to the graphene sheets, the nanocomposite still has an excellent magnetic response. It is indicating that the graphene-Fe_3O_4 has easy recycling attributes and useful for magnetic separation application [15], [16].

3.4. Electrochemical Properties

Cyclic voltammetry has done to study the electrochemical properties. Figure 3.a displayed a CV diagram of the bare CPE and CPE/graphene-Fe_3O_4 electrode in 0.01 M $K_3[Fe(CN)_6]$/ $K_4[Fe(CN)_6]$-0.1 M NaCl at the scan rate of 100 mVs^{-1}. A couple of redox peaks corresponding to the reduction and oxidation of $[Fe(CN)_6]^{3-/4-}$ were observed for all sample electrode indicating a reversible electron transfer process. Both anodic (I_{pa}) and cathodic(I_{pc}) peak current were increased after the bare CPE was modified. The I_{pa} was increased until 76.58 % and I_{pc} also increase until 346.96 % from the bare CPE. The data saw that applying nanomaterial as the electrode surface can enhance the redox currents. Besides that, the peak-potential separations (ΔE_p) getting smaller and the I_{pa} and I_{pc} moving closer to each other reveal to the better transfer electron media.

The effective surface area (A) was determined using Randles-Servcik equation for a reversible process as follow [9]:

$$Ip = \left(2.69 \times 10^5\right) n^{3/2} A \cdot D^{1/2} \cdot v^{1/2} \cdot C \qquad (1)$$

where I_p (A) is the peak current, n is the number of electron transfer in the reaction, A (cm^2) is the effective surface area of electrode, D (cm^2s^{-1}) is the diffusion coefficient of the reduced, C (mol cm^{-3}) is the bulk concentration of the reduced form, and v(V s^{-1})is the scan rate. For this calculation, $n = 1$, $D = (7.6 \times 10^{-6})$ cm^2s^{-1}, $C = 10^{-5}$ $molcm^{-3}$. It was calculated that the effective surface area of CPE/graphene-Fe_3O_4 (0.0436 cm^2) > bare CPE (0.0247 cm^2). The result indicated that the use of nanomaterial for modified surface electrode could improve the electroactive surface area of the electrode.

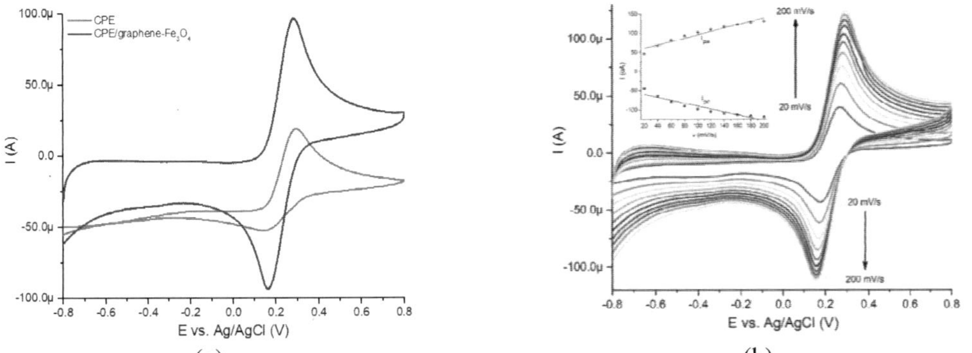

(a) (b)

Figure 3. (a) CVs of the bare CPE and CPE/graphene-Fe_3O_4 electrode and (b) effect of scan rate from CPE/graphene-Fe_3O_4 modified electrode at scan rate (20, 40, 60, 80, 100, 120, 140, 160, 180, and 200) $mV{\cdot}s^{-1}$, and (inset) plot of the relationship between anodic and cathodic peak current with scan rate

The effect of scan rate from CPE/graphene-Fe_3O_4 modified electrode shown in Figure 3.b. From the CV diagram, the anodic (I_{pa}) and cathodic (I_{pc}) peak current were increased linearly by increasing in the scan rate (v) from 20 mVs^{-1} until 200 mVs^{-1}. The plot of the relationship between anodic and cathodic peak current with scan rate inserted in Fig 4b. The linear line equation of I_{pa} (μA) = 0.4452 v + 50.712 (mVs^{-1}), R = 0.9411, and I_{pc} (μA) = –0.3754 v –52.086 (mVs^{-1}), R = 0.8953. It may suggest that the mass transfer of electrochemical kinetics in the electrode mainly diffusion controlled [17]. It can also declare that the combination of Fe_3O_4 and graphene for the modified surface electrode provided an excellent mediated transfer between the electrode and electrolyte, thus leading to the promotion of the electron transfer rate. The graphene-Fe_3O_4 nanocomposite is offering better performance when applied as a working electrode in the electrochemical sensor.

4. Conclusions

Graphene-Fe_3O_4nanocomposite synthesized by a simple, effective sonochemical method at room temperature. The nanocomposite has a good crystallinity with well size and shape of Fe_3O_4 nanoparticles decorated graphene nanosheets. It also exhibits an excellent magnetic response. The electrochemical studies demonstrated the graphene-Fe_3O_4 nanocomposite could improve the electroactive surface area of the electrode and mediated electron transfer between electrode and electrolyte. It was due to the integration of the unique properties of Fe_3O_4 and graphene. Graphene-Fe_3O_4 nanocomposite can lead to the promotion of the electron transfer rate and enhancement the redox currents, so it was generated higher sensitivity of output signal. Accordingly, graphene-Fe_3O_4 nanocomposite would provide a great promising material to apply in the development of high-performance electrochemical sensors, including biosensor based on electrochemical procedure.

References

[1] A. Chaubey and B. Malhotra, "Mediated biosensors," *Biosens. Bioelectron.*, vol. 17, no. 6–7, pp. 441–456, 2002.
[2] C. I. Justino, T. A. Rocha-Santos, and A. C. Duarte, "Advances in point-of-care technologies with biosensors based on carbon nanotubes," *TrAC Trends Anal. Chem.*, vol. 45, pp. 24–36, 2013.
[3] N. L. W. Septiani, B. Yuliarto, H. K. Dipojono, and others, "Multiwalled carbon nanotubes–zinc oxide nanocomposites as low temperature toluene gas sensor," *Appl. Phys. A*, vol. 123, no. 3, p. 166, 2017.
[4] Yuliarto B, Ramadhani M, Nugraha F, Septiani N L W, and Hamam K A, "Enhancement of SO_2 gas sensing performance using ZnO nanorod thin films: the role of deposition time," *Springer US*, vol. 54, pp. 4543–4554, 2017.

[5] P. Bollella *et al.*, "Beyond graphene: electrochemical sensors and biosensors for biomarkers detection," *Biosens. Bioelectron.*, vol. 89, pp. 152–166, 2017.
[6] Y. Song, Y. Luo, C. Zhu, H. Li, D. Du, and Y. Lin, "Recent advances in electrochemical biosensors based on graphene two-dimensional nanomaterials," *Biosens. Bioelectron.*, vol. 76, pp. 195–212, 2016.
[7] T. A. Rocha-Santos, "Sensors and biosensors based on magnetic nanoparticles," *TrAC Trends Anal. Chem.*, vol. 62, pp. 28–36, 2014.
[8] F. He, Z. He, J. Xie, Y. Li, and others, "IR and Raman Spectra Properties of Bi_2O_3-ZnO-B_2O_3-BaO Quaternary Glass System," *Am. J. Anal. Chem.*, vol. 5, no. 16, p. 1142, 2014.
[9] H. Teymourian, A. Salimi, and S. Khezrian, "Fe_3O_4 magnetic nanoparticles/reduced graphene oxide nanosheets as a novel electrochemical and bioeletrochemical sensing platform," *Biosens. Bioelectron.*, vol. 49, pp. 1–8, 2013.
[10] R. Rahmawati, M. G. Permana, B. Harison, B. Yuliarto, D. Kurniadi, and others, "Optimization of Frequency and Stirring Rate for Synthesis of Magnetite (Fe_3O_4) Nanoparticles by Using Coprecipitation-Ultrasonic Irradiation Methods," *Procedia Eng.*, vol. 170, pp. 55–59, 2017.
[11] R. Rahmawati *et al.*, "Preparation of MWCNT-Fe_3O_4 Nanocomposites from Iron Sand Using Sonochemical Route," in *IOP Conference Series: Materials Science and Engineering*, 2017, vol. 202, p. 012013.
[12] B. Jaleh, A. Khalilipour, S. Habibi, M. Niyaifar, and M. Nasrollahzadeh, "Synthesis, characterization, magnetic and catalytic properties of graphene oxide/Fe_3O_4," *J. Mater. Sci. Mater. Electron.*, vol. 28, no. 6, pp. 4974–4983, 2017.
[13] J. Zhu, G. Zhang, S. Gu, and B. Lu, "SnO_2 Nanorods on ZnO Nanofibers: A New Class of Hierarchical Nanostructures Enabled by Electrospinning as Anode Material for High-Performance Lithium-Ion Batteries," *Electrochimica Acta*, vol. 150, pp. 308–313, Dec. 2014.
[14] M. Deosarkar, S. Pawar, and B. Bhanvase, "In situ sonochemical synthesis of Fe_3O_4–graphene nanocomposite for lithium rechargeable batteries," *Chem. Eng. Process. Process Intensif.*, vol. 83, pp. 49–55, 2014.
[15] K.-K. Cheng *et al.*, "Sugarcane bagasse hemicellulose hydrolysate for ethanol production by acid recovery process," *Biochem. Eng. J.*, vol. 38, no. 1, pp. 105–109, 2008.
[16] D. Yang *et al.*, "In situ synthesized rGO–Fe_3O_4 nanocomposites as enzyme immobilization support for achieving high activity recovery and easy recycling," *Biochem. Eng. J.*, vol. 105, pp. 273–280, 2016.
[17] J. Salamon *et al.*, "One-pot synthesis of magnetite nanorods/graphene composites and its catalytic activity toward electrochemical detection of dopamine," *Biosens. Bioelectron.*, vol. 64, pp. 269–276, 2015.

Effect of Temperature and Duration of Sintering on Perovskite Material $(La_{1-x}Ag_x)_{0.8}Ca_{0.2}MnO_3$

B. Kurniawan[1], S. D. Rosanti[1], R. Kamila[1], N. B. Sahara[1], D. S. Razaq[1], T. Komala[1], and D. R. Munazat[1]

[1]Department of Physics, Faculty of Mathematics and Natural Sciences (FMIPA), Universitas Indonesia, Depok 16424, Indonesia

E-mail: bkuru@fisika.ui.ac.id

Abstract. Lanthanum compounds of $(La_{1-x}Ag_x)_{0.8}Ca_{0.2}MnO_3$ have been prepared for x = 0 and x = 0.5 by using sol-gel method. Bulk samples were sintered by varying temperature and duration of sintering processes. The sintering temperature was varied at two conditions, 600 °C for 5 h and 900 °C for 24 h. Crystal structure of the samples was observed by X-Ray diffraction spectrometer. Rietveld refinement of X-Ray diffraction results showed that the sample with x = 0, both sintered at 600 °C for 5 h and at 900 °C for 24 h are single phase and they have an orthorhombic structure with Pnma space group. The sample with x = 0.5 sintered at 900 °C for 24 h did not show a single phase due to the appearance of Ag peaks, and they have a perovskite structure. Meanwhile, the sample with x = 0.5 sintered at 600 °C for 5 h, also did not show a single phase due to the appearance of Ag and LaO peaks. This results showed that for x = 0.5, temperature and duration of sintering are influenced by nucleation processes in the samples. The higher temperature and more longer duration of sintering will make the nucleation processes in the sample better.

Keywords. Ag-doped, $(La_{1-x}Ag_x)_{0.8}Ca_{0.2}MnO_3$, sol-gel, and sintering.

1. Introduction

Recently, synthesis of materials with multifunctional properties is a field of intense activity, especially regarding technological innovation and development of new materials [1]. The parental material is $LaMnO_3$, the fascinating materials in condensed-matter research.$LaMnO_3$ is an antiferromagnetic (AFM) insulator with the Neel temperature (T_N) of ~140 K. In order to form $La_{1-x}A_xMnO_3$ (La = trivalent rare earth element and A = alkaline earth) compounds, substituting the La site by an alkaline earth or alkaline element (A) exhibits unusual physical effects on their ferromagnetic (FM)–paramagnetic (PM) phase-transition temperature (T_C), colossal magnetoresistance and magnetocaloric effects (MCEs) [2–4].

Substitution an alkaline earth metal Ca^{2+}ion to La^{3+}ion can change the ratio of Mn^{3+}ion and Mn^{4+}ion to form perovskite $La_{0.8}Ca_{0.2}MnO_3$ (LCMO). The application of LCMO is in the fields of magnetic memory devices, refrigeration, sensor, biology, medicine, and catalyst [5]. Substitution the monovalent metal Ag^{+}ion to La^{3+}ion can also change the ratio of Mn^{3+}ion and Mn^{4+}ion to form $La_{1-x}Ag_xMnO_3$ (LAMO). The ionic radii of La^{3+} and Ag^{+}ions are 1.18 Å and 1.13Å, respectively. By using the ionic radii and based on a calculation of the tolerance factor, it reveals that LAMO forms a stable

Content from this work may be used under the terms of the Creative Commons Attribution 3.0 licence. Any further distribution of this work must maintain attribution to the author(s) and the title of the work, journal citation and DOI.
Published under licence by IOP Publishing Ltd

manganite [6]. Physical properties of LCMO may be modified by the chemical composition as well as by various temperature and duration of sintering that is affecting to their microstructure [7, 8].

Modifying LCMO and LAMO can be done by substitution of the Ag^+ ion for La^{3+}ion to become $(La_{1-x}Ag_x)_{0.8}Ca_{0.2}MnO_3$ (LACMO). LACMO was synthesized by a sol-gel process. Doping Ag in La site of LCMO induced an increase of Magnetoresistance (MR), Curie temperature and metal-insulator transition temperature. Furthermore, the interesting experiments have been continued by varying the synthesis process. In order to get more information in the structure, magnetic and electrical properties of LACMO, the variation of temperature and duration of sintering in the synthesis process were done. It suggests that the variation of temperature and duration of sintering in the synthesis process will influence phase and crystal structure of the sample.

2. Materials and methods

The materials of $(La_{1-x}Ag_x)_{0,8}Ca_{0,2}MnO_3$ have been prepared by sol-gel method for x = 0 and x = 0.5. Precursors were used including of La_2O_3, $AgNO_3$, $Ca(NO_3)_2.4H_2O$ and $Mn(NO_3)_2.4H_2O$. Each precursor was dissolved with nitric acid to obtain a transparent solution. Citric acid ($C_6H_8O_7.H_2O$) is used with a mole ratio of citric acid of 1:1.2 from the total of metal ions. The ammonia solution is added to adjust the pH of the solution to be 7. The resulting mixture is homogenized by using magnetic stirrer at a temperature of (80 to 90) °C to form a gel. The resulting gel was heated at a temperature of 120 °C in the oven for 3 h in order to dry the sample to become high viscosity gel. After that the gel was heated at 500 °C for 5 h in the furnace to get a decomposition of the organic precursor. Then the gel was sintered with two combinations treatments at 600 °C for 5 h and 900 °C for 24 h. The obtained samples then to be characterized by using X-ray Diffraction (XRD) spectrometer and to be done for refining process to get the phase and structural analysis of the samples.

3. Results and discussion

The data of XRD of $(La_{1-x}Ag_x)_{0,8}Ca_{0,2}MnO_3$for x = 0 and x = 0.5 showed the characteristic peaks of the compound as can be seen in Figure 1 to Figure 4. Data of XRD signals were refined by using Rietveld method. The system was presented perovskite parameters, and unit cell volume of the LCMO and LACMO samples obtained from XRD patterns are presented in Table 1 and Table 2.

In Figure 1, the sample with x = 0 both sintered at 600 °C for 5 h and at 900 °C for 24 h were in single phases. They showed an orthorhombic structure with *Pnma* space group. No change in orthorhombic structure by changing the sintered temperature from 600 °C to 900 °C and changing the duration of sintering from 5 h to 24 h but the peaks become sharper and clearer. Previous research material of $La_{0.7}Ca_{0.3}MnO_3$, reported by Siwach *et al*, that are sintered at 600 °C and 900 °C for 2 h and 6 h also showed the same properties of an orthorhombic structure [9].

Figure 2 showed the characteristic peaks for LACMO with x = 0.5. The samples did not show single phase system. It showed the appearance of another peak beside the same peaks as for x = 0. We found the discrepancies characteristic between two different sintering temperatures. For sintering temperature at 600 °C, we observed the appearance peaks from Ag and LaO. Meanwhile, we observed only the appearance peak from Ag for sintering temperature at 900 °C. It means that the higher sintering temperature can increase the nucleation process in the samples so the LaO peaks could be disappeared due to the melting temperature of LaO is around 900 °C. The Ag peaks were observed at 2θ = (38.0°, 44.5°, and 64.5°) that are the same positions that are found by Irmak *et al.* in$La_{0.95}Ag_{0.05}MnO_3$ and $La_{0.75}Ag_{0.25}MnO_3$ [10]. Besides, for the Ag peaks, they still appeared due to the breaking solubility limitation of Ag^+ ions from the stoichiometric value [3]. In previous research perform by Huang and Srivastava, It has been explained that the solubility limit of Ag^+ ions should be related to x = 0.15. It means, in order to get single phase system, Ag concentration should be below x = 0.15 [11, 12]. We also observed in our previous reports [ISCPFM Nov 2017] that the single phase system could be achieved for x < 1.5.

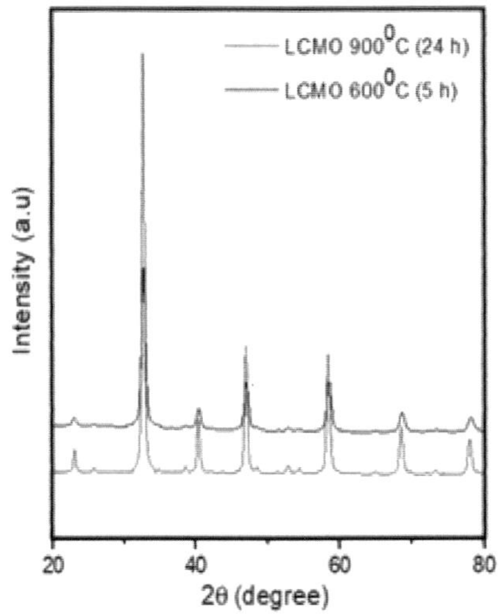

Figure 1. LCMO is sintered at 600°C (5 h) and 900°C (24 h)

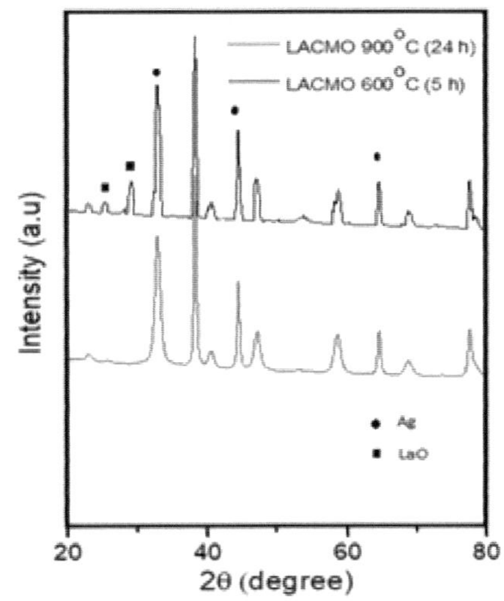

Figure 2. LACMO is sintered at 600°C (5 h) and 900°C (24 h)

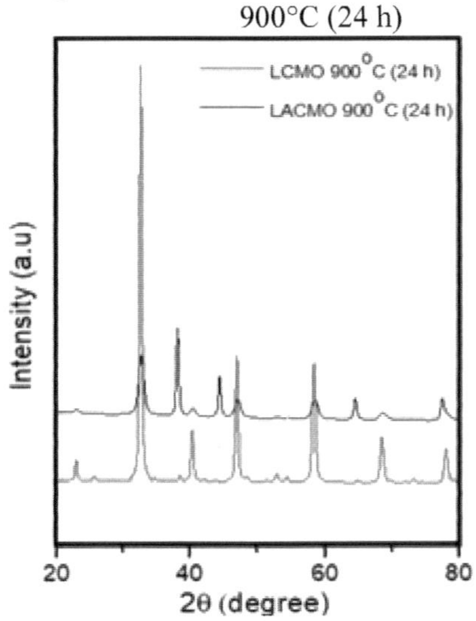

Figure 3. LCMO and LACMO were sintered at 900°C (24 h)

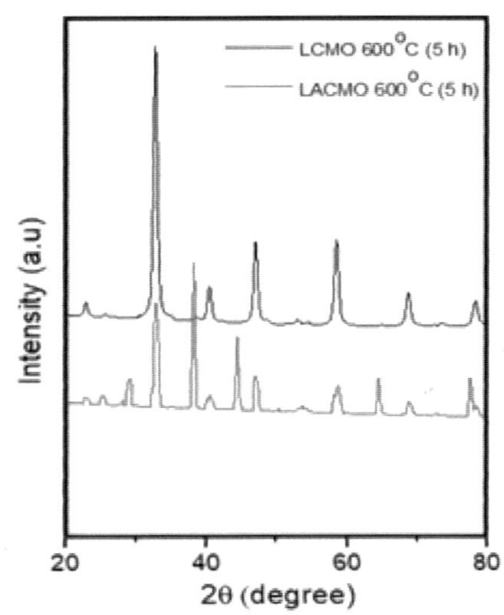

Figure 4 LCMO and LACMO were sintered at 600°C (5 h)

Figure 3 showed the differences characteristic peaks at same sintering temperature 900 °C during 24 h for LCMO and LACMO as also mentioned in Figure 4 at same sintering temperature 600 °C during 5 h. The differences are due to the appearance another phase at LACMO. The lower sintering temperature causes the appearance of more phases.

We also found that the crystallite size for $x = 0$ is 14.92 nm (sintered at 600 °C for 5 h), 25.51 nm (sintered at 900 °C for 24 h) and for $x = 0.5$ are 10 nm (sintered at 600 °C for 5 h) and 10.4 nm (sintered at 900 °C for 24 h). Crystallite size is increased with the increasing temperature and duration of sintering. We got that R profile and weighted R profile on data processing were below 10%. It means that the result is in accordance with experimental analysis which is supported by the chi-squared value of under 1.3.

Table 1. Lattice parameters, unit cell volume and goodness of fit ($\chi 2$) LCMO and LACMO sintered at 900 °C for 24 h

Parameters	$La_{0,8}Ca_{0,2}MnO_3$	$La_{0.7}Ag_1Ca_{0,2}MnO_3$	
	Perovskite	Perovskite	Silver
a (Å)	5.4653	5.4730	4.0770
b (Å)	7.73329	7.7310	4.0770
c (Å)	5.49248	5.4880	4.0770
Volume (Å)	232.1383	230.3828	68.1195
Percentage of each phase (%)	100	74	26
Rp	7.3894	7.4666	
Rwp	9.3343	9.4006	
Goodness of fit	1.09674	1.1935	

Table 2. Lattice parameters, unit cell volume and goodness of fit ($\chi 2$) LCMO and LACMO sintered at 600 °C for 5 h

Parameters	$La_{0,8}Ca_{0,2}MnO_3$ Perovskite	$La_{0.7}Ag_1Ca_{0,2}MnO_3$		
		Perovskite	Silver	Lanthanum Oxide
a (Å)	5.4638	5.4302	4.0845	4.0650
b (Å)	7.7257	7.7016	4.0845	4.0650
c (Å)	5.4595	5.5475	4.0845	6.2645
Volume (Å)	230.4548	232.0018	68.1415	89.6467
Percentage of each phase (%)	100	68.2	18.4	13.4
Rp	7.5812	7.6416		
Rwp	9.5221	9.5795		
Goodness of fit	1.1528	1.2399		

4. Conclusions

$(La_{1-x}Ag_x)_{0.8}Ca_{0.2}MnO_3$ were successfully synthesized by sol-gel process and were varied by temperature and duration of sintering with x = 0 and x = 0.5. Sintering temperature variations were done at 600 °C for 5 h and 900 °C for 24 h. For LCMO, the samples showed a single phase condition at both sintering temperature. For LACMO, the samples did not show a single phase condition due to the appearance another LaO and Ag peaks. The sintering temperature influence to the characteristic of XRD peaks. At higher sintering temperature, the peaks for both LCMO and LACMO become clearer and sharper. At LACMO the higher temperature and longer sintering duration, the samples showed decreasing numbers of the appearance phases.

Acknowledgments

This work is financially supported by Universitas Indonesia under research grant "Hibah PITTA (*Hibah Publikasi Internasional Terindeks untuk Tugas Akhir Mahasiswa Universitas Indonesia*" with grant contract number is 634/UN2.R3.1/HKP.05.00/2017. The authors are also grateful to Department of Physics, Universitas Indonesia for research facilities.

References

[1] Andrade V M, Vivas R J C, Pedro S S, Tedesco J C G, Rossi A L, Coelho A A, and Reis M S 2016 Magnetic and magnetocaloric properties of La06Ca04MnO3 tunable by particle size and dimensionality *Acta Materialia* **102** 49 – 55

[2] Riahi K, Messaoui I, Cheikhrouhou-Koubaa W, Mercone S, Leridon B, Koubaa M, and Cheikhrouhou A 2016 Effect of synthesis route on the structural, magnetic and magnetocaloric properties of La078Dy002Ca02MnO3 manganite: A comparison between

sol-gel, high-energy ball-milling and solid state process *Journal of Alloys and Compounds*, **688** 1028 – 1038

[3] Ye S L, Song W H, Dai J M, Wang K Y, Wang S G, Zhang C L, and Fang J 2002 Effect of Ag substitution on the transport property and magnetoresistance of LaMnO3 *Journal of Magnetism and Magnetic Materials* **2481** 26–33

[4] Debnath J C and Strydom A M 2015 Large low field magneto-resistance and temperature coefficient of resistance in La08Ca02MnO3 epitaxial thin film *Journal of Alloys and Compounds* **621** 7–11

[5] Xia W, Li L, Wu H, Xue P, and Zhu X 2017 Structural, morphological, and magnetic properties of sol-gel derived La07Ca03MnO3 manganite nanoparticles *Ceramics International* **433** 3274–3283

[6] Battabyal M, and Dey T K 2005 Electrical conductivity in La1−xAgxMnO3 pellets between 10 and 350 K *Physica B: Condensed Matter* **367** 40–47

[7] Pekała M, and Drozd V 2008 Magnetocaloric effect in nano- and polycrystalline La08Sr02MnO3 manganites *Journal of Non-Crystalline Solids* **354** 5308–5314

[8] Pkała M, Drozd V, Fagnard J F, and Vanderbemden P 2010 Magnetocaloric effect in nano- and polycrystalline manganites La 05Ca05MnO3 *Journal of Alloys and Compounds* **5072** 350–355

[9] Siwach P K, Goutam U K, Srivastava P, Singh H K, Tiwari R S, and Srivastava O N 2006 Colossal magnetoresistance study in La07Ca03MnO3 manganite *J. Phys. D: Appl. Phys.* **39** 14–20

[10] Ekber-Irmak A, Coskun A, Tasarkuyu E, Akturk S, Unlu G, Samancioglu Y, and Yucel A 2010 The influence of the sintering temperature on the structural and the magnetic properties of doped manganites: La095Ag005MnO3 and La075Ag025MnO3 *Journal of Magnetism and Magnetic Materials* **3228** 945–951

[11] Huang S, Deng L, Zhou K, Hu Z, Sun S, Ma Y, and Xiao P 2012 Effect of Ag substitution on the electromagnetic property and microwave absorption of $LaMnO_3$ *Journal of Magnetism and Magnetic Materials* **324** 3149–3153

[12] Srivastava P, Srivastava A, Srivastava O N, Singh D P, and Siwach P K 2011 Effect of ag doping and admixing on the magnetotransport properties of nanophasic $La_{07}Ca_{03}MnO_3$ *International Journal of Nanoscience* **1004** 1019–1024

IOP Conf. Series: Materials Science and Engineering **367** (2018) 012056 doi:10.1088/1757-899X/367/1/012056

Effects of Molar Ratios and Sintering Times on Crystal Structures and Surface Morphology of $Nd_{1+x}FeO_3$ Oxide Alloy Prepared by using Solid Reaction Method

E. H. Sujiono[1], J. Agus[1], S. Samnur[1] and K. Triyana[2]

[1]Laboratory of Materials Physics, Department of Physics, Universitas Negeri Makassar, Parang Tambung campus, Makassar 90224, Indonesia
[2]Department of Physics, Universitas Gadjah Mada, Sekip Utara BLS 21 Yogyakarta 55281, Indonesia

E-mail: e.h.sujiono@unm.ac.id

Abstract. The effects of molar ratios and sintering times on crystal structures and surface morphology on $NdFeO_3$ oxide alloy have been studied. $NdFeO_3$ oxide alloy formed by chemical preparation with solid reaction method using raw oxide Fe_2O_3 (99.9 %) and Nd_2O_3 (99.9 %) powders. In this article we reported the effects of molar ratios x = (–0.1, –0.2 and –0.3) and sintering times for 15 h and 20 h on crystal structures and surface morphology of $Nd_{1+x}FeO_3$ synthesized by solid-state reaction method. The results indicate that variation of molar ratio and sintering time has influenced the FWHM, crystalline size and grain size. The $Nd_{1+x}FeO_3$ have a major phase is $NdFeO_3$, and other minor phases are Fe_2O_3, Nd_2O_3 and $Nd(OH)_3$. The dominant intensity of *hkl* (121) with a value in FWHM, crystallite size, and grain size an indication the results will be applied as a gas sensor material as the focus of the further study.

Keywords. Crystal structure, $Nd_{1+x}FeO_3$, molar ratio, sintering time, surface morphology, and solid state reaction method.

1. Introduction

Metal particles smaller than 100 nm in primary particle diameter are generally considered as nanoparticles. Such metal nanoparticles often exhibit very interesting electronic, magnetic, optical, and chemical properties [1]. For example, their high surface-to-volume ratios have large fractions of metal atoms at surface available for catalysis [2, 3]. The phase majority of catalysts used in modern the chemical industry is based on mixed metal oxides including perovskite oxides ABO_3 [4]. The nano-perovskite oxides ABO_3 (*A*: La, Nd, Sm, and Gd; *B*: Fe, Co and Ni; and O: oxygen) have high catalytic activities and high sensitivity with CO and HCs. Their applications in gas sensors were studied especially for its properties in electrical and response for CO gas [5]. $NdFeO_3$ is a perovskite transition metal oxide with an orthorhombic structure and space group of Pbnm possesses insulator properties at room temperature [6].

One of $NdFeO_3$ synthesis method is solid reaction method [7]. Solid reaction method is the synthesis of solid materials by reacting with another solid at high temperatures [8]. Solid reaction method is having advantages such as the resulting crystals having good purity and crystallized.

Content from this work may be used under the terms of the Creative Commons Attribution 3.0 licence. Any further distribution of this work must maintain attribution to the author(s) and the title of the work, journal citation and DOI.
Published under licence by IOP Publishing Ltd

IOP Conf. Series: Materials Science and Engineering **367** (2018) 012056 doi:10.1088/1757-899X/367/1/012056

However, the synthesis results obtained in this method produce particles of a large size and irregular morphology [9]. Based on the interesting phenomena developing of oxide materials using the solid-state method, the authors have the conduct of previous research for fabrication oxide material such as $YBa_2Cu_3O_y$, $NdBa_2Cu_3O_y$, $NdFe_xBa_{2-x}Cu_3O_y$ and $Nd_{1+x}FeO_3$, the results have been reported elsewhere [10–14].

In this study will be assessed the effect of molar ratio and sintering time on crystalline structure and surface morphology of $Nd_{1+x}FeO_3$ prepared by solid-state reaction method. Heating temperature, heating time and molar ratio are important because it can be adjusted to crystallinity, grain size, and homogeneity. Therefore, the development of $NdFeO_3$ oxide material needs to be realized to obtain the best crystal structure and surface morphology for gas sensor application.

2. Materials and methods

Preparation of $Nd_{1+x}FeO_3$ material was made 6 samples, 3 samples with a variation of molar ratio indicate with sample #1, #2 and #3, and 3 samples with a variation of sintering time indicate with sample #4, #5 and #6. In this study, $Nd_{1+x}FeO_3$ oxide Alloy Material were synthesized from Nd_2O_3 (99.9 %) and Fe_2O_3 (99.9 %) powder using solid state reaction method. Stoichiometric calculation of raw material mass used Proust law with molar ratio value are x = –0.1, x = –0.2 and x= –0.3 for sample #1, #2 and #3, respectively and x = 0 for sample #4, #5 and #6. The mixed was grinded by mortar and pastel for ± 3 hours to maximize the reaction and then were calcined using the furnace. The sample #1, #2 and #3 was calcined at 950 °C for 50 h, sample #6 at 600 °C for 40 h. Sample #4 and #5 without calcination process. After calcination process is complete, the sample then prepared into a pellet with 1.25 cm diameter under 16 t pressure. The pellet of sample #1, #2 and #3 then sintered at temperature 950 °C for 67 h, sample #4 and #6 at 400 °C for 20 h and sample 5 at 400 °C for 15. All of the samples were characterized using XRD to identify the crystalline phase and using SEM to identify the surface morphology.

3. Results and discussion

3.1. Variation of molar ratio

3.1.1. Chemical Composition

The spectrum showed that increased of peaks intensity and widened of peaks width when the value of the molar ratio is added. However, there are some irregularities where the intensity of the peak sample #2 (x = –0.2) higher than sample #1 (x = –0.1) at certain positions.

Figure 1 shows the addition of the molar ratio, the peak intensity continues to increase, and the peak width becomes wider. However, there are some deviations in which the intensity of the peak x = -0.2 is higher than x = –0.1 at certain positions, e.g., at 32.6° corresponding to the *hkl* (121). The XRD characterization result obtained that the sintering temperature influence of the FWHM. The FWHM is 0.11° for the sample #1, related to the intensity of 984.56 counts. The similar results have been reported on reference [2, 4, 11, 14] though with different research method.

The crystals size of $NdFeO_3$ was calculated on the basis of Scherrer's equation at the highest peak and obtained the crystals size from the smallest to the largest is associated with x = –0.2 (94.0 nm), x = –0.1 (109.7 nm) and x = –0.3 (131.6 nm). While XRD diffraction pattern obtained of FWHM value for molar ratio varied of x = –0,1, x = –0,2 and x = 0, at 32.6° corresponding to the *hkl* (121) are 0.11°, 0.13° and 0.12° and the intensity for each molar ratio is 834 counts, 979 counts and 474 counts, respectively.

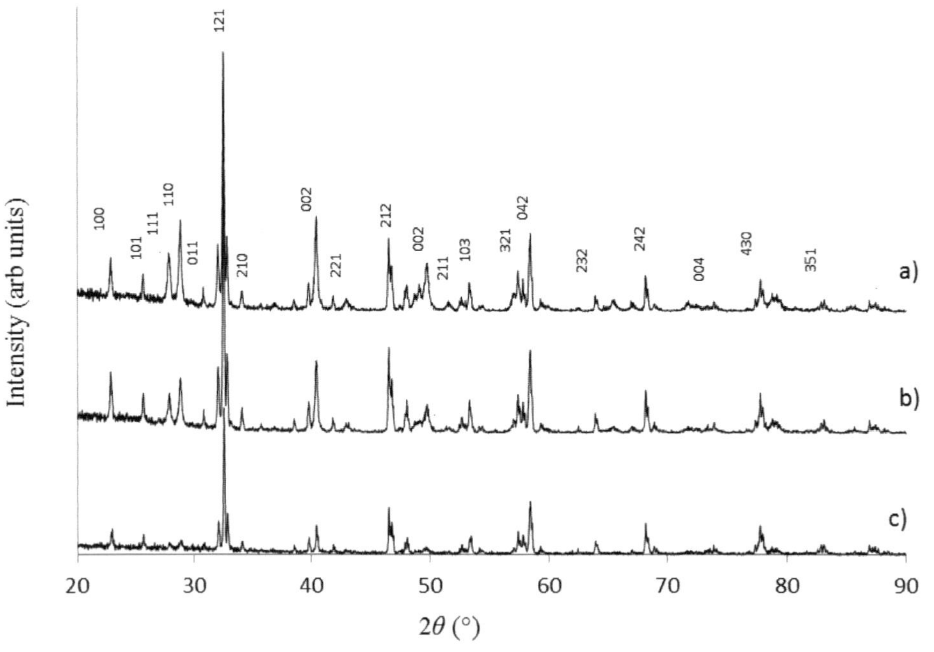

Figure 1. XRD spectra of a) sample #1 , b) sample #2 and c) sample #3, respectively.

3.1.2. Microstructure and Microanalysis

Figure 2 shown SEM image of the $Nd_{1+x}FeO_3$ alloy was prepared with different parameter process. As shown in the figure, it can be seen that the morphology in Figure 2.a and Figure 2.b is more homogeneous although there are still grain boundaries. This data indicated that the result from phase composition was obtained is confirmed. On the other hand, based on Figure 2.c shows that many of grain size shape were observed and this is caused by appearing of grain boundaries. Therefore, we confirmed that the sintering temperature also has influence formed of grain size and grain boundary [3, 9, 11]. Based on the SEM image of the samples above, the variation of the molar ratio has a clumpy surface morphology in which the grain boundary appears on the sample looks very thin and is almost invisible, and the crystalline grain appears much unified.

Crystalline phases analysis using the software *Match!* shows each sample contains four phases is $NdFeO_3$, $Nd(OH)_3$, Fe_2O_3 and Nd_2O_3. It can be seen in Table 1, increasing of molar ratio gradually increase the percentage of $NdFeO_3$ phase. Similar results have been reported in previous studied [13], and there is a minority phase of $Nd(OH)_3$ due to the absorption of hydrogen and oxygen by neodymium [15].

Figure 2. SEM images of a) sample #1, b) sample #2 and c) sample #3 with mag. of 20,000x.

Table 1. Phase analysis of the $Nd_{1+x}FeO_3$ samples.

Samples	$NdFeO_3$ (%)	Nd_2O_3 (%)	Fe_2O_3 (%)	$Nd(OH)_3$ (%)
#1	83	4	-	13
#2	83	4	5	8
#3	78	4	7	11

3.2. Variation of sintering times

3.2.1. Chemical Composition

The XRD graph shown in Figure 3, it is can be seen that if $Nd(OH)_3$ phase formed almost dominant, in contrast, that the oxide alloy material has not formed $NdFeO_3$ phase. It can be seen clearly that there are many peaks with a variation of intensity value as shown in Figure 3.a sample #4 and Figure 3.b sample #5. These results also confirmed that the synthesis process does not show the formation of $NdFeO_3$ phase. However, based on the results of sample #5 the oxide alloy material still forms the $Nd(OH)_3$ phase, although the sintering time is 5 h longer than sample #4 which is sintered for 15 h. This structure can also be seen that the most dominant peak as an indication of $NdFeO_3$ phase still does not observe as consequences of the sintering process at a low temperature of 400 °C and without calcination. In contrast Figure 3.c sample #6 show the formation of $NdFeO_3$ phase indicate the appearance of a peak at 32.6° corresponding to the *hkl* (121). In fact, this result confirmed of a strong indication that calcination process at 600 °C for 40 h following sintering process at a low temperature of 400 °C still produce oxide alloy with $NdFeO_3$ phase. It means the calcination process has the more dominant effect of the crystal structure if compare than the sintering time at low temperature.

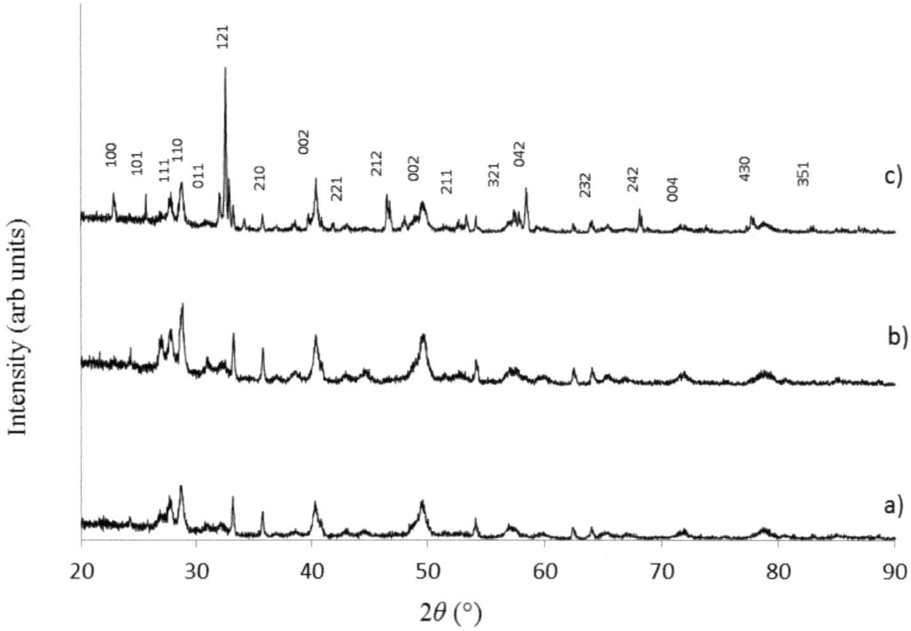

Figure 3. XRD spectra of a) sample #4, b) sample #5 and c) sample #6, respectively.

Calculation of grain size using Debye-Scherrer's equation on the highest intensity obtained crystalline size value are (163.97 ± 0.01) nm, (167.21 ± 0.01) nm and (255.47 ± 0.01) nm for the sample#4, sample #5, and sample #6, respectively. While based on XRD diffraction pattern obtained FWHM value at 2θ of 32.6° for sample #4, sample#5 and sample #6 are 0.19°, 0.19° and 0.12° and the intensity for each sample are 57 counts, 76 counts, and 1227 counts, respectively. This result is similar as has reported by Lou Xiangdong et al. with temperature 800 °C [8], Yabin Wang et al. with temperature 1000 °C [9] and they get similar results that phase $NdFeO_3$ exist at $2\theta = 32.56^\circ$ with *hkl* value is (121). V. Zharvan *et al.* [13] reported the crystalline size of samples ranging at 137.0 – 152.0 nm.

3.2.2. Microstructure and Microanalysis

The SEM image of $Nd_{1+x}FeO_3$ samples was shown in Figure 4. It can be seen that all of the samples have non-homogeneous morphology. There are still many agglomerates made due to temperature and mechanical treatment.

Figure 4.a sample #4 and Figure 4.b sample #5 shows an unfinished grain growth forming of $NdFeO_3$ oxide alloy. Correspond to the crystal structure as have been explained, the surface morphology of the sample #4 and sample #5 indicate with granular patterns that are very difficult to identify. Given the variation of sintering time in these two samples, it does not show any significant difference grain size. However, if we observe more closely at Figure 4.b sample #5, there are some parts that have formed a grain. In contrast, the surface morphology of the sample#6 as shown in Figure 4.c which was calcined for 40 h at a temperature of 400 °C then sintered for 20 h at a temperature of 600 °C indicates a better morphology than the two samples which is sample #4 and sample #5 without calcination process. The size of the grains is large and in some parts of the surface appear grain growth is not perfect due to the low sintering and calcination temperature. Similiar results were obtained by Khorasani-Motlagh M. *et al.* that synthesis nanocrystals NdFeO3 materials with has good porosity and the crystalline size influence of time and temperature heat treatment [16].

Figure 4. SEM image of a) sample #4, b) sample #5 and c) sample #6 with mag. of 20.000x.

Phases analysis using the software *Match!* show dominant phases of $NdFeO_3$ and contains phase in each sample are $Nd(OH)_3$, Fe_2O_3 and Nd_2O_3, respectively. The highest intensity corresponds to the $NdFeO_3$ phase with polycrystalline material and orthorhombic structure.

Table 2. Phase analysis of the $Nd_{1+x}FeO_3$ samples.

Samples	$NdFeO_3$ (%)	Nd_2O_3 (%)	Fe_2O_3 (%)	$Nd(OH)_3$ (%)
#4	55	8	20	17
#5	53	9	19	17
#6	74	2	14	10

It can be seen in Table 1 and Table 2, the percentage of $Nd(OH)_3$ are quite large due to the absorption of hydrogen and oxygen by neodymium [15]. Its means to produce high quality of $NdFeO_3$ oxide alloy the temperature of calcination and sintering process for synthesis should be higher than 950 °C or increasing the heating time that the reaction of the raw material occurs perfectly and containing lattice -OH has missing.

4. Conclusions

The molar ratio greatly affects the crystal structure of $Nd_{1+x}FeO_3$ oxide alloy, which is indicated intensity differences, FWHM value, and crystal size. The XRD analyzed was obtained FWHM value is 0.11, 0.13, 0.12, 0.19, 0.19 and 0.12, and crystalline size value of $NdFeO_3$ are (294.86 ± 0.01) nm, (248.30 ± 0.01) nm, (288.17 ± 0.01) nm, (163.97 ± 0.01) nm, (167.21 ± 0.01) nm and (255.47 ± 0.01) nm for sample #1, #2, #3, #4, #5 and #6, respectively. There are four phases that contain each sample:

$NdFeO_3$, $Nd(OH)_3$, Fe_2O_3 and Nd_2O_3. The highest intensity of $NdFeO_3$ giving information that samples are a polycrystalline material with dominant phase formed is orthorhombic correspond to *hkl* (121).

Acknowledgements
This research was funded by Directorate Research and Community Services, Directorate General of Research and Development, Ministry of Research, Technology, and Higher Education, Republic of Indonesia, under research scheme of *Hibah Kompetensi* fiscal year 2016/2017.

References

[1] Masoud Salavati-Niasari, Fatemeh Davar and Maryam Shaterian 2008 Preparation of cobalt nanoparticles from [bis(salicylidene)cobalt(II)]–oleylamine complex by thermal decomposition *Journal of Magnetism and Magnetic Materials* **320** 3-4 pp 575–578
[2] Siegel R. W, Nastasi M, Parkin D. M and Gleiter H. 1993 Synthesis and properties of nanophase materials *Materials Science and Engineering*. **168** (2) pp 189–197
[3] Siegel R W 1994 Nanostructured materials –mind over matter *Nanostructured Materials* **4** (1) pp 121–138
[4] Pena M A and Fierro J L G 2001 *Chem. Rev* **101** (7) pp 1981–2018
[5] Truong G H *et al* 2011 *Adv. Nat. Sci: Nanosci. Nantotechnol* **2** p 015012
[6] Anhua W *et al* 2009 Preparation of $ReFeO_3$ nanocrystalline powders by auto-combustion of citric acid geL *Asia-Pac. J. Che. Eng*. **4** pp 518–521.
[7] Yabin W *et al* 2011 Growth rate dependence of the NdFeO3 single crystal grown by float-zone technique *Journal of Crystal Growth* **318** pp 927–931
[8] Truong Giang Ho 2011 Nanosized perovskite oxide NdFeO3 as material for a carbon-monoxide catalytic gas sensor *Advances In Natural Sciences: Nanoscience And Nanotechnology* **2** p 015012
[9] Xiangdong Lou, X Jia, and J Xu 2005 *Journal of Rare Earths* **23** (3) pp 328–331
[10] Sujiono E H 2017 ID Patent No. P00200800471 Reg. No. 44179
[11] Sujiono E H *et al* 2001 Crystal Structure and Morphology Analysis of $Nd_{1+x}Ba_{2-x}Cu_3O_7$ Oxide Alloy Surface Developed by Solid State Reaction Method *Physica Status Solidi (A) Applied Research* **187** pp 471–479.
[12] Sujiono E H, Arifin P and Barmawi M 2002 $YBa_2Cu_3O_{7-\delta}$ thin films deposited by a vertical MOCVD reactor *Materials Chemistry and Physics* **73** pp 47–50.
[13] V. Zharvan *et al* 2017 The Effect of Molar Ratio on Crystal Structure and Morphology of $Nd_{1+x}FeO_3$ (X=0.1, 0.2, and 0.3) Oxide Alloy Material Synthesized by Solid State Reaction Method *IOP Conference: Materials Science and Engineering* **202** pp 012072
[14] Sujiono E H *et al* 2011 *Jurnal Teknologi dan Aplikasi* (ITS*)* **7** pp 166–173
[15] Pedro V S *et al* 2016 *Materials Research* **19 (2)** pp 389–393
[16] Khorasani-Motlagh M *et al* 2013 Chemical Synthesis and Characterization of Perovskite NdfeO3 Nanocrystals via a Co-Precipitation Method *Int. J. Nanosci. Nanotechnol* **9 (1)** pp 7–14
[17] Xinshu N *et al* 2003 *Journal of Rare Earth* **21** (6) pp 630–632
[18] Ru Zhang *et al* 2010 Electrical and CO-sensing properties of $NdFe_{1-x}Co_xO_3$ perovskite system *Journal of Rare Earths* **28** (4) pp 591–595
[19] Shujuan Y *et al* 2011 The magnetic properties **and** specific heat of $NdFeO_3$ single crystal were systematically studied in the temperature range from 2 to 300 K *Journal of Applied Physics* **109** p 07E141
[20] Zhang et al 2009 Lanthanide hydroxide nanorods and their thermal decomposition to lanthanide oxide nanorods *Materials chemistry and physics* **114** (1) pp 160-167

Influence of PT Vale's Granulated Gradation Nickel Slag Aggregate on Compressive Strength of Concrete

E. H. Sujiono[1], V. Zharvan[1], S. N. Yunita[1] and S. Samnur[1]

[1]Laboratory of Materials Physics, Department of Physics, *Universitas Negeri Makassar*, Makassar 90224, Indonesia

E-mail: e.h.sujiono@unm.ac.id

Abstract. Influence of granulated gradation nickel slag aggregate on compressive strength of concrete has been investigated. In this study, the influence of granulated gradation nickel slag aggregate with aggregate size LS 9.50 mm, LS 5.60 mm, LS 2.36 mm, LS 2.00 mm and the combination of those fourth aggregate size on compressive strength and microstructure of concrete has reported. The concrete was prepared according to SNI 03-1972-1990. Samples then characterized using universal testing machine controller respect to ASTM C39M-01 method to obtain the compressive strength value and scanning electron microscope to obtain morphology of all samples. Results show that sample with nickel slag aggregate size at 5.60 mm has highest compressive strength value at (44.05 ± 1.93) MPa. Characterization of microstructure shows a good interface between aggregate and cement paste. The morphology after compressive testing shows that cracks were occurred in the cement paste, not between aggregate and paste interface. It can be concluded that the aggregate size of nickel slag was directly proportional to the compressive strength of the concrete depending on the bond between cement paste and the size of nickel slag aggregate.

Keywords. Aggregate, compressive strength, concrete, morphology, and nickel slag.

1. Introduction

Concrete is a composite material made of aggregate that bonded together by liquid cement. Concrete can be visualized as a multi-phase composite material that made up of three phases: mortar, mortar/aggregate interface and the coarse aggregate phase [1, 2, 3]. Its major components are cement, water, and aggregates with 50 % to 60 % of the total volume depending on the mix proportion [2]. The coarse aggregate in normal concrete is from rock fragments that characterized by high strength [3]. The larger percentage of coarse aggregate in the concrete mix makes it contribute to the strength of concrete [4].

PT Vale Indonesia produces about 3000 t of nickel slag in solid form every week [5]. On the other hand, those waste materials generated a big problem to the environment. Slag is a waste produced during manufacturing of pig iron and steel [6]. Previous research showed that chemical composition of nickel slag contains majority oxides such as SiO_2 38.85 %, MgO 23.60 %, FeO 29.75 % and Al_2O_3 4.38 % in wt%. Besides, concrete with 100 % aggregate of nickel slag has compressive strength value 36.7 MPa [7]. According to that fact, it can be concluded that concrete with nickel slag as an aggregate can be used in industry.

Content from this work may be used under the terms of the Creative Commons Attribution 3.0 licence. Any further distribution of this work must maintain attribution to the author(s) and the title of the work, journal citation and DOI.
Published under licence by IOP Publishing Ltd

In this paper, the obtain influence of aggregate size on concrete that built from Portland cement, sand and nickel slag from PT. Vale as aggregate is presented. Furthermore, analyses of mechanical properties and morphology of samples were reported.

2. Materials and methods

The raw materials including Portland's cement, sand, water and nickel slag (grain size: 9.50 mm) were mixed together according to SNI 04-1972-1990 composition for making high-quality K-350 concrete (compressive strength value: 31.20 MPa). The paste then molds into three cubes with dimension (5 × 5 × 5) cm^3 is shown in Figure 1 and let them dried. All of the concrete then cured by dipping into the water for seven days according to SNI 03-2847-2002. The same procedure was repeated by varying the aggregate size of nickel slag at 5.60 nm, 2.36 nm and 2.00 mm also the combination of those fourth aggregate size (9.50 mm, 5.60 mm, 2.36 mm and 2.00 mm).The compressive strength test of concrete was carried out according to ASTM C39M-01 using universal testing machine controller (*Testometric*) and then characterized using SEM-EDS (*Tescan Vega3*) to investigate surface morphology and the interface zone.

Figure 1. Concrete with nickel slag (aggregate size:9.50 mm) as aggregate.

3. Results and discussion

3.1. Compressive strength of concrete

The result of the compressive strength of concrete is shown in Figure 2. Based on Figure 2, it can be seen that compressive strength of concrete linearly increase as increasing of aggregate size and reach a maximum value of compressive strength, (44.05 ± 1.93) MPa, at concrete with aggregate size 5.60 mm. These increases could be attributed to the differences in the wetted surface in concrete batches. The smaller aggregates in concrete have a larger wetted area than larger aggregates, and when concrete in the curing process it leaves pores when microcracks start that makes low compressive strength in concrete with smaller sized aggregates [1]. Furthermore, the strong bond between cement paste and aggregate makes this material to maintain fracture until to maximum limit before the nickel aggregate crushed [8]. Table 1 shows the values of compressive strength for each sample.

Table 1. Compressive strength values of concrete

Concrete Sample	Compressive Strength (MPa)
LS 2.00	19.23 ± 0.37
LS 2.36	22.01 ± 1.32
LS 5.60	44.05 ± 1.93
LS 9.50	33.72 ± 1.77
MIX	37.80 ± 1.10

The compressive strength then decreases when the size of the aggregate increase. Concrete that has a mix of aggregate size has the highest compressive strength value at (37.80 ± 1.10) MPa compare with high-quality K-350 concrete with compressive strength value 31.20 MPa. Vilane et al. also report the effect of aggregate size on the compressive strength of concrete and he found that aggregate at size 9.50 mm has compressive value at 15.34 MPa [1] which is lower than our result (33.72 ± 1.77) MPa due to the mechanical characteristic of nickel slag aggregate [7].

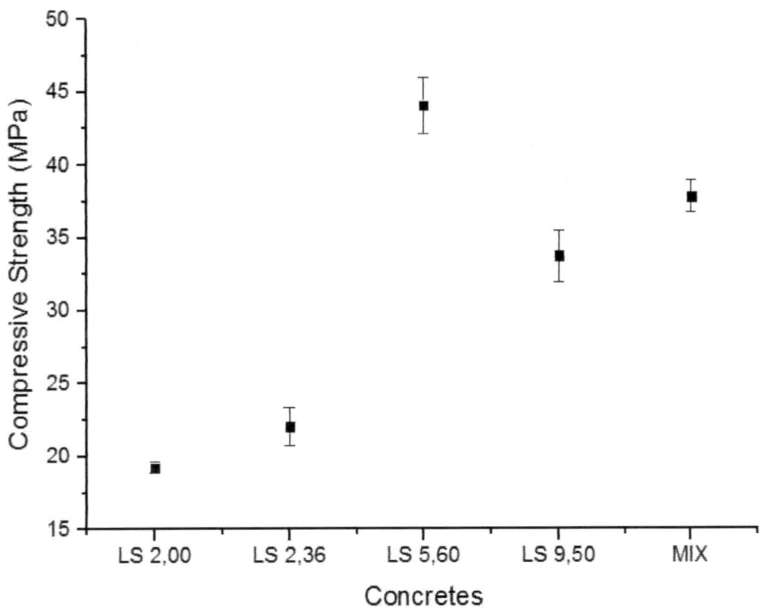

Figure 2. The compressive strength of concrete as a variation of nickel slag aggregate size.

Increasing of compressive strength is attributed to large interface distance among particles and when the addition of aggregate size increase leading to the formation a brittle material that it stables to microcracks, hence the fracture of material was occurred [9]. Besides, increasing the aggregate size makes the bond between matrix-aggregate became weaker that makes the value of compressive strength decrease [7].

3.2. Morphology of concrete

Morphology of concrete before and after the compressive test can be seen in Figure 3. Figure 3(a) shows the morphology of concrete before compressive strength test. It gives information that there is a good bond between cement paste and aggregate of nickel slag. Also, there was no void due to the presence of high SiO_2 on nickel slag. It also shows the solid-smooth surface indicated as nickel slag aggregate [5].

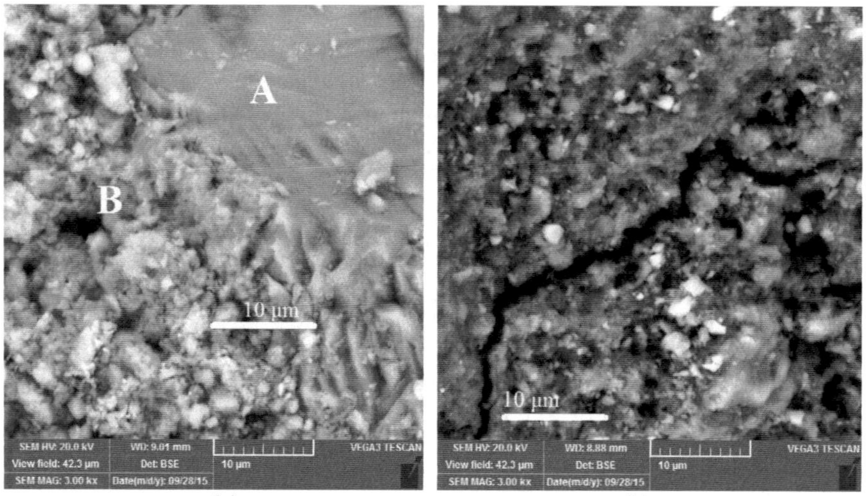

(a) (b)

Figure 3. SEM micrograph of concrete (a) before and (b) after compressive strength test (A= nickel slag, B= cement paste).

Morphology of concrete after compressive strength test was shown in Figure 3(b). It can be seen that there was a crack, about 1 μm width, present on the concrete due to the effect of compressive testing. Figure 3(b) also gives information that crack did not present in the interface zone between matrix (cement paste) and aggregate (nickel slag) but in the matrix only.

The result of EDS is shown in Table 2. It gives information that SiO_2 has a high percentage of the concrete about 35.74 %, then CaO 32.12 % and Al_2O_3 16.62 %. The highest percentage of SiO_2 on the concrete due to present of nickel slag aggregate [10] that makes a strong bond between nickel slag aggregate and Portland cement [8].

Table 2. EDS results of concrete

Compound	Percentage (wt %)
SiO_2	35.74
CaO	32.12
Al_2O_3	16.62
FeO	4.88
MgO	3.66
Na_2O	3.30
K_2O	1.26
TiO_2	0.79
Cr_2O_3	0.35
MnO	0.24
P_2O_5	0.16

4. Conclusions

The influence of nickel slag grain size aggregate on compressive strength and morphology of concrete has been investigated. It was found that concrete has highest compressive strength value at 44.05 MPa with aggregate size 5.60 mm. Morphology of concrete shows that the interface zone between mortar and slag nickel aggregate has a good bond due to the presence of silica (SiO_2) in nickel slag that binding with cement paste.

Acknowledgments
This research was fully funded by Directorate General of Higher Education, Ministry of Research, Technology and Higher Education, through Priority Research Scheme (MP3EI) fiscal year 2013/2015.

References

[1] Vilane B R T and Sabelo N 2016 *Journal of Agricultural Science and Engineering* **2** 66 – 69
[2] Beshr H, Almussalam A A, and Maslehuddin M 2003 Effect of coarse aggregate quality on the mechanical properties of high strength concrete *Construction and Building Materials* **17** 97 – 103
[3] Abdullahi M 2012 Effect of Aggregate Type on Compressive Strength of Concrete *International Journal of Civil and Structural Engineering* **2** 791 – 800
[4] Ajamu S O and Ige J A 2015 Effect of coarse aggregate size on the compressive strength and the flexural strength of concrete beam *International Journal of Engineering Research and Applications* **5** 67 – 75
[5] Sugiri S 2005 *J. Infras. Built Environ* **1**
[6] Singh G, Sangwan S, and Usman M 2015 Experimental study of blast furnace slag concrete *International Journal of Engineering Sciences and Research Technology* 475 – 480
[7] Sujiono E H *et al* 2016 The influence of nickel slag aggregate concentration to compressive and flexural strength on fly ash-based geopolymer composite *AIP Conf. Proc* **1725** 020083
[8] Sujiono E H *et al* 2017 Deformation Pattern of Nickel Slag Bonding on the Development of Concrete Construction *IOP Conf. Series: Journal of Physics: Conf. Series* **846** 012017
[9] Radhi S H, Mohammed M R, and Hadi A N 2013 *International Journal of Advanced Research* **1** 372 – 374
[10] Samnur S *et al* 2016 Study on physical-chemical properties of furnace-nickel-slag powder for geopolymer application *Jurnal Pendidikan Fisika Indonesia* **12** 177 – 182

IOP Conf. Series: Materials Science and Engineering **367** (2018) 012058 doi:10.1088/1757-899X/367/1/012058

Analysis of AISI 304 Tensile Strength as an Anchor Chain of Mooring System

I. Hamidah[1], R. Wati[1], and R. A. Hamdani[1]

[1]Department of Mechanical Engineering Education, Universitas Pendidikan Indonesia, Jl. Dr. Setiabudhi No.229 Bandung, Indonesia, 40154

E-mail: idahamidah@upi.edu

Abstract. The background of this research is the use of mild steel (i.e., St37) as anchor chain that works on the corrosive environment of seawater which is possible to decrease its tensile strength. The longer soaked in seawater, the more significant the lowering of its tensile strength. Anchor chain needs to be designed by considering its tensile strength and corrosion resistance, so it's able to support mooring system well. The primary purpose of this research is obtaining the decreasing of stainless steel 304 (AISI 304) tensile strength which is corroded by seawater as anchor chain of the mooring system. It is also essential to obtain the lifetime of AISI304 and St37 as anchor chain with the same load, the corrosion rate of AISI 304, and St 37 in seawater. The method which was employed in this research is an experiment with four pieces of stainless steel AISI 304, and of St 37 corrosion testing samples, six pieces of stainless steel 304, and six pieces of St 37 for tensile testing samples. The result of this research shows that seawater caused stainless steel AISI 304 as anchor chain has decreased of tensile strength about 1.68 % during four weeks. Also, it indicates that AISI 304 as anchor chain has a lifetime about 130 times longer than St 37. Further, we found that the corrosion rate of stainless steel 304 in seawater is 0.2042 mpy in outstanding category, while the St 37 samples reached up to 27.0247 mpy ranked as fair category. This result recommends that AISI 304 more excellence than St 37 as anchor chain of the mooring system.

Keywords. AISI 304, corrosion rate, St 37 and tensile strength.

1. Introduction

Indonesia's cruise accident tends to increase for every year. During 2010 – 2016, it had occurred 54 cruise accidents that caused 337 people died or missed, and 474 people injured. Cruise accident can be drowning, burning or exploding, crashing, and grounding [1]. This matter needs to get attention about the importance of designing of an anchor chain for supporting mooring system safety. Anchor chain is a tool that is used between anchor and ship so the movement of the ship can be decreased. Anchor chain has a vital role in mooring system. The performance of mooring system depends on the strength of anchor grip in the seabed, and the strength of anchor chain to hold the ship speed.

Anchor chain is usually designed from mild steel with tensile strength about (37 to 55) $kg \cdot mm^{-2}$ and strain about (20 to 30) %. St37 contains carbon of 0.17 %, it has tensile strength of (36.7 to 52) kg/mm^2 and strain of (16 to 24) % [2]. Corrosion testing of carbon steel in seawater shows that corrosion resistance of carbon steel is very low [3]. Compared to the AISI 304, the corrosion resistance of St 37 is lower than corrosion resistance of AISI 304. The AISI 304 consist of (18 to 20)

Content from this work may be used under the terms of the Creative Commons Attribution 3.0 licence. Any further distribution of this work must maintain attribution to the author(s) and the title of the work, journal citation and DOI.
Published under licence by IOP Publishing Ltd

IOP Conf. Series: Materials Science and Engineering **367** (2018) 012058 doi:10.1088/1757-899X/367/1/012058

% chromium [4]. Corrosion rate testing of AISI 304 in corrosive solutions such as NaOH, KOH, and NaCl shows the corrosion resistance of (0 to 0.0234) mpy in the excellent category [5]. AISI 304 testing of welded strength in seawater shows that it has occurred decreasing of tensile strength about 2.9 % [6].

Decreasing of anchor chain's tensile strength may cause the breaking of anchor chain while anchoring. The breaking of anchor chain may cause ship can't be anchoring, or mooring system is failing. The breaking of anchor chain also caused economically disadvantage because when anchor chain is breaking, then anchor will be drowned through the anchor chain. If anchor chain is well designed, the mooring system can be supported well too. Anchor chain is designed for considering the loads, safety factor, corrosion resistance, and allowed of tensile strength. One of the purposes of this research is obtaining the strengths of St 37 and AISI 304. By identifying the strengths ratio of St 37 and AISI 304, it could be expected that AISI 304 can be used to replace St 37 as an anchor chain.

2. Materials and methods

ST 37 and AISI 304 were used as samples. There are two methods to characterize the samples: mass loss method and tensile testing. Mass loss method was used to measure the corrosion rate, and tensile testing was used to measure the strengths of ST 37 and AISI 304. Detailed information on the parameters used for testing the tensile strength of St 37 AISI 304 is shown in Table 1. The data of corroded tensile testing samples were compared with tensile testing of non-corroded samples. Corrosion testing data were used to determine the corrosion rate and its lifetime. Meanwhile, tensile testing data was used to analyze the decreasing of tensile strength.

Table 1. Parameters used in the testing of St 37 and AISI 304 tensile strength in seawater

Condition	Value
Exposure times	168 h to 672 h
Temperature	25 °C
Relative humidity	72 %
Air pressure	960 mbar

3. Results and discussion

The corrosion rate of AISI 304 in seawater is about (0.1804 to 0.2491) mpy (see Table 2). The average corrosion rate of AISI 304 is about 0.2042 mpy or 0.000518 mmpy. The corrosion resistance of AISI 304 in seawater belongs to the outstanding category. This result is in line with the other study [7]. The existence of (18 to 20) % chromium, caused AISI 304 resists to the corrosive environment [4]. The chromium in AISI 304 can bind spontaneously with oxygen, and it forms a passive film that can protect AISI 304 from the environment [8].

Table 2. Corrosion rates of AISI 304 on seawater

Sample	Time (hour)	Surface Area (in^2)	Weight Loss (mg)	Corrosion Rate	
				mpy	mm/y
#A1	168	0.8123	0.4	0.1949	0.00050
#A2	336	0.7679	0.7	0.1804	0.00046
#A3	504	0.8909	1.3	0.1925	0.00049
#A4	672	0.7944	2.0	0.2491	0.00063

As seen in Table 3, St 37 corrosion rates in seawater is about (24.3341 to 31.8751) mpy. The average corrosion rate of St 37 is 27.0247 mpy or 0.068642 mmpy. For this value, the corrosion resistance of St 37 belongs to a fair category, which caused by the 0.17 % carbon content in St 37. The carbon content of St 37 is 47.06 % bigger than AISI 304 [2]. The role of carbon in the material is to

IOP Conf. Series: Materials Science and Engineering **367** (2018) 012058 doi:10.1088/1757-899X/367/1/012058

increase the mechanic strength, but it can decrease the corrosion resistance [9]. Different results about the comparison of corrosion rates of St 37 and AISI 304 were obtained when both materials were immersed in molten zinc bath [10]. Available high carbon in steel resulted in forming thick layers on the surface of St 37 and reducing the corrosion. In addition, St 37 has no chromium content so it cannot create a passive film to protect the material from the corrosive environment. High carbon content and no chromium content in St 37 caused corrosion resistance of this material lower than AISI 304. From the other study, the calcium carbonate precipitate observed after immersion in seawater probably cover the surface of material more effectively and may, therefore, be a better barrier against from corrosive environment [11].

Table 3. Corrosion rates of St 37 on seawater

Sample	Time (hour)	Surface Area (in^2)	Weight Loss (mg)	Corrosion Rate	
				mpy	mm/y
#B1	168	0.7990	62.9	31.8751	0.08096
#B2	336	0.7885	104.8	26.9081	0.06835
#B3	504	0.7921	146.6	24.9817	0.06345
B4	672	0.7921	190.4	24.3341	0.06181

The tensile strength of St 37 and AISI 304 is shown in Table 4. It was observed that AISI 304 that corroded in seawater has decreasing of tensile strength about 1.68 %. Meanwhile, St 37 has reduced of tensile strength about 17.79 %. The lowering of tensile strength in AISI 304 is lower than St 37, so AISI 304 will has a lifetime longer than St 37. The difference of the decreasing of tensile strength is affected by variation of material corrosion rate. Furthermore, the tensile strength reduced is directly proportional to corrosion rate [12].

Table 4. The tensile strength of St 37 and AISI 304

Material	Not Corroded σ_{max} (kg/mm^2)	Corroded in Seawater σ_{max} (kg/mm^2)	Difference (%)
St 37	32.83	26.93	17.97
AISI 304	68.20	67.05	1.68

The function of anchor chain in many types of the ship is to hold the movement of the ship. Anchor chain is made with particular specification considering the prevailing standard. The scales of tensile strength that has to be held by anchor chain are based on the type of material and its diameter. Based on carrying loads and breaking loads data issued by Bureau Veritas, one of the anchor chains has to carry the 51 kN maximum load.

The force that works in anchor chain is a shear force; then it needs to be determined the maximum shear strength that can be received by each material. Maximum shear strength can be determined by [13]:

$$t_{max} = 0.80\sigma_{max} \tag{1}$$

The allowed of shear strength has to be determined to design the dimensions of the anchor chain. The allowing of shear strength can be determined using Equation 2 [13]:

$$\tau_{allow} = \frac{t\max}{sf} \tag{2}$$

if breaking load of anchor chain is 51 kN or 5198.77 kg, then the diameter of chain can be determined using Equation 3 [13]:

$$A = \frac{F}{\tau_{allow}} \tag{3}$$

By using the Equation 1 to Equation 3, the maximum shear strength, the permission of shear strength, and the chain diameter of St 37 and AISI 304, can be shown in Table 5.

Table 5. The tensile strength of St 37 and AISI 304

Material	σ_{max} (kg/mm^2)	τ_{allow} (kg/mm^2)	diameter (mm)
St 37	26.26	15.72	20.50
AISI 304	54.56	32.67	11.02

From the calculation above, it is known that to hold the loads about 51 kN, it is needed minimum diameter about 20.50 mm for St 37 and 11.02 mm for AISI 304. The diameter of AISI 304 is 53.75 % smaller than the diameter of St 37 for same loads. Anchor chain is designed to hold certain load with certain lifetime. To design the lifetime, it needs a corrosion allowance that compensates for corroded material. If the corrosion allowance is 0.05 mm of each material, then the lifetime of each material as anchor chain can be determined by Equation 4 [14]:

$$CA_{AISI304} = CR \times LT \times 2 \quad (4)$$

where CR is corrosion rate, LT is a lifetime.

The lifetime can be obtained by using Equation (4) based on the data of corrosion rate of Table 2 and Table 3. The tensile strength and the lifetime of both samples are depicted in Table 6.

Table 6. The tensile strength of St 37 and AISI 304

Material	CR (mmpy)	LT (year)
St 37	0.067356	15.72
AISI 304	0.000518	0.37

It is observed that lifetime of AISI 304 is longer about 130.43 times than St 37. Therefore, AISI 304 exhibits an excellent performance compared to St 37 as anchor chain. Lifetime prediction of steel structure in the marine environment is a challenging task for corrosion engineers either for existing or new structures to be constructed [15].

4. Conclusions

Seawater caused AISI 304 corroded about 0.2042 mpy for four weeks, and it creates the decreasing of its tensile strength of about 1.68 %. The lifetime of AISI 304 is nearly 130 times longer than St 37. The corrosion rate of AISI 304 in seawater is 0.2042 mpy belongs to the outstanding category. Meanwhile, the corrosion rate of St 37 in seawater is 27.0247 mpy belongs to the good category. It suggests that AISI 304 may be considered as an anchor chain to replace St37 for some reasons.

Acknowledgements

Authors are grateful for the financial support provided by Directorate General of Higher Education, Ministry of Research, Technology, and Higher Education, Republic of Indonesia through Grant of Penelitian Unggulan Perguruan Tinggi.

References

[1] Komite Nasional Keselamatan Transportasi (KNKT) 2016 *Data Investigasi Kecelakaan Pelayaran Tahun 2010 – 2016* (Jakarta : KNKT)
[2] Bebon *Carbon and Low Alloy Steel EN10025 ST37–2 Steel Plate* (Bebon, China)
[3] Pattireuw K J and Rauf F A 2013 *Lumintang RCA Jurnal Online Poros Teknik Mesin* **2** 1 – 10
[4] Steel AK 2007 *Poduct Data Sheet 304/304L Stainless Steel* (West Chester, AK Steel)
[5] Hamidah I, Solehudin A, and Setiawan A 2016 *ARPN J. of En.g and Appl. Sci.* **11** 972 – 977
[6] Nindhia TGT 2015 *Jurnal Energi dan Manufaktur* **7** 155 – 158
[7] Syed S 2012 The effect of atmospheric pollution on materials damage *Materials and Corrosion* **63** 1 – 12

[8] ASM International 2005 *ASM Handbook Volume 1, Properties and Selection: Irons, Steels, and High–Performance Alloy* (ASM International, USA)
[9] Outokompu 2013 *Handbook of Stainless Steel, Stainless Steel Categories, and Grades* (Outokompu, Sweden)
[10] Saeed B and Ahmad S 2013 *Advanced Process in Materials* **7** 55–63
[11] Moller H, Boshoff ET, and Froneman H 2006 *The J. of South African Institute of Mining and Metallurgy* **106** 585–592
[12] Ismail A and Adan NH 2014 *Am. J. of Eng Research* **3** 64–67
[13] Norton R L 2006 *Machine Design New Jersey* (Pearson Prentice Hall)
[14] Davis J R 2000 *Corrosion Understanding the Basics USA* (ASM International)
[15] Paul S 2010 *Canadian Metallurgical Quarterly* **49** 1–8

IOP Conf. Series: Materials Science and Engineering **367** (2018) 012059 doi:10.1088/1757-899X/367/1/012059

Preliminary Study: Local Iron Sand Characterization of Titanomagnetite Type

Y E Gunanto[1], Maya P Izaak[1], L Cahyadi[1], E Jobiliong[1], and W A Adi[2]

[1]Department of Physics Education, Faculty of Education, University of Pelita Harapan, Karawaci, Tangerang 15811, Indonesia
[2]Center for Science and Technology of Advanced Materials, BATAN, Tangerang Selatan 15314, Indonesia

E-mail: yohanes.gunanto@uph.edu

Abstract. In this study, we have characterized the natural iron sand type of titanomagnetite taken from the coast of West Lampung. Prior to characterization, the natural iron sand type of titanomagnetite was milling with high energy milling (HEM) for 10 h. Scanning electron microscopy (SEM) and energy dispersive spectroscopy (EDS) were used to characterize surface morphology and analysis of constituent elements of iron sands respectively. X-ray diffraction (XRD) was used for phase analysis and crystal structure. The morphology of the surface of the natural iron sand type of titanomagnetite had a variety of shapes and sizes of particles and contrasting colors. This is consistent with the result of characterization using energy dispersive spectroscopy where the element of iron sand consists of Fe (66.57 %), Mg (1.37 %), Si (0.57 %), Al (1.66 %), Ti (8.03 %), La (1.17 %) and O (20.64 %). These constituent elements did not stand on their own but bond to each other forming certain phases. Analysis by X-ray diffraction is generated that the phases formed on the iron sand of types titanomagnetite, consist of ilmenite ($FeTiO_3$), hematite (α-Fe_2O_3) and titanomagnetite ($Fe_{3-x}(MT,Ti)_xO_4$) phases with the content respective of 9.63 %, 19.09 % and 71.28 %, where MT = metal transition.

Keywords. Hematite, iron sand, ilmenite, and titanomagnetite.

1. Introduction

In this competitive globalization era, Indonesia has been intensively developing infrastructure in order to improve the economic level and to be a self-reliant country that does not rely on imports. Thus, supporting materials such as steel, cement and others are needed to develop the in infrastructure. This has an impact on the increasing need for basic materials in the form of iron sand as the base material for making steel or cement. The year 2025 is estimated to meet the steel industry required about 250 million tons of iron ore and 110 million tons of iron sand [1].

In Indonesia, natural iron sand resources totaling more than 2 billion tons are spread along the southern coast of Java, Sumatra, West Nusa Tenggara, and so on with the main content of iron oxide, titania, silica, and alumina [2]. Indonesia has great iron mineral resources, comprising primary iron ore (17 %), iron sand (8 %) and lateritic iron ore (75 %) [3]. Materials contained in the iron sand are magnetic and non-magnetic. Magnetic materials in the iron sand are drawn strongly by magnets, and

Content from this work may be used under the terms of the Creative Commons Attribution 3.0 licence. Any further distribution of this work must maintain attribution to the author(s) and the title of the work, journal citation and DOI.
Published under licence by IOP Publishing Ltd

some are weakly pulled. Strongly withdrawn consisted of titanomagnetite (Ti ~ 10%) and hematite (Fe:Ti ~ 1:1) content, while weakly pulled, the Ti content reached nearly 50 %.

In this study, natural iron sand of titanomagnetite type is produced from natural iron sand by using a separator of a weak permanent magnet around 1000 Oe. This stage is an important parameter that not conducted by previous researchers due to they still used the magnetic separator with magnet field high so that not obtained the titanomagnetite type perfectly. Thus only the iron sand of titanomagnetite type will attach to the permanent magnet. Some characterization will be done such as scanning electron microscope (SEM) is used to analyze surface morphology. To analyze the elements contained in the iron sand of titanomagnetite type used energy dispersive spectroscopy (EDS) and to analyze the phases use X-ray diffractometer (XRD).

2. Materials and methods

Iron sand was retrieved from Lampung, Indonesia. 100 g of iron sand was prepared and separated using magnetic separator with a low magnetic field about 1000 Gs to eliminate major impurity, so that obtained iron sand with Fe rich or called the type of titanomagnetite. The iron sand powders of titanomagnetite type were milled by using high-energy milling (HEM) type PW1000 in for 10 h at room temperature to obtain fine particles to facilitate the characterization and further processing. The analysis of phase qualitative and quantitative was carried out using X-ray diffractometer equipped (Pan Analytical) with a tube provided with the copper anode (CuKα). The Rietveld analysis was done by applying for GSAS program with the pseudo-Voigt function used to approximate of diffraction line profiles. The elemental composition analysis and surface morphology were observed by using dispersive spectroscopy (EDS) and scanning electron microscope (SEM), respectively

3. Results and discussion

Figure 1 shows the observed microstructure on iron sands of titanomagnetite type using scanning electron microscope. The surface morphological observations showed that the sample had a variety of shapes, particle size, and color contrast. This indicated that the suspected sample had a very diverse phase. The average size of the grains is between (5 to 10) μm. These results were much smaller than the results obtained Rusianto *et al.*, which is about 212 μm [4]. However, a photograph of this surface morphology cannot determine the phases contained in this titanomagnetite sandstone iron sample. Thus further analysis is needed to show that this type of titanomagnetite iron sandstone has a diverse phase, using x-ray diffraction analysis. To analyze the phases with x-ray diffraction, it is necessary to support the elementary data of this titanomagnetite iron sandstone by using energy dispersive spectroscopy (EDS).

Figure 2 shows the results of elementary analysis using energy dispersive spectroscopy on titanomagnetite iron sand rocks. The result of energy spectrum showed that the most dominant elements were oxygen (O), titanium (Ti) and iron (Fe) which were consecutively at 0.525 keV, 4.508 keV and 6.398 keV with wavelength Kα. This meant that these rocks contain iron-rich elements (Fe), but there were still some other elements contained in it in a relatively small amount. The elements contained in this titanomagnetite iron sandstone rock is presented in detail in Table 1.

Table 1 shows that the total iron content of iron sandstone of this type of titanomagnetite is 66.57 %. These results indicate that the iron content in the sand sandstone type titanomagnetite Lampung is more potential and economical to be further processed. The element content in this titanomagnetite iron sandstone is very complex. Some of the elements contained in the iron sand of titanomagnetite type did not stand alone most likely instead they bind to each other to form certain minerals according to the largest content of this iron sand of titanomagnetite type. Thus a structural analysis is needed to identify the phases present in the iron sand of titanomagnetite type. Mufti *et al.* obtain Fe content in the natural iron sand about 41.2 % before being extracted with a magnet [5].

IOP Conf. Series: Materials Science and Engineering **367** (2018) 012059 doi:10.1088/1757-899X/367/1/012059

Figure 1. The surface morphology of iron sandstone samples of the titanomagnetite type

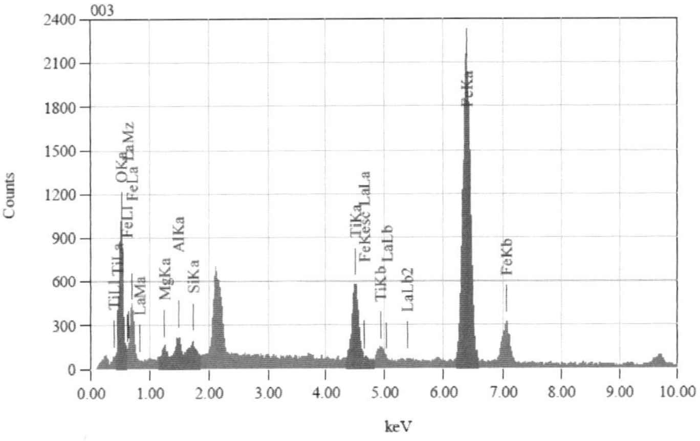

Figure 2. elementary analysis using energy dispersive spectroscopy

Table 1. The result of elementary analysis using energy dispersive spectroscopy

No.	Element	Content (% weight)
1.	Iron (Fe)	66.57 ± 0.35
2.	Magnesium (Mg)	1.37 ± 0.22
3.	Silicon (Si)	0.57 ± 0.16
4.	Aluminum (Al)	1.66 ± 0.18
5.	Titanium (Ti)	8.03 ± 0.22
6.	Lanthanum (La)	1.17 ± 0.61
7.	Oxygen (O)	20.64 ± 0.21
8.	Calcium (Ca)	ND
9.	Chromium (Cr)	ND
10.	Vanadium (V)	ND

*) ND: not detected and limit detection around 1 wt.%

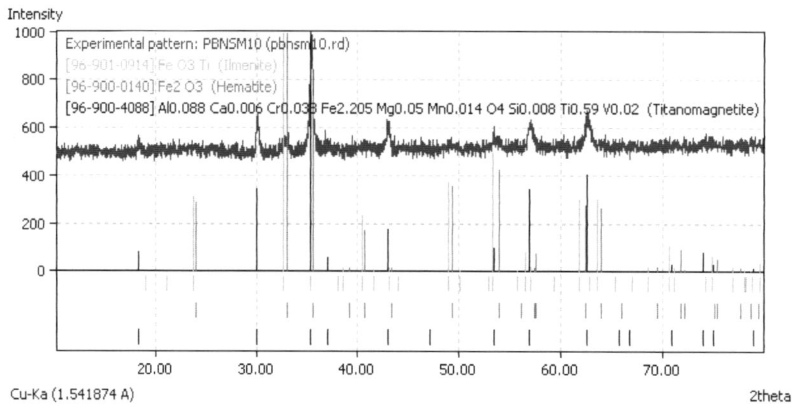

Figure 3. X-ray diffraction pattern on the iron sand of titanomagnetite type samples

Figure 3 shows the results of the measurement and identification of x-ray diffraction patterns of titanomagnetite iron sandstone samples. This revealed that there are formed Bragg diffraction peaks which were suspected to be multi-phase. The phase identification refers to the results of the Yamanaka study, ICDD 96-901-0914 [6], Blake, 96-900-0140 [7], Stout, ICDD 96-900-4088 [8], which are the ilmenite, hematite, and titanomagnetite. Thus, based on the results of the analysis of the X-ray diffraction pattern shows that the amount of mineral phase content contained in this iron sand of titanomagnetite type consist of ilmenite ($FeTiO_3$), hematite (α-Fe_2O_3) and titanomagnetite ($Fe_{3-x}(MT,Ti)_xO_4$) phases with the content respective of 9.63 %, 19.09 % and 71.28 %, where MT (metal transition) as shown in Table 2.

Based on the results of x-ray diffraction analysis in Table 2, then the next steps can be done for processing of iron sandstone of this type of titanomagnetite. For further research, it will focus on a local natural resource that can be used as a permanent magnetic material ingredient. And based on the initial identification of mineral content in this titanomagnetite iron sandstone, the most efficient use is processing for purification into the Fe_3O_4 raw material to reach more than 99 % Fe_3O_4. The Fe_3O_4 material can be used as a permanent ferrite based magnet material.

In general, mineral analysis results can be classified into two parts, namely magnetic materials and non-magnetic materials. Minerals classified as magnetic materials are maghemite minerals, titanomagnetite iron sands, and goethite. Meanwhile, the minerals belonging to non-magnetic materials are minerals staurolite, quartz, and lime. However, to get the raw material of Fe_3O_4 up to reach more than 99 % level, the extraction process of iron sandstone of this type of titanomagnetite should be made so that the phases other than iron can be separated and expected to be obtained by a material with relatively high iron content. The results obtained by Jalil *et al.*, Fe_3O_4 content in iron sand taken from the coast of Syiah Kuala about 85 % [9].

Table 2. The mass fraction of the mineral phases contained in the titanomagnetite iron sandstone rock

No.	Mineral	Empirical Formula	Mass Fraction (%)
1.	Ilmenite	$FeTiO_3$	9.63 ± 0.004
2.	Hematite	α-Fe_2O_3	19.09 ± 0.005
3.	Titanomagnetite	$Fe_{3-x}(MT,Ti)_xO_4$ (MT: Transition Metal)	71.28 ± 0.006

*) MT : Mn, Cr, V, Ni, etc.

4. Conclusions

Iron sand of titanomagnetite type has been obtained from natural iron sand using a weak magnetic separator of 1000 Oe. To increase Fe content, iron sand of titanomagnetite type was milled for 10 h using high energy milling (HEM). The result of surface morphology by using SEM found that the average size of the grains is about (5 to 10) μm. The elements contained in natural iron sand were obtained by energy dispersive spectroscopy (EDS) analysis, consisting of: Fe (66.57 %), Mg (1.37 %), Si (0.57 %), Al (1.66 %), Ti (8.03 %), La (1.17 %) and O (20.64 %). The results of the analysis using X-ray diffraction (XRD) have been obtained that the phase contained in the natural iron sand of the type of titanomagnetite, namely ilmenite ($FeTiO_3$), hematite (α-Fe_2O_3) and titanomagnetite ($Fe_{3-x}(MT,Ti)_xO_4$) phases with the content respective of 9.63 %, 19.09 % and 71.28 %, where MT (metal transition).

Acknowledgements
This work was supported by Higher Education Research Grants 2017 No. DIPA: SP DIPA-042.06.1.401516/2017, Development of Iron Oxide Nanoparticles from Iron Sand Processing in Support of National Industry

References
[1] Performance Report of the Ministry of Industry Year 2015, 2016.
[2] Ministry of Industry Republic of Indonesia, Profile of Steel Industry, 2014
[3] Rochani S et al 2008 The current status of iron minerals in Indonesia *Indonesian Mining Journal* **11** 1 – 17
[4] Rusianto T *et al* 2012 The potential of iron sand from the coast south of Bantul Yogyakarta as raw ceramic magnet materials *Journal of Technology* **5** 62 – 69
[5] Mufti, N., Munfarriha, U., Fuad, A., & Diantoro, M. (2016, February). Synthesis and photocatalytic properties of Fe_3O_4@TiO_2 core-shell for degradation of Rhodamine B. In AIP Conference Proceedings (Vol. 1712, No. 1, p. 050009). AIP Publishing.
[6] Yamanaka T, Komatsu Y, and Nomori H 2007 Electron density distribution of $FeTiO_3$ ilmenite under high pressure analyzed by MEM using single crystal diffraction intensities Locality: synthetic Sample: P = 3.6 Gpa *Physics and Chemistry of Minerals* **34** 307 – 318
[7] Blake R L *et al* 1966 Refinement of the hematite structure *American Mineralogist* **51** 123 – 129
[8] Stout M Z and Bayliss P 1975 Crystal structure of a natural titanomagnetite *The Canadian Mineralogist* **13** 86 – 88
[9] Jalil Z *et al* 2017 Magnetic behavior of natural magnetite (Fe_3O_4) extracted from beach sand obtained by mechanical alloying method *AIP Conf. Proc.* 1862 **030023** 1 – 4

Photoelectrochemical Performance of TiO_2/Ti Electrode for Organic Compounds

M. Maulidiyah[1], I. B. P. Wijawan[1], D. Wibowo[1], A. Aladin[2], B. Hamzah[3], M. Nurdin[1]

[1]Department of Chemistry, Faculty of Mathematics and Natural Sciences, Universitas Halu Oleo, Jl. HEA Mokodompit Kampus Baru, Kendari 93232, Indonesia.
[2]Department of Chemical Engineering, Universitas Muslim Indonesia, Jln. Urip Sumoharjo, KM 5. Makassar 90231, Indonesia.
[3]Department of Mathematics and Natural Sciences Education, Faculty of Teacher Training and Education, Universitas Tadulako, Kampus Bumi Tadulako, Jl. Sukarno-Hatta Km 9 Tondo, 94118 Palu, Indonesia.

E-mail: mnurdin06@gmail.com

Abstract. Photoelectrochemical performance of TiO_2/Ti electrode was investigated by using organic compounds. The TiO_2/Ti electrode was prepared by anodic oxidation at a potential bias of 25 V for 4 h then calcined for 450 °C to obtain the anatase polymorph. Subsequently, it was characterized by X-Ray Diffraction (XRD) and Scanning Electron Microscope-Energy Dispersive X-ray (SEM-EDX). The XRD pattern showed that TiO_2 has anatase phase as confirmed by 2θ peaks at 37.93°; 63.00°, and 83.00°. Using SEM-EDX data the TiO_2 layer was formed on Ti plate with the composition of Ti (4.5 keV) and O (0.5 keV) elements. Furthermore, the photoelectrochemical sensing on the three organic compounds (ascorbic acid, glucose, and titan yellow) with the electrolyte addition showed that the linearity of TiO_2/Ti electrode were 0.937, 0.968, and 0.938, meanwhile without the electrolyte were 0.998, 0.989, and 0.923, respectively.

Keywords. Electrode, organic compounds, photoelectrochemical, and TiO_2/Ti.

1. Introduction

The organic compound is a class of compounds composed of carbon and hydrogen atoms and can contain other elements such as nitrogen, oxygen, phosphor, and sulfur. Organic compounds which often cause a variety of environmental pollution problems, especially disposal of organic wastes, are ascorbic acid, glucose, and titan yellow. One of the alternatives in the process of overcoming environmental pollution due to organic compounds in the process of organic compounds degradation or decomposition using titanium dioxide (TiO_2) photocatalyst [1, 2].

The last few decades, the utilization of TiO_2 has been widely developed among others as the process of the photocatalyst, sensing, photoelectrolysis, photovoltaic, solar cells, filtration and biomedicine [3, 4]. The utilization of the TiO_2 compound in the photocatalyst field is generally to degrade organic pollutant perfectly that produces carbon dioxide (CO_2) products in aquatic environment [5, 6]. Maulidyah *et al* (2016) reported the utilization of TiO_2 photoelectrocatalysis to

Content from this work may be used under the terms of the Creative Commons Attribution 3.0 licence. Any further distribution of this work must maintain attribution to the author(s) and the title of the work, journal citation and DOI.
Published under licence by IOP Publishing Ltd

degrade the amino acid waste and Zhao *et al* (2004) reported the improved performance of TiO_2 photocatalyst with the ability to degrade various types of organic substances and surface capacity of TiO_2 in absorbing organic substances [7, 8].

Common methods used to produce TiO_2 materials include hydrothermal, sonochemistry, and coprecipitation methods. Another method is an electrochemical method known as anodizing method [9, 10]. This method can form a nanotube structure on thin layer of titanium metal [11]. Along with the development of photocatalyst utilization in degrading organic compounds, it still has deficiency such as relatively low efficiency because it can lead to electron-hole recombination [12, 13].

One of the alternatives that can prevent electron-hole recombination is the electric field formation in the electron-hole area [14, 15]. The electric field can occur in the presence of potential difference. This potential difference will be formed when there is electron current in an electrochemical cell that will occur by utilizing electrons formed from the degradation process of organic compounds [16, 17]. Therefore, to improve the performance of photocatalysis, the combination between the process of photocatalyst and electrochemical, which is called as photoelectrocatalysis process, is required [18]. It utilizes photoelectrocatalysis as the main factor in the manufacture of COD sensors. Sensitivity of the sensors to the water-soluble organic compound is the sensitivity effect of thin TiO_2 layer on the indium tin oxide (ITO) glass used as the working electrode.

Based on the above description, then an electrochemical sensitivity test was performed using TiO_2/Ti nanotube electrodes and photocurrent response test with ascorbic acid, glucose, and titan yellow photometry profile testing. The results of this study are expected to be applied to determine the electrochemical profile of various organic compounds electrochemically.

2. Materials and methods

2.1. Fabrication of TiO_2/Ti electrode

The fabrication of TiO_2/Ti electrode has been made by using Maulidiyah *et al* (2015) method. The Ti plate was cleaned and shined. Subsequently, it was inserted in the probe which was filled with the electrolyte solution in the form of 0.27 M NH_4F and distilled water with a 98 % glycerol. The electrolyte solution was produced by dissolving 0.99 g of NH_4F using 4 mL of distilled water and 96 mL of 98 % glycerol. Anodizing process was done by placing the Ti plate as anode and Cu plate as a cathode, using a magnetic stirrer. A potential bias of 25 V was applied by connecting to a power supply. This anodizing process was carried out for four hours. The final stage was the calcinations of Ti plate for 1.5 h at a temperature of 500 °C to obtain TiO_2 anatase crystal which has better activity than other types of TiO_2 crystals [5].

2.2. Degradation test of organic compounds

The measured photocurrent in this study was identified from the photoelectrochemical circuit using the bath system. This procedure was a simple system in which three electrodes i.e. working, reference and comparison electrodes. The electrodes were immersed in a beaker containing test solution in the form of organic compounds such as ascorbic acid, glucose, and titan yellow with the addition of $NaNO_3$ as an electrolyte. The photocurrent was measured using a potentiostat with a potential bias of 0.5 V and a measurement time of 60 s. At the beginning of the measurement, UV lamp was off and started up at 10 s. Measurement for 60 s was carried out by using UV radiation [3, 19].

Dye degradation test of organic compounds for (1, 3, 5, and 7) ppm was added by 0.1 M $NaNO_3$ then analyzed using the Multi-Pulse Amperometry (MPA) with a ten-minute duration and a potential bias of 0.5 V in a dark and UV irradiation.

3. Results and discussion

3.1. Characterization of TiO_2/Ti electrode

3.1.1. X-Ray Diffraction (XRD)

The crystallinity and phase of TiO_2/Ti were characterized using XRD. The ability of diffraction methods is used for the determination of crystalline structure, quantitative phase, qualitative analysis, phase transformation, crystal texture, and crystal size [20]. The crystallinity and phase of crystalline TiO_2/Ti are shown in the diffractogram in Figure 1. It is shown that there are peaks at 2θ of (37.93°, 63°, and 83°) which indicates the formation of crystalline TiO_2 with anatase phase. Based on the works of Han, et al. (2004) and Khanna *et al* (2007), anatase TiO_2 can be observed from the peaks at 2θ = (25.3°, 37.9°, 48.4°, 54°, 63°, 70°, 75°, and 83°) [21, 22].

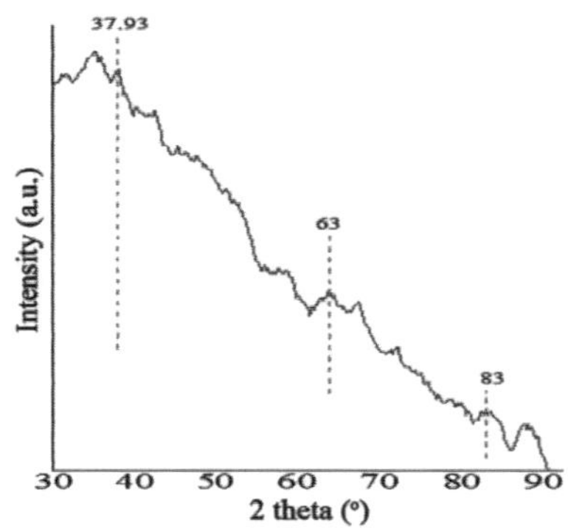

Figure 1. XRD pattern of TiO_2/Ti electrode

3.1.2. Scanning electron microscope-energy dispersive x-ray (SEM-EDX)

Characterization using SEM-EDX aims to determine morphology, particle size, and material composition. The characterization results of SEM-EDX of TiO_2/Ti anodizing results can be observed in Figure 2. The morphology of TiO_2/Ti anodizing results indicates the presence of 10 μm particles (Figure 2.a), which are fairly evenly on all surfaces indicating the dispersed TiO_2 particles on the Ti plate surface. SEM-EDX also generates additional data in the form of X-ray energy graph. The peak position of an element in EDX depends on the energy of the element. The element will be detected at its typical energy point with intensity value on Y axis. Y-axis shows quantity values of elements in the whole sample which is later confirmed in wt%. Based on Figure 2.b the TiO_2/Ti anodizing results has the composition of O (oxygen) element of 22.34 wt% or 46.27 at% and Ti (titanium) element of 77.66 wt% or 53.73 at%.

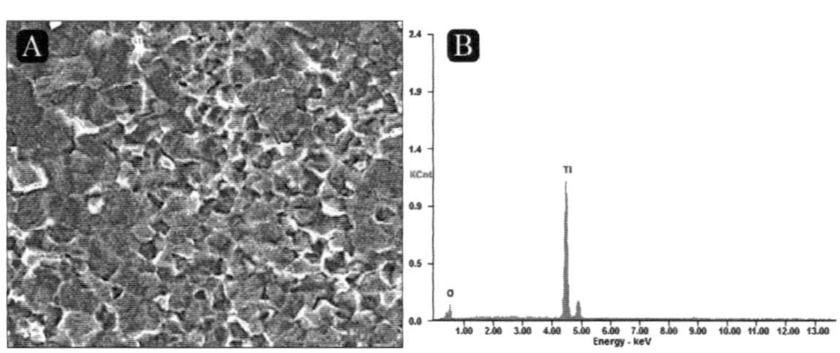

Figure 2. Analysis by using SEM-EDX; (a) morphology of TiO_2/Ti electrode (b) EDX spectrum of TiO_2/Ti electrode

IOP Conf. Series: Materials Science and Engineering **367** (2018) 012060 doi:10.1088/1757-899X/367/1/012060

3.2. Photocurrent response against organic compounds

In this study, ascorbic acid is an organic compound with the molecule formula of $C_6H_8O_6$ as test compound caused it has three hydroxyl groups allowing readily adsorbed on electrode surface and can produce a high photocurrent response. This may occur as a result of negatively charged hydroxyl group on ascorbic acid will attract with positively charged TiO_2 surface in solution. The ascorbic acid photocurrent response can be seen in Figure 3.

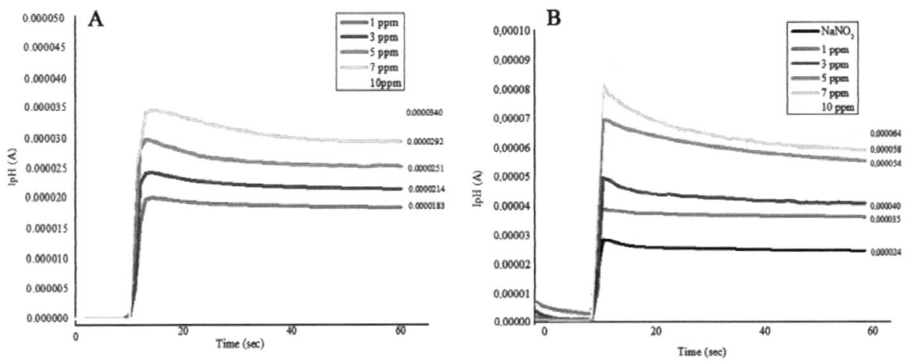

Figure 3. Photocurrent response of ascorbic acid, (A) non-electrolyte (B) addition of electrolyte

The photocurrent responses of ascorbic acid as shown in Figure 3.a and Figure 3.b show the differences in photocurrent intensities. The current response generated by both non-electrolyte and electrolyte ascorbic acid show greater current responses than $NaNO_3$ blank solution response. This is due to ascorbic acid is more readily adsorbed on the surface of photocatalyst, but it can be seen that the curve of electrolyte ascorbic acid response does not coincide with the curve of $NaNO_3$ [10]. This indicates that ascorbic acid is not completely degraded by photocatalysts.

The differences in photocurrent intensity show differences in non-electrolyte and electrolyte glucose response, as shown in Figure 4. The current response produced by electrolyte glucose is greater than the current response produced by non-electrolyte glucose. This is due to the $NaNO_3$ electrolytes that serve as conductors of electrons.

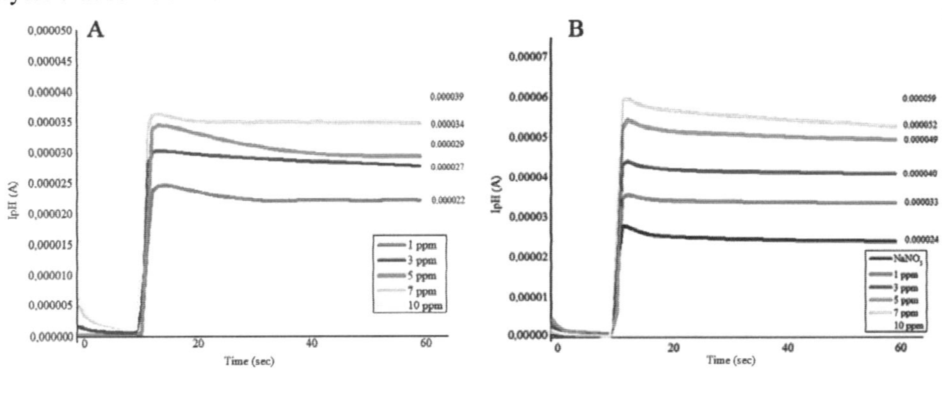

Figure 4. Photocurrent response of glucose, (a) non-electrolyte (b) addition of electrolyte

The current difference occurs because electrolyte glucose is more readily adsorbed on the photocatalyst surface than non-electrolyte glucose. Thus, on electrolyte glucose current response curve with $NaNO_3$ blank solution curve shows no coincidence. This indicates that glucose is not decomposed or degraded completely by the photocatalyst.

The photocurrent responses of non-electrolytic and electrolyte titan yellow are shown in Figure 5.a and Figure 5.b. The photocurrent responses of electrolyte titan yellow seen as observed in Figure 5.a shows higher current responses at each concentration variation than non-electrolyte titan yellow responses (Figure 5.a). This tendency occurs because the electrolyte $NaNO_3$ used in analyte plays an important role in the transfer of electrons in the solution resulting in more optimal current responses. The curve of photocurrent response in Figure 5.b indicates the degradation activity of the titan yellow at concentrations above 5 ppm does not occur perfectly. This is clearly seen on the curve of yellow titan in concentrations of 7 ppm and 10 ppm do not coincide with the $NaNO_3$ curve which is the blank in the measured test solution. The imperfect process of degradation is due to the saturation on the surface of the photocatalyst thereby inhibiting the degradation process.

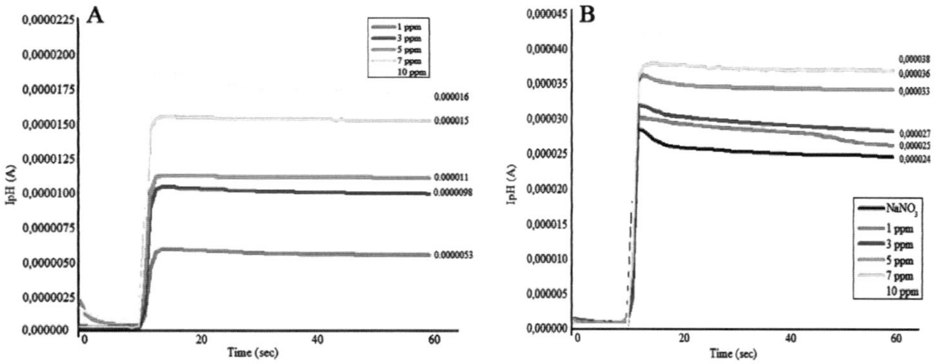

Figure 5. Photocurrent response of titan yellow, (A) non-electrolyte (B) addition of electrolyte

3.3. Compared linearity of electrolyte against non-electrolyte

In this study, linearity was obtained by plotting concentration (C) and photocurrent response (Iph) values generated by the test compounds. The linearity of some test compounds with electrolytes is related to the resulting current response in some concentrations. As observed on the graph of Figure 6, the linearity of titan yellow is 0.938 that higher than the ascorbic acid, indicating that the average increase in photocurrent for each titan yellow concentration is more stable than ascorbic acid.

The ascorbic acid curve shows increases in current responses in the concentration range of 0 ppm to 10 ppm. This indicates the higher concentration of ascorbic acid, the higher the resulting current response. It is also experienced by some other test compounds that show increases in current responses as the concentrations increase. Photocurrent responses of glucose and titan yellow also increase along with the increasing concentration but the resulting current response at each concentration variation is lower than on ascorbic acid. The linearity of glucose and titan yellow as shown in Figure 6 are 0.968 and 0.938, respectively.

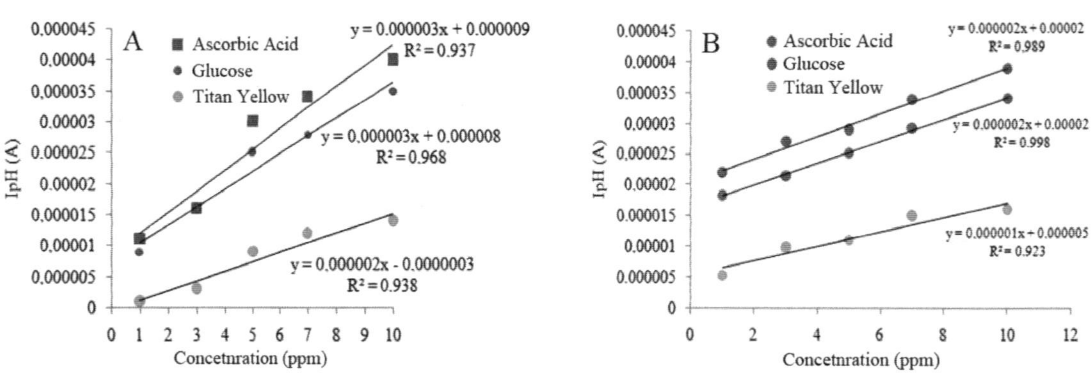

Figure 6. Comparison of photocurrent response (a) using electrolyte, and (b) non-electrolyte

The correlation graph between current response produced and the variation concentration of several non-electrolyte test compounds is shown in Figure 6.a and Figure 6.b. The photocurrent response of ascorbic acid shows increase in each concentration variation with a linearity value of 0.998.

The current response produced by ascorbic acid at each concentration variation is lower than the current response produced by glucose. It shows that non-electrolyte glucose can decompose optimally than non-electrolytic ascorbic acid with a linearity value of 0.998. Unlike the curve of non-electrolyte titan yellow that shows lower photocurrent response value than glucose and ascorbic acid for each concentration variation. It also shows that titan yellow is more difficult to degrade than glucose and ascorbic acid. In addition, the linearity value of non-electrolyte titan yellow of 0.923 is also lower than in the linearity values of glucose and ascorbic acid.

4. Conclusions

Based on the results of this study, it can be concluded that the XRD pattern showed that TiO_2 in TiO_2/Ti electrode has anatase crystal phase as confirmed by 2θ peaks at 37.93°, 63.00°, and 83.00°. From SEM-EDX data the TiO_2 layer was formed on Ti plate with the composition of Ti (4.5 keV) and O (0.5 keV) elements. The photoelectrochemical sensing on the three organic compounds (ascorbic acid, glucose, and titan yellow) with the electrolyte addition showed that the linearity of TiO_2/Ti electrode were 0.937, 0.968, and 0.938, meanwhile without the electrolyte were 0.998, 0.989, and 0.923, respectively.

Acknowledgments

We acknowledge the financial support of the DRPM - Ministry of Research, Technology and Higher Education, the Republic of Indonesia to Research of Competency-based, contract number 065/ADD/SP2H/LT/DRPM/VIII/2017.

References

[1] Nurdin M *et al* 2016 *Int. J. ChemTech. Res.* **9** 483–491
[2] Nurdin M *et al* 2016 *Oriental J. Chem.* **32** 2713–2721
[3] Nurdin M *et al* 2016 *J. Mater. & Environ. Sci.* **7** 3334–3343
[4] Maulidiyah *et al* 2017 *Asian J. Chem.* **29** 204–2508
[5] Maulidiyah *et al* 2015 *Int. J. Pharma. & Pharmaceu. Sci.* **7** 141–146
[6] Maulidiyah *et al* 2017 *Environ. Nanotechnol. Monit. & Manag.* **8** 109–111
[7] Maulidiyah *et al* 2016 *Anal & Bioanal Electrochem.* **8** 761–776
[8] Zhao H, Jiang D ,Zhang S, Catterall K, and John R 2004 *Anal Chem.* **76** 155–160
[9] Nurdin M *et al* 2017 *J. Coatings & Technol. Res.*
[10] Nurdin M *et al* 2017 *Anal Bioanal Electrochem.* **9** 480–494
[11] Wibowo D, Ruslan, Maulidiyah, and Nurdin M 2017 *IOP Conf. Series: Mater. Sci. Eng.* **267** p 012007
[12] Ruslan, Mirzan M, Nurdin M, and Wahab A W 2016 *Int. J. Appl. Chem.* **12** 399–409
[13] Maulidiyah *et al* 2015 *Oriental J. Chem* **31** 2337–2342
[14] Hikmawati A H *et al* 2017 *IOP Conf. Series: Mater. Sci. Eng.* **267** p 012005
[15] Maulidiyah *et al* 2015 *Biosci. Biotechnol. Res. Asia* **12** 985–1989
[16] Maulidiyah *et al* 2015 *Int. J. ChemTech. Res.* **8** 645–653
[17] Arham Z, Nurdin M, and Buchari B 2016 *Int. J. ChemTech. Res.* **9** 113–120
[18] LO Mursalim A M *et al* 2017 *IOP Conf. Series: Mater. Sci. Eng.* **267** p 012006
[19] Maulidiyah *et al* 2015 *Int. J. ChemTech. Res.* **8** 416–423
[20] Maulidiyah *et al* 2015 *ARPN J. Eng. & Appl. Sci.* **10** 6250–6256
[21] Han H, Zan L, Zhong J, Zhang L, and Zhao X 2004 *Material Science and Engineering B* **110** 227–232
[22] Khanna P K, Singh N, and Charan S 2007 *Materials Letters* **61** 4725–4730